RECONSTRUCTIVE PHASE TRANSITIONS

In Crystals and Quasicrystals

RECONSTRUCTIVE PHASE TRANSITIONS

In Crystals and Quasicrystals

Pierre Tolédano
University of Amiens, France

Vladimir Dmitriev
University of Rostov on the Don, Russia

World Scientific
Singapore • New Jersey • London • Hong Kong

Published by

World Scientific Publishing Co. Pte. Ltd.

P O Box 128, Farrer Road, Singapore 912805

USA office: Suite 1B, 1060 Main Street, River Edge, NJ 07661

UK office: 57 Shelton Street, Covent Garden, London WC2H 9HE

Library of Congress Cataloging-in-Publication Data

Tolédano, Pierre.
Reconstructive phase transitions : in crystals and quasicrystals / Pierre Tolédano, Vladimir Dmitriev.
p. cm.
Includes bibliographical references and index.
ISBN 9810223641. -- ISBN 9810228406 (pbk.)
1. Quasicrystals. 2. Crystals. 3. Phase transformations (Statistical physics) I. Dmitriev, Vladimir. II. Title.
QC173.4.Q36T65 1996
530.4'14--dc20 96-9517
CIP

British Library Cataloguing-in-Publication Data

A catalogue record for this book is available from the British Library.

This book is printed on acid-free paper.

Printed in Singapore by Uto-Print

To Zoya

To Olga

FOREWORD

Reconstructive phase transitions constitute the most fundamental and widespread type of structural transitions existing in Nature, forming a large majority of the transitions found in elemental crystals, alloys, minerals and complex fluids. Despite this fact they have been persistently considered by theorists as the "uninteresting case" (P. W. Anderson, Pierre Curie Colloquium, 1980). Such a conclusive statement is mainly based on the presupposed opinion that the absence of group-subgroup relationship between the symmetries of the phases surrounding a reconstructive transition forbids the definition of an order-parameter, which is the primary step for a theoretical approach to a phase transition.

In some respect this pessimistic view results from the popularity of Landau's theory of second-order transitions (the "interesting case") in which the author examines the situation of continuous or almost continuous transitions between group-subgroup related phases. It also stems from the fact that models of phase transitions in statistical physics such as the Ising model, lead to a lowering of symmetry between the high and low temperature phases. The Landau theory and the complementary theory of Critical Phenomena offer the possibility of working out the universal properties of systems undergoing phase transitions on the sole basis of symmetry considerations. Along this line some classes of structural transitions have been thoroughly investigated during the past decades, such as ferroelectric and ferroelastic transitions, or transitions in thermotropic liquid crystals.

Reconstructive transitions have been left aside from the preceding theoretical framework. The aim of our book is to demonstrate that these transitions can indeed be the subject of a full theoretical treatment providing a strong revision and a better understanding of the Landau concepts, and most importantly on the condition that one leaves the realm of pure symmetry for the world of concrete structures and effective transition mechanisms.

For a long time reconstructive transitions have been a center of interest for metallurgists and crystallographers. They have investigated comprehensively the phase diagrams involving such "transformations" and the corresponding displacive and ordering processes. They also established a nomenclature and proposed some classifications. However, while the specific features of the various subclasses of reconstructive transitions have been clearly disclosed experimentally, a unifying theoretical picture could not emerge. This is mainly because the nature and symmetry of the transition order-parameters have not been well understood and related organically to the critical variables and to the thermodynamic functions.

It is the purpose of the theory presented in this book to show the intimate connections between the set of critical degrees of freedom (atomic displacements or probability of occupancy of selected positions by the atoms), the order-parameters, the phase diagrams, and the static and dynamical properties of the two main types - displacive and ordering - of reconstructive transitions. On that goal the structures are expressed under the form of lattice functions corresponding to a sum of stationary density waves. The increment $\delta\psi$ of the lattice function at the transition provides a natural definition and the explicit form of the order-parameter as a function of the critical degrees of freedom. Thus, the order-parameter corresponds to the phase or to the amplitude of $\delta\psi$ for, respectively, displacive and ordering mechanisms.

Striking differences appear in our theoretical description of displacive reconstructive transitions with respect to the standard Landau approach to displacive transitions. Thus, the order-parameters are non-linear periodic functions of the critical displacements instead of linear functions. Therefore the thermodynamic potentials have the meaning of truncated Fourier series, instead of Taylor expansions. Their minimization, which is performed with respect to the critical displacements, yield two sorts of stable states: the Landau states which are group-subgroup related to the initial phase, and *additional states for which the group-subgroup relationship is lost*.

For ordering reconstructive transitions there exists, as in the current description of continuous order-disorder transitions, a linear relationship between the order-parameters and the critical variables. We show that such transitions generally occur between differently ordered close packed structures which are derived from a disordered parent structure for definite occupancies of the atomic sites and specific crystallogeometrical requirements. The fulfillment of these conditions creates new symmetry operations in the ordered phases and prevent the possibility of a group-subgroup relationship between them.

Accordingly, for both displacive and ordering reconstructive transitions the absence of a group-subgroup relationship between the symmetries of the phases is the result of specific values taken by the critical variables, i.e. the phases correspond to *limit states*. This essential property is shown to have significant consequences on the static and dynamical features of reconstructive transitions, as well as on the critical behavior of the relevant physical quantities.

This book is a monograph. Its subject is however related to the wider field of discontinuous phase transitions. We have therefore included a long introductory chapter devoted to a unified presentation of the phenomenological theory of first-order phase transitions. In this chapter the results concerning the solutions of the radial problem in the Landau theory, which have been essentially obtained by Gufan and his coworkers, are detailed. They provide a useful basis for working out the phase diagrams of reconstructive phase transitions. Along the same line, a presentation of the concepts of the theory of Singularities is given. These concepts allow to place on a rigorous basis some empirical rules used for the construction of thermodynamic potentials and of the corresponding phase diagrams.

In Chapter II the density wave theory of reconstructive phase transitions is formulated and illustrated by a number of working examples. The applicability of the theory to real crystalline systems is shown in Chapters III and IV. In these Chapters we have preferentially, but not exclusively, considered reconstructive transitions in crystals of the elements since the simplicity of their structures allows to show more clearly the connection between the transition order-parameters and the critical displacements and ordering of the atoms. Only few examples of minerals have been discussed because their transitions generally possess an implicit morphotropic character. The description of morphotropic transitions, also predominantly found in complex fluid systems, requires further theoretical considerations which are beyond the subject of this book.

The predictive character of the density wave approach to phase transitions is well demonstrated in the examples of crystal-quasicrystal transitions examined in Chapter V. The approach permits a determination of the atomic positions in the quasicrystal structure providing their positions in the crystal structure of reference are known. The loss of the crystal's translational symmetry and the onset of a non-crystallographic point-group symmetry, which characterize a crystal-quasicrystal transition, are described in terms of combined displacements and disordering of the atoms.

This book materializes many years of a theoretical investigation of concrete examples of reconstructive transitions in crystals and quasicrystals. We feel indebted to the colleagues and friends who collaborated with us during this period. We are especially grateful to Yuri Gufan and Serguei Rochal who took part to the most important steps in our effort to insert the interpretation of reconstructive transitions into a general theoretical framework. We have also benefited from the collaboration of Vladimir Lorman and Bruno Mettout. We thank Victor Torgashev for many helpful discussions and Evgeni Kut'in who allowed us to quote some of the unpublished results of his thesis. In many occasions Jean-Claude Toledano helped us to clarify the formulation of our theoretical approach. At last, we want to thank Vladimir Dvorak who offered us the possibility to write the last pages of our book in the beautiful city of Prague.

P. T. and V. D., March 1996

CONTENTS

CHAPTER II. DENSITY-WAVE THEORY OF RECONSTRUCTIVE PHASE TRANSITIONS

CHAPTER IV. ORDERING TYPE RECONSTRUCTIVE PHASE TRANSITIONS

CHAPTER I

PHENOMENOLOGICAL THEORY OF FIRST-ORDER PHASE TRANSITIONS

1. Introduction

Two main classes of phase transitions occur in crystals, which involve a modification of the crystal's atomic configuration. A first class is characterized by the fact that the structural changes which take place across the transition do not break the chemical bonds, but only slightly alter their lengths and orientations. When the transition can be described in terms of displacements, it means more concretely that the atomic shifts from one phase to another are generally of the order of 0.01-0.1 A°. At the experimental level, this category of structural transitions corresponds to latent heat smaller than a few calorie/gram, and to small discontinuities of the physical quantities at the transition point. Most importantly: a group-subgroup relationship between the invariance groups of the phases surrounding the transition, can be inferred from the evolution of the X-ray and neutron diffraction spectrum across the transition. This type of transition is found in most of the insulators, in some alloys, and in a small number of crystals of pure elements. A minority of these structural transitions is found to be second-order (continuous), and in most cases they are slightly first-order.

To a second class of structural transitions pertain most of the allotropic transformations of the elements, the majority of the transitions found in alloys, and also a few well established examples of transitions which take place in insulating materials. These transitions, often denominated as *reconstructive*, since they involve a breaking of part of the chemical bonds of the initial structures, exhibit drastic changes at the transition point, with large latent heat and thermal hysteresis. The corresponding atomic displacements are of the order of the lattice parameters, and no group-subgroup relationship is found, between the symmetry of the phases. The transitions have generally a strongly first-order character.

The fact that there is no apparent symmetry relationship between the crystal structures of the phases enframing reconstructive transitions, has led to the presupposed opinion that their mechanisms and physical properties, could not be conveniently accounted by a unifying theoretical scheme. The remaining chapters of this book, are devoted to the demonstration that this opinion is largely erroneous. In contrast, the apparent simplicity of the symmetry connection between the structures in the first class of structural phase transitions has motivated, since the seminal work of Landau [1], a widespread effort at working out their general properties, on the level of the phenomenological theories. The aim of this first chapter is to make a general survey of this phenomenological approach to structural transitions in which the phases are group-subgroup related, as the theoretical methods used for their interpretation, and some of their general properties, will serve for the description of reconstructive phase transitions.

The chapter focuses on first-order phase transitions. This is not only because the subject of this book concerns the theory of a class of first-order transitions, but also because most of the extensions of the original Landau theory apply to the interpretation of first-order transitions, which constitute, by far, the more common situation among structural systems. Even when considering only structural transitions, first-order transitions constitute a very heterogeneous family. Obviously, the thermodynamic classification of phase transitions, which distinguishes only between second- and first-order transitions, is insufficient, as it denominates similarly weekly discontinuous transitions and reconstructive ones. In spite of the fact that the various types of first-order transitions possess a number of common characteristics (existence of a latent heat, coexistence of the phases at the transition point, discontinuity of the order-parameter and of other physical quantities etc.) it is possible to differentiate them *qualitatively* by a complete set of measurements. Thus *it should be possible to differentiate them also on a theoretical ground*, and not only *quantitatively*. It is one of the aims of this book to show that the *full* symmetry of the order-parameter, which includes *the intrinsic symmetry of the concrete transition mechanism*, may serve as a basis for distinguishing between the various categories of first-order transitions.

Before dealing with first-order transitions, we need to introduce the basic concepts of the Landau theory of second-order phase transitions, and the various general procedures which allow the phenomenological description of a phase transition.

2. Basic concepts of the Landau theory of phase transitions

2.1. Aim of the Landau theory

In the study of phase transitions, crystallographers are often confronted to an ambiguous situation. Small changes in atomic positions determine structures having different space groups, and the data are generally compatible with several possible symmetry assignments. In order to make a choice, the crystallographer must be able to rely on some theoretical substrate which will allow him to discard certain of the possible assignments. The relevant theoretical framework in this field is constituted by the thermodynamic and symmetry considerations which form the so-called Landau Theory of Phase Transitions.

The Landau theory is a *phenomenological* theory. It does not aim to establish that a phase transition exists in a given system. The existence of a transition is an *experimental fact* considered as a starting point of the theory. The explanatory power of the theory is to establish the overall consistency of the microscopic characteristics of the transition (space-symmetry and structural changes, anomalies in the phonon spectrum etc...) and the results of the measurement of various relevant macroscopic quantities (thermal, dielectric, optical, elastic ...).

The *continuous* ("second-order") nature of the transition plays an essential role in the working out of the general foundations of the Landau theory. However, though the *strict field of validity* of the theory is that of continuous transitions, the theory satisfactorily applies to most of the first-order transitions, for which a group-subgroup relationship exists between the phases surrounding the transition (i.e. the first class of structural transitions, mentioned in the Introduction).

The Landau theory defines two basic concepts: the *order-parameter* and the *transition free-energy*. From the symmetry properties of these two quantities, it is possible to infer, on the one hand, a certain number of observable symmetry characteristics of the system: degeneracy of the "low-symmetry" phase (i.e. the number of energetically equivalent domain orientations in this phase), enumeration of the possible symmetries of the "low-symmetry" phase, for a given symmetry of the "parent" high-symmetry phase. On the other hand, macroscopical quantities can be classified as function of their symmetries with respect to the order-parameter from which can be deduced a classification of the type of anomalous behavior induced by the occurrence of the phase transition.

In order to give an intuitive approach to the basic arguments of the Landau theory, and to its use, we will first analyze a simplified working example of structural phase transition.

2.2. An introductory example

2.2.1. Crystallographic description and assumptions

Fig. 1 represents the unit-cell of a speculative crystalline structure with simple tetragonal Bravais lattice, in which a phase transition is assumed to take place. Negative ions occupy the vertices of the tetragonal cell, which has the lattice constants $a=b\neq c$. A positive ion M^+ is at the center of the cell. This configuration is assumed to be the equilibrium state of the system *above* the temperature T_c of the transition. Below the transition, equilibrium is assumed to correspond to a structure which differs from the preceding *parent* structure, by the fact that M^+ lies out of the center of the cell in an unspecified direction. Hence the latter equilibrium is characterized by the magnitude and direction of the displacement $\vec{v}_0=(v_x^0\ ,\ v_y^0\ ,\ v_z^0)$ of the central ion. At high temperature the equilibrium corresponds to $\vec{v}_0=0$.

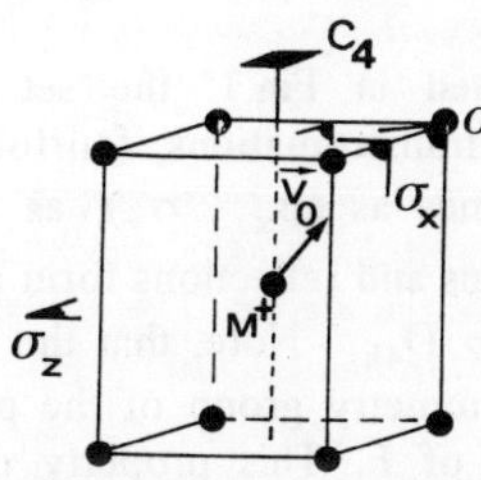

Fig.1. Unit-cell reflecting the tetragonal D_{4h} symmetry of the structure considered in the introductory example. $\vec{v}_0$ shows the displacement of the ion M^+.

Let us determine the displacement $\vec{v}_0$ as a function of temperature. This can be done by a *variational method.* One considers an *arbitrary* displacement $\vec{v}$ of the M^+ ion. For given temperature T and pressure P (or volume V), and specified values of the components of $\vec{v}$, there is a definite value $F(T,P,v_x,v_y,v_z)$ for the *free energy* of the system. This function is a *variational free-energy* and not the equilibrium free energy of the system, since it is calculated for an arbitrary displacement. The equilibrium displacement $\vec{v}_0(T,P)$ is defined as the displacement which minimizes the variational free energy F. The equilibrium free energy of the system is $F_{eq}(T,P)=F(T,P,\vec{v}_0)$. Note that, in the case of a change in the volume, one would have to consider a variational Gibbs function, F+PV, in order to determine the equilibrium of the system.

One cannot go further as long as one does not know the form of the variational free energy as function of the components of $\vec{v}$. The second step of Landau's theory is to show that, given general assumptions, one is able to determine simply the form of $F(T,P,\vec{v})$ in the required range of values of the functions arguments. The *basic assumption* is that of *continuity of the phase transition* in two different meanings: on the one hand, one assumes that the equilibrium values $v_i^0(T,P)$ *vary continuously across the transition at* T_c. On the other hand, one assumes that F is a *continuous* and *derivable function* of the parameters T, P, $\vec{v}$, which can be expanded in the form of a Taylor expansion, as function of these arguments. Let us stress that, as on either sides of T_c, $|\vec{v}_0|$ is small, one can restrict the determination of the functional form of $F(T,P,\vec{v})$ to *small values* of (v_x,v_y,v_z) and of $|T-T_c|$. F will then be equal to the *sum of the first relevant terms of a Taylor series in the preceding variables.*

2.2.2. *Symmetry constraints: Form of the free-energy*

The central property of the variational free-energy which allows to specify its form is a *symmetry property.* F is a function of (v_x,v_y,v_z) which is *invariant by the symmetry transformations of the high-temperature (parent) equilibrium structure*. In other terms, an arbitrary displacement $\vec{v}$ and the displacement $\vec{v}'$ obtained from $\vec{v}$ by applying one of the latter symmetry transformations correspond to the same value of the free-energy. Indeed, both displacements determine an identical set of mutual distances between the positive and negative ions of the system, and the free energy only depends on this *internal* configuration of the ions.

In the example represented in Fig.1, the set of symmetry transformations comprises, aside from the lattice transformations, fourfold rotations (C_4) around the Z axis, mirror symmetries into planes as $(\sigma_x$, $\sigma_y)$, as well as the products of these transformations. The set of rotations and reflections form a *group* G_0 of order 16, which is the crystallographic *point group* D_{4h} . Note, that the symmetry property of the free energy, to be invariant by the symmetry group of the parent structure, holds for each degree of the Taylor expansion of F. This property derives from the fact that the geometric transformations of G_0 act *linearly* on the components of $\vec{v}$. Hence, terms of different degrees belonging to the expansion of F will not mix, and must be separately invariant. It is easy to check on Fig.1, that by successive application of the mirror symmetries σ_x , σ_y , σ_z , to the components of $\vec{v}$, no linear combination of these components is invariant by G_0: each of the three former symmetry transformations reverses one component of $\vec{v}$ and preserves the two others. *Linear terms are therefore absent from the expansion.*

As for second degree terms, the same symmetry transformations preclude the existence of configurations of bilinear products of the type $(v_x v_y)$. Actually, one finds that the fourfold symmetry imposes that the most general form of the second degree contribution to the variational free-energy is a linear combination of v_z^2 and $(v_x^2 + v_y^2)$. Hence, the Taylor expansion of F, restricted to its lowest degree terms in (v_x, v_y, v_z) is:

$$F(T,P,v_x,v_y,v_z) = F_0(T,P) + \frac{\alpha_1(T,P)}{2} v_z^2 + \frac{\alpha_2(T,P)}{2}\left(v_x^2 + v_y^2\right). \tag{1.1}$$

2.2.3. The order parameter

Let us now derive the key result of the Landau theory: *either the component* v_z *, or the pair of components* (v_x, v_y) *will take non-zero values below* T_c *(but not both)*. The derivations of this result relies on the fact that one, and one only of the two coefficients α_i in Eq.(1.1) vanishes and changes sign at T_c , and that the other coefficients remain positive in the neighborhood of T_c . The demonstration is the following: at T_c the equilibrium values v_i^0 of the components of $\vec{v}_0$ are zero. Therefore, at this temperature, the variational free-energy (1.1) is minimum for v_i=0. Considering the form (1.1) of F, this property implies that we have $\alpha_i(T_c) \geq 0$ for i=1, 2 . These inequalities cannot be *strict* for *both* coefficients α_i , because their positiveness would hold on either sides of T_c , in the vicinity of this temperature. Consequently, the minimum of F would correspond to v_i=0 on either sides of the transition, while the situation assumed is only compatible with this result *above* T_c . Using the converse argument that the equilibrium values of the v_i are not all equal to zero *below* T_c , leads easily to the conclusion that *one of the two coefficients* α_i *, at least,*

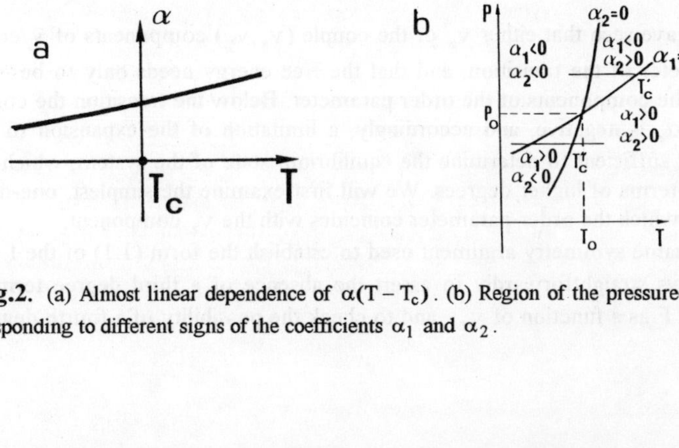

Fig.2. (a) Almost linear dependence of $\alpha(T - T_c)$. (b) Region of the pressure-temperature plane corresponding to different signs of the coefficients α_1 and α_2 .

vanishes at T_c *, and becomes negative below this temperature*, as represented in Fig.2a.

Let us now show that the two coefficients α_i *cannot vanish simultaneously* at T_c . This result relies on the assumption that the two coefficients are *different functions* of temperature and pressure. Fig.2b shows, in the (T, P) plane, the lines corresponding to the vanishing of the two functions α_i . The simultaneous vanishing occurs at an isolated point (T_0, P_0). Let us consider, for instance, the situation depicted in Fig.2b. For P above P_0 , on lowering the temperature, α_1 vanishes at T_c' and α_2 remains positive in the neighborhood of T_c' . Hence, the equilibrium value of the set (v_x, v_y) remains equal to zero, on either side of T_c' . A transition at this temperature will only concern a possible change in v_z^0 . Likewise, for P below P_0 a transition at T_c'' only concerns a possible change of the set of components (v_x^0 , v_y^0), the third component v_z^0 remaining equal to zero on either side of T_c''. Hence, an infinitesimal change of the pressure (for instance a small fluctuation of the atmospheric pressure) from above P_0 to below P_0 will modify qualitatively the nature of the phase transformation. This is a singular situation not encountered in nature. Rather, the standard situation corresponds to pressures away from P_0 , for which a slight change of the pressure does not modify significantly the direction of the displacement. In this case, *one coefficient* α_i *only vanishes and changes sign at the transition temperature*.

If one α_i coefficient is strictly positive, then the minimum of F with respect to the components v_i appearing in the second degree term multiplying this coefficient, *has zero equilibrium value in the vicinity of* T_c *above and below this temperature*. Thus, depending of the α_i which remains positive, either v_z or (v_x, v_y) can be omitted, in the first place, from the free-energy expansion. *The remaining set of components constitutes the order-parameter of the transition.* At this stage the order-parameter is defined as the set of degrees of freedom whose coefficient of the second degree contribution to F *vanishes and changes sign at* T_c. The number of independent components of the order-parameter (one in the case of v_z , two in the case of (v_x, v_y)) is the *dimension of the order-parameter*. In reference to Fig. 1, it means that the displacement of the M^+ ion below T_c cannot occur in arbitrary direction of space. It is *either* along the Z axis, *or* in the (X,Y) plane.

2.2.4. Stable state below T_c *and physical anomalies at* T_c

We have seen that either v_z or the couple (v_x, v_y) components of $\vec{v}$ constitute the order parameter of the transition, and that the free energy needs only to be expanded as function of the components of the order parameter. Below the transition the corresponding coefficients α_i is negative, and accordingly, a limitation of the expansion to the second degree is not sufficient to determine the equilibrium state of the system, which requires to add positive terms of higher degrees. We will first examine the simplest, one-dimensional, situation for which the order-parameter coincides with the v_z component.

The same symmetry argument used to establish the form (1.1) of the Landau free-energy, allows straightforwardly to assert the absence of a third degree term v_z^3 in the expansion of F as a function of v_z , and to check the possibility of a fourth degree term. If we assume that this simplest form of expansion is sufficient to determine the equilibrium state of the system, the coefficient of the fourth degree term must be positive in the neighborhood of T_c. Up to the latter degree, the form of the relevant contributions to the free-energy is therefore:

$$F(T,P,v_z) = F_0(T,P) + \frac{\alpha(T-T_c)}{2} v_z^2 + \frac{\beta}{4} v_z^4 . \tag{1.2}$$

In this expression, α_1, which is an odd function of $(T-T_c)$ has been expanded up to its *lowest* degree term. α being a positive constant. Likewise, the lowest degree expansion of $\beta(T-T_c)$ is a positive constant β in the vicinity of T_c. The function $F_0(T,P)$, which is the zeroth degree term in the expansion, represents the normal "background" part of the free energy. It behaves smoothly across the transition since it does not depend on the order-parameter. A plot of $F(v_z)$-F_0, for three characteristic temperatures is indicated in Fig.3.

The minima of F, determined by the set of conditions

$$\frac{\partial F}{\partial v_z} = 0 \ , \quad \frac{\partial^2 F}{\partial v_z^2} > 0 \tag{1.3}$$

occur, above T_c, for v_z^0=0, as expected. For $T < T_c$ they occur for :

$$v_z^0 = \pm \left[\frac{\alpha(T_c - T)}{\beta} \right]^{1/2} . \tag{1.4}$$

This behavior has a general validity: the order-parameter of a transition, in the framework of Landau's theory, should possess a square-root dependence as a function of temperature. Note that one finds two minima corresponding to the same value of the free-energy, and opposite values of v_z^0. The corresponding upward and downward displacements of the

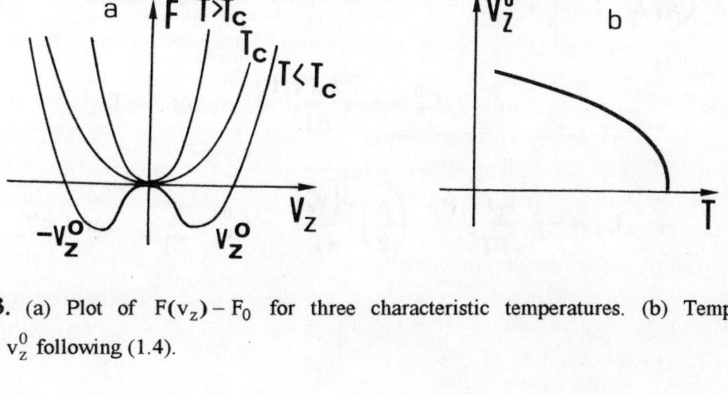

Fig.3. (a) Plot of $F(v_z) - F_0$ for three characteristic temperatures. (b) Temperature dependence of v_z^0 following (1.4).

M^+ ion (Fig. 1) are distinct states of the system possessing the same stability.

Other physical consequences of the form (1.2) of the free energy can be drawn: *absence of latent heat* associated with the crossing of the transition, anomalous behavior of the *specific heat*, behavior of the *susceptibility* related to the order-parameter.

The latent heat is $L=T_c\Delta S$, where ΔS is the difference of entropy between the two phases, at T_c . We can derive S in each phase from the equilibrium free energy $F\left[T,P,v_z^0(T,P)\right]$ using the standard definition: $S=-\left(\frac{\partial F_{eq}}{\partial T}\right)_{P=const}$. We can write:

$$\frac{\partial F_{eq}}{\partial T}=\left(\frac{\partial F}{\partial T}\right)_{v_z}+\left(\frac{\partial F}{\partial v_z}\right)_T\cdot\left[\frac{d(v_z)}{dT}\right]_{v_z^0} \quad . \tag{1.5}$$

However, since F is minimum for $v_z=v_z^0$, the second contribution in (1.5) vanishes. Hence:

$$S=-\left(\frac{\partial F}{\partial T}\right)_{v_z^0}=-\frac{\alpha}{2}\left(v_z^0\right)^2-\frac{\partial F_0(T,P)}{\partial T} \quad . \tag{1.6}$$

Since both v_z^0 and $\left(\frac{\partial F_0}{\partial T}\right)$ are continuous at T_c , there is no entropy jump ($\Delta S=0$) at T_c , and thus, *there is no latent heat for the transition.*

Several specific heats can be considered for a system, depending on the quantity which is maintained constant. In the above example, the displacement v_z of a positive ion determines the occurrence of an electric dipole (or a macroscopic polarization P_z). The quantity ε_z which is thermodynamically conjugated to v_z is therefore proportional to an electric field (the conjugation between quantities X and Y, is expressed by the fact that infinitesimal work on the system has the form YdX). Let us show that the specific heat at constant electric field has a characteristic type of anomaly. The specific heat is expressed by: $C_\varepsilon=T\left(\frac{\partial S}{\partial T}\right)_\varepsilon$. Using Eq.(1.6), we find the following expressions:

$$C_\varepsilon^0=-\frac{\partial^2 F_0(T,P)}{\partial T^2} \qquad \text{for } T>T_c$$

and

$$C_\varepsilon=-\frac{\partial^2 F_0(T,P)}{\partial T^2}-\left(\frac{\alpha}{2}\right)\frac{d\left(v_z^0\right)^2}{dT}=C_\varepsilon^0+\frac{\alpha^2}{2\beta} \qquad \text{for } T<T_c\,. \tag{1.7}$$

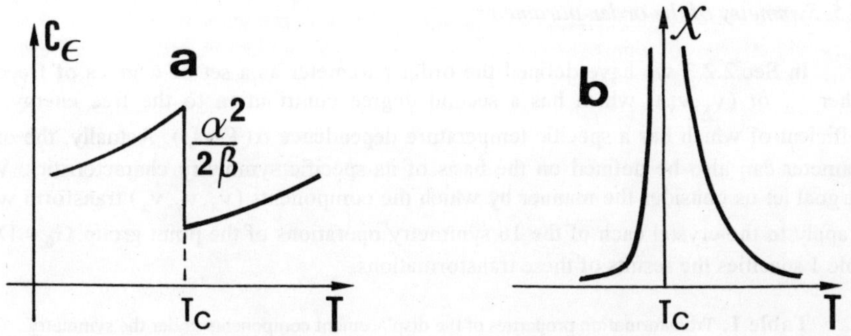

Fig.4. Critical behavior of (a) the specific heat following Eq.(1.7), (b) the susceptibility following Eq.(1.9).

Accordingly, above and below T_c, the specific heat is a different smoothly varying function of temperature, determined by the background free-energy $F_0(T,P)$, and by the smooth variation of the β coefficient. Fig.4a reproduces the anomaly of the specific heat, which has the form of an upward step on cooling through T_c.

The anomaly of the susceptibility χ, in the considered case, is proportional to the dielectric susceptibility of the material. It is defined as: $\chi = \lim_{\varepsilon_z \to 0} \left(\frac{\partial v_z}{\partial \varepsilon_z}\right)_{v_z^0}$. Its calculation requires to examine the behavior of the system in the presence of a small field ε conjugated to the order-parameter. In this case, the appropriate thermodynamic potential whose minimum determines the equilibrium of the system is not F but $G=F-v_z\varepsilon_z$. Minimizing G with respect to v_z leads to:

$$\varepsilon_z = v_z\left[\alpha(T-T_c)+\beta v_z^2\right]. \tag{1.8}$$

For small values of ε_z, the solution of this equation must tend towards the equilibrium values of v_z^0. Deriving these solutions with respect to ε_z, we obtain:

$$\chi(T>T_c) = \frac{1}{\alpha(T-T_c)} \quad ; \quad \chi(T<T_c) = \frac{1}{2\alpha(T-T_c)} \tag{1.9}$$

i.e. the susceptibility goes to infinity when $T \to T_c$ from either sides of the transition (Fig.4b).

The set of anomalies described in this paragraph represent the basic effects of temperature on quantities which are affected by a phase transition. They constitute the "canonical" signature of second-order phase transitions.

2.2.5. Symmetry of the order-parameter

In Sec.2.2.3 we have defined the order parameter as a set of degrees of freedom (either v_z or (v_x,v_y)) which has a second degree contribution to the free energy, the coefficient of which has a specific temperature dependence $\alpha(T-T_c)$. Actually, the order parameter can also be defined on the basis of its specific symmetry characteristics. With that goal let us consider the manner by which the components (v_x,v_y,v_z) transform when we apply to the crystal each of the 16 symmetry operations of the point group $G_0 = D_{4h}$. Table 1 specifies the results of these transformations.

Table 1. Transformation properties of the displacement components under the symmetry operations of D_{4h}.

G_0	C_1	C_4	C_2	C_4^3	σ_x	σ_y	σ_{xy}	$\sigma_{\bar{x}y}$
v_z	v_z	v_z	v_z	v_z	v_z	v_z	v_z	v_z
v_x	v_x	v_y	$-v_x$	$-v_y$	$-v_x$	v_x	v_y	$-v_y$
v_y	v_y	$-v_x$	$-v_y$	v_x	v_y	$-v_y$	$-v_x$	v_x

G_0	I	S_4^3	σ_z	S_4	U_x	U_y	U_{xy}	$U_{\bar{x}y}$
v_z	$-v_z$	$-v_z$	$-v_z$	$-v_z$	$-v_z$	$-v_z$	$-v_z$	$-v_z$
v_x	$-v_x$	$-v_y$	v_x	v_y	v_x	$-v_x$	v_y	$-v_y$
v_y	$-v_y$	v_x	v_y	$-v_x$	$-v_y$	v_y	v_x	$-v_x$

Let us first note that v_z is transformed either into itself or into $-v_z$. Considering this coordinate as the basis vector of a one dimensional vector space, we can conclude that this vector space is invariant by all the transformations of G_0 (a vector space being defined as the set of vectors which are linear combinations of the basis). Such a space, containing no space of smaller dimension, is called *an irreducible invariant space, relative to the group* G_0. On the other hand, each of the components (v_x,v_y) is not transformed into a proportional component by all the elements of G_0. Half of these elements transform v_x into $\pm v_y$, and vice versa. Hence, v_x and v_y are not bases for one-dimensional irreducible invariant spaces. However, their set generates a two-dimensional vector space which has the property to be invariant and irreducible by all the transformations of G_0. Note that the set of the three components (v_x,v_y,v_z) carries a three-dimensional vector space which also has the property to be invariant by all the transformations of G_0. However this vector space contains the two invariant spaces carried respectively by v_z and by (v_x,v_y). Hence it is *reducible*.

In summary, from a symmetry point of view, the order-parameter of a phase transition is a set of degrees of freedom which carries an irreducible vector space (an *irreducible representation*) with respect to the action of the group G_0, which is the symmetry group (point group) of the high-symmetry (parent) phase.

2.2.6. Degeneracy of the low symmetry phase

We have noted hereabove that in the stable state of the system, above T_c, the structure is invariant by G_0. When $\vec{v} \neq 0$ the structure becomes invariant by a smaller set of transformations. Let us enumerate these transformations for each possible stable state of the system below T_c.

When the order-parameter coincides with v_z, we determined below T_c two stable states $v_z^0 = \pm\left[\frac{\alpha(T_c - T)}{\beta}\right]^{1/2}$. The crystalline structures corresponding to the displacements of the M^+ ion parallel to the Z axis are both invariant by the same set of 8 symmetry transformations. These comprise the cyclic group of order 4 generated by the fourfold rotation around Z, and by the reflections in planes containing this axis. This set is the group C_{4v} *subgroup of* G_0. *The transition induces a lowering of the symmetry of the system.* Also note that the two states $\pm v_z^0$ are transformed into each other by some of the symmetry operations such as the mirror-symmetry σ_z "lost" below T_c. These two states necessarily possess the same free energy; they are equally stable. Indeed, application to the system of σ_z preserves the value of the free energy (which is invariant by all the transformations belonging to G_0) while the state $+v_z^0$ is not preserved, and is transformed into $-v_z^0$.

We have not worked out explicitly the stable states of the system in the case of a two-dimensional order-parameter (v_x, v_y). A simple discussion along the line developed for the one-dimensional order parameter v_z would show that the relevant form of the free energy is:

$$F(T,P,v_x,v_y) = F_0(T,P) + \frac{\alpha(T-T_c)}{2}\left(v_x^2 + v_y^2\right) + \frac{\beta_1}{4}\left(v_x^4 + v_y^4\right) + \frac{\beta_2}{2}\left(v_x^2 v_y^2\right) \quad (1.10)$$

and the possible stable states below T_c are:

I) $v_x = \pm\left[\frac{\alpha(T-T_c)}{\beta_1}\right]^{1/2}$, $v_y=0$; **II)** $v_y = \pm\left[\frac{\alpha(T-T_c)}{\beta_1}\right]^{1/2}$, $v_x=0$;

III, IV) $v_x = \pm v_y = \pm\left[\frac{\alpha(T-T_c)}{\beta_1+\beta_2}\right]^{1/2}$.

Similarly to the case of v_z, there is a lowering of the crystal's symmetry below T_c. In the enumerated cases one finds that the respective point groups of the structures are: I) $G_1 = C_{2v}^{X}$, II) $G_1' = C_{2v}^{Y}$, III) $G_2 = C_{2v}^{XY}$, IV) $G_2' = C_{2v}^{\overline{X}Y}$. States I and II correspond to each other through one of the lost transformations of G_0 (e.g. a fourfold axis), and the pair of states III and IV correspond to the same value of the equilibrium free energy. Note that the symmetry groups associated to equally stable states are *conjugate* relative to G_0=D_{4h}, i.e. they satisfy the relationship: $G_i' = g \cdot G_i \cdot g^{-1}$ (i=1,2), with g belonging to G_0.

2.3. Generalization of the preceding example: Formulation of the Landau theory

2.3.1. Summary of the results obtained in the introductory example

In the specific model described in the preceding section, we have shown that an order parameter (e.g. v_z or (v_x, v_y)) is a set of scalar degrees of freedom which allows to describe the symmetry and physical changes accompanying the phase transition in a crystal. The equilibrium values of the n-components of the order-parameter are zero for $T \geq T_c$, and *not all zero* for $T < T_c$. The n-components define a *vector space* ε_n which is an irreducible invariant space by the group G_0, which is the crystallographic symmetry group of the high-temperature parent phase. A variational free-energy F associated with the transition, can be expanded as a function of the components of the order-parameter. The second degree term of this expansion has a coefficient which vanishes and changes sign at the transition temperature T_c. The form of F is determined by the symmetry properties of the order parameter, i.e. by the mode of transformation of the components of the order parameter into each other, when the operations of G_0 are applied to them. The specific form of F determines the symmetries of the potentially stable phases, below T_c, the degeneracy of these phases and the temperature dependences of relevant physical quantities.

In this section we will briefly outline the arguments used in order to formulate the preceding approach in a general framework. These arguments rely in part on the properties of the irreducible representations of a group. The reader can report to the Appendix I, for an introductive presentation of irreducible representations. We will considered here representations of a group G_0 carried by a set $(\eta_1, \eta_2, \ldots, \eta_n)$ of degrees of freedom, and use essentially the following properties:

i) If the set is irreducible and non-totally symmetric, there is no linear combination $f_1(\eta_i) = \alpha_1\eta_1 + \ldots + \alpha_n\eta_n$ of the η_i, which is *invariant* by all the elements of G_0.

ii) If the set is irreducible, there is a single homogeneous polynomial $f_2(\eta_i)$ of degree *two*, which is invariant by G_0. Its form is: $f_2(\eta_i) = \sum_{i=1}^{n} \eta_i^2$.

iii) If $g(\vec{r})$ is an arbitrary function of the space coordinates, one can always write:

$$g(\vec{r}) = \sum_{\gamma}\sum_{j}\varphi_{\gamma j}(\vec{r}) \tag{1.11}$$

where each set of functions $\varphi_{\gamma j}$ (j=1,2,...,p) defines an irreducible invariant space by the group G_0, corresponding to the irreducible representation Γ_γ of G_0.

2.3.2. Description of a second-order phase transition

In order to generalize the considerations developed in Section 2.2, we have to define independently from any specific structure the basic ingredients of the theory: definition of a variational set of degrees of freedom, construction of a free energy, determination of the stable states on either sides of a transition temperature T_c. In a first step, we describe an arbitrary atomic configuration of a crystal by the set of densities $\rho_i(\vec{r})$ of the particles of type i. This set constitutes variational degrees of freedom, which can be used to construct the variational free energy of the crystal $F[T,\rho_i(\vec{r})]$. An equilibrium state of the crystal is defined by the set of functions $\rho_i^{eq}(T,\vec{r})$ which minimize the free energy.

The symmetry of the system at given temperature and pressure, is defined as the set of geometrical transformations which leave all the ρ_i^{eq} invariant. This set forms a group. We are then in a position to define a second-order (continuous) transition at T_c by two conditions: 1) The $\rho_i^{eq}(T,\vec{r})$ are functions whose form change continuously across T_c. 2) The symmetry group of the system just above T_c is different from the symmetry just below T_c. Accordingly, the phase transition though associated to a *continuous change of the spatial configuration of the atoms* is associated to a *sudden change of the symmetry.*

2.3.3. The transition order-parameter

As compared to the set (v_x, v_y, v_z) used in the example of Section 2.2, the variational function $\rho_i(\vec{r})$ have the drawback of *not being small and scalar quantities*, thus making an expansion of the free energy questionable. Without loss of generality, let us restrict to a single type of particles. In order to remove the mentioned difficulty, we put:

$$\rho(\vec{r}) = \rho_0(\vec{r}) + \sum_{\gamma}\sum_{j}\eta_{\gamma j}\varphi_{\gamma j}(\vec{r}) \tag{1.12}$$

where $\rho_0(\vec{r}) = \rho^{eq}(T_c,\vec{r})$ is the equilibrium density at T_c. Let us denote G_0 the symmetry group of this equilibrium density. The sum in the second member of Eq.(1.12) *is a small increment since the transition is continuous.* This increment has been expanded, as in Eq.(1.11) as function of irreducible functions with respect to the group G_0. Moreover, each function has been expressed as the product of a *normalized function* φ and of a *small scalar parameter* η. It is easy to convince oneself that it is possible, at choice, to consider that under the action of G_0, either the φ_j transform into each other, the η_j being fixed coefficients, *or the* η_j *transform into each other, the functions* φ_j *being fixed.* We adopt the second convention.

The variational free energy can now be written in the form: $F\left[T, \eta_{\gamma j}, \varphi_{\gamma j}(\vec{r})\right]$. At each temperature the characteristics of the system are specified by the following conditions: (i) the equilibrium states correspond to the values $\eta^0_{\gamma j}$ which make the free energy minimum. These values define through Eq.(1.12) the equilibrium density of particles in the system. (ii) The symmetry of the system is the group of invariance of the determined equilibrium density. Note that in $(\rho - \rho_0)$, the contribution of the degrees of freedom $\eta_{\gamma j}$ spanning a totally symmetric representation can be ignored in the first place, as one can show that such a contribution would have the same symmetry on either sides of the transition. Therefore it is not crucially associated with the symmetry change defining the phase transition.

Following the ideas introduced in Section 2.2, the free energy form is obtained as a Taylor expansion as function of the small parameters $\eta_{\gamma j}$. Besides, each polynomial term of given degree in this expansion, is invariant by G_0. Indeed, F being a scalar is unchanged by any rotation or reflection. Among the transformations, those belonging to G_0 have the additional property of leaving invariant the density ρ_0, which is, besides T and $\eta_{\gamma j}$, an argument of F.

The group theoretical rules recalled in Section 2.3.1, determine the absence of an invariant linear term in the expansion. Moreover, these rules specify the form of the second degree contribution. We have:

$$F\left[T, \eta_{\gamma j}, \varphi_{\gamma j}(\vec{r})\right] = F_0(T) + \sum_{\gamma} \alpha_\gamma \left(\sum_j \eta_{\gamma j}^2 \right). \tag{1.13}$$

On the other hand the increment of the equilibrium density at the transition is:

$$\delta\rho_{eq}(\vec{r}) = \rho^{eq}(\vec{r}) - \rho_0(\vec{r}) = \sum_{\gamma} \sum_{j} \eta^0_{\gamma j} \varphi_{\gamma j}(\vec{r}). \tag{1.14}$$

Let us first consider the system at T_c. The equilibrium values of all the $\eta_{\gamma j}$ are zero. Hence on the basis of Eq.(1.13), we conclude that all the α_γ satisfy the condition $\alpha_\gamma \geq 0$ (the second degree expansion must be minimum at the origin). Note that *this condition cannot be strict for all* γ. Otherwise, these coefficients would also be positive in the vicinity of T_c , on either sides of this temperature. As a consequence, the equilibrium values of the η_j would all be zero and the symmetry would be unchanged. Hence, *one at least of the* α_γ *has to vanish and change sign at* T_c. An argument already invoked in Section 2.2 allows to assert that *one* α_γ *coefficient only has this property*. The $\eta_{\gamma j}$ corresponding to the other indices γ keep zero equilibrium values in the vicinity of T_c, and can be ignored in the first place. In Eq.(1.13), we can therefore drop all degrees of freedom except the ones associated with the α_γ which vanishes at T_c. This set $(\eta_{\gamma j})$ (j=1,2,...,n) constitutes *the n-dimensional order-parameter of the considered transition.* As it comprises all the degrees of freedom contributing to a single second degree term in the free energy, it necessarily constitutes a basis for an irreducible vector space with respect to G_0, following the group theoretical rules recalled above.

2.3.4. Stable states and symmetry in the vicinity of T_c

Above T_c, due to the positivity of α (we drop the γ coefficient) the equilibrium values of the η_j are zero and the symmetry of the system is G_0, identical to the symmetry at T_c. Below T_c, α is negative and the minimum of F occurs away from the origin in the η_j space. The symmetry of the system comprises all the transformations leaving invariant the density:

$$\rho^{eq} = \rho_0 + \sum_j \eta_j^0 \varphi_j(\vec{r}). \qquad (1.15)$$

Since the η_j^0 contribution to the second member *is small*, these transformations have to be selected among those belonging to the invariance group of ρ_0. As the space $\{\eta_j\}$ defines a non-trivial representation of the latter group the increment of density in Eq.(1.15) cannot be invariant by all the transformations of G_0: *the symmetry group of the system below* T_c *is a subgroup* G *of* G_0. Note that the group-subgroup relationship $G \subset G_0$ requires to assume that the η_j^0 are small. For large η_j^0, Eq.(1.15) does not imply *necessarily* a group-subgroup relationship between the invariance groups of ρ^{eq} and ρ_0 (see Fig.16 of Chapter II).

As pointed out in Section 2.2, in order to determine the minimum of F below T_c, it is necessary to expand the free energy to degrees higher than two. The relevant expression of the free energy is then:

$$F(T,P,\eta_j) = F_0(T,P,\rho_0) + \frac{\alpha(T-T_c)}{2}\left(\sum_j \eta_j^2\right) + f_3(\eta_j) + f_4(\eta_j) + ... \qquad (1.16)$$

where we have developed the coefficient of the second degree term to the lowest degree in $(T-T_c)$. It can be shown that the existence of a *third degree term* $f_3(\eta_j)$ depends exclusively on the nature of the representation Γ_n associated with the order parameter. If the symmetry of the order-parameter is such that a third degree term is not symmetry-forbidden, the transition will be *necessarily discontinuous*. This symmetry-predicted type of first-order transition will be discussed in Section 4.2.

For any symmetry of the order parameter fourth degree terms will always be present in the free energy expansion (if only the square of the second degree term). No further general statement can be made. Depending on the form and coefficients of this term, and eventually of higher degree terms (see Section 3), a continuous (or discontinuous) transition will be possible towards one or several distinct low-symmetry phases. The form of the f_4 term is determined by searching the most general 4th degree polynomial, invariant by the set of transformations belonging to G_0.

2.3.5. Summary of the Landau scheme for the study of a phase transition

In the light of the considerations developed in this section the investigation of a phase transition, in the framework of the Landau theory, proceeds on the following line:
1) Search, as a starting information on the system, the symmetry group G_0 of the more symmetric phase surrounding the transition, and the nature of the irreducible representation Γ_n associated with the order-parameter.
2) Construct the free energy by determining the form of the invariant polynomials of the required degrees.
3) Determine, as a function of the coefficients of the free-energy expansion, the absolute minima of F.
4) For each minimum, determine the invariance group of the density ρ^{eq}, i.e. the "low-symmetry" groups of the system.
5) Derive the temperature dependence of the quantities related to the order parameter.

A further step is to consider the coupling of the order parameter to other relevant degrees of freedom, and derive the temperature dependence of these "secondary" order-parameters.

2.4. Secondary order-parameters

2.4.1. Primary and secondary order-parameters

Starting from Eq.(1.13), we have considered that the free-energy F of the system and the particle density $\rho(\vec{r})$ describing its structure, were functions of a single irreducible set of degrees of freedom of the system: the n components η_j of the order-parameter. The rejection of other degrees of freedom corresponding to other irreducible representations of G_0 (equivalent or unequivalent) was based on the form of the second degree expansion of F, which imposed that only the η_j can take non-zero equilibrium values below T_c. *This conclusion is no longer valid if the expansion of F to terms of degree higher than two is considered.*

In this section we analyze the general conditions which preside over the onset, below T_c, of non-zero equilibrium values for degrees of freedom distinct from the order-parameter components η_j. These degrees of freedom will be termed *secondary order-parameters*, while the set of η_j can be termed the *primary order-parameter*. We will examine the form of the contribution of the *secondary* order-parameter to the free-energy and show that the symmetry characteristics of the transition (e.g. the possible symmetries of the phases stable below T_c) are entirely determined by the primary order-parameter. Before establishing this result on a general ground (Sec.2.4.5) we will examine the influence of a particular secondary order-parameter in the example discussed in Sec.2.2.

2.4.2. Return to the introductory example

The only relevant degree of freedom assumed in the example of Sec.2.2, is the set of three components of the displacement $\vec{v}$. Let us allow the possibility that the c-parameter in Fig.1, equal to the distance along Z between the planes of the negative ions, can change with temperature and pressure. This change corresponds to a deformation of the unit-cell along Z, which coincides with the component e_{zz} of the strain tensor. In the same manner as for the components of $\vec{v}$, the equilibrium value of e_{zz} can be determined by minimizing the variational free-energy F assumed to be a function of e_{zz} as well $F(T,P,\vec{v},e_{zz})$.

The form of the contribution of e_{zz} to F is determined by noting that e_{zz} is invariant under the application of all the transformations of $G_0=D_{4h}$. The part of the expansion of F, depending exclusively on e_{zz} is therefore of the form:

$$\lambda_1 e_{zz} + \frac{\lambda_2}{2} e_{zz}^2 + \frac{\lambda_3}{3} e_{zz}^3 + ... \qquad (1.17)$$

as terms of any degree in e_{zz} are allowed to be present in F. Besides, any power of e_{zz} multiplied by any invariant polynomial of the components of $\vec{v}$ are invariant by G_0 and may be included in F. The *lowest degree* term of this type is:

$$e_{zz}\left[\delta_1(v_x^2 + v_y^2) + \delta_2 v_z^2\right]. \qquad (1.18)$$

In order to examine the effect of the presence of e_{zz}-dependent terms, one has to minimize F with respect to e_{zz}. It yields:

$$\lambda_1 + \lambda_2 e_{zz} + \lambda_3 e_{zz}^2 + \delta_1(v_x^2 + v_y^2) + \delta_2 v_z^2 = 0. \qquad (1.19)$$

As $\lambda_1(T,P)$ is a parameter which, in general, *is not* vanishingly small, e_{zz} cannot either be vanishingly small. λ_1 is actually attached to the thermal expansion of the unit-cell along Z, which is not related to the phase transition. We can eliminate the resulting component of e_{zz} by setting: $e_{zz}=e_{zz}^0+e_{zz}^c$, where e_{zz}^0 is the non-zero value of e_{zz} at T_c, defined by:

$$\lambda_1 + \lambda_2 e_{zz}^0 + \lambda_3 (e_{zz}^0)^2 + ... = 0$$

and e_{zz}^c is the *spontaneous* contribution to e_{zz}, *resulting from the transition*. As both λ_1 and e_{zz}^0 vary smoothly across the transition, Eq.(1.19) can be rewritten as:

$$e_{zz}^{c} = -\frac{1}{C}\left[\delta_1(v_x^2+v_y^2)+\delta_2 v_z^2\right] \tag{1.20}$$

where the coefficient C, which depends on e_{zz}^{0}, is the *effective* coefficient of the quadratic contribution in e_{zz}^{c} to the free energy.

If α_1 or α_2 vanish at T_c, C necessarily remains strictly positive in the neighborhood of T_c. We can then deduce from (1.20) that e_{zz}^{c} vanishes above T_c and becomes non-zero below T_c. Hence, the order-parameter (v_z or (v_x, v_y)) *is not the only quantity possessing non-zero values for* $T<T_c$ *and zero value for* $T>T_c$. Quantities such as e_{zz}^{c} also possess this property and are *secondary order-parameters*. They can be distinguished from the primary order-parameters by the fact that the α_i coefficient associated with it vanishes at T_c, while the coefficient C remains strictly positive nearly T_c. The onset of a non-zero value for e_{zz}^{c} is related to the occurrence of a *coupling term* in the free-energy. If the coefficient (δ_1 or δ_2) of the coupling term is zero, Eq.(1.20) shows that e_{zz}^{c} will remain equal to zero.

2.4.3. Influence of the secondary order-parameter on the properties of the system

Assuming $\alpha_1(T_c,P_c)=0$, i.e. that the primary order-parameter of the transition is v_z, we can eliminate, using Eqs.(1.2) and (1.20), e_{zz}^{c} from the free energy and obtain an expansion of F, which depends only on v_z. This expression is:

$$F = F_0(T,P)+\frac{(\alpha_1 - e_{zz}^{c}\delta_2)}{2}\cdot v_z^2+\frac{(\beta-2\delta_2^2)}{4}\cdot v_z^4 \tag{1.21}$$

which shows that F has the same functional form as (1.2) *but different coefficients*. The effect of coupling between primary and secondary order-parameters is thus *to modify the coefficients of the free energy associated with the primary order-parameter* by amounts increasing with the magnitude δ_2 of the coupling. Eq.(1.21) establishes that none of the conclusions reached in Sec.2.2 is changed: the possibly stable phases and the form of the anomalies associated with the transition, which only depend on the functional form of F, are the same as in the absence of coupling. Only *the boundaries* of the phases *are displaced*, and *the amplitudes* of *the anomalies* are modified. For instance, the transition temperature corresponds to the vanishing of $(\alpha_1-2e_{zz}\delta_2)$, instead of α_1, and will take place at $(T_c+\frac{2e_{zz}^{c}\delta_2}{\alpha})$, instead of T_c. Furthermore, this point is not necessarily significant, as one can consider that (1.2) is the free-energy obtained when the coupling to secondary degrees of freedom, such as e_{zz}^{c}, have been taken into account, and already eliminated through procedures similar to the one leading to (1.21).

Let us also emphasize that the *symmetry* of the various phases *are the same*, as found in Sec.2.2.5, as e^c_{zz} *preserves* the symmetry G_0, and also the breaking of symmetry resulting from the onset of v_z: *the symmetry characteristics of the transition is exclusively determined by the primary order-parameter.*

2.4.4. Physical anomalies related to secondary order-parameters

Along the same line, we can show that the primary order-parameter is also essential for working out the anomalies related to the secondary order-parameters. This is a consequence of the fact that *the existence of a coupling* between the secondary degrees of freedom and the primary order-parameter, is a *necessary* condition for the singular behavior of the additional quantities. Eq.(1.20) can be used to deduce the temperature dependence of e^c_{zz}. One has e^c_{zz}=0 above T_c, and, as below T_c v_z is proportional to $(T_c - T)$, it yields:

$$e^c_{zz} = -\frac{1}{C}(T_c - T) \tag{1.22}$$

corresponding to the *linear dependence* shown in Fig.5a, which differs from the square-root dependence of the primary order-parameter in function of $T_c - T$, represented in Fig.3b.

The quantity which is thermodynamically conjugated to e^c_{zz} is the stress-tensor component σ_{zz}. The susceptibility χ_e associated with e^c_{zz}, defined by $\chi_e = \lim_{\sigma_{zz}\to 0}\left(\frac{\partial e^c_{zz}}{\partial \sigma_{zz}}\right)$, has the meaning of an *elastic compliance*. Minimizing the thermodynamic potential

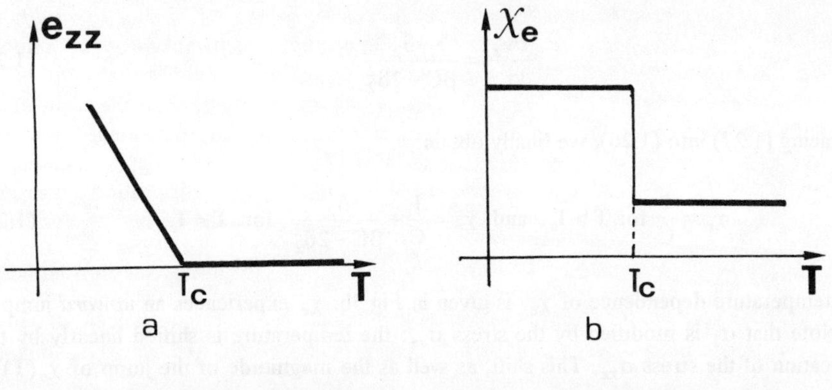

Fig.5. Temperature dependence of (a) the secondary order-parameter $e_{zz} \propto T_c - T$; (b) the elastic susceptibility χ_e following Eq.(1.28).

$F-\sigma_{zz}e_{zz}$ with respect to σ_{zz}, one gets as a substitute to (1.22):

$$e_{zz}^{c}=\frac{\sigma_{zz}}{C}-\frac{\delta_2}{C}v_z^2. \tag{1.23}$$

Likewise, instead of (1.21), one gets the following expression:

$$F-\sigma_{zz}e_{zz}=\frac{1}{2}\left[\alpha_1-2\delta_2 e_{zz}^{0}+\frac{\sigma_{zz}\delta_2}{C}\right]\cdot v_z^2+\frac{1}{4}\left(\beta-\frac{2\delta_2^2}{C}\right)\cdot v_z^4. \tag{1.24}$$

If we assume that the mechanical susceptibility χ_e corresponds to conditions of zero applied stress σ_{zz}, we can write the minimum of (1.24) with respect to v_z:

$$v_z\left\{\left[\alpha_1-2\delta_2 e_{zz}^{0}+\frac{\sigma_{zz}\delta_2}{C}\right]+\frac{\left(\beta-2\delta_2^2\right)}{C}v_z^2\right\}=0. \tag{1.25}$$

From (1.23) we draw the expression of χ_e:

$$\chi_e=\frac{1}{C}-\frac{\delta_2}{C}\cdot\left(\frac{\partial v_z^2}{\partial\sigma_{zz}}\right) \tag{1.26}$$

where $\left(\frac{\partial v_z^2}{\partial\sigma_{zz}}\right)$ is obtained, below T_c, by minimizing the term into bracket in (1.25) with respect to σ_{zz}:

$$\frac{\partial v_z^2}{\partial\sigma_{zz}}=\frac{-\delta_2}{\beta C-2\delta_2^2}. \tag{1.27}$$

Replacing (1.27) into (1.26), we finally obtain:

$$\chi_e=\frac{1}{C}\ \text{for}\ T>T_c,\ \text{and}\ \chi_e=\frac{1}{C}+\frac{\delta_2^2}{\beta C-2\delta_2^2}\ \text{for}\ T<T_c. \tag{1.28}$$

The temperature dependence of χ_e is given in Fig.5b. χ_e experiences an *upward* jump at T_c. Note that α_1 is modified by the stress σ_{zz}: the temperature is shifted linearly by the application of the stress σ_{zz}. This shift, as well as the magnitude of the jump of $\chi_e(T)$ at T_c, is controlled by the magnitude δ_2 of the coupling between e_{zz}^{c} and by the *primary* order parameter.

2.4.5. General properties of secondary order-parameters

The properties disclosed for secondary order-parameters in the preceding working example, can be established within a general framework. Let η_j (j=1,...,n) and ζ_k (k=1,...,m) be the respective components of the primary and secondary order-parameters, and assume that they transform as two *different irreducible representations* Γ_1 and Γ_2 of the parent phase symmetry group G_0. Because Γ_1 and Γ_2 have different symmetries, there is no *bilinear* polynomial of the form:

$$\sum a_{jk}\eta_j\zeta_k \tag{1.29}$$

invariant by G_0, and the second degree invariant polynomials $\sum_j \eta_j^2$ and $\sum_k \zeta_k^2$ are *decoupled*. For instance, the fourth degree homogeneous polynomial:

$$\left(\sum_j \eta_j^2\right)\left(\sum_k \zeta_k^2\right) \tag{1.30}$$

is invariant by G_0, since it is the product of two invariants terms. More generally, one can find *mixed invariants* $g_{p,q}(\eta_j,\zeta_k)$ which are homogeneous polynomials of the set of variables η_j and ζ_k. These invariants are linear contributions of monomials of η_j and ζ_k, which, unlike (1.30), are not necessarily products of polynomials of η_j and ζ_k *separately* invariant by G_0, and have a degree p in the η_j, and a degree q in the ζ_k. The free-energy of the system, in which the primary and secondary order-parameters are involved can be written under the general form:

$$F(T,P,\eta_j,\zeta_k) = F_0(T,P) + \frac{\alpha}{2}\eta^2 + f_4(\eta_j) + \frac{b}{2}\zeta^2 + h_3(\zeta_K) + h_4(\zeta_K) + \\ + ... + \sum_{p,q} g_{p,q}(\eta_j,\zeta_k) \tag{1.31}$$

where $\eta^2 = \sum_j \eta_j^2$ and $\zeta^2 = \sum_k \zeta_k^2$. f_4, h_3 and h_4 are homogeneous invariants of the η_j and ζ_k of respective degrees 4, 3 and 4. $\alpha = \alpha_0(T - T_c)$ and b>0. Taking into account the degrees of the various polynomials in Eq.(1.31), one can write the ζ_k contribution to F as:

$$\zeta^2\left[\frac{b}{2} + \zeta h_3\left(\frac{\zeta_k}{\zeta}\right) + \zeta^2 h_4\left(\frac{\zeta_k}{\zeta}\right) + ... + \sum \zeta^{q-2}\eta^p g_{p,q}\left(\frac{\eta_j}{\eta}, \frac{\zeta_k}{\zeta}\right)\right] \tag{1.32}$$

where, due to their homogeneous character, the polynomials $h_3\left(\frac{\zeta_k}{\zeta}\right)$, $h_4\left(\frac{\zeta_k}{\zeta}\right)$ and $g_{p,q}\left(\frac{\eta_j}{\eta},\frac{\zeta_k}{\zeta}\right)$ remain *finite* when $\eta \to 0$ and $\zeta \to 0$.

If q>2, the coefficient of ζ^2 in (1.32) tends to b/2>0 when η *and* $\zeta \to 0$. Consequently, F has a minimum for $\zeta = 0$, on either sides of a *continuous* transition at T_c: *the* ζ_k *components keep a zero equilibrium value below* T_c. By contrast, if the mixed invariant $g_{p,q}$ of *smallest degree* in the ζ_k, is a *linear function* of the ζ_k, the preceding argument does not hold: the effective coefficient of ζ^2 is determined by the limit of the mixed invariant:

$$\frac{\eta^p}{\zeta} g_{p,1}\left(\frac{\eta_j}{\eta},\frac{\zeta_k}{\zeta}\right). \tag{1.33}$$

Since the *total degree* (p+1) of this invariant is *strictly larger than 2*, as the bilinear term (1.29) is discarded, the exponent p satisfies the condition: $p \geq 2$.

In conclusion, a *necessary condition* for certain components ζ_k to acquire a *non-zero equilibrium value below a continuous transition* at T_c, is the possibility to construct from the set η_j and ζ_k components a *mixed homogeneous polynomial* $g(\eta_j,\zeta_k)$ invariant by G_0, *linear in the* ζ_k *and of degree* $p \geq 2$ *in the* η_j. The projection method which allows to work out these mixed polynomials, and the possible symmetries of the secondary order-parameters for different values of the integer p, with respect to the symmetry of the primary order-parameter, can be found in Ref. [17].

Let us now show that *the contribution of secondary order-parameter to the free-energy can be restricted*, just below the transition, *to two terms*: the *quadratic invariant* $\frac{b}{2}\zeta^2$ and the mixed invariant $\sum \zeta_k g_{p_0}(\eta_j)$, where p_0 is the *smallest degree* involved in mixed invariants of the type:

$$g_{p,q}(\eta_j,\zeta_k) = \sum_k \zeta_k g_k(\eta_j) \tag{1.34}$$

where the $g_k(\eta_j)$ are homogeneous polynomials of degree $p \geq 2$. With that goal, let us substitute (1.34) to the sum $\sum_{p,q} g_{p,q}(\eta_j,\zeta_k)$, in the free energy (1.31). By minimization with respect to η_j and ζ_k we obtain:

$$\frac{\partial F}{\partial \eta_j} = \eta_j\left[\alpha + \frac{1}{\eta_j}\cdot\frac{\partial f_4}{\partial \eta_j} + \ldots + \sum_k \frac{\zeta_k}{\eta_j}\cdot\frac{\partial g_k}{\partial \eta_j}\right] = 0, \tag{1.35}$$

$$\frac{\partial F}{\partial \zeta_k} = b\zeta_k + \frac{\partial h_3}{\partial \zeta_k} + \frac{\partial h_4}{\partial \zeta_k} + ... + g_k(\eta_j) = 0. \tag{1.36}$$

These two equations determine the extrema of F. Since $\frac{\partial h_3}{\partial \zeta_k}$ and $\frac{\partial h_4}{\partial \zeta_k}$ are both of degree larger or equal to 2 in the ζ_k, Eq.(1.34) yields for small η_j and ζ_k: $\zeta_k \cong -\frac{g_k(\eta_j)}{b}$. Replacing ζ_k in Eq.(1.35), one gets

$$\eta_j\left[\alpha + \frac{1}{\eta_j}\cdot\frac{\partial f_4}{\partial \eta_J} - \frac{1}{2b\eta_j}\cdot\sum_k \frac{\partial g_k^2}{\partial \eta_j}\right] = 0\,. \tag{1.37}$$

In Eq.(1.37), $\frac{1}{\eta_j}\cdot\frac{\partial f_4}{\partial \eta_J}$ is of degree 2, while $\frac{1}{\eta_j}\cdot\frac{\partial g_k^2}{\partial \eta_j}$ is of degree ($2p_0$-2). Accordingly two different situations can be met:

1) p_0 is *strictly* larger than 2. For small η_j, the ($2p_0$-2) degree term is negligible compared to the degree 2 term, and Eq.(1.37) is identical to the equation providing the equilibrium values of the η_j in the absence of secondary order-parameter: *the equilibrium is not modified by the* ζ_k.

2) p_0=2. The expression inside bracket in Eq.(1.37) is modified by the existence of the coupling. The modification is equivalent to the replacement in the primary order-parameter expansion, of $f_4(\eta_j)$ by $\left(f_4 - \sum_k \frac{g_k^2}{2b}\right)$, or in other words, to a change in the coefficients of the fourth degree terms in the preceding expansion. If this change preserves the *positive* character of $f_4(\eta_j)$ a continuous transition will still occur for α=0, at the same temperature T_c, as in the absence of coupling term. If the fourth degree term becomes *negative*, a continuous transition will no longer be possible. Placing ourselves in the first case, one has $\zeta_k^0 \cong -\frac{g_{p_0}(\eta_j^0)}{b}$, i.e. *the equilibrium values* ζ_k^0 *are determined by the equilibrium values* η_j^0. If all functions $g_k(\eta_j)$ vanish identically for η_j=η_j^0, the secondary order-parameter does not acquire a non-zero value below T_c, . If, on the contrary, certain $g_k(\eta_k^0)$ are non-zero below T_c, the equilibrium values ζ_k^0 will obey the relation:

$$\zeta_k^0 \propto (\eta_j^0)^{p_0} \tag{1.38}$$

with $p_0 \geq 2$. The ζ_k^0 are therefore small quantities as compared to the η_j^0, *under the conditions that the* η_j^0 *are small*. For example, the terms $h_3(\zeta_k^0)$, $h_4(\zeta_k^0)$ are respectively of order $3p_0 \geq 6$, and $4p_0 \geq 8$, and thus *can be neglected*. Thus, in addition to the coupling term $\sum_k \zeta_k g_{p_0}(\eta_j)$ only the quadratic contribution of the secondary order-parameter will be significant.

At last, let us show that the symmetry of the phase stable below T_c is entirely determined by the symmetry properties of the order-parameter. Following (1.14), we can express the density increment describing the equilibrium structure of the system below T_c, as:

$$\delta\rho_{eq}(T,P,\vec{r}) = \sum_j \eta_j^0 \cdot \varphi_j^1(\vec{r}) + \sum_k \zeta_k^0 \cdot \varphi_k^2(\vec{r}) \tag{1.39}$$

where $\varphi_j^1(\vec{r})$ and $\varphi_k^2(\vec{r})$ are normalized functions spanning respectively the n- and m-dimensional vector spaces ε_n and ε_m, in which Γ_1 and Γ_2 are defined.

Let G be the invariance group of the "vector" η_j^0 in ε_n. The action of G leaves each η_j^0 unchanged, and Eq.(1.38) shows that likewise, G will also leave the "vector" ζ_k^0 invariant in ε_m. As a consequence, the full-invariance group of the primary order-parameter increment $\delta\rho_{eq} = \sum_j \eta_j^0 \cdot \varphi_j^1(\vec{r})$ is also the full-invariance group of the total increment (1.39).

Accordingly, it is not necessary to consider the secondary order-parameters for the study of the nature of the symmetry changes. Note that the increment $\delta\rho_{eq} = \sum_k \zeta_k^0 \cdot \varphi_k^2(\vec{r})$ possess at least the symmetry elements of the group G, but may have a higher-symmetry than G, which coincides at the utmost with G_0. In this limit case, Γ_2 identifies to the totally symmetric irreducible representation of G_0

As a conclusive remark, let us mention the particular (*pseudoproper*) case, where Γ_1 and Γ_2 are *identical* irreducible representations. Here, the relevant mixed invariant (1.29) is bilinear in the components of the primary and secondary order-parameters. In Chapter III and IV we will meet for example situations in which the secondary order-parameter is a macroscopic strain component, which has the same symmetry as the "improper" set of displacements forming the primary order-parameter. In the corresponding transition free-energy, one has to take into account invariants of the same degrees with respect to both variables η_j and ζ_k, namely, the ζ^2 and $h_4(\zeta_k)$ terms in Eq.(1.31).

3. The angular and radial problems in the Landau theory of structural transitions

3.1. Introduction

The general considerations exposed in the preceding section apply to any type of phase transitions in which a group-subgroup relationship exists between the symmetries of the phases: structural transitions in crystals, but also magnetic and superconducting transitions, transitions in liquid crystals etc.... For each type of transition, one has to take into account the specific symmetries of the system and its irreducible degrees of freedom, which may imply particular procedures in applying the theory. For example, magnetic systems require the use of magnetic space-groups, and their corepresentations [2], and to follow the approach described by Dzialoshinskii [3], which separates in the thermodynamic potential the terms representing exchange and relativistic forces. In the normal - to superconducting transition, the symmetry of the phases include the rotations in real and spin space, the gauge symmetry, and time-reversal [4], whereas in liquid crystals, continuous groups in real space and their representations have to be used [5].

Let us examine the particular ingredients needed to apply the Landau theory to transitions between crystalline phases. In this case, the invariance group, G_0 or G, coincides with a *crystallographic space* group, the main properties of which are given in Appendix I. Let us only recall here, that a space group is a group of infinite order formed by elements which can be written $\{R|\vec{t}\}$ where R is a point-symmetry operation and $\vec{t}$ is a translation. The symmetry operations R generate the point-group of the crystal. On the other hand, among the translations $\vec{t}$, there is an infinite group $\vec{T}$ of "primitive" translations, generating the three-dimensional Bravais lattice of the crystal.

As illustrated in Appendix I, there is an infinite set of nonequivalent irreducible representations of a space group G_0. They cannot be tabulated in a systematic manner as the more known representations of finite groups, and have to be constructed starting from simplest representations. Namely each representation has a double index: the first index is a $\vec{k}$-vector in reciprocal space, belonging to the first Brillouin zone of the space. The former vector defines a subgroup $G(\vec{k})$ of G_0 which is the set of elements $\{R|\vec{t}\}$ of G_0 whose elements R leaves $\vec{k}$ unchanged, or transforms it into an equivalent vector, i.e. differing from $\vec{k}$ by a reciprocal lattice vector. This group $G(\vec{k})$ has irreducible representations $\hat{\tau}_i(\vec{k})$ of dimension n_i which are tabulated [6-8]. A representation of G_0 can be denoted $\Gamma_{k\gamma}$. It is constructed according to systematic rules using $\hat{\tau}_i(\vec{k})$. Its dimension is $n = n_i \times r$, where r is the number of vectors in *the star* $\vec{k}^*$ of $\vec{k}$. This star is the set of vectors nonequivalent to $\vec{k}$, having the same modulus as $\vec{k}$, and obtained from $\vec{k}$ by application of all the point symmetry elements R.

As shown in Sec.2.3.3 the difference between two stable phases surrounding a transition is specified by the increment of equilibrium density (1.12) which, following the preceding consideration, can be written for a structural transition:

$$\delta\rho^{eq}(\vec{r}) = \rho(\vec{r}) - \rho_0(\vec{r}) = \sum_{kni} \eta^i_{kn} \cdot \varphi^i_{kn}(\vec{r}) \; . \tag{1.40}$$

The symmetry properties of $\delta\rho^{eq}(\vec{r})$ determine the form of the non-equilibrium variational free-energy $F(T, \eta^i_{kn}, \varphi^i_{kn})$. As underlined in Sec.2.3.2, because the transformation properties of the η^i_{kn} and φ^i_{kn} are identical, F can be considered as function only of the η^i_{kn}. When the transition occurs between strictly periodic structures, and is associated to a single irreducible representation, the indices k and i can be omitted, and the set (η_j) (j=1-n) constitute the order-parameter components. The equilibrium values (η^e_j) are determined from the condition that F is minimal with respect to the order-parameter components:

$$\frac{\partial F}{\partial \eta_j} = 0 \quad (j=1\text{-}n) \tag{1.41}$$

and

$$\left\| \frac{\partial^2 F}{\partial \eta_j \partial \eta_k} \right\| \geq 0 \quad (j,k=1\text{-}n). \tag{1.42}$$

(1.42) expresses the positiveness of the eigenvalues of the $n \times n$ matrix of the second derivatives of F, with respect to the order-parameter components.

In the example discussed in Sec.2.2 the number of components (η_j) is one (v_z), or two (v_x, v_y), and the expansion $F(T, \eta_j)$ was restricted to the minimal fourth power, which may give rise to a stable low-symmetry structure, across a *second-order* transition. When the number n of η_j components increases (and there exist many examples of structural transitions with n=3, 4, 6, 8,...), or when F is expanded to higher powers (and this will be shown to be necessary in the description of some first-order transitions, or for disclosing low-symmetry structures), the minimization of F presents mathematical difficulties. Systematic investigations of the order-parameter symmetries of structural transitions [9-11], and theoretical methods considering the more essential symmetries involved in the Landau approach [12, 13], have allowed to overcome most of these difficulties. The corresponding methods, mainly proposed and applied by Gufan and his collaborators [10, 13], have shown that the problems related to the description of structural transitions break into two categories: the *angular problem* and the *radial problem*.

The angular problem involves the working out of the characteristics of the system which are determined only by the transformation properties (the symmetry) of the order-parameter, but not by the specific form of the thermodynamic potential: classification of the solutions of the equations of states (1.41), enumeration of the possible low-symmetry phases, number of domains in each phase, symmetry of the external interactions which act on the domains etc.. The introduction of simplifying concepts as the ε-space, the image-group, the complete rational basis of invariants etc. has made possible to obtain a geometrically exact solution of the angular problem.

The radial problem consists in finding the equilibrium values of the order-parameter, and subsequently of the other physical quantities, inducing anomalies at the transition, and related to the order-parameter. It includes also the construction of the phase diagram, with the description of the boundaries between the phases, the coexistence regions etc.. The problem cannot be solved without model assumptions, namely without knowing the specific form of the thermodynamic potential, and the power at which it is truncated. Many progresses have also been made in this field, using the results obtained in the theory of singularities [14].

In this section we will present successively the main results obtained in the solutions of the angular problem (Sec.3.2), and of the radial problem (Sec.3.3).

3.2. The angular problem and its solutions

3.2.1. Image of G_0 *in the order-parameter representation*

Consider the set of matrices $M(g_i)$ defining a representation Γ of G_0 in a given space ε (irreducible or not). These matrices are not necessarily distinct. For instance in Table 1, the matrices of D_{4h} spanned by the one-dimensional irreducible subspace z are equal to 1 for half of the g_i and -1 for the other half. In the two-dimensional irreducible subspace (x,y), the matrices $M(C_1)$ and $M(\sigma_z)$ are both equal to $\begin{bmatrix} 1 & 0 \\ 0 & 1 \end{bmatrix}$. Γ is a *faithful* representation of G_0, when all its matrices are distinct. In this case, the correspondence between the g_i and the $M(g_i)$ is an *isomorphism*. If Γ is not faithful, the correspondence is an *homomorphism*.

The set of distinct matrices $M(g_i)$ form a *group*, which is called the image (denoted I_0) of G_0 in the homomorphism. The *kernel* K of the homomorphism is the set $g_i \in G_0$, such as the matrices $M(g_i)$ coincide with the identity matrix in the space ε. K is an *invariant* subgroup of G_0 (i.e. $g_i K = K g_i$ for all $g_i \in G_0$), and G_0 can be expressed as a sum of left or right *cosets* associated to K:

$$G_0 = K + g_2 K + g_3 K + ... \tag{1.43}$$

where the g_i *do not belong to* K. The set of cosets form the *quotient group of* G_0 *by* K, which is denoted G_0/K. This group is isomorphous to the image G_0, and we can write $G_0 = I_0 \times K$. For a *finite* group G_0, the orders (n_G) of the three groups G_0, I_0 and K satisfy the equality: $n_{G_0} = n_{I_0} \cdot n_K$. Reciprocally, *the order of* I_0 *can be finite even if* G_0 *is an infinite group.*

The preceding considerations apply to the n-dimensional irreducible representations carried by the n-components of the transition order-parameter. The set of distinct matrices of this irreducible representation constitutes a group I_0, which is the image of the group G_0 of invariance of the high symmetry phase. Each matrix in I_0 is *orthogonal*, since it is real and unitary. Its action on the components (η_j) of the order-parameter can therefore be considered as a geometrical transformation in the n-dimensional space ε_n of the order-parameter, which has the property to preserve the origin in this space, *as well as the length of vectors*. In the 3-dimensional space such transformations comprise rotations about an axis passing through the origin, reflections about a plane passing through the origin, as well as products of these transformations. The rotations are the orthogonal transformations whose matrices have a determinant equal to +1. The set of all rotations about any axis passing through the origin form a group denoted SO(3), while the set of all orthogonal transformations is a group denoted O(3). More generally, in the n-dimensional space, the set of all orthogonal transformations with determinant +1 forms a group denoted SO(n) while the set of all orthogonal transformations is the group denoted O(n).

The image I_0 of G_0 in the order-parameter representation is isomorphous to a subgroup of O(n) since it is composed of $n \times n$ orthogonal matrices. I_0 also has an additional peculiarity, which is related to the irreducibility of the order-parameter space, under the action of I_0: it can only be a subgroup of O(n) which has no invariant subspace of the n-dimensional space. The subgroups of O(n) possessing this property are termed *irreducible subgroups of* O(n). Another possible terminology is based on the remark that the n-dimensional space ε_n is spanned by the components of a vector ($\eta_1,\ldots,\eta_n$). Thus ε_n spans the "*vector representation*" of the n-dimensional space. The preceding requirement can be expressed by stating that the vector representation of the group I_0 be *irreducible*. For instance if n=3, the group D_{4h} in Table 1 leaves the $\vec{z}$ direction and the (x,y) plane separately invariants. This group *is not* an irreducible subgroup of O(3). By contrast, the restriction of D_{4h} to $\vec{z}$ or to (x,y) are irreducible subgroups of O(1) or O(2), respectively.

3.2.2. Use of the image-group in the application of the Landau theory

The action of G_0 in the order-parameter space is effected by means of the elements of its image I_0. Hence, once the order-parameter symmetry and the related image I_0 are known, it is possible on the basis of this sole knowledge to determine the form of the free-energy, and to work out the possible symmetries of the low symmetry phase.

Each degree in the free-energy expansion, will be constituted by the most general polynomial of this degree, in the η_j components, which is invariant by I_0. As I_0 is generally a more simple group than G_0 (e.g. a finite group instead of an infinite one), the consideration of I_0 will simplify the procedure of determination of the successive terms of the Landau free-energy. Besides, consideration of the images shows that the determination of the forms of the free-energies which arise for any phase transition associated with a n-component order-parameter, is a mathematically well-defined and solvable task. This task consists in enumerating the irreducible subgroups of O(n), and for each subgroup, in working out the form of the invariant P-th degree polynomials constructed from the set η_j of coordinates in the n-dimensional space.

A similar simplification will occur in the derivation of the symmetry group of the low-symmetry phase, which is a subgroup of G_0 leaving invariant a given vector in the space ε of the order-parameter. In this space, G is represented by its image I_G which is the subgroup of I_0 leaving invariant the considered vector. If I_G is known, G can be determined as its reciprocal image: if $M(g_i)$ is an element of I_G, all the elements $g_i K$ belong to G. Hence, in order to determine G, it is sufficient to determine the simpler group I_G and use the relation $G = I_G \times K$. Examples of this procedure are treated in Sec.3.2.7.

The enumeration of the irreducible subgroups of O(n) has been performed for $n \le 4$ [13, 15]. Thus one has access to a full list of the order-parameter symmetries for order-parameter dimensions up to 4. Let us summarize the corresponding results.

1) *n=1.*

The group O(1) is reduced to the identity and to the inversion in the order-parameter space. It has only one strict subgroup coinciding with the identity. Accordingly, there are two possible symmetries for n=1. For O(1)=(E,I) (C_i in the Schoenflies notation), the resulting free-energy is:

$$F = \frac{\alpha}{2}\eta^2 + \frac{\beta_1}{4}\eta^4 + \frac{\beta_2}{6}\eta^6 + \ldots \qquad (1.44)$$

while for the identity subgroup C_1, all powers of η are allowed:

$$F = a\eta + \frac{\alpha}{2}\eta^2 + \frac{B_1}{3}\eta^3 + \frac{\beta_1}{6}\eta^4 + \frac{B_2}{5}\eta^5 + \ldots \qquad (1.45)$$

2) *n=2.*

The group O(2) acts in a plane. It comprises the set of all rotations about an axis perpendicular to the plane and passing through the origin, as well as the reflections about any plane containing the preceding axis. Besides SO(2), O(2) possesses an infinite set of irreducible subgroups which can be classified in two categories: the groups C_n ($n \ge 3$), which are cyclic abelian groups generated by the rotation $2\pi/n$, and the groups

$C_{nv} = C_n + \sigma C_n$, where σ is an *arbitrary* reflection, as different σ will lead to conjugated, physically equivalent, groups.

Hence, the relevant images comprise two infinite groups O(2) and SO(2), and an infinite set of finite groups C_n and C_{nv}. Choosing polar coordinates (ρ, θ) in the order-parameter plane of components (η_1, η_2), and since O(2) and SO(2) contain rotations by an arbitrary angle θ, the polynomials of (η_1, η_2) which are invariant by these two groups are only functions of $\rho^2 = \eta_1^2 + \eta_2^2$. The free energy adapted to these images is:

$$F = \frac{\alpha}{2}\rho^2 + \frac{\beta_1}{4}\rho^4 + \frac{\beta_2}{6}\rho^6 + \dots \quad (1.46)$$

The form of the invariant polynomials and free-energies associated with the irreducible subgroups C_n and C_{nv}, of O(2), as well as the case n=3, can be more elegantly worked out by introducing the concept of integrity bases [10(a,b), 16].

3.2.3. Integrity bases

For a *finite group*, as C_n or C_{nv}, any invariant polynomial can be obtained as linear combination of the powers of a limited number of basic polynomials, forming the *integrity basis of the group* [10(a,b), 16]. This mathematical property has generated methods of finding the integrity bases for certain types of groups [17]. We will restrict ourselves to summarize the corresponding results.

In the case of the groups C_n and C_{nv}, the integrity bases are , for C_n:

$$I_1 = \rho^2, \quad I_2 = \rho^n \mathrm{Cos}(n\theta), \quad I_3 = \rho^n \mathrm{Sin}(n\theta) \quad (1.47)$$

and for C_{nv}:

$$I_1 = \rho^2, \quad I_2 = \rho^n \mathrm{Cos}(n\theta). \quad (1.48)$$

For instance for n=3 the basis comprises ρ^2 and $\rho^3\mathrm{Cos}(3\theta)$ for C_{3v}, while for C_3 it contains the additional polynomial $\rho^3\mathrm{Sin}(3\theta)$. The corresponding free-energy will be written, for example, for C_{3v}:

$$F = a_1 I_1 + a_2 I_2 + a_3 I_1^2 + a_4 I_2^2 + a_{12} I_1 I_2 + \dots \quad (1.49)$$

Analogously, one has for C_{4v}, $I_1 = \rho^2$, $I_2 = \rho^4\mathrm{Cos}(4\theta)$ and:

$$F = a_1 I_1 + a_2 I_1^2 + a_3 I_2 + a_4 I_1^3 + a_{12} I_1 I_2 + \dots \quad (1.50)$$

3) *n=3.*

The irreducible subgroups of O(3) form an infinite set and, as in the case n=2, we can identify the subgroups which are conjugate to each other as representing equivalent physical situations. One then finds [17] that besides SO(3), there are 7 irreducible groups among the subgroups of O(3): the 3 subgroups of SO(3), denoted T, O and Y, and the 3 subgroups of O(3), denoted T_h, O_h and Y_h, which are obtained by adjoining to (T, O, Y) the symmetry about the origin. The seventh group, T_d, results from the combination with T of the reflection about a symmetry plane of the regular tetrahedron.

The two continuous group O(3) and SO(3) give rise again to the free-energy (1.46), where $\rho^2 = \eta_1^2 + \eta_2^2 + \eta_3^2$. Table 2 lists the integrity bases of the invariant polynomials corresponding to the 5 finite subgroups T, T_h, T_d, O and O_h. The integrity basis of the invariants polynomials for the irreducible groups Y and Y_h, can be found in Ref.[17].

Table 2. Integrity bases for invariant polynomials corresponding to five irreducible subgroups of O(3).

T	$I_1 = \eta_1^2 + \eta_2^2 + \eta_3^2$; $I_2 = \eta_1\eta_2\eta_3$; $I_3 = \eta_1^4 + \eta_2^4 + \eta_3^4$; $I_4 = (\eta_1^2 - \eta_2^2)(\eta_2^2 - \eta_3^2)(\eta_3^2 - \eta_1^2)$
T_h	I_1; I_3; I_4; $I_5 = \eta_1^2\eta_2^2\eta_3^2$
T_d	I_1; I_2; I_3
O	I_1; I_3; I_5; $I_2 \times I_4$
O_h	I_1; I_3; I_5

4) *n=4.*

For n=4, the irreducible subgroups of O(4), and the corresponding integrity bases have been worked out and can be found in Ref.[18]. For $n \geq 4$, the enumeration of irreducible subgroups of O(n), and of their invariant polynomials is only partial at present. Systematic results corresponding to definite classes of phase transitions have been established, in particular for some categories of structural transitions [19], for liquid crystal systems [20], and for "unconventional" superconductors [21].

3.2.4. Isotropy groups and other simplifying mathematical concepts

In order to determine the absolute minima of the Landau free-energy, one has to search the values of the order-parameter components η_j, which satisfy Eqs.(1.41) and (1.42). The standard minimization process presents no major difficulty for small dimensionalities of the order-parameter, and for order-parameter expansions truncated to the lowest powers of the η_j. For a large number of order-parameter components or (and) when higher degree expansions are taken into account, there may exist strong technical difficulties. A number of methods, more elaborated than the standard one, have been described, which take advantage of the symmetry properties of the minima η_j^{eq}, satisfying (1.41) and (1.42), and use some symmetry concepts, which will be summarized in this section.

a) *Isotropy group.*

In Sec.3.2.2, we have seen that the space $\varepsilon_n=(\eta_1,\eta_2,\ldots,\eta_n)$ is irreducible under the action of the image $I_0 \in O(n)$. No subspace of ε_n is invariant by I_0. However every vector direction $\vec{\eta}$ in ε_n is left invariant by some subgroup I_η of I_0. I_η, which is defined as the complete set of elements of I_0 leaving $\vec{\eta}$ unchanged, is called the *isotropy-group* of $\vec{\eta}$ (or *little group* of $\vec{\eta}$). Along this line, the symmetry-group G of the phase stable below T_c, is related to the isotropy group $I_{\eta^{eq}}$ of the vector $\vec{\eta}^{eq}=(\eta_1^{eq},\eta_2^{eq},\ldots,\eta_n^{eq})$ corresponding to the absolute minimum of the free-energy, by $G=I_{\eta^{eq}}\times K$, where K is the kernel of the homomorphism $G_0 \rightarrow I_0$.

b) *Orbit, stratum.*

Since $\vec{\eta}$ is not invariant by all the elements of I_0, when applying the elements of the group I_0 to a given vector $\vec{\eta}$, we generate a set of distinct vectors $\vec{\eta}(p)$ which forms the *orbit* (or the *star*) of $\vec{\eta}$. The isotropy groups $I_\eta^{(p)}$ and $I_\eta^{(q)}$ of any two vectors $\vec{\eta}(p)$ and $\vec{\eta}(q)$ of the orbit are *conjugated* in I_0. This result applies to the orbit of the absolute minimum $\vec{\eta}^{eq}$ of F and has the consequence that the groups $G^{(p)}$, such as $G^{(p)}/K=I_\eta^{(p)}$ are conjugated in G_0.

Two vectors whose isotropy groups are conjugated in I_0, do not necessarily belong to the same orbit. If they belong to different orbits, these distinct orbits are said to be of the *same type*. The set of all orbits of a given type (i.e. characterized by isotropy groups which are conjugated in I_0) is called *a stratum*. Distinct orbits of a stratum contain the same number of vectors, i.e. they are associated with the same number of minima of the free-energy. However the distinct orbits do not correspond to the same *value* of the free-energy.

c) *Minimal and Maximal isotropy-groups.*

The isotropy-group I_η of the vector $\vec{\eta}$ can leave invariant other vectors in the ε-space. These vectors and $\vec{\eta}$ generate a subspace ε_η of ε, each vector of which is invariant by I_η. This subspace is termed the *isotropy manifold* of I_η. If I_0 is finite, ε_η belongs to the stratum of $\vec{\eta}$. The isotropy groups I_η associated to all the vector directions in ε possess a common subgroup I_k, possibly reduced to the identical transformation. I_k is an invariant subgroup of I_0, and of all the I_η. Consequently (as it is self-conjugate) it is the isotropy group of all the vectors of a stratum. This stratum contains a continuous set of distinct vector directions. If a vector $\vec{\eta} \in \varepsilon$ does not belong to this stratum, it constitutes a *symmetry direction* in ε: its isotropy group will contain I_k as a *strict subgroup*. The isotropy manifold of I_η will be a *symmetry manifold.*

A partial order exists between the isotropy-groups I_η, up to a conjugation in I_0. Thus, for two directions $\vec{\eta}$ and $\vec{\eta}'$, the isotropy-groups I_η and $I_{\eta'}$, can satisfy an inclusion relation $I_\eta \subset I_{\eta'}$, up to a conjugation (i.e. I_η belongs to a group conjugated to $I_{\eta'}$). If the direction $\vec{\eta}$ is the subgroup of no other isotropy-group I_η, I_η is called a *maximal-isotropy group*, and $\vec{\eta}$ can be termed a *maximal-symmetry direction* in ε.

3.2.5. *Illustration of the preceding concepts*

Let us apply the preceding mathematical concepts to the example treated in Sec.2.2, considering only the two-dimensional order-parameter space $\varepsilon_2 = (v_x, v_y)$ of the group $G_0 = D_{4h}$. The set of distinct matrices of D_{4h} in the order-parameter representation (v_x, v_y) which expresses the action of D_{4h} in ε_2 is:

$$\begin{matrix} C_1,\sigma_z & C_2,I & C_4,S_4 & C_4^3,S_4^3 & \sigma_x,U_y & \sigma_y,U_x & \sigma_{xy},U_{\bar{x}y} & \sigma_{\bar{x}y},U_{xy} \\ \begin{bmatrix} 1 & 0 \\ 0 & 1 \end{bmatrix} & \begin{bmatrix} -1 & 0 \\ 0 & -1 \end{bmatrix} & \begin{bmatrix} 0 & 1 \\ -1 & 0 \end{bmatrix} & \begin{bmatrix} 0 & -1 \\ 1 & 0 \end{bmatrix} & \begin{bmatrix} -1 & 0 \\ 0 & 1 \end{bmatrix} & \begin{bmatrix} 1 & 0 \\ 0 & -1 \end{bmatrix} & \begin{bmatrix} 0 & -1 \\ -1 & 0 \end{bmatrix} & \begin{bmatrix} 0 & 1 \\ 1 & 0 \end{bmatrix} \end{matrix} . \qquad (1.51)$$

One can verify on the first row of (1.51), that it corresponds to the image $I_0 = C_{4v}$. As D_{4h} and C_{4v} are groups of order 16 and 8, respectively, the kernel of the homomorhism $G_0 \to I_0$ is a group K of order 2 comprising the identity C_1, and the reflection σ_z. The invariance group G of every direction (v_x, v_y) in ε_2 are given in Table 3. The isotropy groups of these directions, subgroups of I_0, can be derived by taking the quotient groups (G/K). It is straightforward to work them out directly by simple inspection of Fig.6. Three types of orbits can be distinguished (i.e. three strata):

(1) the orbit of the vector $(v_x, 0)$ which comprises three additional vectors: $(-v_x, 0)$, $(0, v_x)$ and $(0, -v_x)$. Their isotropy group is, up to a conjugation, the group of order 2:

Table 3. Invariance groups of the various directions in ε_2.

Direction	G
$(v_x,0)$	C_{2v}^{x}
$(0,v_y)$	C_{2v}^{y}
(v_x,v_x)	$C_{2v}^{\bar{x}y}$
$(-v_x,v_x)$	C_{2v}^{xy}
(v_x,v_y)	C_s^{z}

$I_1=(C_1,\sigma_y)$. The isotropy manifold of I_1 is reduced to the $\pm(v_x,0)$ direction: it is a line, which corresponds to a *maximal symmetry direction* in ε_2, i.e. the corresponding isotropy-group is *maximal*.
(2) The orbit of the vectors (v_x,v_x) comprises the additional vectors $(v_x,-v_x)$, $(-v_x,-v_x)$, and $(-v_x,v_x)$. The isotropy group is also of order 2, $I_2=(C_1,\sigma_{\bar{x}y})$. Here again the isotropy manifold is a line coinciding with one of the bisectors, which is a maximal symmetry direction in ε_2, corresponding to a maximal isotropy group.
(3) The orbit of the vector (v_x,v_y) with $v_x\neq v_y\neq 0$. There are 8 distinct vectors in the orbit: $(\pm v_x,\pm v_y)$ and $(\pm v_y,\pm v_x)$. The isotropy-group of any vector of the orbit is C_1, which is obtained for all the orbits whose representative vector (v_x,v_y) is in a general direction of the (v_x,v_y) plane. The stratum corresponding to the isotropy group is composed of all the directions in the plane, except the directions corresponding to the two first orbits. Such directions correspond to the *minimal isotropy-group*.

In the considered example, we have determined the isotropy groups of every direction in the 2-dimensional space, by checking in each case, which ones of the 8 elements of $I_0=C_{4v}$ leave the considered vector direction invariant. An alternate

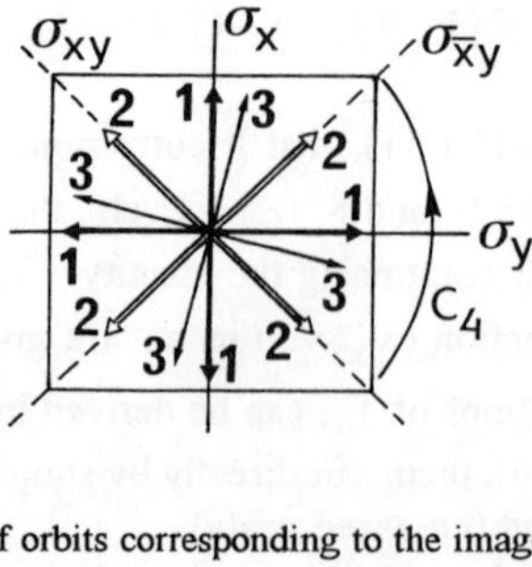

Fig.6. The three types of orbits corresponding to the image C_{4v}: $[\pm(v_x,0),\pm(0,v_x)]$ (orbit 1), $[\pm(v_x,\pm v_x)]$ (orbit 2), and $(\pm v_x,\pm v_y)$ (orbit 3).

method which does not require the detailed knowledge of the action of I_0 on the order-parameter space can be used. This method, which is termed the subduction criterion [22], relies on the knowledge of the characters of the representation of I_0, carried by the space ε. It is exposed in Refs. [17, 22].

3.2.6. Minima of F and maximal isotropy-groups

In the present section and in the next one we briefly describe the method which can be used in order to determine in certain cases the minima of the expansion F, without solving the minimization equations (1.41) and (1.42).

In the example considered in the preceding section, we can notice that the two phases corresponding to vector directions in the order-parameter space, *whose isotropy groups are maximal*, namely $[(\pm v_x, 0), (0, \pm v_x)]$ or $(\pm v_x, \pm v_x)$ are the same that have been found in Sec.2.2.6 *to be stable below a second-order transition*. By contrast the general direction in the order-parameter space $(v_x \neq v_y \neq 0)$ which has the *minimal isotropy-group* does not describe one of the relevant stable phases. Generalizing, one can state: *the vector directions which correspond to the absolute minimum of F, below a second-order transition are along maximal directions in the order-parameter space*. This property which holds strictly for order parameters of dimension $n \leq 3$, is *not always* true for $n \geq 4$, and loses its validity for first-order transitions.

One can express the relation between the absolute minimum of F and the maximal isotropy directions in two different formulations:

Formulation 1: Below a continuous transition the symmetry of the stable phase necessarily corresponds to a vector $\vec{\eta}$ in the order-parameter space, whose isotropy group $I_\eta \subset I_0$ (I_0 being the image of G_0) is maximal.

Let I'_0 be the largest subgroup of O(n) which leaves invariant the fourth degree free-energy expansion F, associated with the n-component order-parameter. In reference to I'_0, we can define isotropy groups for every $\vec{\eta}$-direction in the order-parameter space: I'_η is the largest subgroup of I'_0 which leaves $\vec{\eta}$ invariant. This gives the alternate formulation:

Formulation 2: The absolute minimum of F occurs in a direction $\vec{\eta}$ whose isotropy-group I'_η is maximal.

Proof: The connection between the location of the extrema of F and the isotropy-groups of the various directions in the order-parameter space relies on the fact that since F is invariant under the action of $g \in I'_0$, the set of derivatives $\left(\frac{\partial F}{\partial \eta_1}, \ldots, \frac{\partial F}{\partial \eta_n}\right) = \mathrm{grad}_\eta F(\vec{\eta})$ transforms under the action of g in the same way as the components of the vector $\vec{\eta}$ of the order-parameter space. Let $\vec{\eta}_0$ be a direction in the order-parameter space ε and I'_η the isotropy-group referred to the invariance group I'_0 of the free-energy $F(\vec{\eta})$. Let us place ourselves in the case where the subspace of ε, invariant by I'_η is reduced to the single vector direction $\vec{\eta}_0$ (the proof can be extended [23] to the general situation where the isotropy-manifold associated with I'_η is not restricted to a single direction). In this case, $\text{grad}_\eta F(\vec{\eta}_0)$ is parallel to $\vec{\eta}_0$ since the two vectors have the same symmetry properties, and consequently $\text{grad}_\eta F(\vec{\eta}_0)$ belongs to the isotropy-space of I'_η, which is reduced to the direction $\vec{\eta}_0$. Accordingly, the occurrence of an extremum (or a saddle point) along $\vec{\eta}_0$ can be deduced from the variation of the *one variable function* $F(\eta_0)$, equal to the restriction of the free-energy $F(\vec{\eta})$ to the $\vec{\eta}_0$ axis. $F(\eta_0)$ is a fourth degree polynomial which has a local maximum for $\eta_0=0$, and a minimum for some finite, non zero value, of η_0. This minimum is either a local minimum or a saddle point of $F(\vec{\eta})$. As, on the other hand, one dimensional isotropy spaces are necessarily maximal symmetry directions, the preceding arguments establish that maximal symmetry directions correspond to extrema of the free-energy.

Let us stress that Formulation 2 of the claimed property, refers to the symmetry group I'_0 of the *fourth* degree expansion of the order-parameter, which is in general a larger group than I_0 ($I_0 \subseteq I'_0$) as the image of G_0 leaves the full (non truncated) expansion invariant, while I'_0 only leaves invariant the 4-th degree expansion. Let us also emphasize that there exist counter-examples to both the statements 1 and 2 [24]. Nevertheless these statements are useful, despite their lack of absolute validity as: (i) they are strictly valid for $n \leq 3$, (ii) even for $n \geq 4$ they are satisfied by *most* of the relevant cases, (iii) general theoretical considerations given in the hereabove proof, show that directions $\vec{\eta}$ having maxima isotropy groups I'_η are necessarily extrema of the order-parameter expansion, whatever its degree. Actually, the mentioned counter examples to the statement 2 show that the maximal symmetry directions are not the *only extrema* of the fourth degree expansion, and that *additional extrema*, can constitute the absolute minimum of F.

Another method has been proposed by Michel and Mozrzymas [25], which allows in certain cases to determine the possible minima of the Landau free-energy, without solving the minimization equations. The method makes use of the Morse inequalities [25]:

$$\sum_{k=0}^{n} \mu_k \leq (P-1)^n + 1 \qquad (1.52)$$

where μ_k is the number of extrema, the Morse index k corresponding to the number of negative eigenvalues of the n×n matrix (1.42). P is the maximal power considered in the order-parameter expansion. These inequalities express the facts that if a polynomial of a given degree has a number μ_0 of maxima, there are constraints on its number of saddle points and maxima, and that the total number of its extrema is bounded. Besides the method uses the property that each type of extremum is *degenerate* in a way which is specified by the symmetry of the orbits in the order-parameter space. The procedure allowing to apply this method is exposed in Ref.[17, 25] and will not be detailed here, as it mainly permits to determine the stable orbits with respect to a fourth degree Landau expansion for a small number n of order-parameter components (n=2, 3) (the Morse inequalities become insufficiently restrictive for n≥4).

3.2.7. Minima of F and decomposition into orbits of the integrity basis of I_0

It has been first noted by Gufan [10(a,b)] that the expansion of F in terms of the integrity basis of invariant polynomials leads naturally to a classification of the extrema of F based on their symmetry, i.e. on the type of orbit associated to them. Developments along this line have been pointed out by various authors [23,26,27]. The method aims at obtaining results concerning the extrema of the free-energy, which *is not necessarily restricted to the fourth degree* and also, results which are *independent from the degree of the expansions*. It is thus well adapted to the description of first-order transitions, and will be exposed in some details.

a) Gufan's method.

The method relies on two elements. On the one hand, it uses the possibility to expand F as a function of a limited number of basic polynomials I_m which are the members of the integrity basis relative to the image I_0 of the invariance group of the system, in the order-parameter representation. On the other hand, it implies a prior enumeration of the maximal isotropy groups and of their isotropy spaces. The use of these elements allow a separation of the order-parameter components corresponding to an extremum into two sets: (1) a set of components which is determined by symmetry considerations and which is consequently independent both on the temperature and of the degree of the free energy; (2) the remaining set for which an actual calculation of the extremum must be performed. This set concerns temperature dependent order-parameter components, whose behavior depend on the degree and on the coefficients of the free-energy.

Let $(I_1, I_2, \ldots, I_k)$ be the k elements of the integrity basis of polynomials relative to the image I_0, as defined in Sec.3.2.3. We denote $I_1 = \sum_{j=1}^{n} \eta_j^2 = \rho^2$ the second-degree invariant polynomial which is always part of the basis for any group I_0. As any term in the order-parameter expansion can be expressed as a function of the I_m, we can write:

$$F = F(I_1, I_2, \ldots, I_k) \tag{1.53}$$

where each I_m is a homogeneous polynomial of degree p_m in the order-parameter components η_j. Let us put, $\eta_j = \rho\gamma_j$, where the γ_j specify a *direction* in the order-parameter space. On the other hand, as J_m is homogeneous, we have:

$$I_m(\eta_1,...,\eta_n) = \rho^{p_m} \cdot D_m(\gamma_1,...,\gamma_n). \tag{1.54}$$

Substituting ρ^2 and the D_m (m=2,...,k) in (1.53), we obtain:

$$F = F_1(\rho^2, D_2,...,D_k). \tag{1.55}$$

The extrema of F are determined by the set of equations:

$$\frac{\partial F}{\partial \rho} = 2\rho \frac{\partial F_1}{\partial \rho^2} = 0, \tag{1.56}$$

$$\frac{\partial F}{\partial \gamma_j} = \sum_{m=2}^{k} \frac{\partial F_1}{\partial D_m} \cdot \frac{\partial D_m}{\partial \gamma_j} = 0. \tag{1.57}$$

Note that there are only (n-1) independent equations of the type (1.56). Let us consider a symmetry direction (γ_r^0). It is a direction whose isotropy-group I_η has an invariant manifold reduced to the sole direction (γ_r^0), and as noted in Sec.3.2.6, we can observe that the gradient (with respect to the η_j or the γ_j) of any function invariant by I_0, at a point of the axis γ_j^0, is parallel to this axis. Hence, all the basic invariants D_m necessarily satisfy the set of conditions:

$$\left(\frac{\partial D_m}{\partial \gamma_j}\right)_{\gamma_j^0} = 0. \tag{1.58}$$

The equations (1.57) are identically satisfied along the direction (γ_j^0). In order to determine the location of the extrema of F_1 along this direction we have to solve Eq.(1.56), and specify the temperature dependent value of the modulus $\rho_0(T)$. In this case, the actual computation is reduced to the determination of $\rho_0(T)$. It is only the result of the latter computation which depends on the coefficients of F or on its degree. The values of the γ_j^0 derive from the decomposition of the order-parameter space into orbits with respect to I_0, or, alternately, from a set of equations (1.58) bearing on a small number of polynomials in the γ_j (i.e. the polynomials of the integrity basis). These polynomials have well defined numerical coefficients and their extremalization does not determine a temperature dependence for the γ_j^0.

We can generalize the preceding situation by considering the case where the isotropy-group I_η has an isotropy-space of dimension 1<n'<n. Let us chose in the order-parameter space, the n' first coordinate axes ($\gamma_1,...,\gamma_{n'}$) within the isotropy manifold of I_η. Using, once more the argument quoted in 3.2.6, that the gradient of any invariant function D_m, at a point lying within the manifold is a vector belonging to the isotropy-manifold of I_η, we conclude that the components of the gradient perpendicular to the considered manifold, vanish identically. This implies:

$$\left(\frac{\partial D_m}{\partial \gamma_{j'}}\right)_{\gamma_j^0} = 0 \qquad (1.59)$$

with j'=n'+1, n'+2,...,n and $\gamma_j^0=0$ for j=n'+1, n'+2,...,n. In (1.57), (n-n') equations vanish identically. The remaining "angular" equations (1.57) and the "radial" equations (1.56) determine a temperature-dependent direction (γ_j^0) (j=1,...,n') within the isotropy-manifold of I_η.

It is worth pointing out that the n'-dimensional manifold considered can contain symmetry directions, whose isotropy-groups are larger than I_η. The extrema of F_1 corresponding to these directions (of the types (1.57)) must be part of the extrema existing in the n'-dimensional manifold. Thus, if one has solved beforehand Eqs.(1.58) for these directions, it is possible to separate the corresponding solutions out of Eqs.(1.56), (1.57) and lower the degree of the latter equations.

In order to locate the minima among the extrema precedingly determined, one can make use, again, of the form (1.55) of the free-energy, end express the second derivatives of F with respect to ρ and γ_j, in the form:

$$\left(\frac{\partial^2 F}{\partial \rho^2}\right)_{\rho_0,\gamma_j^0} = 4\rho_0^2\left(\frac{\partial^2 F_1}{\partial(\rho^2)^2}\right)_{\rho=\rho_0}, \qquad (1.60)$$

$$\left(\frac{\partial^2 F}{\partial \rho \partial \gamma_j}\right)_{\rho_0,\gamma_j^0} = 2\rho_0 \sum_m \frac{\partial^2 F_1}{\partial(\rho^2)\partial D_m}\left(\frac{\partial D_m}{\partial \gamma_j}\right)_{\rho_0,\gamma_j^0}, \qquad (1.61)$$

$$\left(\frac{\partial^2 F}{\partial \gamma_j \partial \gamma_{j'}}\right)_{\rho_0,\gamma_j^0} = \left[\sum_{m,m'} \frac{\partial^2 F_1}{\partial D_m \partial D_{m'}} \cdot \frac{\partial D_m}{\partial \gamma_j} \cdot \frac{\partial D_{m'}}{\partial \gamma_{j'}}\right]_{\rho_0,\gamma_j^0} + \left[\sum_m \frac{\partial F_1}{\partial D_m} \cdot \frac{\partial^2 D_m}{\partial \gamma_j \partial \gamma_{j'}}\right]_{\rho_0,\gamma_j^0} . \qquad (1.62)$$

In particular, if we consider a symmetry direction for which condition (1.58) is satisfied, (1.61) is identically zero, and the second member of (1.62) reduces to its second term. Finally, if one wishes to associate the worked-out minima of F with possible low-symmetry phase below a continuous transition, one must check that the conditions defining these minima remain valid for infinitesimal values of ρ.

b) An illustrative example of Gufan's method.

Following Gufan and Sakhnenko [10(b)], let us consider as a first illustrative example of Gufan's method, the case of the 2-dimensional order-parameter with the symmetry corresponding to the image $I_0 = C_{nv}$, with $n \geq 4$. The integrity basis is composed of 2 polynomials given by Eq.(1.47): $I_1 = \rho^2$ and $I_2 = \rho^n \mathrm{Cos}(n\theta)$. On the other hand, one can easily check (Figs.7a and 7b) that depending if n is odd or even there are *2 or 3 types of isotropy groups*, unequivalent up to a configuration in C_{nv}.

In the case where *n is odd*, the first type $I_\eta^{(1)}$ corresponds to directions $\vec{\eta}$ lying within the symmetry planes. These isotropy-groups, of order 2, are composed of the elements (C_1, σ_v) and their manifold is reduced to a single symmetry direction. The second type $I_\eta^{(2)}$ corresponds to the general direction, away from symmetry planes, in the order-parameter space, and is reduced to the identity (C_1). Its manifold is 2-dimensional. If *n is even*, two types of isotropy-groups $I_\eta^{(1)}$ and $I_\eta^{(2)}$ correspond to directions $\vec{\eta}$ lying in either of two families of symmetry planes rotated by the angle (π/n) with respect to eachother. Their respective manifolds are reduced to a single symmetry direction. The third isotropy-group $I_\eta^{(3)}$ is, again, the invariance group of the general direction in the order-parameter space. It consists of the sole identity C_1 and its manifold is 2-dimensional.

In agreement with (1.55), let us write $F = F_1(\rho^2, D = \mathrm{Cos}(n\theta))$ and use the variable θ, instead of the set (γ_1, γ_2), to specify a direction in the order-parameter space. Eq.(1.58) takes the form:

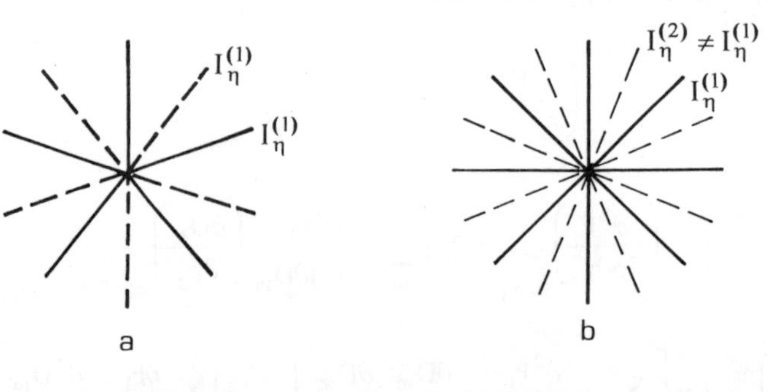

Fig. 7. Symmetry directions and isotropy groups corresponding to $I_0 = C_{nv}$ with (a) n odd, (b) n even.

$$\mathrm{Sin}(n\theta)=0. \tag{1.63}$$

Its solutions $\theta=(k\pi/n)$ correspond to the directions lying within the symmetry planes of the image C_{nv}, as expected from the considerations following Eq.(1.58). The value of the modulus ρ_0 along these directions is determined by Eq.(1.56). Below T_c, we have:

$$\frac{\partial F_1}{\partial(\rho^2)}=\frac{1}{2\rho}\left(\frac{\partial F}{\partial \rho}\right)_{D=\pm 1}=0. \tag{1.64}$$

These extrema are minima if Eqs.(1.60)-(1.62) determine a positive matrix of second derivatives. The conditions are expressed as:

$$\frac{\partial^2 F}{\partial \rho^2}>0 \text{ and } \mp\frac{\partial F_1}{\partial D}>0 \tag{1.65}$$

where the $\mp$ signs correspond to D=±1. We can see that the second of the conditions (1.65) can only be satisfied if F is expanded up to terms of n-th degree (otherwise F is independent of $I_2=\rho^n\cdot\mathrm{Cos}(n\theta)$ and $\partial F_1/\partial D_2=0$), in order to ensure the stability of the low symmetry phases. Let us note that this circumstance does not preclude the possibility of a continuous transition, as the incriminated condition $(\partial F/\partial D)>0$ pertains to the "angular stability" of the considered symmetry directions. As for the "radial" stability, given by (1.64), it is controlled by a condition independent of the angular stability for n>4 and $\rho \to 0$. Indeed, we can restrict to the 4-th degree expansion which has the standard form ((1.46) with $\gamma=0$) which is clearly compatible with a continuous transition.

In the case n>4, the continuity of the transition is ensured by the radial terms, which differentiate the high-symmetry phase from the low-symmetry one. The angular

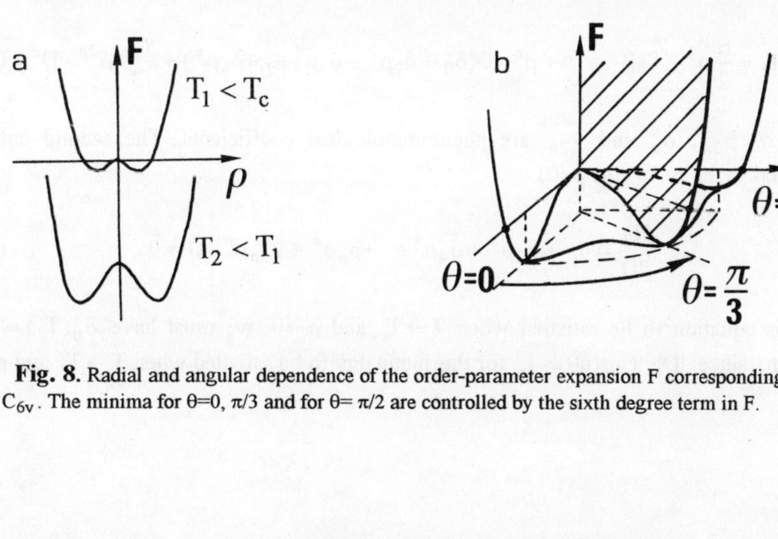

Fig. 8. Radial and angular dependence of the order-parameter expansion F corresponding to the image C_{6v}. The minima for θ=0, π/3 and for θ= π/2 are controlled by the sixth degree term in F.

stability condition, which can be expressed independently, ensures the preference of the system for one set of symmetry directions, as shown in Fig.8. Thus a continuous transition is possible between the high-symmetry phase and either of the phases corresponding to the isotropy group (C_1, σ_v).

Let us now consider the second type of solutions in (1.56) and (1.57). They are determined by the set of equations:

$$\frac{\partial F}{\partial \rho} = 0, \quad \frac{\partial F_1}{\partial D} = 0. \tag{1.66}$$

In this case, an expansion up to the n-th degree is insufficient to determine the direction θ in the order-parameter space, since, up to this degree, F_1 depends linearly on D. The minimum required degree is 2n (as a 2n-th degree expansion contains terms D^2). Eqs.(1.66) then provide a value of D, which is a function of ρ, unlike Eq.(1.63). Hence, the corresponding θ value will specify a general direction in the order-parameter space, whose isotropy-group coincides with the identity C_1. Besides, this direction will be temperature dependent. A necessary condition for the stability of this phase is provided by the positiveness of (1.62). The corresponding inequality reduces to:

$$\frac{\partial^2 F_1}{\partial D^2} > 0 \tag{1.67}$$

which also requires expanding F up to the 2n-th degree, at least. Here a continuous transition between the high-symmetry phase and this second type of low-symmetry phase is impossible. This can be viewed by considering the expansion F up to the 2n-th degree term, and examine the behavior of the solutions of (1.66) when $\rho \to 0$. We have for the first equation (1.66):

$$F_1 = \frac{\alpha}{2}\rho^2 + \ldots + \beta_{2n}\rho^{2n} + \rho^n \cdot D(\delta_0 + \delta_2\rho^2 + \delta_4\rho^4 + \ldots + \delta_n\rho^n) + \frac{\gamma_{2n}}{2}\rho^{2n} \cdot D^2 \tag{1.68}$$

where α, β_{2n}, δ_n and γ_{2n} are phenomenological coefficients. The second equation (1.66) takes the form (for $\rho \neq 0$):

$$\frac{\partial F_1}{\partial D} = \delta_0 + \delta_2\rho^2 + \delta_4\rho^4 + \ldots + \delta_n\rho^n + \gamma_{2n}\rho^n \cdot D = 0. \tag{1.69}$$

For this equation to be satisfied when $T \to T_c$ and $\rho \to 0$, we must have $\delta_0(T_c) = 0$. In addition, since $|D| = |Cos(n\theta)| < 1$, for this inequality to be satisfied when $T \to T_c$ and $\rho \to 0$,

Table 4. Symmetry directions and isotropy groups for the images C_n, O, T_h and O_h.

Image of G_0	Direction in the order-parameter space	Isotropy group
C_n	General	C_1
O		C_4
T_h	$\gamma_1 = 1,\ \gamma_2 = \gamma_3 = 0$	C_{2v}
O_h		C_{4v}
O		C_3
T_h	$\gamma_1 = \gamma_2 = \gamma_3 = 1/\sqrt{3}$	C_3
O_h		C_{3v}

we must either have $\delta_2(T), \delta_4(T_c) = 0$... (the value of D then tends towards $(-\delta_n / \gamma_{2n})$) or have $\gamma_{2n}(T) = 0$. Since the transition at T_c also implies $\alpha(T_c)=0$, three coefficients, at least, of the expansion (1.68) must vanish at T_c, a condition which in general cannot be met at a continuous transition.

An analogous analysis of the image C_n, as well as of the three-component order-parameter images, which may give rise to continuous transitions, has been performed by various authors [10-17]. The corresponding results are summarized in Table 4.

c) Kim's addition to Gufan's method.

The method exposed in a) emphasizes the role of the integrity basis in sorting the "symmetry extrema" from the other extrema. They also provide simplifications in discussing the nature of the extrema (minima, maxima, saddle points) independently from the specific degree of the expansion. However, the determination of the *absolute* minimum still requires carrying out lengthy calculations (see Sec.3.3). Kim [27] has described a refinement of Gufan's method which provides a simpler working-out of the absolute minimum. The method relies on a further use of the integrity basis, in order to locate geometrically the absolute minimum. It concerns two aspects of Gufan's method.

In the first place the various directions in the order-parameter space are specified by the values of the $D_m(\gamma_0)$ invariants. The correspondence $\gamma_i \rightarrow D_m(\gamma_i)$ is a *mapping* of the unit-circle in the order-parameter space into a contour traced in the space spanned by the D_m. Actually, each point of the contour represents all the directions of an orbit (since D_m is invariant by I_0, while each vector of an orbit is transformed to the other vectors of the orbit). Besides, the contour in the D_m-space possesses cusps corresponding to the symmetry-directions in the order-parameter space (since in the corresponding directions one has $\partial D_m / \partial \gamma_i = 0$). Fig.9a shows such a contour in the case of the image $I_0 = C_{4v}$. In this case, there is a single D=Cos(4θ). The contour is thus a segment ±1 on the coordinate axis D. The extremities θ=0 (mod.π/2) and θ=π/4 (mod.π/2) correspond to the symmetry directions (with isotropy group (C_1, σ_v), while the internal points correspond to

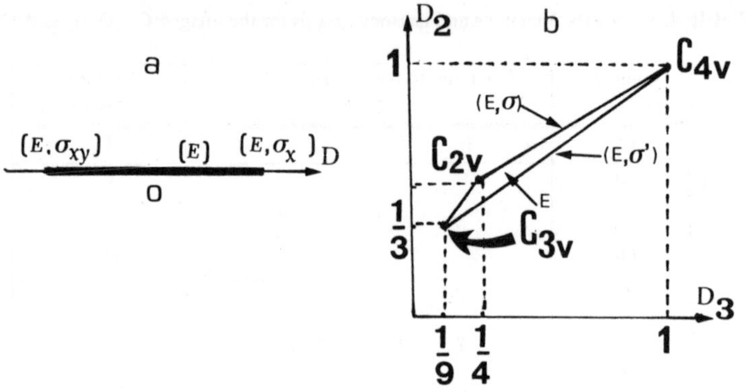

Fig. 9. Illustration of Kim's refinement of Gufan's method. (a) Contour in the case of the image $I_0 = C_{4v}$. (b) Contour relative to the image O_h.

the general directions in the order-parameter space (with isotropy group C_1).

Fig.9b shows the contour relative to the image O_h, corresponding to a 3-component order-parameter, from Ref.[27]. In this case, Table 2 shows that there are, in addition to ρ^2, two invariants I_3 and I_5 in the integrity basis. I_5 can be replaced by $(\eta_1^6 + \eta_2^6 + \eta_3^6)$, which is a linear combination of ρ^6, I_5, and $\rho^2 \cdot I_3$. The contour is traced in the space spanned by $D_2 = (\gamma_1^4 + \gamma_2^4 + \gamma_3^4) = I_3/\rho^4$, and $D_3 = (\gamma_1^6 + \gamma_2^6 + \gamma_3^6)$. The isotropy-groups of the various orbits are indicated for the cusps, and the lines joining them.

The second innovation of Kim's method is to search for the extrema of F in a given direction of the D_m-space. These extrema yield values $\rho_0(D_m)$ of the modulus ρ of the order-parameter. Using these values, we can express the extremal value of the free-energy as a function $F_0(D_m)$, and study the variations of $F_0(D_m)$ in the D_m-space. Examination of the location of the contour, precedingly defined, with respect to these variables allows to specify the nature of the absolute minimum of F. It is worth pointing out that the variations of $F_0(D_m)$ depend on the degree of the order-parameter expansion, and these variations will become more difficult to study as measure as this degree is increased.

As an illustration of Kim's considerations, let us again consider the two-component order-parameter with the image $I_0 = C_{4v}$, and study a free-energy expanded to the 4-th degree:

$$F = \frac{\alpha}{2}\rho^2 + \frac{\beta_1 + \beta_2 D}{4}\rho^4 \qquad (1.70)$$

with D=Cos(4θ). For fixed D, the extremum of F occurs for $\rho^2 = -\dfrac{\alpha}{\beta_1+\beta_2 D}$, and the function $F_0(D)$ takes the form:

$$F_0(D) = -\frac{\alpha^2}{\beta_1+\beta_2 D}. \tag{1.71}$$

The positiveness of ρ^2 requires $(\beta_1+\beta_2 D)>0$. When D varies between -1 and +1, $F_0(D)$ varies *monotonously* from $-\dfrac{\alpha^2}{\beta_1-\beta_2 D}$ to $-\dfrac{\alpha^2}{\beta_1+\beta_2 D}$. This monotonous variation implies (whatever the sign of β_2) that the minimum of F_0 is at one of the extremities of the segment ±1 in the D-space. Considering the contour in Fig.9a, we see that this minimum will correspond to either of the orbits with maximal isotropy groups. We can see that, in order to obtain the absolute minimum of F in a general direction in the order-parameter space (i.e. *inside* the interval ±1, in the D-space), $F_0(D)$ must have a non-monotonous variation on the interval ±1. Such a variation can arise in particular, if we expand F up to a degree such as certain coefficients of ρ^{2p}, depend non-linearly on D. For instance, in the present example, it can be checked, that a sixth degree expansion of F has coefficients depending linearly on D, and that the absolute minimum of F remains at D=±1. By contrast an eight degree expansion involves a non-linear dependence of the coefficients on D and is compatible with an absolute minimum *within* the interval (D=±1) for certain ranges of the expansion coefficients. However, such a minimum is not stable just below a continuous transition, since it is not obtained for the 4-th degree expansion.

3.2.8. Determination of the low-symmetry space groups

If the absolute minimum of the order-parameter expansion, below T_c, corresponds to the set of order-parameter components ($\eta_1^0,\ldots,\eta_n^0$), the invariance group G of the low symmetry phase is the largest subgroup of G_0 which leaves invariant the vector $\vec{\eta}_0=(\eta_1^0,\ldots,\eta_n^0)$ of the order-parameter space. G=$I_G\times$K, where I_G is the isotropy-group of the $\vec{\eta}_0$ direction and K is the kernel of the homomorphism $G_0\rightarrow I_0$. In the case of structural transitions, the advantage of making use of the image I_0 to determine G is clear: I_0 is a finite group, which will be simpler to handle than the infinite group G_0. G is one of the 230 space groups. Accordingly, once the vector $\vec{\eta}_0$ is specified, the determination of G can be performed in three main steps: 1) Determination of *K and* I_0; 2) Determination of the *isotropy-group*; 3) Determination of the *space-group* G. The last step requires the identification of the point-group $\hat{G}$, of the *Bravais-lattice* of G, and of the screw-axes and glide-planes contained in G.

To illustrate this procedure, we shall consider the irreducible representation, denoted $\Gamma_1(\vec{k}_8^*)$ [6], of the O_h^5 space-group, associated with the wave-vector $\vec{k}_8 = \frac{3}{4}\vec{a}^* + \frac{1}{2}\vec{b}^* + \frac{1}{4}\vec{c}^*$, where $\vec{a}^*$, $\vec{b}^*$, and $\vec{c}^*$ are the reciprocal lattice vectors of the face-centered cubic (FCC) Brillouin-zone.

a) Image of $\Gamma_1(\vec{k}_8^*)$.

The wave-vector $\vec{k}_8$ ends at the W-point of the FCC Brillouin zone. As $G_0 = O_h^5$ is a symmorphic space-group, the *small* representations $\hat{\tau}_i(\vec{k}_8)$ coincide with ordinary irreducible representations of the point group $\hat{G}(\vec{k}_8) = D_{2d}$ [6, 7]. Among the five irreducible representations $\hat{\tau}_i$ of D_{2d}, we will consider the *totally symmetric* representation $\hat{\tau}_1$. As the ratio of the orders of G_0 (48) and $\hat{G}(\vec{k}_8)$ (8) is *six*, the *star* $\vec{k}_8^*$ has six arms: $\vec{k}_8^{(1)}$, $\vec{k}_8^{(2)} = -\vec{k}_8^{(1)}$, $\vec{k}_8^{(3)} = C_3(\vec{k}_8^{(1)})$, $\vec{k}_8^{(4)} = -\vec{k}_8^{(3)}$, $\vec{k}_8^{(5)} = C_3^2(\vec{k}_8^{(1)})$, $\vec{k}_8^{(6)} = -\vec{k}_8^{(5)}$. Thus , the irreducible representation $\Gamma_1(\vec{k}_8^*)$ constructed from $\hat{\tau}_1$ is six-dimensional. On the other hand, G_0 is isomorphous to the product $\hat{G}_0 \times \vec{T}$, where $\hat{G}_0$ is the point- group associated to G_0, and $\vec{T}$ is the group of primitive translations. Consequently the image I_0 of G_0 for the representation $\Gamma_1(\vec{k}_8^*)$ can be written $I_0 = I_T \times I_{\hat{G}_0}$, where I_T is the image of $\vec{T}$, and $I_{\hat{G}_0}$ is the image of the group of elements $\{R|0\}$ isomorphous to $\hat{G}_0$. I_T is a group of order 32 generated by the three 6×6 matrices:

$$\vec{a} = \begin{bmatrix} 0 & 1 & 0 & 0 & 0 & 0 \\ -1 & 0 & 0 & 0 & 0 & 0 \\ 0 & 0 & 0 & -1 & 0 & 0 \\ 0 & 0 & 1 & 0 & 0 & 0 \\ 0 & 0 & 0 & 0 & -1 & 0 \\ 0 & 0 & 0 & 0 & 0 & -1 \end{bmatrix}, \vec{b} = \begin{bmatrix} 0 & -1 & 0 & 0 & 0 & 0 \\ 1 & 0 & 0 & 0 & 0 & 0 \\ 0 & 0 & -1 & 0 & 0 & 0 \\ 0 & 0 & 0 & -1 & 0 & 0 \\ 0 & 0 & 0 & 0 & 0 & 1 \\ 0 & 0 & 0 & 0 & -1 & 0 \end{bmatrix}, \vec{a}+\vec{b}+\vec{c} = \begin{bmatrix} -1 & 0 & 0 & 0 & 0 & 0 \\ 0 & -1 & 0 & 0 & 0 & 0 \\ 0 & 0 & -1 & 0 & 0 & 0 \\ 0 & 0 & 0 & -1 & 0 & 0 \\ 0 & 0 & 0 & 0 & -1 & 0 \\ 0 & 0 & 0 & 0 & 0 & -1 \end{bmatrix}. \quad (1.72)$$

I_2 is a group of order 48, isomorphous to $\hat{G}_0 = O_h$, generated by the matrices:

$$A = \begin{bmatrix} 0 & 0 & 0 & 0 & 1 & 0 \\ 0 & 0 & 0 & 0 & 0 & 1 \\ 1 & 0 & 0 & 0 & 0 & 0 \\ 0 & 1 & 0 & 0 & 0 & 0 \\ 0 & 0 & 1 & 0 & 0 & 0 \\ 0 & 0 & 0 & 1 & 0 & 0 \end{bmatrix}, B = \begin{bmatrix} 1 & 0 & 0 & 0 & 0 & 0 \\ 0 & -1 & 0 & 0 & 0 & 0 \\ 0 & 0 & 0 & 0 & 1 & 0 \\ 0 & 0 & 0 & 0 & 0 & 1 \\ 0 & 0 & 1 & 0 & 0 & 0 \\ 0 & 0 & 0 & -1 & 0 & 0 \end{bmatrix}, C = \begin{bmatrix} 1 & 0 & 0 & 0 & 0 & 0 \\ 0 & 1 & 0 & 0 & 0 & 0 \\ 0 & 0 & 0 & 0 & 1 & 0 \\ 0 & 0 & 0 & 0 & 0 & 1 \\ 0 & 0 & 1 & 0 & 0 & 0 \\ 0 & 0 & 0 & 1 & 0 & 0 \end{bmatrix}, D = \begin{bmatrix} 1 & 0 & 0 & 0 & 0 & 0 \\ 0 & -1 & 0 & 0 & 0 & 0 \\ 0 & 0 & 1 & 0 & 0 & 0 \\ 0 & 0 & 0 & -1 & 0 & 0 \\ 0 & 0 & 0 & 0 & 1 & 0 \\ 0 & 0 & 0 & 0 & 0 & -1 \end{bmatrix}. \quad (1.73)$$

So, the image I_0 possesses 32×48=1536 elements.

b) Order-parameter expansion.

Let us calculate the integrity basis corresponding to I_0, using a simple procedure adapted to the specific *form* of I_0. With that goal, it can be noted that all the matrices (1.72) and (1.73) consist of 3 *diagonal* blocks, acting separately on the three sets of order-parameter components (η_1,η_2), (η_3,η_4) and (η_5,η_6) spanning the 6-dimensional space ε_6 of $\tau_1(\vec{k}_8^*)$. Accordingly, we can form the required invariant polynomials of two variables (η_i,η_{i+1}) (i=1, 3, 5) which are left invariant by the set of 2×2 matrices :

$$\begin{bmatrix} 0 & 1 \\ -1 & 0 \end{bmatrix}, \begin{bmatrix} 1 & 0 \\ 0 & -1 \end{bmatrix}, \begin{bmatrix} -1 & 0 \\ 0 & -1 \end{bmatrix}.$$

These matrices generate a group, identical to the group of matrices (1.51), i.e. isomorphous to C_{4v}, the basic invariants of which are ρ_i^2 and $\rho_i^4 \mathrm{Cos}(4\theta_i)$. Any function of these invariants (with i=1, 3, 5) will be invariant by the subgroup of I_0 considered. We are interested in the polynomials of n-th degree which are also invariant by the matrices (1.73). Clearly, the set generated by these matrices operates a permutation of the invariants of index i. Hence the solutions to the investigated problem are polynomials of ρ_i^2 and $\rho_i^4 \mathrm{Cos}(4\theta_i)$, which are invariant by permutation of the i-indices. Polynomials of *third* degree are absent, and polynomials of fourth degree are easily determined, allowing to write the order-parameter expansion, related to I_0, as:

$$F(T,\rho_i,\theta_i) = F_0(T) + \frac{\alpha}{2}\left(\sum_{i=1}^{3}\rho_i^2\right) + \frac{\beta_1}{4}\left(\sum_{i=1}^{3}\rho_i^4\right) + \frac{\beta_2}{2}\left(\sum_{\substack{i=1 \\ i\neq j}}^{3}\rho_i^2\rho_j^2\right) + \frac{\beta_3}{4}\left(\sum_{i=1}^{3}\rho_i^4 \mathrm{Cos}(4\theta_i)\right). \tag{1.74}$$

c) Kernel of the homomorphism $G_0 \rightarrow I_0$*, and isotropy-group* I_G.

As stressed above, the image I_0 contains a subgroup $I_{\hat{G}_0}$ isomorphous to $\hat{G}_0$. Hence the kernel of the homomorphism $G_0 \rightarrow I_0$ necessarily coincides with a subgroup of the translation group $\vec{T}$ of G_0, as G_0 is symmorphic. The image of $\vec{T}$ being of order 32, the kernel K is a subgroup of $\vec{T}$ of index 32. In other words the unit-cell of the infinite translation group $\vec{T}$ is 32 times larger than the unit-cell $(\vec{a},\vec{b},\vec{c})$ of the FCC lattice. The corresponding basic vectors are:

$$\vec{a}_k = 2(\vec{a} - \vec{b} + \vec{c}) = 2a(100), \quad \vec{b}_k = 2(\vec{a} + \vec{b} - \vec{c}) = 2a(010),$$
$$\vec{c}_k = 2(-\vec{a} + \vec{b} + \vec{c}) = 2a(001) \tag{1.75}$$

where the basic translations of the FCC lattice are:

$$\vec{a} = \frac{a}{2}(110), \ \vec{b} = \frac{a}{2}(011), \ \vec{c} = \frac{a}{2}(101). \tag{1.76}$$

One can check that $\left(\vec{a}_k \cdot (\vec{b}_k \times \vec{c}_k)\right) = 32\left(\vec{a} \cdot (\vec{b} \times \vec{c})\right)$.

In the 6-dimensional space spanning the matrices (1.72) and (1.73), let us assume that for a certain range of the coefficients, the absolute minimum of the free-energy corresponds to the direction (η $-\eta$ 0 0 0 0) of the order-parameter space (using Eqs.(1.41) and (1.42), it is straightforward to show that this range is defined by $\beta_3 > 0$, $\beta_2 > \beta_1 - \beta_3 > 0$). The isotropy-group I_G is composed of the elements of I_0 which preserve the manifold (η_1, η_2) of the order-parameter space, and which, in the manifold, leave the first bisector invariant . As I_0 acts, within this manifold, as the group C_{4v}, among the elements of this group, only two elements (C_1 and $\sigma_{\bar{x}y}$) leaves the [11] direction unchanged, as they are represented by the 2×2 matrices $\begin{bmatrix} 1 & 0 \\ 0 & 1 \end{bmatrix}$ and $\begin{bmatrix} 0 & 1 \\ 1 & 0 \end{bmatrix}$ (see (1.51)). Inspection of the matrices (1.72) and (1.73) shows that I_G is generated by the matrices which possess as first diagonal block one of the two former matrices. Thus, only the matrices denoted C in (1.73), or C×B and C×D belong to I_G. Hence, I_G is generated by the subgroup of order 16 of I_0, formed by the matrices B, C, D, and by the subgroups of order 8 of I_T, whose first diagonal block is $\begin{bmatrix} 1 & 0 \\ 0 & 1 \end{bmatrix}$. I_G is constituted by the product of these two sets: its order is 8×16=128, i.e. it is a subgroup of order 12 of G_0.

d) Point-group $\hat{G}$. Bravais lattice of G.

In the isomorphism between the group generated by the matrices (1.73) and O_h, the matrices A, B, C, D correspond respectively to the 3-fold rotation C_3^{xyz}, to the 4-fold rotation C_4^x, to the two-fold rotation $U_{\bar{y}z}$, and to the inversion *I*. The isotropic group I_G of the considered direction in the order-parameter space contains the matrices denoted C, C×B and C×D. Accordingly, the low-symmetry group contains the reciprocal images of these matrices, namely $\{U_{\bar{y}z}|0\} \times K$, $\{C_4^x|\vec{a} + \vec{b} + \vec{c}\}$, and $\{I|\vec{a} + \vec{b} + \vec{c}\}$. Since K is the group of primitive translations (1.75), and since the reciprocal images of the other generators of I_G are also primitive translations of G_0, we can infer that the *point group* $\hat{G}$ is generated by the elements *I*, C_4^x and $U_{\bar{y}z}$. It is the crystallographic point-group D_{4h}^x of order 16, the 4-fold axis of which is along the x-axis.

D_{4h} belongs to the tetragonal system and the translation group $\vec{T}_G$ of the crystal in the low-symmetry phase of space group G, defines one of the Bravais lattice (primitive P or body-centered I) of this system. The specification of the relevant Bravais lattice involves two steps:

(i) The first step consists in working out the basic translations of $\vec{T}_G$. Taking into account the fact that the index of $\vec{T}_G$ in $\vec{T}$ is 4 (since the index of G in G_0 is 3), the basic translations of $\vec{T}_G$ must define a unit-cell 4 times larger than $(\vec{a},\vec{b},\vec{c})$. One set of basic translations fulfilling this property is:

$$\vec{a}'_G = -\vec{a}+\vec{b}+\vec{c} = a(001),\quad \vec{b}'_G = \vec{a}+\vec{b}-\vec{c} = a(010),\quad \vec{c}'_G = \vec{a}+\vec{c} = a\left(1\frac{1}{2}\frac{1}{2}\right). \tag{1.77}$$

(ii) The second step is the identification of the set (1.77) of basic translations with the standard set associated with the relevant Bravais lattice, as given for example in Tables [6, 7]. When the 4-fold axis is along X, the two Bravais lattices of the tetragonal system are defined by the following basic translations:

$$\text{P-lattice}: C(1\ 0\ 0),\ A(0\ 1\ 0),\ A(0\ 0\ 1),$$

$$\text{I-lattice}: \left(\frac{C}{2}\ \frac{A}{2}\ \frac{A}{2}\right),\ \left(-\frac{C}{2}\ \frac{A}{2}\ \frac{A}{2}\right),\ \left(\frac{C}{2}\ \frac{A}{2}\ -\frac{A}{2}\right). \tag{1.78}$$

In order to identify the Bravais lattice generated by the translations $\vec{a}'_G,\vec{b}'_G,\vec{c}'_G$ given by (1.77), we must find a combination of vectors (1.77) which define a unit-cell of same volume as $\vec{a}'_G,\vec{b}'_G,\vec{c}'_G$, and display the same geometrical configuration as one of the sets of vector (1.78). One can see that the combinations:

$$\vec{a}_G = \vec{c}'_G = \left(a\ \ \frac{a}{2}\ \ \frac{a}{2}\right),\quad \vec{b}_G = \vec{a}'_G+\vec{b}'_G-\vec{c}'_G = \left(-a\ \ \frac{a}{2}\ \ \frac{a}{2}\right),$$

$$\vec{c}_G = \vec{b}'_G-\vec{c}'_G = \left(-a\ \ \frac{a}{2}\ -\frac{a}{2}\right) \tag{1.79}$$

form a set of translations identical to the standard translations of the I-Bravais lattice in (1.78) provided C=2a, and A=a. Thus, the translations $\vec{a}_G,\vec{b}_G,\vec{c}_G$ are the three basic translations of the I-lattice of the low-symmetry phase.

e) Identification of the space-group G.

G contains the elements $\{U_{\bar{y}z}|0\}$, $\{C_4^x|\vec{c}\} = \{C_4^x|\frac{C}{4}0\frac{A}{2}\}$, $\{I|\vec{c}\} = \{I|\frac{C}{4}0\frac{A}{2}\}$, as well as the product of these elements. Finding the space group G (its symbol) with $\hat{G}=D_{4h}$ requires specifying the nature of the 4-fold axis (pure or helicoidal rotation) and the nature of the 3 types of reflection-planes σ_x,σ_y and $\sigma_{\bar{y}z}$ (pure or glide-reflection). One finds here the following elements:

$$\{C_4^x|\tfrac{C}{4}0\tfrac{A}{2}\},\quad \{\sigma_x|\tfrac{C}{4}0\tfrac{A}{2}\},\quad \{\sigma_y|0\},\quad \{\sigma_{\bar{y}z}|\tfrac{C}{4}0\tfrac{A}{2}\}$$

which are written in a non standard setting. For a standard setting, one has to perform the transformation (x→z, z→y, y→x), which gives the elements:

$$\left\{C_4^z\middle|\frac{A}{2}\,0\,\frac{C}{4}\right\},\ \left\{\sigma_z\middle|\frac{A}{2}\,0\,\frac{C}{4}\right\},\ \left\{\sigma_x\middle|0\right\},\ \left\{\sigma_{\bar{x}y}\middle|\frac{A}{2}\,0\,\frac{C}{4}\right\}$$

showing that the fourfold axis is a *screw-axis* whose non-primitive translation is one-fourth of the period C. σ_z is a *glide-plane* with a translation A/2, σ_x is a pure reflection, and $\sigma_{\bar{x}y}$ is a glide-plane with a translation C/4. This allows, using the space-groups Tables [6-8] to find that $G = D_{4h}^{19}\,(I4_1/amd)$.

3.3. The radial problem and its solutions

3.3.1. Arbitrariness of the phenomenological models: Specific form of the Landau potential

The solutions of the radial problem consists in finding the dependence of the equilibrium values of the order-parameter components (η_j) on the phenomenological coefficients of the free-energy (Landau potential), which depend on the external conditions (temperature, pressure, and eventually concentration). These equilibrium values (η_j^{eq}) allow to evaluate the critical behaviour (anomalies) of various relevant physical quantities: susceptibilities, specific heat etc... They also lead to the construction of the phase diagram in the space of the external parameters, which includes the transition lines bounding the phases, the shape and width of the regions of coexistence of the phases (for first-order transitions) and, possibly, various types of singular points. One can also deduce the effect of external (conjugated) fields on the phase diagram and on the transition anomalies, as well as the dynamics of the order-parameter, using equations of motion such as the Landau-Khalatnikov [28] and the Landau-Lifshitz [29] equations.

Although the *general form* of the Landau potential $F(T,P,\eta_j)$ has been shown in Sec.3.2. to be entirely determined by the symmetry of the order-parameter, the equilibrium values of the order-parameter components, as functions of the phenomenological coefficients, depend on the *specific model* chosen for F, namely: 1) On the maximal power of the (η_j) at which F is truncated, 2) On the sign and dependence on the external parameters assumed for the phenomenological coefficients, 3) On the influence of secondary order-parameters, and on the maximal degree considered for the corresponding coupling terms, which, as shown in Sec.2.4. lead to a renormalization of the coefficients of the primary order-parameter expansion.

Usually the choice of the model of Landau potential is made with some arbitrariness, and is essentially determined by empirical considerations: one takes the *minimal "necessary"* power for the η_j and coupling terms which allow to describe the more realistically the full set of available experimental data, namely the observed sequence of phases and the critical anomalies. Along the same line the simplest (linear or constant) dependence on the external variables is assumed for the phenomenological coefficients.

Such procedure is coherent in a phenomenological approach, since the most "economic" model, which contains the smallest number of phenomenological variables, is indeed the most predictive one. In the examples discussed in this section we will show that the degree of arbitrariness in the choice of the Landau potential, can be reduced using a number of simple rules.

In the present section we present the main results which have been obtained in working out the solutions of the radial problem for a number of selected models. Since the features of these models are chiefly determined by the number n of order-parameter components, and by the corresponding image-groups, we shall order our presentation as in Secs.3.2.2. and 3.2.3. i.e. following the increasing values of n.

3.3.2. Solutions of the radial problem for one-component order-parameter expansions

a) Canonical model of first-order phase transition.

In Sec.2.2.4. the simplest one-component order-parameter expansion (1.2) truncated to the fourth degree, was shown to describe a single second-order phase transition between a high- and low-symmetry phase at T_c. Another canonical example is the sixth-degree expansion:

$$F(T,P,\eta) = F_0(T,P) + \frac{\alpha}{2}\eta^2 + \frac{\beta}{4}\eta^4 + \frac{\gamma}{6}\eta^6 \qquad (1.80)$$

which describes, when assuming α=a(T-T_c), β<0 and γ>0, a first-order transition between the high- and low-symmetry phases. The equations minimizing F are:

$$\frac{\partial F}{\partial \eta} = \eta\left(\alpha + \beta\eta^2 + \gamma\eta^4\right) = 0, \qquad (1.81)$$

$$\frac{\partial^2 F}{\partial \eta^2} = \alpha + 3\beta\eta^2 + 5\gamma\eta^4 \geq 0. \qquad (1.82)$$

The discussion of the conditions (1.81) and (1.82) is visualized in Figs.10a and 10b. It can be summarized as follows: For $T > T_2 = T_c + \beta^2/4\gamma a$ only one phase is stable, namely the parent phase (I) corresponding to η_I^{eq}=0. At T=T_2 phase II, corresponding to:

$$\eta_{II}^{eq} = \pm\left[\frac{-\beta + (\beta^2 - 4\alpha\gamma)^{1/2}}{2\gamma}\right]^{1/2} \qquad (1.83)$$

appears as a metastable state (i.e. corresponding to secondary minima in the non-equilibrium curve F(η)-F_0). As the temperature is lowered, the stability of phase II

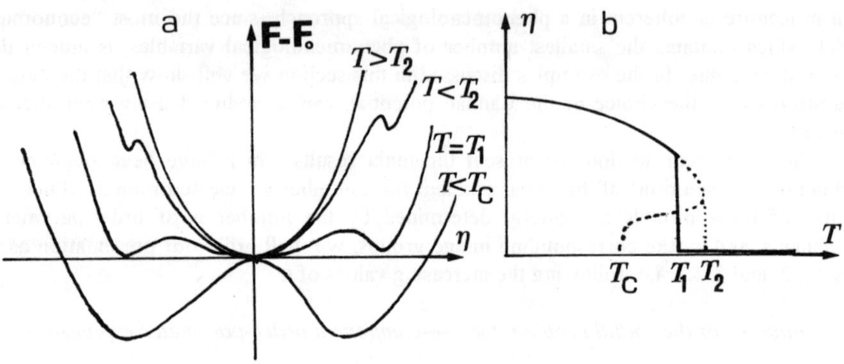

Fig. 10. (a) Variation of the non-equilibrium potential $F-F_0$ given by (1.80), as a function of temperature. (b) Temperature dependence of the order-parameter $\eta(T)$.

increases. Phases I and II are equally stable when the additional condition $F(\eta)-F_0=0$ is fulfilled, at:

$$T_1 = T_c + \frac{3\beta^2}{16\gamma a}. \tag{1.84}$$

Below T_1, phase I becomes less stable than phase II, and remains as a metastable state until α changes sign for $T=T_c$.

From the preceding discussion, the main distinctive properties of first-order transitions can be inferred. These transitions are characterized:
i) by *a region of coexistence of the two phases* I and II, within the interval:

$$\Delta T = T_2 - T_c = \frac{\beta^2}{4\gamma a} \tag{1.85}$$

the two phases being alternately metastable for $T>T_1$ and $T<T_1$. Although the transition may in reality take place at any temperature within the interval ΔT, the most probable transition temperature is T_1, which corresponds to an equal stability for the two phases. As a consequence a *thermal hysteresis* will generally be observed, since the measured transition temperature will not be the same on heating and cooling.
ii) by a discontinuous jump of the order-parameter at the transition which is given by:

$$\eta(T_1) = \left(-\frac{\beta}{4\gamma}\right)^{1/2}. \tag{1.86}$$

Below T_1 η increases with decreasing temperature following the law (1.83).
iii) by a *latent heat* L which results from the discontinuity of $\eta(T)$ at the transition. One has:

$$L = T_1 \Delta S(T_1) = \frac{a}{2} T_1 \eta^2(T_1) = -\frac{a\beta T_1}{8\gamma}. \qquad (1.87)$$

In the same way, a discontinuous variation should be found at T_1 for all the physical quantities proportional to the first derivatives of the thermodynamic potential, e.g. the volume, the lattice constants, the induced strains or polarization etc...
iv) by specific discontinuities at T_1 for the physical quantities related to the second derivatives of the thermodynamic potential. For example, if ξ is the field conjugated to η, minimization of $F(\eta)$-$\eta\xi$ with respect to η and ξ provides the following form for the susceptibility relative to the order-parameter:

$$\chi_\eta = \lim_{\xi \to 0} \left(\frac{\partial \eta}{\partial \xi} \right) = \begin{cases} \frac{1}{a}(T - T_c)^{-1} & \text{for } T > T_1, \\ \frac{16\gamma}{3\beta^2} & \text{for } T = T_1, \\ (\alpha + 3\beta\eta^2 + 5\gamma\eta^4)^{-1} & \text{for } T < T_1. \end{cases} \qquad (1.88)$$

χ_η is represented in Fig.11a. It differs from the standard Curie-Weiss law (Fig.4b and Eq.(1.9)) obtained for a second-order transition, as no divergence of χ_η occurs at T_1, but a *finite jump*.

Although at T_1 the specific heat curve $C_\eta(T)$ undergoes an upward step on cooling proportional to a^2/β as for a second-order transition (Fig.4a and Eq.(1.7)), the

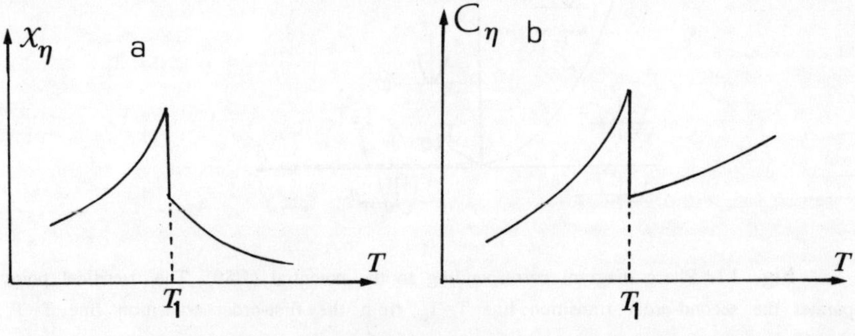

Fig. 11. Temperature dependence of (a) the susceptibility χ_η, (b) the specific heat C_η, for a first-order transition corresponding to the potential (1.80).

influence of higher degree terms in the thermodynamic potential will manifest itself in the temperature dependence of $C_\eta(T)$ below T_1. In this region $C_\eta(T)$ varies as:

$$C_\eta(T) = C_0 + a^2\left(\beta^2 - 4\alpha\gamma\right)^{-1/2} \qquad (1.89)$$

and accordingly the decrease of $C_\eta(T)$ below T_l will be sharper for a first-order transition than for a second-order one, as shown in Fig.11b.

In the preceding discussion only the temperature has been considered as an external variable. Assuming a dependence of the phenomenological coefficients α and β in (1.80) on *two* external variables (e.g. temperature and pressure), and a constant value for γ, one gets the phase diagram in the (α, β) plane shown in Fig.12. In this diagram, one can see that the line of first-order transitions defined by:

$$\alpha = \frac{3\beta^2}{16\gamma} \quad (\beta<0) \qquad (1.90)$$

goes over the second-order transition line:

$$\alpha=0 \quad (\beta>0) \qquad (1.91)$$

at the tricritical point T_{tc} located at the origin ($\alpha_{tc} = \beta_{tc} = 0$). Close to this point, for $\beta<0$, β can be neglected in Eq.(1.81), and one has the approximate equilibrium value for η:

$$\eta^{eq} \cong \left(-\frac{\alpha}{\gamma}\right)^{1/4} \qquad (1.92)$$

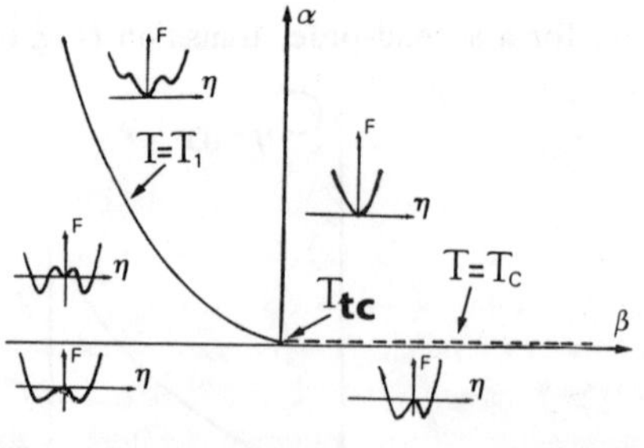

Fig. 12. Phase diagram corresponding to the potential (1.80). The tricritical point T_{tc} separates the second-order transition line $T=T_c$ from the first-order transition line $T=T_1$. The curves $(F-F_0)(\eta)$ are shown in some significant regions of the phase diagram.

For $\beta>0$, close to the tricritical point, one can assume a power law $\beta = A^u$, where A is a positive constant, and $0 \leq u \leq 1/2$. Thus, the order-parameter close to the second-order transition line $T=T_c$, has the form:

$$\eta^{eq} \cong \left(-\frac{\alpha}{\beta}\right)^{1/2} \cong \left(-\frac{\alpha}{A}\right)^{1/2-u} \tag{1.93}$$

when approaching the tricritical point, the critical exponent associated with the order-parameter varies in the interval $\left[\frac{1}{4}, \frac{1}{2}\right]$.

b) General features of the phase diagrams associated with one-component order-parameter expansions, with the image-group C_i.

The considerations developed for the sixth degree potential (1.80), and previously for the fourth degree potential (1.2), can be generalized to any one-component order-parameter expansion of higher degree containing even powers of the order-parameter η:

$$F(T,P,\eta) = F_0(T,P) + a_1\eta^2 + a_2\eta^4 + a_3\eta^6 + a_4\eta^8 + \dots \, . \tag{1.94}$$

As indicated in Sec.3.2.2 such potentials correspond to the image-group C_i, with only one basic polynomial ($I=\eta^2$) in their integrity basis. The equation of state and stability conditions are:

$$\frac{\partial F}{\partial \eta} = 2\eta\left(a_1 + 2a_2\eta^2 + 3a_3\eta^4 + 4a_4\eta^6 + \dots\right) = 0 \tag{1.95}$$

$$\frac{\partial^2 F}{\partial \eta^2} = 2\left(a_1 + 2a_2\eta^2 + 3a_3\eta^4 + 4a_4\eta^6 + \dots\right) + 4\eta^2\left(2a_2 + 6a_3\eta^2 + 12a_4\eta^4 + \dots\right) \geq 0\,. \tag{1.96}$$

If 2m is the highest degree assumed for (1.94), i.e. if m is the maximal number of non-zero coefficients a_i, Eq.(1.95) possesses at the utmost (m-1) positive roots for the variable η^2, which are associated with (m-1) stable low-temperature phase *having identical symmetries*. Hence, if m is even the phase diagram will display a maximal number of m/2 successive isostructural transitions, whereas the maximal number of such transitions is (m-1)/2 for m odd. In Fig.13 one can verify that for m odd *all the transitions are first-order*, while for m even, *an initial second-order transition is followed by (m/2-1) first-order transitions*. We will now discuss in more details the phase diagrams and physical anomalies corresponding to the potentials with 2m=8 and 2m=10, which have been studied by Gufan [13] and Gufan and Larin [30, 31].

c) Phase diagram for an eight-degree expansion (2m=8).

As one can see in Fig.13, the 8-th degree expansion is the lowest degree which allows to describe two low-symmetry phases. These phases have identical symmetries but

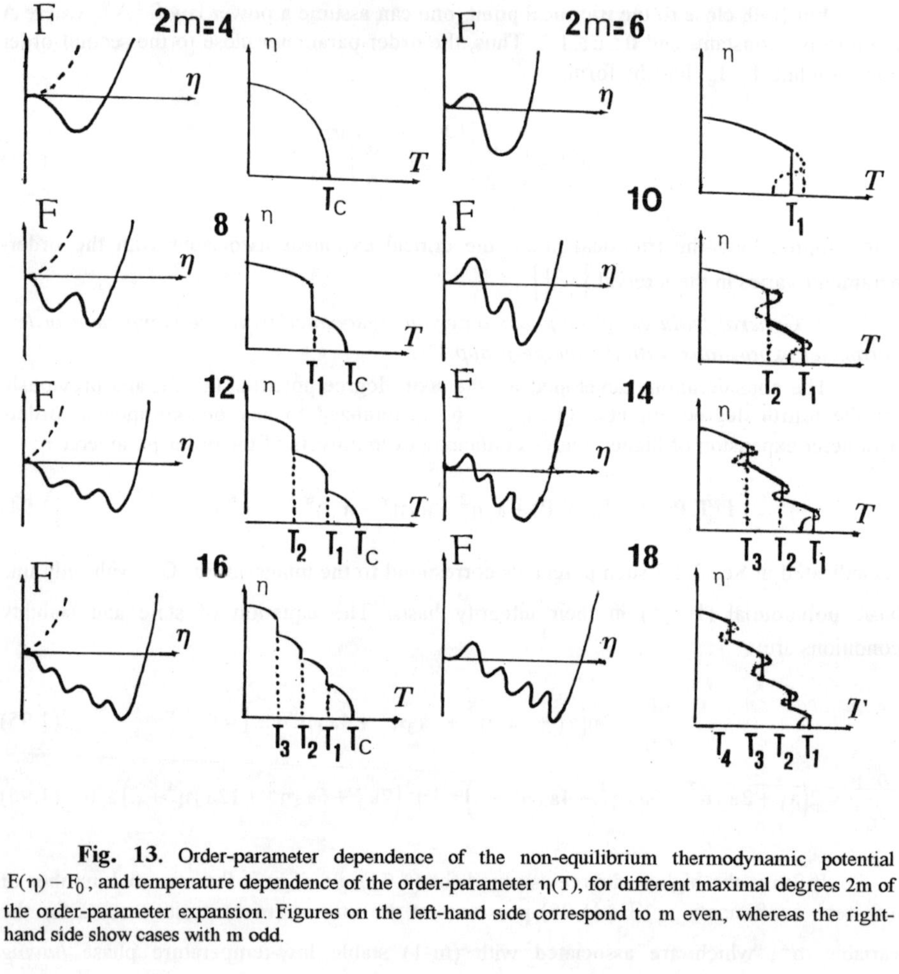

Fig. 13. Order-parameter dependence of the non-equilibrium thermodynamic potential $F(\eta) - F_0$, and temperature dependence of the order-parameter $\eta(T)$, for different maximal degrees 2m of the order-parameter expansion. Figures on the left-hand side correspond to m even, whereas the right-hand side show cases with m odd.

differ by the magnitude of the order-parameter η. The properties of the phase diagram, which includes a second-order transition line followed by a first-order transition line (or a single first-order transition line) can be worked out from the equation of state:

$$\eta\left(a_1 + 2a_2\eta^2 + 3a_3\eta^4 + 4a_4\eta^6\right) = 0 \tag{1.97}$$

and from the limit of stability lines, which are: $a_1 = 0$ for the parent phase I, obtained for $\eta=0$, and:

$$\eta^2\left(2a_2 + 6a_3\eta^2 + 12a_4\eta^4\right) = 0 \tag{1.98}$$

for the low-temperature phases denoted II and III. The detailed mathematical procedure for solving Eqs.(1.97) and (1.98) can be found in Ref.[12]. We will restrict ourselves hereafter, to discuss its essential steps.

In addition to the solution $\eta=0$, Eq.(1.97) possesses either one or three positive solutions for η^2, depending on the sign of a_3. As shown in Fig.14a in the three-dimensional coefficient space (a_1, a_2, a_3) one has for $a_3>0$ only one low-temperature stable phase (as in the case 2m=6), while for $a_3<0$ and $a_4>0$, one finds two distinct regions of stability for the low-temperature phase. Assuming these latter conditions for the signs of a_3 and a_4, one can represent the complete set of phases for 2m=8, using as variables the coefficients a_1 and a_2, which are supposed to vary linearly as functions of external parameters (e.g. T and P). Fig.14b represents the lines of transitions bounding the various phases in the plane $(\tilde{a}_1, \tilde{a}_2)$, where $\tilde{a}_1 = \left|a_3^3\right|\frac{a_4^2}{a_1}$, $\tilde{a}_2 = \frac{a_3^2}{a_4}$. Thus the isostructural transition line II-III is a straight-line segment of equation:

$$\tilde{a}_1 = -\frac{1}{2}\tilde{a}_2 + \frac{1}{8} \tag{1.99}$$

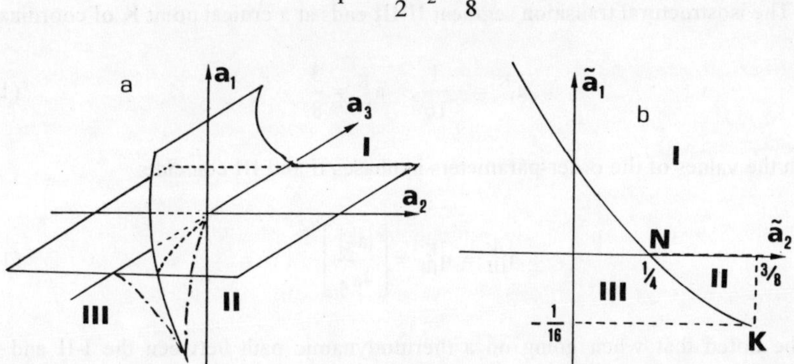

Fig. 14. Topology of the phase diagrams corresponding to an eight degree potential (1.94): (a) in the space (a_1, a_2, a_3), (b) in the $(\tilde{a}_1, \tilde{a}_2)$ plane for $a_3 < 0$. Full and dashed lines in (b) represent first- and second-order transition lines, respectively.

at which the jump of the order-parameter, given by $F(\eta_{II}) = F(\eta_{III})$ is evaluated by:

$$\left[\eta_{III}^2 - \eta_{II}^2\right]_{II-III} = \frac{|a_3|}{2a_4}\left(3 - 8\tilde{a}_2\right)^{1/2}. \qquad (1.100)$$

The first-order transition segment (1.99) starts at a three-phase point N of coordinates $\tilde{a}_1^N = 0$, $\tilde{a}_2^N = 1/4$ where two other transition lines converge: The second-order transition line I-II ($\tilde{a}_1 = 0$, $\tilde{a}_2 > 1/4$) and the line of first-order transitions I-III, of equation:

$$\tilde{a}_1 = -\frac{1}{3}\tilde{a}_2 + \frac{2}{3}\left(\frac{1}{3} - \tilde{a}_2\right)^{3/2} + \frac{2}{27} \qquad (1.101)$$

which is tangent to the II-III line at the N point. The jump of the order-parameter on the I-III line defined by $F(\eta_{III}) = F(\eta_I = 0)$, corresponds to:

$$\left(\eta_{III}^2\right)_{I-III} = \frac{|a_3|}{3a_4}\left[1 + \left(1 - 3\tilde{a}_2\right)^{1/2}\right]. \qquad (1.102)$$

At the N-point the order-parameter discontinuities (1.100) and (1.102) coincide, and one has:

$$\left[\left(\eta_{III}^N\right)^2 - \left(\eta_{II}^N\right)^2\right]_{II-III} = \left(\eta_{II}^N\right)^2_{I-II} = \frac{|a_3|}{2a_4}. \qquad (1.103)$$

The isostructural transition segment II-III ends at a critical point K of coordinates:

$$\tilde{a}_1^K = -\frac{1}{16}, \quad \tilde{a}_2^K = \frac{3}{8} \qquad (1.104)$$

at which the values of the order-parameters in phases II and III coincide:

$$\eta_{II}^K = \eta_{III}^K = \left[\frac{|a_3|}{4a_4}\right]^{1/2}. \qquad (1.105)$$

It can be noted that when going on a thermodynamic path between the I-II and II-III transition lines, close to the critical point K, the temperature dependence of the order-parameter varies as $(T_c - T)^{\beta}$, where β goes from $\beta=1/2$ to $\beta=1/3$ close to the K point. Figs.15a and 15b, summarize the form of the non-equilibrium potential (1.94), up to

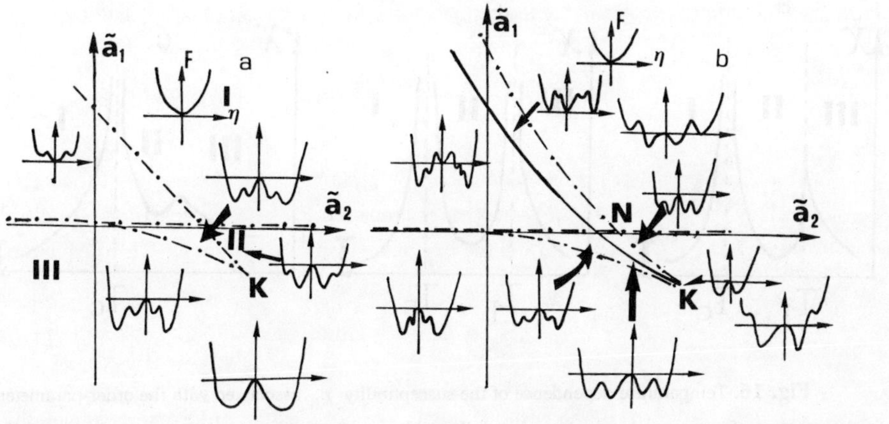

Fig. 15. Phase diagram corresponding to an eight-degree potential with $a_3 < 0$ and $a_4 > 0$, in the coefficient plane ($\tilde{a}_1, \tilde{a}_2$). In (a) only the limit of stability of the phases are represented in dashed-dotted lines, whereas in (b) the lines of transitions are also represented in full (first-order) and dashed (second-order) lines.

the eighth degree, as a function of the order-parameter, in different regions of the ($\tilde{a}_1, \tilde{a}_2$) plane (Fig. 15a) and along the lines of equal energies (Fig. 15b).

Using Eqs. (1.99) to (1.105) one can determine the form of the physical anomalies which take place at the various transitions shown in Figs. 14 and 15. For the second-order I-II transition and for the first-order I-III transition the anomalies have the same shape as represented in Fig. 5 and Fig. 11, respectively. The region of the phase diagram corresponding to $1/4 \le \tilde{a}_2 \le 3/8$ displays an interesting succession of physical anomalies, which are represented in Fig. 16 for the susceptibility χ_η. Thus for $1/4 \le \tilde{a}_2 \le 3/8$ a Curie-Weiss law (as in Fig. 5b) is followed by a first-order transition type anomaly (as in Fig. 11a), whereas at the critical point K ($\tilde{a}_2 = 3/8$) one will observe a succession of two Curie-Weiss divergences of χ_η. In the vicinity of the K point ($\tilde{a}_2 \ge 3/8$) the jump in $\Delta\chi$ should decrease progressively with the increasing values of $\tilde{a}_2$.

d) Phase diagram for a tenth-degree expansion (2m=10).

The minimal degree of $F(\eta)$ required to describe two consecutive first-order transitions is 2m=10 (see Fig. 13) if assuming $a_4 < 0$ and $a_5 > 0$. The mathematical discussion of the corresponding equation of state:

$$\eta\left(a_1 + 2a_2\eta^2 + 3a_3\eta^4 + 4a_4\eta^6 + 5a_5\eta^8\right) = 0 \qquad (1.106)$$

and limit of stability condition:

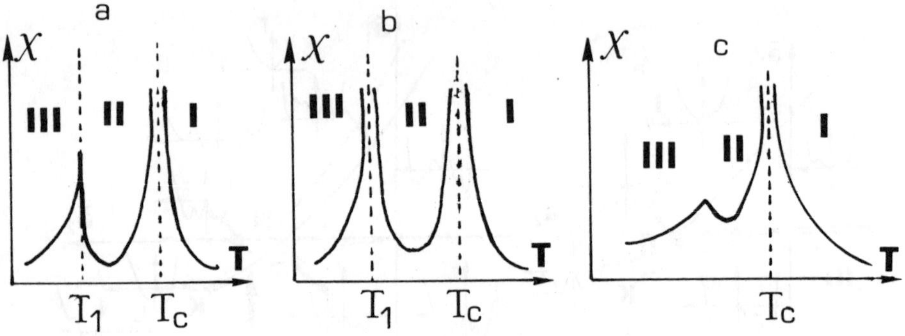

Fig. 16. Temperature dependence of the susceptibility χ_η associated with the order-parameter, for a succession of one second-order transition followed by an isostructural transition. a) $1/4 < \tilde{a}_2 < 3/8$, b) $\tilde{a}_2 = 3/8$ (at the critical point K), c) $\tilde{a}_2 > 3/8$ in the vicinity of the K-point.

$$a_1 + 6a_2\eta^2 + 15a_3\eta^4 + 28a_4\eta^6 + 45a_5\eta^8 = 0 \qquad (1.107)$$

has been performed by Gufan [12]. The resulting phase diagrams shown in Fig.17a,b,c are topologically more complex than in the case 2m=8. They display in particular: 1) Two possible critical end points K_1 and K_2 for the isostructural transition between phases II and III. 2) A triple point T at which merge the three phases (I, II, III), phase II possessing two distinct regions of stability meeting at the T-point. 3) A crossing point L for the two limit of stability lines of phases II and III, with respect to phase I.

Increasing the (even) power of F(η) to twelve (2m=12) gives (as shown in Fig.13) the possibility of stabilizing an additional low-symmetry phase across a second isostructural transition line. It leads to the multiplication of critical points and regions of phase coexistence. The algebraic minimization of potentials with $2m \geq 12$, and the working out of the corresponding phase diagrams is a difficult mathematical task, which has not been performed up to now.

e) One-component order-parameter expansions with the image-group C_1.

As noted in Sec.3.2.2 another class of one-component order-parameter expansions exists, associated with the image-group C_1. It transforms as the identity representation of any of the 230 space groups (at the Γ-point of the Brillouin-zone), and contains all powers (even and odd) of the order-parameter η, as expressed by Eq.(1.45). The corresponding integrity basis is thus reduced to the linear invariant I=η. The equation of state and stability condition associated with expansion (1.45) are:

$$a + \eta\left(\alpha + B_1\eta + \beta_1\eta^2 + B_2\eta^3 + \ldots\right) = 0, \qquad (1.108)$$

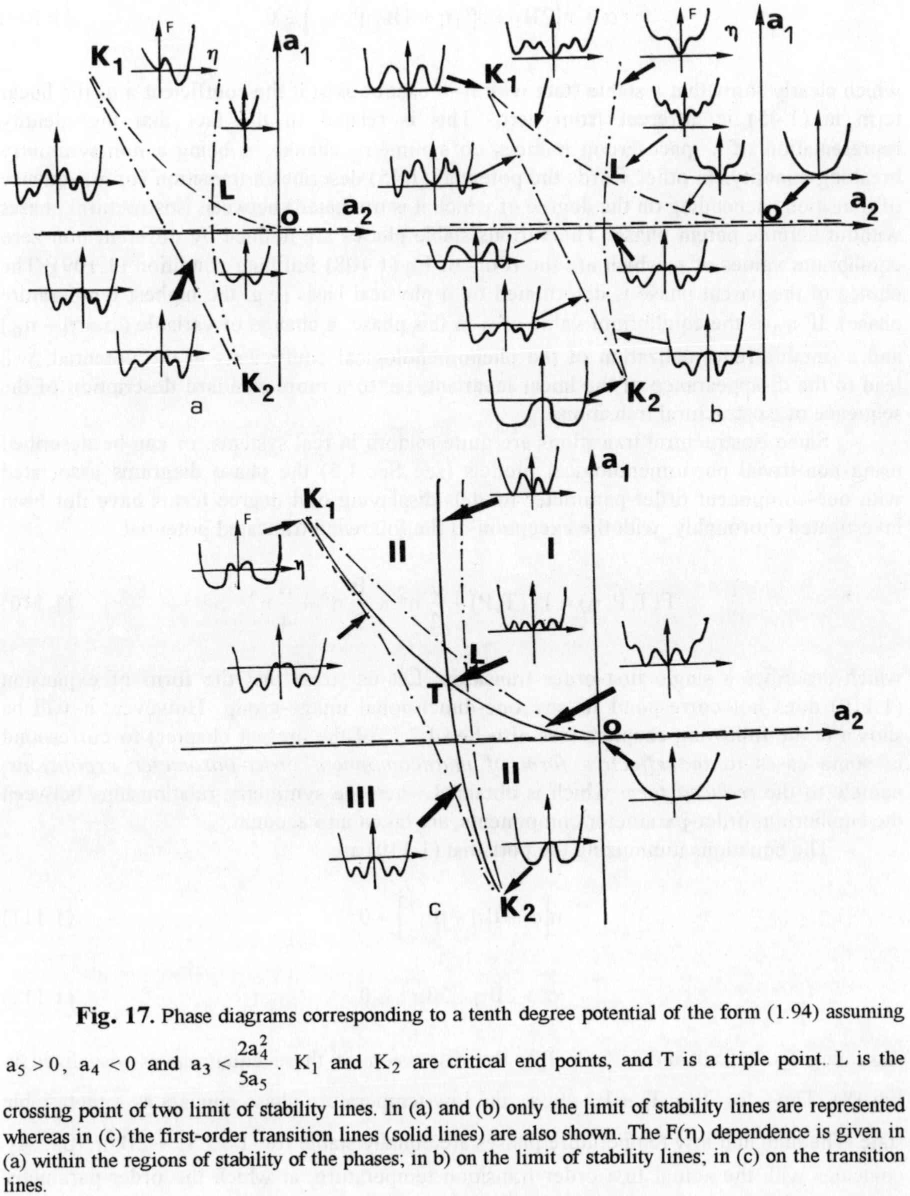

Fig. 17. Phase diagrams corresponding to a tenth degree potential of the form (1.94) assuming $a_5 > 0$, $a_4 < 0$ and $a_3 < \frac{2a_4^2}{5a_5}$. K_1 and K_2 are critical end points, and T is a triple point. L is the crossing point of two limit of stability lines. In (a) and (b) only the limit of stability lines are represented whereas in (c) the first-order transition lines (solid lines) are also shown. The F(η) dependence is given in (a) within the regions of stability of the phases; in b) on the limit of stability lines; in (c) on the transition lines.

$$\alpha + \eta\left(2B_1 + 3\beta_1\eta + 4B_2\eta^2 + \ldots\right) \geq 0 \tag{1.109}$$

which clearly show that a stable state with $\eta=0$ cannot exist if the coefficient a of the linear term in (1.45) is different from zero. This is related to the fact that the identity representation of a space-group induces no symmetry change, η being a non-symmetry breaking quantity. In other words the potential (1.45) describes a transition (or a sequence of transitions depending on the degree of which it is truncated) between isostructural phases without definite parent phase. The various stable phases are defined by different non-zero equilibrium values of η which are the roots of Eq.(1.108) fulfilling condition (1.109). The choice of the parent phase is determined on a physical basis (e.g. the highest temperature phase). If η_0 is the equilibrium value of η in this phase, a change of variable ($\rho = \eta - \eta_0$) and a suitable renormalization of the phenomenological coefficients in the potential, will lead to the disappearance of the linear invariant, i.e. to a more standard description of the sequence of isostructural transitions.

Since isostructural transitions are quite seldom in real systems, or can be described using non-trivial phenomenological models (see Sec.4.5) the phase diagrams associated with one-component order-parameter models displaying odd degree terms have not been investigated thoroughly, with the exception of the following truncated potential:

$$F(T,P,\eta) = F_0(T,P) + \frac{\alpha}{2}\eta^2 + \frac{B}{3}\eta^3 + \frac{\beta}{4}\eta^4 \tag{1.110}$$

which describes a single first-order transition. Let us stress that the form of expansion (1.110) does not correspond to any one-dimensional image-group. However, it will be shown in the following chapters (see also Sec.4.2.1. of the present chapter) to correspond in some cases to the *effective form of multicomponent order-parameter expansions*, namely to the *reduced form* which is obtained when the symmetry relationships between the equilibrium order-parameter components, are taken into account.

The equations minimizing the potential (1.110) are:

$$\eta\left[\alpha + B\eta + \beta\eta^2\right] = 0, \tag{1.111}$$

$$\alpha + 2B\eta + 3\beta\eta^2 \geq 0. \tag{1.112}$$

Assuming $B<0$, $\beta>0$ and $\alpha=a(T-T_c)$, the discussion of these equations is visualized in Fig.18a. Thus, for $T_1 = T_c + B^2/9\beta a$, the low-temperature phase appears as a metastable state. The high and low temperature phases are equally stable for $T_2 = T_c + 2B^2/9\beta a$. T_2 coincides with the actual first-order transition temperature, at which the order-parameter undergoes the jump: $\eta(T_2)=-B/3\beta$, then increases following the law:

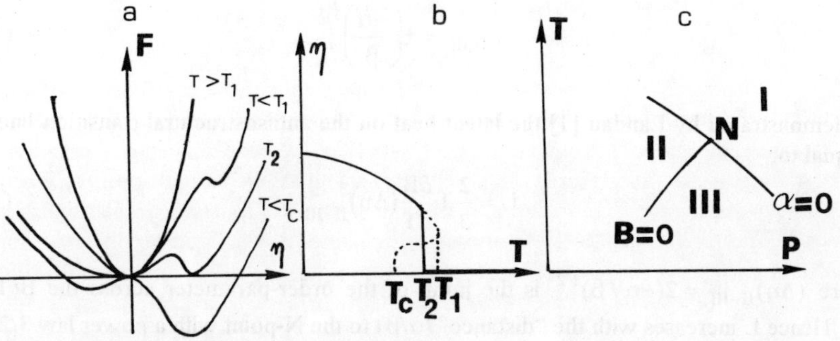

Fig. 18. (a) Order-parameter dependence of the potential (1.110) with B<0 for different temperatures. (b) Temperature dependence of the order-parameter, given by (1.113) for $T < T_2$. (c) Neighborhood of the Landau critical point N. Solid lines are first-order transition lines.

$$\eta(T) = \frac{-B + (B^2 - 4\alpha\beta)^{1/2}}{2\beta} \tag{1.113}$$

which is represented in Fig. 18b. Let us note that assuming B>0 does not modify the preceding description and has only the result of exchanging the left and right-hand side of Fig. 18a.

The preceding discussion resembles to the one performed for the potential (1.80), and most of the features characterizing first-order transitions, as described in Sec. 3.3.2.a, are also found here: region of coexistence for the high and low temperature phases, discontinuous jump at the transition for the order-parameter, thermal hysteresis, latent heat etc. However the phase diagram corresponding to the potential (1.110) involves only a first-order transition regime and no tricritical point. This is due to the existence of the cubic invariant. As stressed by Landau [1] the vanishing of the coefficient of the preceding invariant determines on the line of first-order transitions $\alpha(T_2, P)=0$, a *three-phase point* N (so-called Landau-point) given by:

$$\alpha(T_2, P)=0 \quad \text{and} \quad B(T_2, P)=0. \tag{1.114}$$

As shown in Fig. 18c, N is an isolated point of continuous transition ending a line of first-order transitions of equation $B(T,P)=0$, which separates two *antiisostructural* phases II and III, i.e. two phases of identical symmetries having opposite values for the order-parameter:

$$\eta_{II,III} = \pm\left(\frac{-\alpha}{\beta}\right)^{1/2}. \tag{1.115}$$

As demonstrated by Landau [1] the latent heat on the antiisostructural transition line II-III is equal to:

$$L = \frac{2}{3}T\frac{\partial B}{\partial T}(\Delta\eta)^3_{II-III} \tag{1.116}$$

where $(\Delta\eta)_{II-III} = 2(-\alpha/\beta)^{1/2}$ is the jump of the order-parameter across the B(T,P)=0 line. Hence L increases with the "distance" (α/β) to the N-point with a power law 3/2.

3.3.3. Phase diagrams for two-component order-parameter expansions

As mentioned in Sec.3.2.3 the image groups associated with two-component *irreducible* order-parameters are C_{nv} or C_n, with n=3, 4, 6, 8 and 12 (the image groups for $C_{1v} = C_s$, C_{2v}, and C_2 coincide with reducible order-parameters), and their integrity bases are given by (1.47) and (1.48) respectively. The phase diagrams corresponding to the groups generated by reflections C_{3v}, C_{4v}, C_{6v}, and to the cyclic groups C_3, C_4 have been discussed by Gufan [12], Gufan and Sakhnenko [10b], Gufan and Larin [31], and in a more general way in the unpublished dissertations of Kut'in [32] and Lorman [33].In the present section we will successively describe the phase diagrams related to the image groups: a) C_{4v} and C_4, b) C_{3v} and C_3, before summarizing the general results which hold for any C_{nv} and C_n.

a) Phase diagrams associated with the image-groups C_{4v} and C_4.

The basic invariants forming the integrity basis for the image group C_{4v} are:

$$I_1 = \eta_1^2 + \eta_2^2, \quad I_2 = \eta_1^2\eta_2^2. \tag{1.117}$$

Accordingly, the order-parameter expansion invariant by the C_{4v} group and restricted to the fourth degree contains the monomials I_1, I_1^2 and I_2. One can use equivalently the order-parameter expansion:

$$F(\eta_1,\eta_2) = a_1(\eta_1^2 + \eta_2^2) + b_1(\eta_1^4 + \eta_2^4) + b_2\eta_1^2\eta_2^2 \tag{1.118}$$

since $\eta_1^4 + \eta_2^4 = I_1^2 - 4I_2$ is also invariant by C_{4v}. The equations minimizing (1.118) are:

$$2\eta_1\left(a_1 + 2b_1\eta_1^2 + b_2\eta_2^2\right) = 0, \quad 2\eta_2\left(a_1 + 2b_1\eta_2^2 + b_2\eta_1^2\right) = 0 \tag{1.119}$$

and:

$$\frac{\partial^2 F}{\partial \eta_1^2} \geq 0, \qquad \frac{\partial^2 F}{\partial \eta_1^2} \cdot \frac{\partial^2 F}{\partial \eta_2^2} - \left(\frac{\partial^2 F}{\partial \eta_1 \partial \eta_2} \right)^2 \geq 0 \tag{1.120}$$

with:

$$\frac{\partial^2 F}{\partial \eta_1^2} = 2\left(a_1 + 2b_1\eta_1^2 + 2b_2\eta_2^2\right) + 8b_1\eta_1^2,$$

$$\frac{\partial^2 F}{\partial \eta_2^2} = 2\left(a_1 + 2b_1\eta_2^2 + 2b_2\eta_1^2\right) + 8b_1\eta_2^2,$$

$$\frac{\partial^2 F}{\partial \eta_1 \partial \eta_2} = 4b_2\eta_1\eta_2. \tag{1.121}$$

Simple calculations show that in addition to the trivial solution $\eta_1^e = \eta_2^e = 0$, corresponding to the parent phase (phase I), which is stable for $a_1 > 0$, one has eight solutions for equations (1.119) and (1.120), which are:

$(\mathrm{II}_1, \mathrm{II}_2)$ for $\eta_1^e = 0$, $\eta_2^e = \pm(-a_1 / 2b_1)^{1/2}$;

$(\mathrm{II}_3, \mathrm{II}_4)$ for $\eta_2^e = 0$, $\eta_1^e = \pm(-a_1 / 2b_1)^{1/2}$;

$(\mathrm{III}_1, \mathrm{III}_2, \mathrm{III}_3, \mathrm{III}_4)$ for $\eta_1^e = \pm\eta_2^e = \pm[-a_1 / (2b_1 + b_2)]^{1/2}$.

However, as all the solutions of the type II_i (i=1-4) coincide with the same stability conditions:

$$a_1 < 0, \; b_1 > 0, \; 2b_1 - b_2 < 0 \tag{1.122}$$

whereas the solutions III_i (i=1-4) are all stable for:

$$a_1 < 0, \; b_1 > 0, \; 2b_1 - b_2 > 0, \; 2b_1 + b_2 > 0 \tag{1.123}$$

it is obvious that the four states II_i and III_i constitute respectively four different domains of the same states (denoted hereafter II and III). This can also be foreseen from the fact that equations (1.119) and (1.120) are invariant by the permutations $(\pm\eta_1) \leftrightarrow (\pm\eta_2)$.

Using Eqs.(1.122) and (1.123) one can construct the phase diagram corresponding to expansion (1.118), in which the conditions $b_1 > 0$ and $(4b_1^2 - b_2^2) > 0$ (which are necessary for the stabilization of the phases II and III) are assumed. Figs. 19a and 19b show the phase diagrams in the three-dimensional space (a_1, b_1, b_2) and in the plane (a_1, b_1) with $b_1 = \text{const} \neq 0$. One can see in these phase diagrams that the transitions I-II and I-III, taking place for $a_1 = 0$, are always second-order, while the transition II-III, which occurs on the line $b_2 = 2b_1$ for $a_1 < 0$, is first-order, and corresponds to a jump of the order-parameter given by:

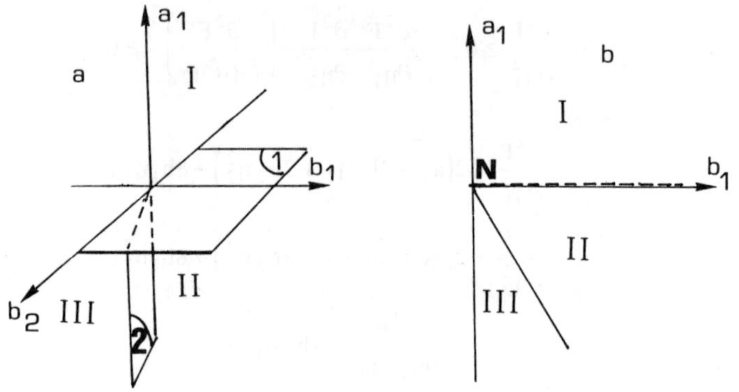

Fig. 19. Phase diagrams corresponding to the image C_{4v} and to the fourth-degree order-parameter expansion (1.118). (a) Phase diagram in the space (a_1, b_1, b_2). Planes 1) and 2) are second-order and first-order transition surfaces. (b) Phase diagram in the plane (a_1, b_1), $b_2 > 0$. Dashed and solid lines are second-order and first-order transition lines.

$$\eta_{II}^2 - \eta_{III}^2 = \frac{a_1}{4b_1}. \tag{1.124}$$

This jump vanishes for $a_1 = 0$ and the three phases are contiguous on the line $a_1 = 0$, $b_2 = 2b_1$ (i.e. at the N point: $a_1 = 0$, $b_1 = 0$ in Fig.19b).

The preceding description clearly reveals the incompleteness of the model using a fourth degree expansion (1.118), as it predicts no region of coexistence between phases II and III, since the limit of stability line ($b_2 = 2b_1$) coincides with the first-order transition line. On the other hand no first-order transition regime is predicted from phase I to the phases II or III. Expanding $F(\eta_1, \eta_2)$ to the sixth degree, with the two additional invariants:

$$c_1(\eta_1^6 + \eta_2^6) \quad \text{and} \quad c_1(\eta_1^4\eta_2^2 + \eta_1^2\eta_2^4) \tag{1.125}$$

yields the equations of state:

$$2\eta_1\left(a_1 + 2b_1\eta_1^2 + b_2\eta_2^2 + 3c_1\eta_1^4 + 2c_2\eta_1^2\eta_2^2\right) = 0$$

$$2\eta_2\left(a_1 + 2b_1\eta_{21}^2 + b_2\eta_1^2 + 3c_1\eta_2^4 + 2c_2\eta_1^2\eta_2^2\right) = 0 \tag{1.126}$$

which possess, in addition to the solutions of (1.119), the additional solutions $\eta_1^e \neq 0$, $\eta_2^e \neq 0$, $\eta_1^e \neq \eta_2^e$, with:

$$(\eta_1^e)^2 + (\eta_2^e)^2 = -\frac{2b_1 - b_2}{3c_1}$$

$$(\eta_1^e)^2 - (\eta_2^e)^2 = \pm\left[\frac{(2b_1 - b_2)^2}{3c_1} - 4\frac{3c_1a_1 - b_2(2b_1 + b_2)}{3c_1 - 2c_2}\right]^{1/2} \cdot \left(\frac{1}{12c_1}\right)^{1/2}.$$

However the corresponding state (phase IV) is unstable, since it coincides with a *maximum* of the sixth-degree expansion. Phase IV possesses a domain of stability in the phase diagram only if additional eighth degree invariants are considered in the order-parameter expansion, namely:

$$d_1(\eta_1^8 + \eta_2^8),\ d_2(\eta_1^6\eta_2^2 + \eta_1^2\eta_2^6),\ d_3\eta_1^4\eta_2^4. \tag{1.127}$$

Accordingly, the lowest degree at which $F(\eta_1, \eta_2)$ should be truncated in order to describe the set of low-symmetry phases II, III and IV is eight. Since phase IV corresponds to a general direction in the order-parameter space ε_2, no other phase of distinct symmetry (no other type of orbit) can be stabilized for the image-group C_{4v}. Thus, the preceding considerations illustrate the general property that *the maximal degree at which the order-parameter expansion must be truncated in order to obtain the maximal number of phases with distinct symmetries is 2N, where N is the highest degree of the invariants forming the integrity basis* (i.e. N=4 for the C_{4v} image). Lower degree expansions *will not* provide a consistent ("structurally stable") model (see Sec.5.2).

A simple way to verify the structural stability of a model is to construct the phase diagram in the *orbit space* which is the space of the integrity basis. Fig.20 shows such a phase diagram for the image group C_{4v}. The diagram has been constructed using expressions (1.117) of I_1 and I_2 and taking into account the equilibrium relationship between the order-parameter components in each phase. Thus, one can see in Fig.20 that phase I coincides with the origin of the orbit space (I_1, I_2), since $\eta_1 = \eta_2 = 0$ yields $I_1 = I_2 = 0$. Phase II $(\eta_1 \neq 0, \eta_2 = 0)$ corresponds to one of the axes $(I_2 = 0,\ I_1 \neq 0)$, whereas phase III $(\eta_1 = \eta_2 = \eta \neq 0)$ is the parabolic branch $I_2 = \frac{1}{4}I_1^2 = \eta^4$. The region

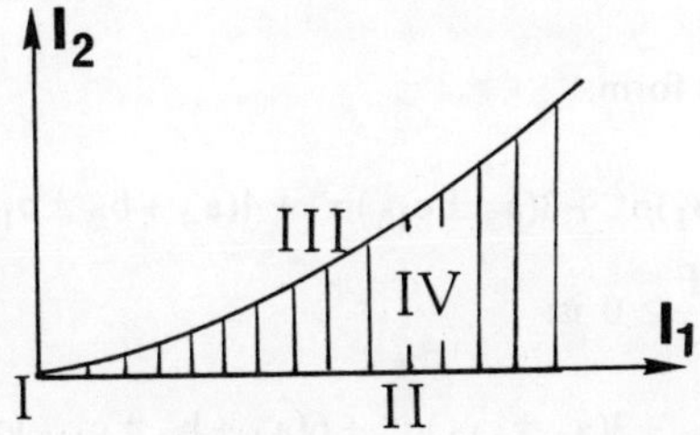

Fig. 20. Phase diagram in the orbit space ("manifold of orbits"), corresponding to the image-group C_{4v}

of stability of phase IV ($\eta_1 \neq \eta_2 \neq 0$) is the portion of the plane (I_1, I_2) located between phases II and III.

Phase diagrams in orbit spaces reveal the full set of possibly stable states for the corresponding image-groups, but also provide indications on the topological properties of the phases in the real (e.g. temperature-pressure) phase diagram. From Fig.20 one can deduce for example that phase IV will always be located *between* phases II and III, or that phase IV can touch phase I only at an isolated point of the phase diagram, (at least when second-order transitions are assumed).

In order to work out the phase diagram associated with an eight-degree expansion $F(\eta_1, \eta_2)$, it is convenient to use polar coordinates ($\eta_1 = \rho\cos\theta$, $\eta_2 = \rho\sin\theta$) in the ε_2-space, and the invariant basis:

$$I_1 = \eta_1^2 + \eta_2^2 = \rho^2, \quad I_2' = I_1^2 - 4I_2 = \rho^4 \cos 4\theta. \tag{1.128}$$

The order-parameter expansion ($F(\eta_1, \eta_2)$=F'(ρ, θ)=F''(I_1, I_2')) has the form:

$$F''(I_1, I_2') = a_1 I_1 + a_2 I_1^2 + a_3 I_1^3 + a_4 I_1^4 + b_1 I_2' + b_2 I_2'^2 + c_{12} I_1 I_2' + c_{112} I_1^2 I_2'. \tag{1.129}$$

Taking $\Phi_1 = \dfrac{\partial F''}{\partial I_1}$, $\Phi_2 = \dfrac{\partial F''}{\partial I_2'}$, the equations of state are:

$$\frac{\partial F'}{\partial \rho} = 2\rho\left(\Phi_1 + 2\rho^2 \Phi_2 \cos 4\theta\right) = 0$$

$$\frac{\partial F'}{\partial \theta} = -4\rho^4 \Phi_2 \sin 4\theta = 0. \tag{1.130}$$

The equilibrium values of the order-parameter components are:

For phases II and III: $\cos 4\theta = \pm 1$ **and** $\Phi_1 \pm 2\rho^2 \Phi_2 = 0$ (1.131)

and for phase IV : $\Phi_1 = 0, \ \Phi_2 = 0.$ (1.132)

Eq.(1.131) has the explicit form:

$$a_1 + 2(a_2 + b_1)\rho^2 + 3(a_3 \pm c_{12})\rho^4 + 4(a_4 + b_2 \pm c_{112})\rho^6 = 0. \tag{1.133}$$

The stability condition $\dfrac{\partial^2 F}{\partial \rho^2} \geq 0$ is:

$$a_2 \pm b_1 + 3(a_3 \pm c_{12})\rho^2 + 6(a_4 + b_2 \pm c_{112})\rho^4 \geq 0. \tag{1.134}$$

Up to a renormalization of the coefficients Eqs.(1.133) and (1.134), which account for the

stabilization of phases II and III, are identical to the equations solved in Sec.3.3.2c for an eighth degree potential with a one-component order-parameter. Using the corresponding results, one obtains the phase diagrams shown in Fig.21, for which the conditions $a_3 = c_{112} = 0$, and $a_2 > 0$ have been assumed. Fig.21a corresponds to the case $(c_{12}^2/4b_2) < a_2 < (c_{12}^2/32a_4)$ and $a_4 < 8b_2$. One can see that the transitions I-II and I-III can be either second or first-order. The first-order transition line I-II becomes a second-order transition line at the tricritical point T_{tc}, whereas the first-order transition line I-III, which separates for $a_1 < 0$ two isostructural phases III^+ and III^-, ends at a critical point K of the liquid-vapor type. Phase IV merge with phases I, II and III at a four-phase point N_1, and is separated from phases II and III by second-order transition lines. Let us note that for large values of b_1 in fig.21a, the transition line II-III crosses the b_1 axis, i.e. the second-order transition II-III takes place within the metastability region of phase I.

Fig.21b represents the phase diagram corresponding to $0 < a_2 < (c_{12}^2/4b_2)$. In this case phase IV cannot be reached directly from phase I, and phases II, III and IV merge at a three-phase point N_2. The II-III transition may occur directly across a first-order transition line N-0.

In the preceding discussion, the condition $c_{12} > 0$ has been assumed. For $c_{12} < 0$ the left and right hand side of the phase diagrams of Figs.21a and 21b are exchanged symmetrically, i.e. the critical point K occurs within phase II, and the transition lines II-IV and III-IV meet in the $a_1 < 0$, $b_1 > 0$ portion of the phase diagram. On the other hand for

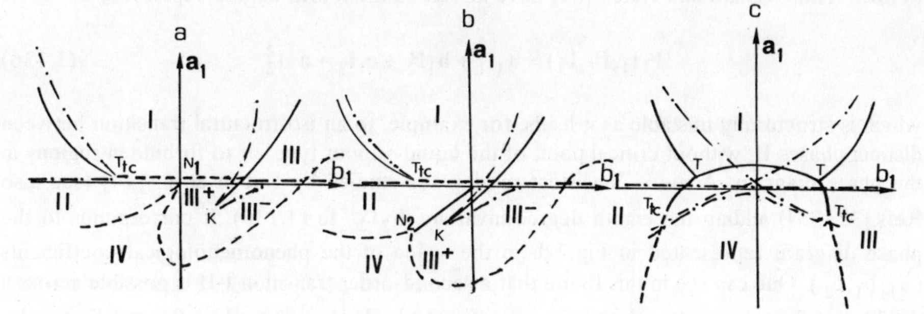

Fig. 21. Phase diagrams corresponding to the image C_{4v} and to the eighth-degree expansion (1.129) for $a_3 = c_{112} = 0$. The diagrams are obtained for the following conditions: (a) $a_4 < 8b_2$, $c_{12} > 0$, $(c_{12}^2/4b_2) < a_2 < (c_{12}^2/32a_4)$; (b) $0 < a_2 < (c_{12}^2/4b_2)$, $c_{12} > 0$; (c) $a_2 < 0$, $4a_2b_2 < c_{12}^2$. Dashed, full and dashed-dotted lines are respectively, second-order, first-order, and limit of stability lines.

$a_2 < 0$ and $4a_2b_2 - c_{12}^2 < 0$, one has the phase diagram represented in Fig.21c, in which Phase IV can be reached directly across a line of first-order transitions, and the II-IV and III-IV transition lines can be either second or first-order, the two regimes being separated by tricritical points. The T-points are triple points.

For the image-group C_4, the order-parameter expansion depends on I_1, I_2', and on the additional invariant $I_3 = \rho^4 \sin 4\theta$. The equations of state, which correspond to the potential $F(I_1, I_2', I_3)$, become:

$$\rho\left[\Phi_1 + 4\rho^2(\Phi_2 \cos 4\theta + \Phi_3 \sin 4\theta\right] = 0,$$

$$4\rho^4\left[-\Phi_2 \sin 4\theta + \Phi_3 \cos 4\theta\right] = 0 \qquad (1.135)$$

with $\Phi_i = \dfrac{\partial F}{\partial I_i}$ (i=1-3). In addition to the trivial solution $\rho=0$ (the parent phase I) there exist only one possible stable solution with $\rho \neq 0$ to the system of equations (1.135), namely: $\eta_1 \neq 0, \eta_2 \neq 0, \eta_1 \neq \eta_2$ (phase II).

Between the invariants forming the integrity basis one has the relationship: $I_1^4 = I_2'^2 + I_3^2$, which determines in the plane (I_2', I_3) a circle of radius I_1^2. Accordingly, in the orbit-space phase II is represented by a cone-type surface, shown in Fig.22a, the apex of which coincides with phase I.

As stressed for the C_{4v} image-group, a consistent model requires to consider at least an eight degree expansion. However, due to the complexity of the complete model, which contains not less than thirteen invariants, only simplified models have been solved up to now. Thus, Galam and Hatch [33] have discussed the fourth-degree expansion:

$$F(I_1, I_2', I_3) = a_1 I_1 + b_1 I_2' + c_1 I_3 + a_2 I_2^2 \qquad (1.136)$$

which is structurally unstable as it leads, for example, to an isostructural transition between distinct phases II, without critical point of the liquid-vapour type, or to forbidden regions in the phase diagram. A more consistent model was studied by Kut'ín et al. [14] (see also Refs.[32, 33]) adding the eighth degree invariant $b_2 I_2'^2$ to (1.136). It corresponds to the phase diagram represented in Fig.22b in the space of the phenomenological coefficients (a_1, b_1, c_1). One can see in this figure that a second-order transition I-II is possible across a disclike surface, located in the plane $a_1 = 0$, which closes a curved surface of first-order transitions between phases I and II. For $a_1 < 0$ one finds a region of metastability into which coexist two isostructural phases, separated by the plane joining two opposite edges of the pyramid. These edges are lines of critical points of the liquid-vapour type. A more simple view of the phase diagram is given in Fig.22c, which represents a cross section of Fig.22b parallel to the (a_1, b_1) plane. Note that the *tricritical circle* in Fig.22b becomes

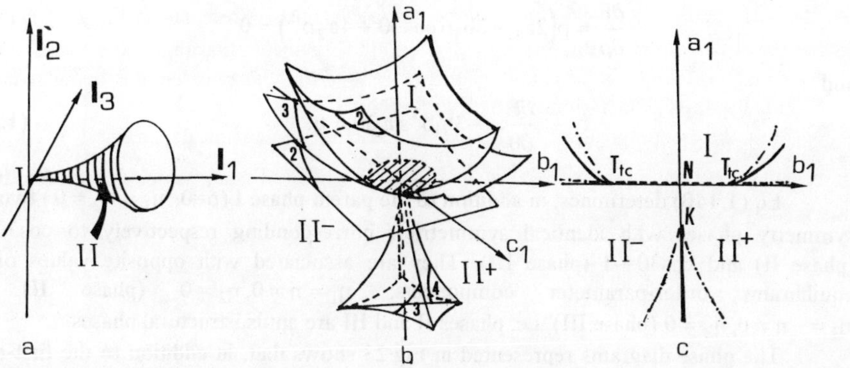

Fig. 22. Phase diagrams corresponding to the image C_4. (a) Location of the phases in the orbit-space. (b) Three-dimensional phase diagram corresponding to an order-parameter expansion with four-degree invariants, and a single ($I_2'^2$) eight degree invariant. The hatched disc 1 is a second-order transition surface. 2 and 3 denote limit of stability and first-order transition surfaces. Plane 4 is a surface of isostructural transitions separating phases II^+ and II^-. (c) Intersection of the phase diagram (b) by a plane parallel to the axes a_1, b_1. Dashed, full and dashed-dotted lines are respectively second-order, first-order, and limit of stability lines. N is a three-phase point. K is a critical point.

two tricritical points in Fig.22c, and that the tricritical end-lines in Fig.22b project on the critical point K.

b) Phase diagrams associated with the image-groups C_{3v} *and* C_3.

The phase diagrams associated with the image-group C_{3v} have been thoroughly investigated since the corresponding integrity basis contains only two basic invariants:

$$I_1 = \eta_1^2 + \eta_2^2 = \rho^2, \quad I_2 = \eta_1^3 - 3\eta_1\eta_2^2 = \rho^3 \cos 3\theta \tag{1.137}$$

having a maximal degree N=3, so structurally stable models involve order-parameter expansions which can be truncated at the sixth (2N) degree, and are thus algebraically solvable. Let us first show, following Kut'in et al. [14], that a fourth degree expansion is indeed insufficient. Up to the fourth degree one has:

$$F(\rho, \theta) = F'(I_1, I_2) = a_1 I_1 + b_1 I_2 + a_2 I_1^2 \tag{1.138}$$

which yields the equations of state:

$$\frac{\partial F}{\partial \rho} = \rho\left(2a_1 + 3b_1\rho\cos 3\theta + 4a_2\rho^2\right) = 0 \qquad (1.139)$$

and

$$\frac{\partial F}{\partial \theta} = -3b_1\rho^3 \sin 3\theta = 0. \qquad (1.140)$$

Eq.(1.140) determines, in addition to the parent phase I (ρ=0, $\eta_1 = \eta_2 = 0$) two low symmetry phases with identical symmetries, corresponding respectively to cos3θ=+1 (phase II) and cos3θ=-1 (phase III). They are associated with opposite values of the equilibrium order-parameter components: $\eta_1 = \eta \neq 0, \eta_2 = 0$ (phase II) and $\eta_1 = -\eta \neq 0, \eta_2 = 0$ (phase III), i.e. phases II and III are antiisostructural phases.

The phase diagrams represented in Fig.23 shows that, in addition to the first-order transitions I-II and I-III, the region of coexistence of phases II and III is reduced to the antiisostructural first-order transition line II-III. In other words, even though the component η undergoes a finite jump at the preceding transition, the boundary between these latter phases coincide. Another inconsistency of the fourth-degree expansion model can be foreseen from the topology of the phases in the orbit space (Fig.24). Thus the invariants I_1 and I_2 possess in phases II and III the following connections: $I_2 = \eta_{II}^3 = I_1^{3/2}$ and $I_2 = -\eta_{III}^3 = -I_1^{3/2}$, i.e. in the (I_2, I_1) plane the two phases only coincide at the origin (phase I) and for $\eta_1 \neq 0$ are separated by a region in which an additional phase (IV) corresponding to $-I_1^{3/2} < I_2 < I_1^{3/2}$ or equivalently to -1<cos3θ<+1, can be stabilized. The construction of the phase diagram associated with the truncated sixth-degree order-parameter expansion:

$$F'(I_1, I_2) = a_1 I_1 + b_1 I_2 + a_2 I_1^2 + b_2 I_2^2 \qquad (1.141)$$

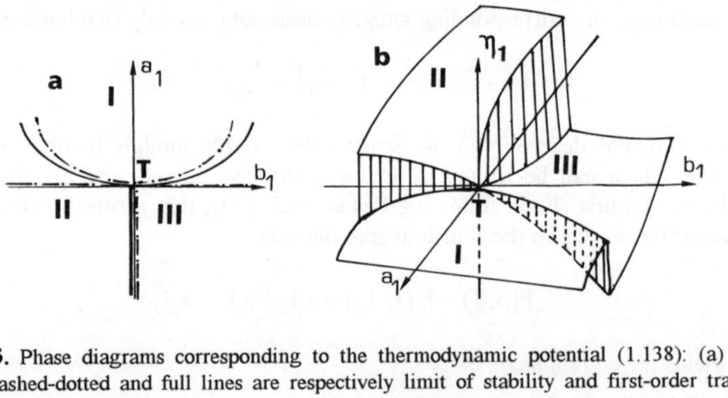

Fig. 23. Phase diagrams corresponding to the thermodynamic potential (1.138): (a) In the (a_1, b_1) plane. Dashed-dotted and full lines are respectively limit of stability and first-order transition lines. T is a triple point. (b) In the (η_1, a_1, b_1) space. Hatched surfaces represent the jumps of the order-parameter $(\eta_1, 0)$ at the transitions.

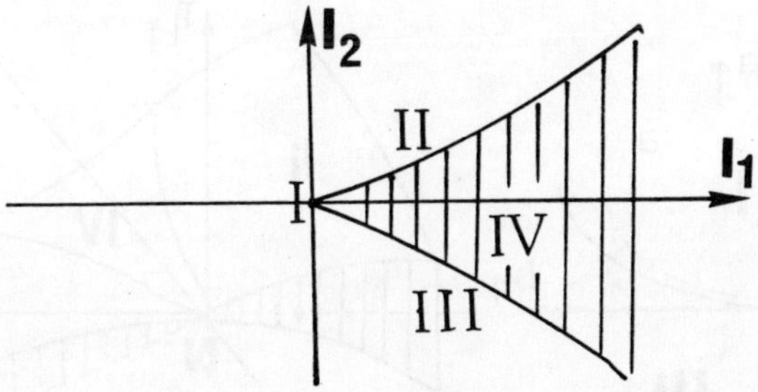

Fig. 24. Topology of the phases corresponding to the C_{3v} image-group, in the orbit-space (I_2, I_1).

clarifies the shortcomings of the fourth-degree model. The equation of state (1.140), becomes:

$$\frac{\partial F}{\partial \theta} = -3\rho^3 \sin 3\theta \left(b_1 + 2b_2\rho^3 \cos 3\theta\right) = 0 \qquad (1.142)$$

which admits the additional solution: $b_1 + 2b_2\rho^3 \cos 3\theta = 0$, corresponding to the stable state IV: (η_1, η_2) with $\eta_1 \neq 0, \eta_2 \neq 0, \eta_1 \neq \eta_2$. The symmetry of phase IV is the maximal common subgroup (of index 2) to the symmetry groups of phases II and III. The phase diagram associated with the expansion (1.141) is shown in Fig.25. Hence, phases II and II are separated by two lines of second-order transitions bounding phase IV. These phases merge with phase I at a four-phase point N of the Landau-type (see Fig.18), as it is an isolated point of second-order transition on a line of first-order transitions.

The model based on the order-parameter expansion (1.141) is structurally stable, but overlooks a number of important features of the phase diagrams which may be constructed for the C_{3v} image-group. These features are obtained when considering the full set of invariants, up to the sixth-degree, in the order-parameter expansion:

$$F'(I_1, I_2) = a_1 I_1 + a_2 I_1^2 + a_3 I_1^3 + b_1 I_2 + b_2 I_2^2 + c_{12} I_1 I_2 \qquad (1.143)$$

which has been comprehensively discussed by Gufan [12]. The corresponding equations of state are:

$$\frac{\partial F}{\partial \rho} = 2\rho\left(a_1 + 2a_2 I_1 + 3a_3 I_1^2 + c_{12} I_2\right) + 3\rho^2 \cos 3\theta \left(b_1 + 2b_2 I_2 + c_{12} I_1\right) = 0 \qquad (1.144)$$

$$\frac{\partial F}{\partial \theta} = -3\rho^3 \sin 3\theta \left(b_1 + 2b_2 I_2 + c_{12} I_1\right) = 0 . \qquad (1.145)$$

In phase II ($\rho \neq 0$, $\cos 3\theta = +1$) the equilibrium value of ρ is given by:

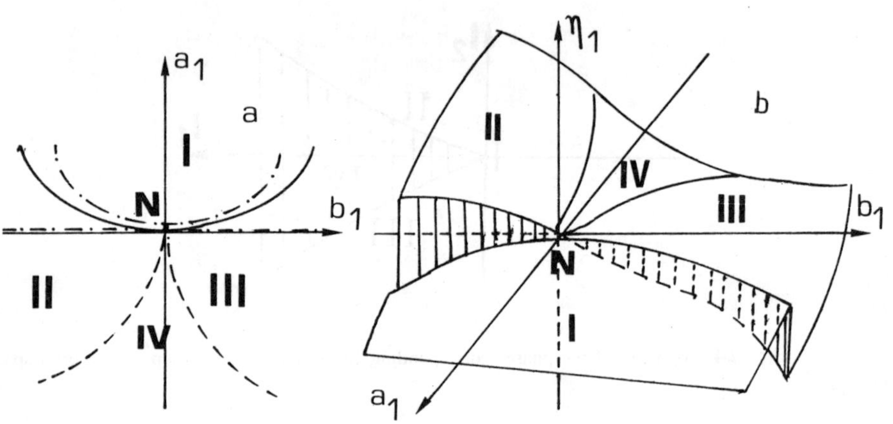

Fig. 25. Phase diagrams corresponding to the thermodynamic potential (1.141): (a) In the (a_1, b_1) plane. Dashed, full, and dashed-dotted lines are respectively, second-order, first-order, and limit of stability lines. N is a four-phase "Landau" point. (b) In the (η_1, a_1, b_1) space.

$$2a_1 + 3b_1\rho + 4a_2\rho^2 + 5c_{12}\rho^3 + 6(a_3 + b_2)\rho^4 = 0$$

and the stability conditions are:

$$\frac{\partial^2 F}{\partial \rho^2} = \rho\left[3b_1 + 8a_2\rho + 15c_{12}\rho^2 + 24(a_3 + b_2)\rho^3\right] \geq 0 \tag{1.146}$$

and

$$\frac{\partial^2 F}{\partial \theta^2} = -9\rho^3\left[-b_1 + c_{12}\rho^2 + 2b_2\rho^3\right] \geq 0. \tag{1.147}$$

Accordingly the limit of stability of phase II with respect to phase I can be expressed under the parametric form:

$$\begin{cases} a_1 - 2a_2\rho^2 - 10c_{12}\rho^3 - 18(a_3 + b_2)\rho^4 = 0 \\ b_1 - \frac{8}{3}a_3\rho + 5c_{12}\rho^2 + 8(a_3 + b_2)\rho^3 = 0 \end{cases}$$

which provides the equation of the line of stability II-I:

$$a_1 = \frac{9}{2^5 a_2} b_1^2 + \frac{3}{2^{10} a_2^3} b_1^3 - \frac{3^6}{2^{17} a_2^5} \left[\frac{2^5 a_2}{3} (a_3 + b_2) - 5^2 c_{12}^2 \right] b_1^4. \tag{1.148}$$

The equation of the first-order transition line I-II, is determined by the equality of the potentials: $F(\rho_I = 0) = F(\rho_{II})$, namely by:

$$\rho^2 \left[a_1 + b_1 \rho + a_2 \rho^2 + c_{12} \rho^3 + 9(a_3 + b_2) \rho^4 \right] = 0$$

in which ρ is given by Eq.(1.144). It yields:

$$a_1 = \frac{1}{2^2 a_2} b_1^2 + \frac{c_{12}}{2^3 a_2^3} b_1^3 + \frac{3c_{12}^2 - 2^2 (a_2 + b_2) a_2}{2^6 a_5} b_1^4. \tag{1.149}$$

The boundary of phase III ($\rho \neq 0$, $\cos 3\theta = -1$) into phase I is obtained by replacing in Eqs.(1.144)-(1.146) b_1 by $-b_1$ and c_{12} by $-c_{12}$. As Eqs.(1.148) and (1.149) are invariant by these changes in sign, the equation of the stability line of phase III with respect to phase I and of the first-order transition line I-III, are deduced from the analogous lines for phase II by a mirror symmetry with respect to $b_1 = 0$.

The limit of stability line between phases II and III can be worked out by introducing the corresponding equilibrium values ρ_{II} and ρ_{III} in Eq.(1.146). One gets:

$$a_1 = \frac{2a_2}{c_{12}} b_1 \pm \Delta \cdot (-b_1)^{3/2} + \frac{3(\Delta \cdot b_2 - a_3 c_{12}^2)}{c_{12}^4} b_1 \tag{1.150}$$

where $\Delta = 4a_2 b_2 - c_{12}^2$. It is easy to show [12] that the mutual limit of stability lines of phases II and III have a common tangent $a_1 = 2a_2 b_1 / c_{12}$ at the origin ($a_1 = b_1 = 0$), and that depending on the sign of Δ, one has between the two phases either two lines of second-order phase transitions bounding phase IV ($\rho \neq 0$, $-1 < \cos 3\theta < +1$) for $\Delta > 0$, or a direct first-order (antiisostructural) transition line II-III for $\Delta < 0$. Figs.26a and 26b represent the two preceding situations in the case $a_2 > 0$. As shown by Gufan [12] (see also Kut'in [32]) the topology of the phase diagrams changes drastically when $a_2 < 0$. In this case, represented in Figs.26c and 26d, one has , instead of a four-phase point N, two three-phase points located on a direct first-order transition line between phases I and IV.

Another important change in the topology of the C_{3v} phase diagram occurs when the coefficients c_{12}, b_2 or a_3 become negative. Higher degree invariants have to be considered in this case in order to insure the stability of the phases. Figs.27a-c represent phase diagrams associated with an eight-degree expansion in which two additional invariants ($c_{112} I_1^2 I_2$ and $a_4 I_1^4$) have been added to the expansion (1.143). One can see that further complexifications of the phase diagrams take place: In Figs.27a and 27b the transition II-III possesses two different, first-order and second-order regimes, with either

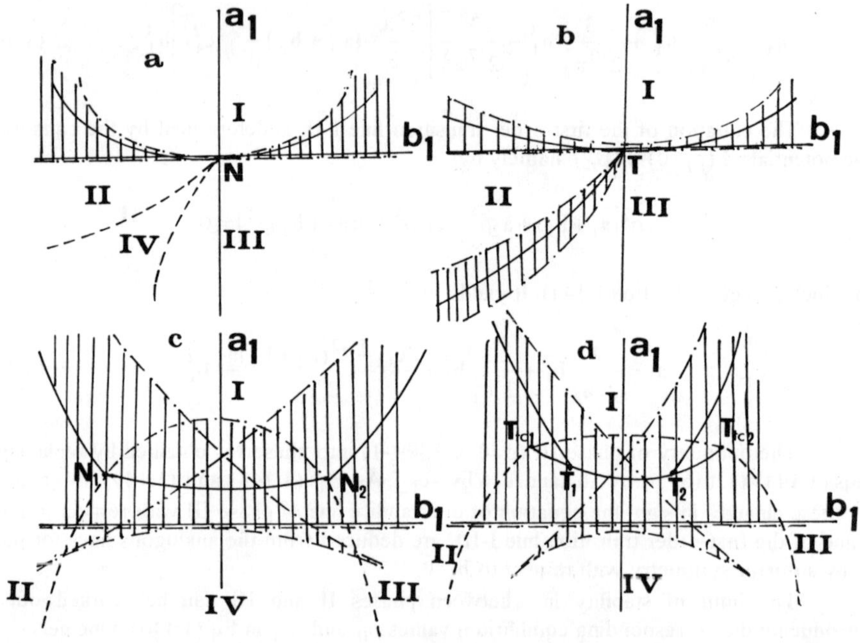

Fig. 26. Phase diagrams corresponding to the image-group C_{3v} for a sixth-degree expansion (1.143). The diagrams are obtained for the following conditions: (a) $a_2 > 0, a_3 > 0, c_{12} > 0$, and $\Delta = 4a_2b_2 - c_{12}^2 > 0$, (b) $a_2 > 0, a_3 > 0, c_{12} > 0, \Delta < 0$, (c) $a_2 < 0, c_{12} = 0, 0 < b_2 < a_3$, (d) $a_2 < 0, c_{12} = 0$, $0 < a_3 < b_2$. Full, dashed and dotted-dashed lines represent first-order, second-order and limit of stability lines, respectively. Hatched surfaces are metastability regions surrounding first-order transition lines. N is a four-phase point. N_1 and N_2 are three-phase points. T, T_1 and T_2 are triple points. T_{tc1} and T_{tc2} are tricritical points.

one or two tricritical points. In Fig.27c phase II has two different regions of stability separated by a line of isostructural transitions, ending at a critical point of the liquid-vapor type [12, 32].

For the image-group C_3 one has with respect to C_{3v} the additional invariant $I_3 = \eta_2^3 - 3\eta_1^2\eta_2 = \rho^3 \sin 3\theta$ in the integrity basis. Using for the order-parameter expansion $F(\rho,\theta)=F'(I_1,I_2,I_3)$ the notation $\Phi_i = \dfrac{\partial F}{\partial I_i}$ and $\Phi_{ij} = \dfrac{\partial^2 F'}{\partial I_i \partial I_j}$ (i,j=1-3), one gets the equilibrium conditions:

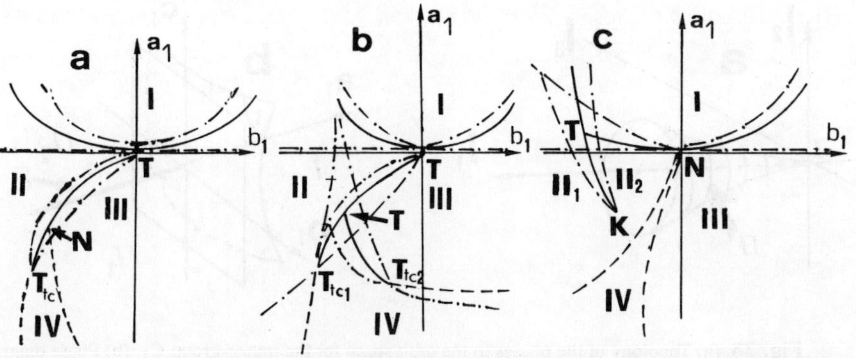

Fig. 27. Phase diagrams corresponding to the image-group C_{3v} for an eighth-degree expansion (see text). The diagrams are obtained for the following conditions: (a) $a_2 > 0$, $a_3 > 0$, $c_{12} < 0$, $b_1 < 0$, $b_2 > 0$, $\Delta > 0$; (b) $a_2 > 0$, $a_3 < 0$, $b_2 < 0$, $a_4 > 0$, $\Delta < 0$; (c) $a_2 > 0$, $b_2 > 0$, $a_3 < 0$, $a_4 > 0$, $\Delta > 0$. Full, dashed and dotted-dashed lines represent first-order, second-order and limit of stability lines. T_{tc1} and T_{tc2} are tricritical points. N is a three phase point in a), and is a four-phase point in c). T, T_1 and T_2 are triple points. K is a critical point of the liquid-vapor type. II_1 and II_2 are two isostructural phases.

$$\frac{\partial F}{\partial \rho} = \rho[\Phi_1 + 3\rho(\Phi_2 \cos 3\theta + \Phi_3 \sin 3\theta)] = 0 \qquad (1.151)$$

$$\frac{\partial F}{\partial \theta} = 3\rho^3[-\Phi_2 \sin 3\theta + \Phi_3 \cos 3\theta] = 0. \qquad (1.152)$$

The corresponding stability conditions involve the second derivatives:

$$\frac{\partial^2 F}{\partial \rho^2} = 4\rho^2 \Phi_{11} + \rho[\Phi_2 \cos 3\theta + \Phi_3 \sin 3\theta] + ...$$

$$\frac{\partial^2 F}{\partial \theta^2} = -9\rho^2[\Phi_2 \cos 3\theta + \Phi_3 \sin 3\theta] + ...$$

$$\frac{\partial^2 F}{\partial \rho \partial \theta} = 6\rho^2[-\Phi_{21} \sin 3\theta + \Phi_{21} \cos 3\theta] + ... \qquad (1.153)$$

in which only the lowest degree terms in ρ have been written. Eqs.(1.152) and (1.153) show that, as for the image-group C_4, only one low-symmetry phase (phase II) can be stabilized. It corresponds in the order-parameter space ε_2 to the "angle" θ^e given by:

$$\mathrm{tg}3\theta^e = \frac{\Phi_2(\rho^e, \theta^e)}{\Phi_3(\rho^e, \theta^e)}$$

and to the "modulus":

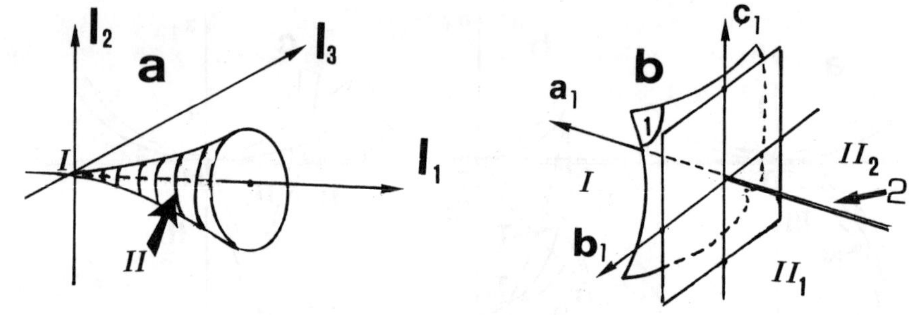

Fig. 28. (a) Topology of the phases in the orbit space for the image-group C_3. (b) Phase diagram of C_3 for the fourth degree expansion (1.154). 1 denotes the surface of first-order transitions I-II, while 2 indicates the isostructural II_1-II_2 transition line.

$$\rho^e = -\frac{\Phi_1(\rho^e,\theta^e)}{3[\Phi_2(\rho^e,\theta^e)\cos 3\theta^e + \Phi_3(\rho^e,\theta^e)\sin 3\theta^e]}$$

which yields the order-parameter components: $\eta_1 = \rho^e \cos\theta^e \neq \eta_2 - \rho^e \sin 0^e$. Fig.28a shows that in the space (I_1, I_2, I_3), phase II is represented by a cone-type surface of equation: $I_2^2 + I_3^2 = I_1^3$, which resembles (although the equation is different) to the surface found for the image-group C_4 (Fig.22a).

From Eqs.(1.151)-(1.153) one can easily deduce that the stability conditions for phase II, namely $\frac{\partial^2 F}{\partial \rho^2} \geq 0$, $\frac{\partial^2 F}{\partial \rho^2} \cdot \frac{\partial^2 F}{\partial \theta^2} \geq \left(\frac{\partial^2 F}{\partial \rho \partial \theta}\right)^2$, can be fulfilled only when $\Phi_1 = \Phi_2 = \Phi_3 = 0$, i.e. *for a finite value of* ρ^e *at the transition point.* In other words, *no second-order transition (even at an isolated point) is possible between phases I and II.* This is a general property, that was already pointed out by Landau [1], of order-parameter expansions involving *two* cubic invariants.

A discussion of the phase diagrams associated with the C_3 image-group has been performed by Kut'in [32] and Lorman [33]. It provides an interesting insight of the respective influence of high-degree invariants on the topology of a phase diagram. Thus, using the (structurally unstable) model based on a fourth-degree expansion:

$$F'(I_1, I_2, I_3) = a_1 I_1 + a_2 I_1^2 + b_1 I_2 + c_1 I_3 \qquad (1.154)$$

leads, as shown in Fig.28b to the stabilization of two isostructural low-symmetry phases (II_1 and II_2), with a region of coexistence reduced to a single line ($b_1 = c_1 = 0, a_3 < 0$) in the (a_1, b_1, c_1) space. In a two-dimensional (e.g. pressure-temperature) phase diagram, it would correspond to an isolated point of first-order II_1-II_2 transition. This unphysical result can be removed by adding to (1.154) the sixth degree invariants: $b_2 I_2^2$ and $c_2 I_3^2$. It yields for $a_2 > 0$ to the phase diagram represented in Fig:29a, in which the isostructural transition plane is surrounded by a pyramidal-type volume of coexistence. The two-dimensional phase diagram of Fig.29b shows no contact between the I-II and II_1-II_2 isostructural transition line. In contrast, when adding also the sixth degree invariant $a_3 I_1^2$ to

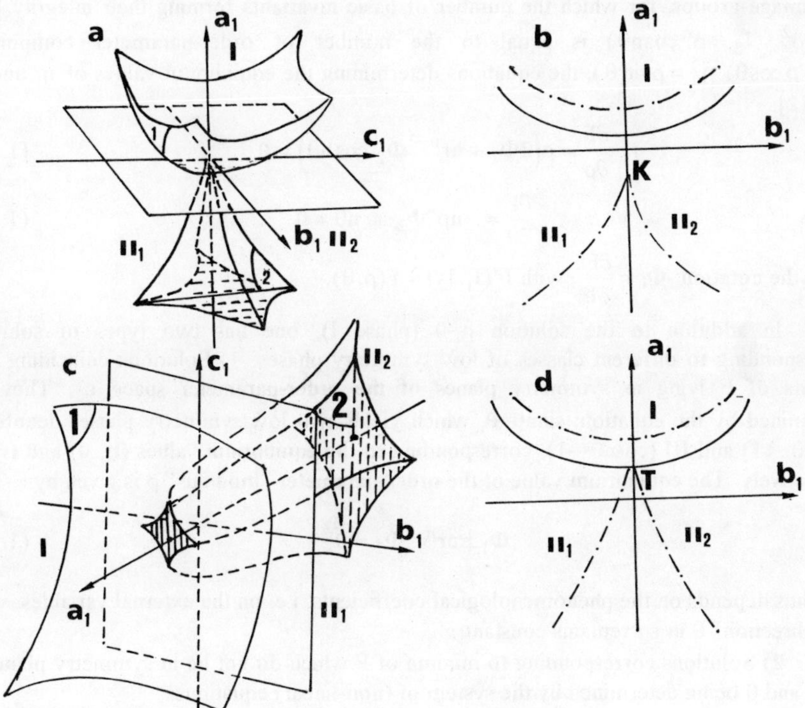

Fig. 29. Phase diagrams corresponding to the image-group C_3 for a sixth-degree expansion of the type (1.154) in which the sixth-degree invariants $b_2 I_2^2$ and $c_2 I_3^2$ are added (Figs. (a) and (b)). In Figs. (c) and (d) the sixth degree $a_3 I_2^3$ invariant is also taken into account (with $a_2 < 0$, $a_3 > 0$). The surfaces denoted 1 and 2 in the three-dimensional diagrams (a) and (c) represent respectively the I-II transition, and the isostructural II_1-II_2 transition. These surfaces become full lines in the two-dimensional diagrams of Figs. (b) and (d) which show cross-sections c_1=const. c_1 is assumed to be small for the cross-section (d). The edges limiting surface 2 are lines of critical points. They correspond to the K-point in (b). T is a triple point. For large values of c_1 T is replaced by a critical point K.

(1.154), with $a_2 < 0$ and $a_3 > 0$, one has a surface of contact between the (II_1+II_2) region and the stability region of phase I in the (a_1, b_1, c_1) space (Fig.29c), i.e. the *critical point K* of Fig.29b, *becomes* (for not too large values of c_1) a triple point, as shown in the two-dimensional diagram of Fig.29d.

c) General features of phase diagrams associated with the image-groups C_{nv} *and* C_n.

The results obtained for C_{4v}, C_{3v}, C_4 and C_3 can be extended to the other image groups C_{nv} and C_n (n=6, 8, 12). Clearly, one has to distinguish the groups generated by reflections (C_{nv}) and the cyclic groups (C_n), which possess different specific features. For C_{nv} image-groups, for which the number of basic invariants forming their integrity basis ($I_1 = \rho^2$, $I_2 = \rho^n \cos n\theta$) is equal to the number of order-parameter components ($\eta_1 = \rho\cos\theta$, $\eta_2 = \rho\sin\theta$), the equations determining the equilibrium values of η_1 and η_2 are [12]:

$$\frac{\partial F}{\partial \rho} = \rho\left(2\Phi_1 + nr^{n-2}\Phi_2 \cos n\theta\right) = 0 \qquad (1.155)$$

$$\frac{\partial F}{\partial \theta} = -n\rho^n \Phi_2 \sin n\theta = 0 \qquad (1.156)$$

using the notation: $\Phi_i = \frac{\partial F'}{\partial I_i}$, with $F'(I_1, I_2) = F(\rho, \theta)$.

In addition to the solution ρ=0 (phase I), one has two types of solutions corresponding to different classes of low symmetry phases: 1) Solutions coinciding with minima of F lying in symmetry planes of the order-parameter space ε_2. They are determined by the equation: sinnθ=0, which yields two low-symmetry phases denoted II (cosnθ=+1) and III (cosnθ=-1), corresponding to the equilibrium values (η, 0) and (η, η) respectively. The equilibrium value of the order-parameter "modulus" ρ is given by:

$$\Phi_1 \pm nr^{n-2}\Phi_2 = 0 \qquad (1.157)$$

and thus depends on the phenomenological coefficients, i.e. on the external variables, while the "direction" θ in ε_2 remains constant.

2) Solutions corresponding to minima of F which do not lie in symmetry planes of ε_2, ρ and θ being determined by the system of (non-linear) equations:

$$\Phi_1 = 0, \quad \Phi_2 = 0 \qquad (1.158)$$

which means that both ρ and θ vary with temperature. There exists actually only one solution for Eq.(1.158), denoted IV and corresponding to the equilibrium values (η_1, η_2) with $\eta_1 \neq \eta_2 \neq 0$.

Accordingly the maximal number of phases which can be found in the phase diagram associated with the image-group C_{nv} is 4. This number is consistent with a theorem by Chevaley, which states that for groups generated by reflections it should be 2^N (when N is the number of order-parameter components). Denoting G_I -G_{IV} the symmetry groups of phases I-IV, one can establish the following results: G_{II} and G_{III} correspond to different symmetries for n≥4 as their respective "directions" θ in ε_2 belong to different classes of conjugate elements. In contrast for C_{3v} all the reflections belong to the same class and consequently, G_{II} and G_{III} have the same symmetry elements. G_{IV} is the maximal common subgroup (of index 2) of G_{II} and G_{III}.

The stability conditions for phases II and III are:

$$\mp\Phi_2 \geq 0, \text{ and } 4\rho^2\Phi_{11} \pm (n-2)n\rho^{n-2}\Phi_2 \pm 4n\rho^4\Phi_{21} + n^2\rho^{n-2}\Phi_{22} \geq 0 \qquad (1.159)$$

where the upper sign corresponds to phase II. From Eqs.(1.155) and (1.159) one can deduce that the boundary between phase II and phase III always correspond to a discontinuous change in ρ, i.e. the II-III transition is always first-order, in agreement with the absence of group-subgroup relationship between the symmetries of the phases. On the other hand phases II and III are always contiguous to phase I, and from the fact that $\lim_{\rho\to 0} \Phi_1 = a_1 = 0$ one can deduce that the transitions I-II and I-III can be of the second order, in agreement with the group-subgroup relationship existing between phase I and, respectively, phase II and III. For n≥4 one has a *line* of second-order transitions, while for n=3, a second-order transition can take place only at a point, which is determined by two conditions: $a_1(P,T)=0$ and $\lim_{\rho\to 0} \Phi_2(P,T,\rho) = 0$. In contrast no second-order transition is possible from phase I to phase IV, whereas the II-IV and III-IV transitions can be second-order.

The preceding properties are well illustrated by the phase diagrams obtained for C_{4v} (Fig.21) and C_{3v} (Fig.27). For n≥6, due to the absence of cubic invariants, the phase diagrams will be similar to those established for C_{4v}. Fig.30 shows for instance the phase diagrams corresponding to C_{6v} [12], for which twelve degree invariants have to be taken into account for a structurally stable model. Along the same line, one needs to consider 16 and 24 degree terms to work out consistently the phase diagrams associated with C_{8v} and C_{12v}, respectively.

Figs.20 and 24 show that the general topology of the phase diagrams of C_{nv} groups in the orbit space (I_1, I_2) is independent from the number n. Only the shape of the I_2=f(I_1) curve changes for phase III ($\eta_1 = \eta_2 = \eta \neq 0$), as one has: $I_2 \propto \eta^n \propto I_1^{n/2}$. Fig.30d summarizes the phase diagrams (I_1, I_2) for the C_{nv} image-groups.

Image-groups C_n arize in ε_2 for physically irreducible two-dimensional representations. For a given n, C_n is a subgroup of index 2 of C_{nv}, and it possesses a larger basis of integrity with the additional invariant $I_3 = \rho^n \sin n\theta$. Thus the dimension of the basis m=3 is larger than the number of order-parameter components. This implies a relationship between the invariants, which is for any n:

$$I_2^2 + I_3^2 = I_1^n \qquad (1.160)$$

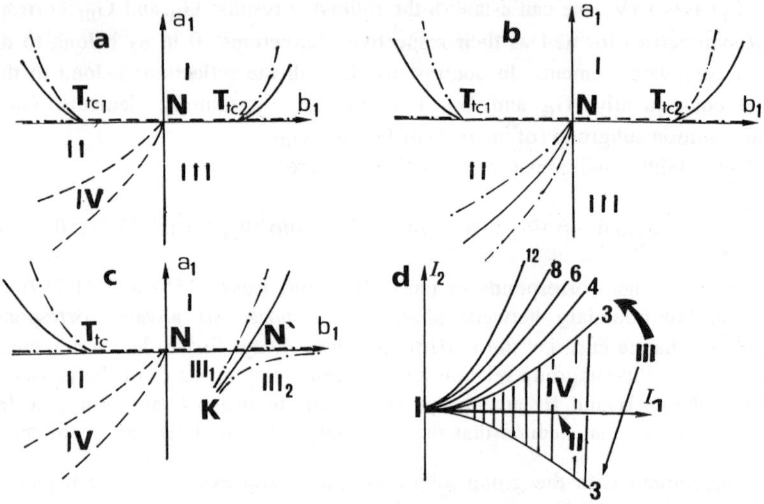

Fig. 30. (a)-(c) Phase diagrams associated with the image C_{6v}, considering the twelvth degree expansion: $F'(I_1,I_2) = a_1I_1 + a_2I_1^2 + a_3I_1^3 + a_4I_1^4 + b_1I_2 + b_2I_2^2 + c_{12}I_1I_2$ with $I_1 = \rho^2$ and $I_2 = \rho^6 \cos 6\theta$. The diagrams correspond to the following conditions: (a) $\Delta = 4a_2b_2 - c_{12}^2 > 0$, $a_2 > 0$, $a_3 > 0$; (b) $\Delta<0$, $a_2 > 0$, $a_3 > a_2$; (c) $\Delta>0$, $a_2 > 0$, $a_3 < b$. Full, dashed and dotted-dashed lines are first-order, second-order, and limit of stability lines. N is a four- or three-phase point. N' is a Landau point. T_{tc1} and T_{tc2} are tricritical points. K is a critical point of the liquid-vapor type. (d) Phase diagram for the C_{nv} image-groups in the orbit space. The numbers 3, 4, 6, 8, 12 denote the exponent $I_2 = I_1^{n/2}$ corresponding to phase III ($\eta_1 = \eta_2 = \eta \neq 0$).

corresponding in the orbit-space (I_1, I_2, I_3) to the same cone-type surface representing phase II in Figs.22a and 28a. This phase which is stabilized for $\eta_1 = \rho\cos\theta \neq \eta_2 = \rho\sin\theta$, is the only distinct low symmetry phase. The associated equilibrium values of ρ and θ are determined by the system of equations:

$$\Phi_1 + n\rho^{n-2}\left(\Phi_2 \cos n\theta + \Phi_3 \sin n\theta\right) = 0 \qquad (1.161)$$

$$-\Phi_2 \sin n\theta + \Phi_3 \cos n\theta = 0 \,. \qquad (1.162)$$

The stability conditions $\dfrac{\partial^2 F}{\partial \rho^2} \geq 0$ and $\left(\dfrac{\partial^2 F}{\partial \rho^2}\right)\left(\dfrac{\partial^2 F}{\partial \theta^2}\right) - \left(\dfrac{\partial^2 F}{\partial \rho \partial \theta}\right) \geq 0$ have the form

$$\frac{\partial^2 F}{\partial \rho^2} = 4\rho^2\Phi_{11} + (n-2)\rho^{n-2}\left(\Phi_2 \cos n\theta + \Phi_3 \sin n\theta\right) + \ldots \geq 0 \quad (1.163)$$

and

$$\frac{\partial^2 F}{\partial \theta^2} = \left[-\rho^n n^2(\Phi_2 \cos n\theta + \Phi_3 \sin n\theta) + \ldots\right] - \left[2\rho^{n-1} n(-\Phi_{21} \sin n\theta + \Phi_{31} \cos n\theta) + \ldots\right]^2 \geq 0. \quad (1.164)$$

For n=3 it has already been stressed that these conditions lead to a first-order I-II transition. By contrast for $n \geq 4$, conditions (1.163) and (1.164) are always fulfilled for small ρ, whatever value takes Φ_1, and consequently the I-II transition may take place as a second-order transition. Accordingly, phase diagrams analogous to the one found for the C_4 image-group (Figs.22b and 22c) will be obtained for C_6, C_8 and C_{12}.

3.3.4. Phase diagrams for three-component order-parameter expansions

As indicated in Sec.3.2.2 there are five image-groups corresponding to structural phase transitions associated with an irreducible three-component order-parameter. These are the cubic point-groups of which the vector representation is irreducible, namely O_h, T_h, O, T_d and T. Their integrity bases have been listed in Table 2. The discussion of the corresponding phase diagrams can be found in a number of works by Gufan and his coworkers [10a-c, 12, 14, 32, 33, 35-37]. We will summarize the main results obtained in these studies, starting from the image-group O_h.

a) Phase diagrams associated with the image-group O_h.

The integrity basis associated with the image-group O_h is (see Table 2):

$$I_1 = \eta_1^2 + \eta_2^2 + \eta_3^2, \qquad I_2 = \eta_1^4 + \eta_2^4 + \eta_3^4, \quad I_3 = \eta_1^2\eta_2^2\eta_3^2. \quad (1.165)$$

Since O_h is a group generated by reflections, it has the property that the dimension m of its basis of integrity is equal to the dimension n of the order-parameter space ε_3. As indicated in the preceding section, one can expect to have 2^n=8 distinct phases, differing by their symmetries, in the phase diagram. Using the notations: $\Phi_i = \frac{\partial F'}{\partial I_i}$, $\Phi_{ij} = \frac{\partial^2 F'}{\partial I_i \partial I_j}$ (i, j=1-3) where $F'(I_1, I_2, I_3) = F(\eta_1, \eta_2, \eta_3)$, one can write the equations of state, under the form:

$$\frac{\partial F}{\partial \eta_i} = 2\eta_i\left(\Phi_1 + 2\eta_i^2\Phi_2 + \eta_j^2\eta_k^2\Phi_3\right) \quad \text{(i, j, k=1-3; } j \neq k \neq i). \quad (1.166)$$

The stability conditions can be calculated using the second derivatives:

$$\frac{\partial^2 F}{\partial \eta_i^2} = 2\Phi_1 + 12\eta_i^2\Phi_2 + 2\eta_j^2\eta_k^2\Phi_3 + 2\eta_i\Phi_{11} + 4\eta_i^3\Phi_{21} + 2\eta_i\eta_j^2\eta_k^2\Phi_{31} \quad (1.167)$$

and

$$\frac{\partial^2 F}{\partial \eta_i \partial \eta_j} = 2\eta_i\Phi_{ij} + 4\eta_i^3\Phi_{jk} + 4\eta_i\eta_j\eta_k^2\Phi_{jk} \quad (i, j, k=1\text{-}3; j\neq k\neq i). \quad (1.168)$$

Table 5 summarizes the equilibrium relations characterizing the eight stable phases which may be found in the phase diagram of the O_h-image group, as deduced from Eqs.(1.166)-(1.168). In particular column (f) of the Table allows to represent the O_h-phase diagram in the orbit-space (I_1, I_2, I_3), which is shown in Fig.31.

Since the maximal degree for the invariants I_i is six (I_3), a structurally stable model describing the full set of phases listed in Table 5 requires to consider an expansion $F(\eta_1, \eta_2, \eta_3)$ of not less than the twelvth degree in the η_i. Let us first consider the sixth degree expansion:

$$F'(I_1, I_2, I_3) = a_1 I_1 + a_2 I_1^2 + a_3 I_1^3 + b_1 I_2 + d_1 I_3 + c_{12} I_1 I_2 \quad (1.169)$$

which has been first discussed by Devonshire [38] in order to describe the structural transitions in $BaTiO_3$. In this case it is easy to deduce from the stability conditions given in Column (e) that only the low-symmetry phases II, III and IV are stable (for example, the condition of stability of phase VI becomes $-\Phi_{12}^2 > 0$ and for phase V one gets the contradictory constraints $\eta_1^2 > \eta_3^2, \eta_1^2 < \eta_3^2, \eta_1 \neq \eta_3$). On the other hand, from the equilibrium conditions listed in column (d) of Table 5 one can conclude that vanishingly small solution of η may occur only along the line of equation: $\Phi_1(P, T, \eta) = 0$. For phases II and III the stability conditions do not impose any restriction for a second-order transition along the preceding line. By contrast, for phase IV the inequalities:

$$\Phi_{11} \geq 0, \quad \Phi_2 \geq 0 \quad \text{and} \quad -2\Phi_2 + \eta^2\Phi_3 \geq 0$$

are inconsistent at finite $\Phi_2(\eta = 0)$ and small η^2. It means that a second-order transition I-II may occur only at one point of the phase diagram given by $\Phi_1(P, T, \eta = 0) = \Phi_2(P, T, \eta = 0) = 0$. Simple calculations [10b] give the following additional features for the sixth-degree model, which are illustrated in Fig.32.

1) The second-order transition I-II occurs for $a_1 = 0$ and $b_1 \geq 0$. Phase II is stable for $a_1 < 0$ when $a_2 > 0$ and $b_1 + a_2 > 0$, or for $a_1 < -(b_1 + a_2)^2 / 3(a_3 + c_{12})$ when $b_1 + a_2 < 0$. The I-III transition can be either second- or first-order the two regimes being separated by a tricritical point.

2) Phase IV exists only when $d_1 > 0$ and is always located between phases II and III (Fig.32a). For $d_1 > 0$ phase IV is stable and the phase diagram contains a direct first-order transition between phases II and III (Fig.32b).

3) For $d_1 > 0$, the limit of stability of phase IV with respect to phase III is given by:

Table 5. Characteristic features of the phase diagrams associated with the image-group O_h. The columns give, for each stable phase, the following informations: (a) Labelling of the phase in reference to the text. (b) Equilibrium value ("direction") of the order-parameter in the ε_3 space. (c) Corresponding isotropy group I_η. (d) Equation of state. (e) Stability conditions. (f) Specific form of the invariants I_1, I_2, I_3.

(a)	(b)	(c)	(d)	(e)	(f)
I	0 0 0	O_h	$\eta_1=\eta_2=\eta_3=0$	$a_1 > 0$	(0 0 0)
II	η 0 0	C_{4v}	$\Phi_1 + 2\eta^2\Phi_2 = 0$	$\Phi_2 \le 0, 2\Phi_2 + \Phi_{11} \ge 0$	$(\eta^2 \;\; 0 \;\; 0)$
III	η η η	C_{3v}	$\Phi_1 + 2\eta^2\Phi_2 + \eta^4\Phi_3 = 0$	$\begin{cases} 2\Phi_2 + \Phi_{11} \ge 0, 2\Phi_2 + 3\Phi_{11} \ge 0 \\ 4\Phi_2^2 + 4\Phi_2\Phi_{11} - 2\eta^2\Phi_3\Phi_{11} \ge 0 \end{cases}$	$(3\eta^2 \;\; 3\eta^4 \;\; \eta^6)$
IV	η η 0	C_{2v}	$\Phi_1 + 2\eta^2\Phi_2 = 0$	$\begin{cases} \Phi_2 \ge 0, \Phi_{11} \ge 0 \\ -2\Phi_2 + \eta^2\Phi_3 \ge 0 \end{cases}$	$(2\eta^2 \;\; \eta^4 \;\; 0)$
V	η_1 η_1 η_3	C_s	$\begin{cases} \Phi_1 + 2\eta_1^2\Phi_2 + \eta_3^4\Phi_3 = 0 \\ \Phi_1 + 2\eta_1^2\Phi_2 + \eta_1^2\eta_3^2\Phi_3 = 0 \end{cases}$	$\begin{cases} \Phi_{11} \ge 0, (\eta_1^2 - \eta_3^2)\Phi_3 \ge 0 \\ [\Phi_{11}\Phi_3 + \eta_1^2\Phi_3^2 + \eta_3^2\Phi_3\Phi_{12} + \\ +4(\eta_1^2 - \eta_3^2)](\eta_3^2 - \eta_1^2) \ge 0 \end{cases}$	$(2\eta_1^2 + \eta_3^2,$ $2\eta_1^4 + \eta_3^4,$ $\eta_1^4\eta_3^4)$
VI	η_1 η_2 0	C_s	$\begin{cases} \Phi_1 + 2\eta_1^2\Phi_2 = 0 \\ \Phi_1 + 2\eta_2^2\Phi_2 = 0 \end{cases}$	$\begin{cases} \Phi_3 \ge 0, \Phi_{11} \ge 0 \\ \Phi_{11}\Phi_{22} - \Phi_{12}^2 \ge 0 \end{cases}$	$(\eta_1^2 + \eta_2^2,$ $\eta_1^4 + \eta_2^4,$ $\eta_1^4\eta_2^2)$
VII	η_3 η_3 η_1	C_s	$\begin{cases} \Phi_1 + 2\eta_3^2\Phi_2 + \Phi_1^4\Phi_3 = 0 \\ \Phi_1 + 2\eta_3^2\Phi_2 + \eta_1^2\eta_3^2\Phi_3 = 0 \end{cases}$	$\begin{cases} \Phi_{11} \ge 0, (\eta_1^2 - \eta_3^2)\Phi_3 \ge 0 \\ [\Phi_{11}\Phi_3 + \eta_3^2\Phi_3^2 + \eta_1^2\Phi_3\Phi_{12} + \\ +4(\eta_3^2 - \eta_1^2)](\eta_1^2 - \eta_3^2) \ge 0 \end{cases}$	$(\eta_1^2 + 2\eta_3^2,$ $\eta_1^4 + 2\eta_3^4,$ $\eta_1^2\eta_3^4)$
VIII	η_1 η_2 η_3	C_1	$\Phi_1 = \Phi_2 = \Phi_3 = 0$	General	$(I_1 \;\; I_2 \;\; I_3)$

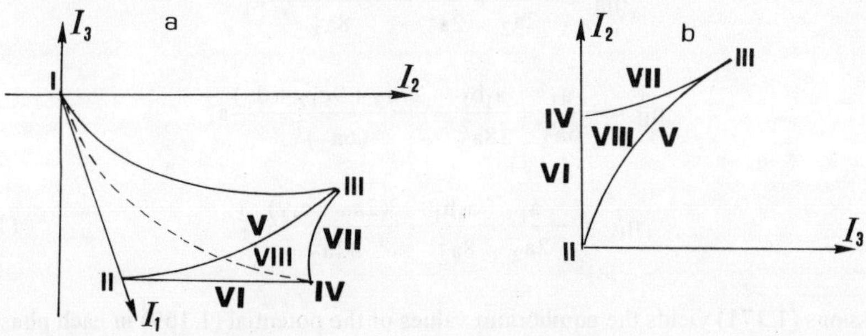

Fig. 31. Topology of the phases corresponding to the phase diagram associated with the image-group O_h (a) in the orbit space (I_1, I_2, I_3), (b) in a cross-section (I_2, I_3). The positions of the phases are deduced from Table 5, column (f).

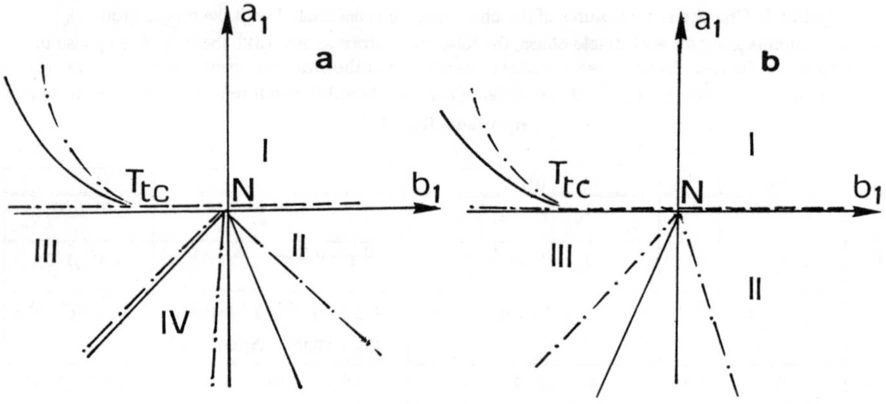

Fig. 32. Phase diagrams corresponding to the image-group O_h for the sixth-degree expansion (1.169), assuming $a_2 > 0$, $\Delta = 4a_2b_2 - c_{12}^2 < 0$ and (a) $d_1 > 0$, (b) $d_1 < 0$. Full, dashed, and dashed-dotted lines are first-order, second-order and limit of stability lines, respectively.

$$a_1 = \frac{2a_2b_1}{c_{12}} \tag{1.170}$$

and with respect to phase II is $a_1 = 8a_2b_1/(4c_{12} - d_1)$. On the other hand phase II possesses a limit of stability within phase IV given by $a_1 = 12a_2b_1/(6c_{12} - d_1)$. The transitions III-IV and II-IV are always first-order. This can be foreseen by the discontinuities of the equilibrium values of the order-parameter components in the three phases. One finds respectively:

$$\eta_{III}^2 = -\frac{a_1}{2a_2} + \frac{a_1b_1}{2a_2^2} - \frac{3(a_3 + c_{12})}{8a_2^3}a_1^2$$

$$\eta_{II}^2 = -\frac{a_1}{6a_2} + \frac{a_1b_1}{18a_2^2} - \frac{27a_3 + 9c_{12} + d_1)}{(6a_2)^3}a_1^2$$

$$\eta_{IV}^2 = -\frac{a_1}{2a_2} + \frac{a_1b_1}{8a_2^2} - \frac{3(2a_3 + c_{12})}{32a_2^3}a_1^2. \tag{1.171}$$

Expressions (1.171) yields the equilibrium values of the potential (1.169) in each phase:

$$F(\eta_{III}) = -\frac{a_1^2}{4a_2} + \frac{b_1 a_1^2}{4a_2^2} - \frac{(a_3 + c_{12})}{8a_2^3} a_1^3$$

$$F(\eta_{II}) = -\frac{a_1^2}{4a_2} + \frac{b_1 a_1^2}{12a_2^2} - \frac{(27a_3 + 9c_{12} + d_1)}{(6a_2)^3} a_1^3$$

$$F(\eta_{IV}) = -\frac{a_1^2}{4a_2} + \frac{b_1 a_1^2}{8a_2^2} - \frac{(2a_3 + c_{12})}{16a_2^3} a_1^3 .$$

from which one can deduce the equations of the first-order transition lines, between phases III and IV ($F_{III} = F_{IV}$):

$$b_1 = \frac{c_{12}}{2a_2} a_1 \tag{1.172}$$

and between phases II and IV ($F_{II} = F_{IV}$):

$$b_1 = \left(c_{12} - \frac{2d_1}{9}\right) \frac{a_1}{2a_2} .$$

Comparing Eqs.(1.170) and (1.172) shows that *the boundary between phases II and IV confounds with the limit of stability between the two phases*, i.e. there is no metastability region between the two phases. This result is indicative of the fact that the sixth degree expansion (1.169) does not provide a consistent description of phase IV. One needs to consider for this purpose an eight degree expansion:

$$F'(I_1, I_2, I_3) = a_1 I_1 + a_2 I_1^2 + a_3 I_1^3 + a_4 I_1^4 + b_1 I_2 + b_2 I_2^2 + d_1 I_3 + c_{12} I_1 I_2 + c_{112} I_1^2 I_2 + c_{13} I_1 I_3 \tag{1.173}$$

which also describes three additional phases listed in Table 5, namely phases V, VI and VII.

The equations of state (column (d) of Table 5) corresponding to phases V-VII can have solutions only at the point $\Phi_1(T,P) = \Phi_2(T,P) = 0$. Stability conditions for phase VI:

$$\Phi_{11} \geq 0, \quad \Phi_3 \geq 0, \quad \Phi_{11}\Phi_{22} - \Phi_{12}^2 \geq 0 \tag{1.174}$$

do not contain any further restriction on the possibility of a second-order transition at the preceding point. One find for the equilibrium values of the order-parameter components in phase VI

$$I_1 = \eta_1^2 + \eta_2^2 = \frac{c_{12} b_1 - 2a_1 b_2}{\Delta} + \frac{c_{12} d_1 - 6a_3 b_2}{\Delta^3} (b_1 c_{12} - 2a_1 b_2)^2$$

$$I_2 = \eta_1^4 + \eta_2^4 = \frac{c_{12} a_1 - 2a_2 b_1}{\Delta} + \frac{3a_3 c_{12} - 2a_2 d_1}{\Delta^3} (b_1 c_{12} - 2a_1 b_2)^2 . \tag{1.175}$$

In order that Eqs.(1.175) possess real solutions for η_1^2 and η_2^2, it is necessary that Eq.(1.174) be fulfilled. This is the case for the region determined by the inequalities: $a_2 > 0$, $\Delta > 0$, $d_1 > 0$, which do not impose any restriction on the nature of the boundaries with the other low symmetry phases. Actually, the boundaries of existence of Eq.(1.174) coincide exactly with the boundaries of phases III and IV, in agreement with Fig.31, which shows that phase VI should be located between the preceding phases. Therefore the transition between phase VI and phase III or IV may be second-order as shown in Fig.33a.

Let us stress that when the situation shown in Fig.33a is realized, the origin ($a_1 = 0$, $b_1 = 0$) which is determined by the double condition $\Phi_1(T,P) = \Phi_2(T,P) = 0$, corresponds to a *five-phase point* N, the sequence of phases around N being I-III-VI-IV-II-I. The existence of five stable phases at the same point, does not contradict Gibbs rule, which forbids the coexistence of more than three phases at one point (triple point) in the plane of the external variables. Gibbs rule holds when all phases merging at the triple point are assumed to be independent, i.e. their thermodynamic potentials, which are equal at the triple point, give rise to independent relationships between the external variables. In the case shown in Fig.33a, the thermodynamic potentials corresponds to phases which are also separated by second-order transitions, and thus consist in extrema, related to each other, of the same non-equilibrium potential, and are thus expressed in terms of the same phenomenological coefficients.

For phase V (or phase VI), which is topologically close to phase III (see Fig.31), near the line $\Phi_1(T,P) = 0$, the contradictory stability conditions $\pm(\eta_1^2 - \eta_3^2)\Phi_3 > 0$, which hold for small (η_1, η_3), disappear when taking into account higher order terms in η, i.e. in

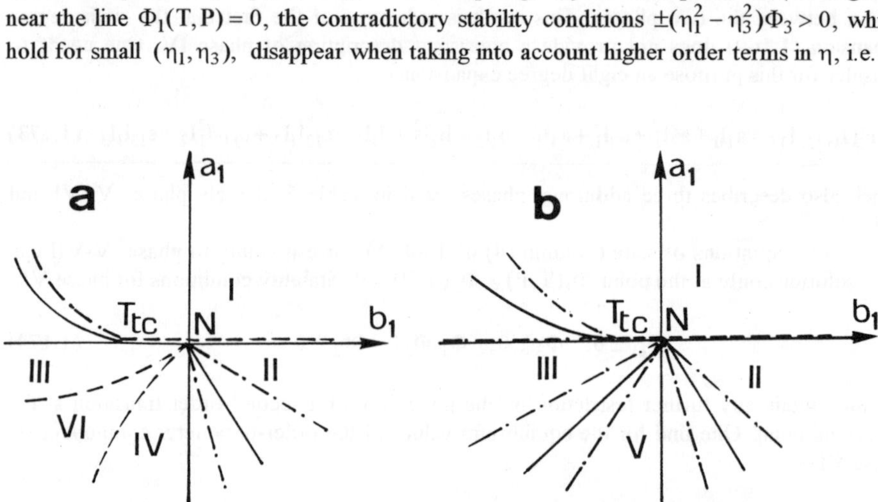

Fig. 33. Phase diagrams for the image-group O_h assuming an eight degree expansion (1.173): (a) $a_2 > 0$, $\Delta > 0$, $d_1 > 0$; (b) $a_2 < 0$, $\Delta < 0$, $d_1 < 0$. Full, dashed and dashed-dotted lines are first-order, second-order and limit of stability lines, respectively.

the region where $\eta_1^2 - \eta_3^2 \propto \eta^4$. Thus, the region of existence of phase V shown in Fig.33b, corresponds to solutions which vanish on the line $\Phi_1 = 0$, and become narrower in proportion to η^4 as the point ($\Phi_1 = \Phi_2 = 0$) is approached. Such region can only be near the boundary of the region of stability of phase II where the potentials of II and V (or VII) differ by an amount of I_1^3. Furthermore, phase V is metastable with respect to phase IV, the potential of which differs from the potential of phase II, only by an amount of the order of I_1^2.

Solution VIII in Table 5 describes a phase with three independent components, and with the lowest symmetry. The solution of the corresponding system of equations of state ($\Phi_1 = \Phi_2 = \Phi_3 = 0$) allows a contact with phase I only across a first-order transition. A consistent model describing phase VIII requires to take into account a twelvth degree potential at least. In this case, illustrated by the phase diagrams of Kut'in [32] represented in Fig.34a all the eight phases listed in Table 5 possess a domain of stability. Additional isostructural transitions may also be stabilized for high-degree expansion models, as shown in the phase diagram of Fig.34b worked out by Lorman [33].

b) Phase diagrams associated with the image-groups T_d, T_h, O, T.

The possible stable phases corresponding to the other irreducible three-dimensional images T_d, T_h, O and T are listed in Table 6. One can see that the image-group T_d, which is also generated by reflection displays (as O_h) 2^3=8 stable states, whereas the three other image-groups have only five possibly stable phases in their phase diagrams. Phases II (η00) and III ($\eta\eta\eta$), which are common to all the three-dimensianal image-groups can be reached from phase I (as for O_h), across lines of second-order transitions, for the image-groups O and T_h. In contrast, due to the existence of a cubic invariant ($I_2 = \eta_1\eta_2\eta_3$) in the integrity basis of the image-groups T_d and T, the preceding phases are separated from phase I by first-order transition lines, which become second-order at only one point N

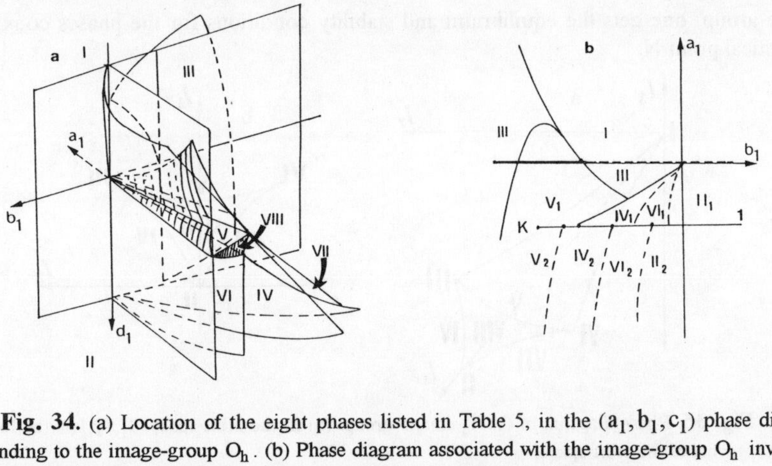

Fig. 34. (a) Location of the eight phases listed in Table 5, in the (a_1, b_1, c_1) phase diagram corresponding to the image-group O_h. (b) Phase diagram associated with the image-group O_h involving an isostructural transition line (line 1) ending at a critical point of the liquid-vapor type.

Table 6. Equilibrium values of the order-parameter components (column (a)) and corresponding isotropy groups (column(b)) for the stable phases associated with the three-dimensional image-groups T_d, T_h, O and T.

T_d				T_h		O		T	
(a)	(b)	(a)	(b)	(a)	(b)	(a)	(b)	(a)	(b)
I. 0 0 0	T_d	V. $\eta_2\ \eta_1\ \eta_1$ $\eta_1>\eta_2$	C_{2v}	I. 0 0 0	T_h	I. 0 0 0	O	I. 0 0 0	T
II. η 0 0	C_{2v}	VI. $-\eta_1\ \eta_1\ \eta_1$	C_{3v}	II. η 0 0	C_{2v}	II. η 0 0	C_4	II. η 0 0	C_2
III. $\eta\ \eta\ \eta$	C_{3v}	VII. $-\eta_1\ \eta_2\ \eta_2$ $\eta_1>\eta_2$	C_s	III. $\eta\ \eta\ \eta$	C_3	III. $\eta\ \eta\ \eta$	C_3	III. $\eta\ \eta\ \eta$	C_3
IV. $\eta_1\ \eta_2\ \eta_2$ $\eta_1>\eta_2$	C_s			IV. $\eta_1\ \eta_2$ 0	C_s	IV. $\eta\ \eta$ 0	C_1	IV. $-\eta\ \eta\ \eta$	C_1
		VIII. $\eta_1\ \eta_2\ \eta_3$	C_1	V. $\eta_1\ \eta_2\ \eta_3$	C_1	V. $\eta_1\ \eta_2\ \eta_3$	C_1	V. $\eta_1\ \eta_2\ \eta_3$	C_1

defined by $\Phi_1(P,T)_{\eta=0} = \Phi_2(P,T)_{\eta=0} = 0$. This isolated second-order critical point corresponds to the point of tangency of the first-order transition lines bounding phases I and II.

The invariants forming the basis of integrity of T_d are related to the invariants of O_h (see Table 2) by:

$$I_1(T_d) = I_1(O_h),\ I_2^2(T_d) = I_3(O_h)\ \text{and}\ I_3(T_d) = I_2(O_h).$$

The location of the phases pertaining to the T_d phase diagram, in the orbit-space (I_1, I_2, I_3) are shown in Fig.35. Since the highest degree of the invariants is *four* (I_3), a consistent description of the phase diagram in the space of the phenomenological coefficients requires an eighth-degree order-parameter expansion model. Using the same notations as for the O_h image-group, one gets the equilibrium and stability conditions for the phases coexisting at the critical point N:

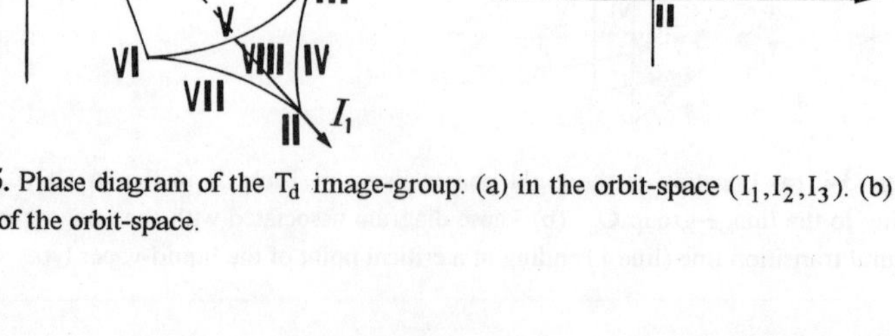

Fig. 35. Phase diagram of the T_d image-group: (a) in the orbit-space (I_1, I_2, I_3). (b) In a cross-section I_1=const of the orbit-space.

1) For phase II: $\Phi_1+2\eta^2\Phi_3=0$, and $4\Phi_1^2-\eta^2\Phi_2^2>0$, $2\Phi_1+4\eta^2(\Phi_{11}+3\Phi_3)\geq 0$, $\Phi_3<0$, $\Phi_1>0$;
2) for phases III and VI: $2\Phi_1\pm\eta\Phi_2+4\eta^2\Phi_3=0$, and $\Phi_3+\Phi_{11}>0$, $\pm 2\eta(-\Phi_2\pm 4\eta\Phi_3)>0$, $-2\Phi_1+4\eta^2(3\Phi_{11}+\Phi_3)>0$, $2\Phi_1+4\eta^2(\Phi_{11}+3\Phi_3)>0$.
It leads to the phase diagram represented in Fig.36a. Fig.36b shows the region of stability of the phases in the space (a_1,b_1,d_1) where a_1,b_1 and d_1 are the coefficients of the invariants I_1,I_2 and I_3, respectively, in an eighth degree expansion. Figs.36c-e show cross-sections (a_1,b_1), in the vicinity of the four-phase point N, and (b_1,d_1) for $a_1<0$ respectively.

With respect to T_d, the group T possesses an additional sixth degree invariant:

$$I_4=(\eta_1^2-\eta_2^2)(\eta_2^2-\eta_3^2)(\eta_3^2-\eta_1^2) \tag{1.176}$$

in its integrity basis. It modifies however appreciably the topology of the corresponding phase diagrams which are shown in Figs.37a,b.

The integrity basis of T_h differs from that of O_h only by the additional sixth degree invariant (1.176). The configuration of the phase diagram of T_h in the orbit space, shown in Fig.37c, is nevertheless topologically different from the corresponding phase diagram of O_h (Fig.31). The same remark holds for the phase diagrams associated with the image-group O, which are represented in Figs.37d,e. Although the integrity basis of O corresponds, with respect to that of O_h, to the simple adjunction of a ninth degree invariant ($I_4'=\eta_1\eta_2\eta_3 I_4$), it modifies in an essential way the features of the phase diagrams (number of phases, their regions of stability etc.). The explanation is that for image-groups such as T_h, O, T the dimension of the orbit space exceeds the dimension of the order-parameter space, while for O_h and T_d the two spaces have the same dimension.

3.3.5. Phase diagrams for image-groups corresponding to an irreducible order-parameter of dimension n≥4

When the number n of order-parameter components is larger than n=3, the corresponding number of image-groups strikingly increases. Thus, even when restricting to phase transitions associated with irreducible representations fulfilling the Lifshitz criterion (see Sec.4.3), the number N of image-groups is N=43 for n=4 (N=22 when considering only "active" irreducible representations [9c, 18]), N=32 for n=6, N=12 for n=8, N=8 for n=12 etc. . Although the images have been identified and classified for n=4 and n=6 [10e,f] no attempt has been made to construct the corresponding phase diagrams except for a minority of specific examples. This can be understood, since the dimension of the corresponding integrity bases, the degrees of the basic invariants, and the number of possible stable states become too large to solve algebraically structurally stable models. Furthermore, in a number of examples with n≥4 discussed in the following chapters, we show that for a qualitative description of concrete examples of phase transition, it is generally sufficient to consider an *effective* order-parameter expansion, i.e. a simplified

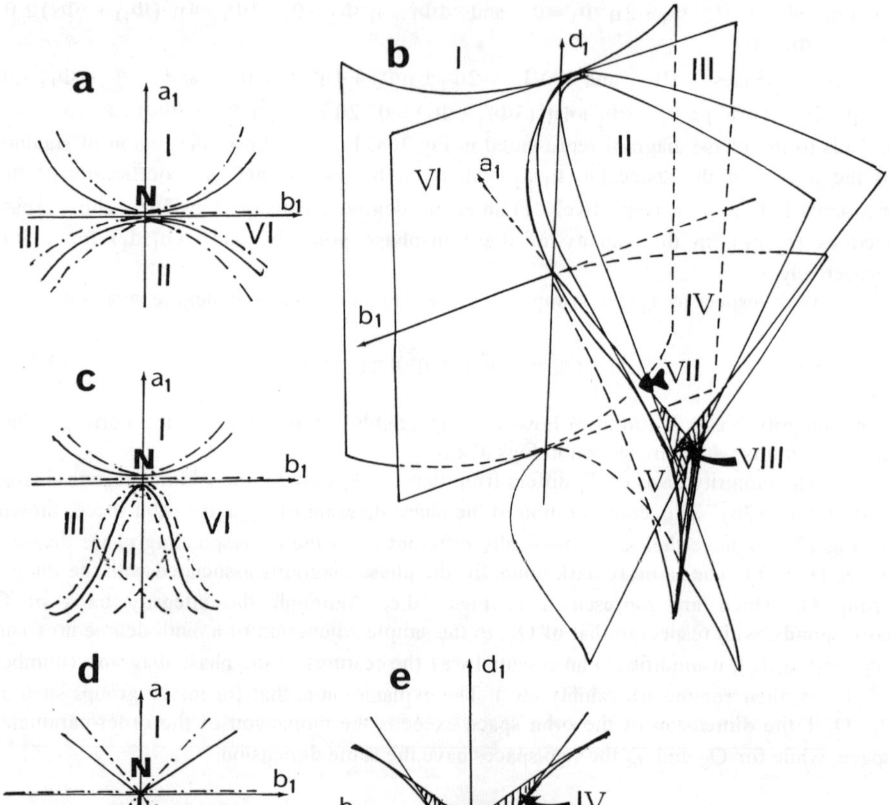

Fig. 36. Phase diagrams associated with the image-group T_d [32]. (a), (c) and (d) represent the four phases I, II, III, VI in the vicinity of the N point for different values of the phenomenological coefficients. Solid, dashed and dashed-dotted lines are first-order, second-order and limit of stability lines, respectively. (b) shows the region of stability of each phase in the (a_1, b_1, d_1) space. (e) is a cross-section (b_1, d_1) for a_1=const<0, precising the location of phases IV, V, VII and VIII.

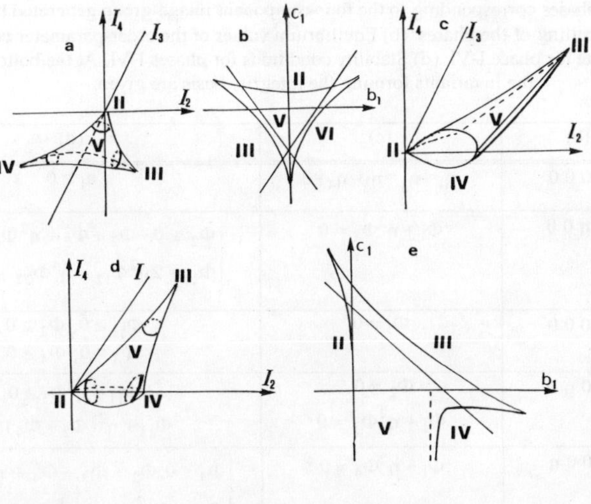

Fig. 37. Phase diagrams associated with the image-groups T, T_h and O. (a), (c) and (d) are cross-sections of the orbit-space for I_1=const, respectively for T, T_h and O. (b) and (e) are cross-sections (a_1=const, d_1=const) in the plane of the phenomenological coefficients (c_1, b_1) for T and O, respectively.

form of the expansion, in which the equilibrium relations between the order-parameter components, which characterize the actually observed phases, are taken into account.

In the following example of a four-component order-parameter image-group, we will illustrate an alternative procedure of discussion of a phase diagram which consists in isolating a region of the phase diagram which contains only part of the phases which may be possibly stable.

Let us consider the four-component order-parameter image-group, which is labelled D32δ in Gufan's classification [10f] and 58.01 in Ref.[9c]. It contains 32 distinct matrices, which are generated by the matrices:

$$\begin{bmatrix} 1 & 0 & 0 & 0 \\ 0 & 1 & 0 & 0 \\ 0 & 0 & -1 & 0 \\ 0 & 0 & 0 & -1 \end{bmatrix} \begin{bmatrix} -1 & 0 & 0 & 0 \\ 0 & -1 & 0 & 0 \\ 0 & 0 & 1 & 0 \\ 0 & 0 & 0 & 1 \end{bmatrix} \begin{bmatrix} -1 & 0 & 0 & 0 \\ 0 & 1 & 0 & 0 \\ 0 & 0 & 1 & 0 \\ 0 & 0 & 0 & -1 \end{bmatrix} \begin{bmatrix} 0 & 1 & 0 & 0 \\ 1 & 0 & 0 & 0 \\ 0 & 0 & 0 & 1 \\ 0 & 0 & -1 & 0 \end{bmatrix} \begin{bmatrix} 0 & 0 & 1 & 0 \\ 0 & 0 & 0 & 1 \\ 1 & 0 & 0 & 0 \\ 0 & 1 & 0 & 0 \end{bmatrix} . \quad (1.177)$$

Using the geometrical method described in Sec.3.2, one finds ten stable directions in the order-parameter space ε_4, corresponding to ten possible stable states, labelled I to X in column (a) of Table 7. Column (b) gives the equilibrium values of the four components of the order-parameter in each phase. Standard projection techniques allow to construct (using the 32 matrices obtained from the generators (1.177)) the basis of integrity of the image-group. It is formed by seven invariants, denoted I_1 to I_7, given at the bottom of Table 7. One can see that the highest power of the preceding invariants (I_7) is 8.

Table 7. Stable phases corresponding to the four-component image-group generated by the matrices (1.177). (a) Labelling of the phases. (b) Equilibrium values of the order-parameter components. (c) Equations of state for phase I-VI. (d) Stability conditions for phases I-VI. At the bottom of the Table, the invariants forming the integrity basis are given.

(a)	(b)	(c)	(d)
I	0 0 0 0	$\eta_1=\eta_2=\eta_3=\eta_4=0$	$a_1 \geq 0$
II	η η 0 0	$\Phi_1+\eta^2\Phi_2=0$	$\begin{cases}\Phi_2 \leq 0, -\Phi_2+\Phi_3+\eta^2\Phi_6 \geq 0 \\ \Phi_{11}+2\eta^2\Phi_{12}+\eta^4\Phi_{22}+\frac{1}{2}\Phi_2 \geq 0\end{cases}$
III	η 0 0 0	$\Phi_1=0$	$\Phi_{11} \geq 0,\ \Phi_2 \geq 0,$ $\Phi_3 \geq 0,\ \Phi_4 \geq 0$
IV	η 0 η 0	$\Phi_2=0,$ $\Phi_1+\eta^2\Phi_3=0$	$\Phi_{11}+\eta^2\Phi_3 \geq 0,$ $\Phi_{11}+\eta^2(\Phi_2+\Phi_4) \geq 0$
V	η 0 0 η	$\Phi_1+\eta^2\Phi_4=0$	$\begin{cases}\Phi_4 \leq 0, \Phi_2+\Phi_3-\Phi_4+\eta^2\Phi_6 \geq 0 \\ \Phi_{11}+2\eta^2(\Phi_{14}+\frac{1}{2}\Phi_{44})+\frac{1}{2}\Phi_4 \geq 0\end{cases}$
VI	η_1 η_2 0 0	$\Phi_1=0,\ \Phi_2=0$	$(\eta_1^2-\eta_2^2)(\Phi_{11}\Phi_{22}-\Phi_{12}^2) \geq 0$
VII	η_1 0 η_2 0	-	-
VIII	η_1 0 0 η_2	-	-
IX	η_1 η_2 η_1 η_2	-	-
X	η_1 η_2 η_3 η_4	-	-
$I_1=\eta_1^2+\eta_2^2+\eta_3^2+\eta_4^2;\ I_2=\eta_1^2\eta_2^2+\eta_3^2\eta_4^2;\ I_3=\eta_1^2\eta_3^2+\eta_2^2\eta_4^2;\ I_4=\eta_1^2\eta_4^2+\eta_2^2\eta_3^2;$ $I_5=\eta_1\eta_2\eta_3\eta_4(\eta_1^2-\eta_2^2+\eta_3^2-\eta_4^2);\ I_6=\eta_1^2\eta_2^2\eta_3^2+\eta_1^2\eta_2^2\eta_4^2+\eta_1^2\eta_3^2\eta_4^2+\eta_2^2\eta_3^2\eta_4^4;$ $I_7=\eta_1\eta_2\eta_3\eta_4(\eta_1^2\eta_3^2-\eta_2^2\eta_4^2)$			

Accordingly, a fully consistent description of the phase diagram, which includes regions of stability for the ten phases, would require a *sixteen-degree* order-parameter expansion. Writing the equations of state and stability conditions for these phases (column (c) and (d) of Table 7), actually shows that the description of phases I-IV can be performed with a sixth degree expansion, whereas phases VI to VIII need at least an eighth-degree expansion. Only phases IX and X necessitate a sixteen power in the η_i for their complete description. Focusing only on phases I-VI one can simplify the form of the order-parameter expansion used in the more economic model, by noting that the equilibrium conditions for phase VI (given in column (c) of Table 7) coincides with the limit of stability of phases II and III (column (d)) and consequently, phase VI should be located between phase II and III, and separated from them by lines of second-order transitions. Analogous topological considerations allow to reduce the minimal degree of the order-parameter expansion to *six* for the sole description of phases I to VI. Retaining only one of the sixth degree invariants (I_6) and neglecting coupling invariants such as I_1I_2, I_1I_3, I_1I_4 one can write:

$$F(\eta_1,\eta_2,\eta_3,\eta_4) = F'(I_1,I_2,I_3,I_4,I_6) = a_1I_1 + a_2I_1^2 + b_1I_2 + c_1I_3 + d_1I_4 + e_1I_6 \quad (1.178)$$

which yields the following non-zero derivatives $\Phi_i = \dfrac{\partial F'}{\partial I_i}$ and $\Phi_{ij} = \dfrac{\partial^2 F'}{\partial I_i \partial I_j}$:

$\Phi_1 = a_1 + 2a_2I_1$, $\Phi_2 = b_1$, $\Phi_3 = c_1$, $\Phi_4 = d_1$, $\Phi_6 = e_1$, $\Phi_{11} = 2a_2$. The stability conditions of column (d) of Table 7 become for:

-Phase II: $a_1 < 0,\ b_1 < 0,\ b_1 \le c_1 + d_1 + \eta^2 e_1$.
-Phase III: $a_2 > 0,\ b_1 > 0,\ c_1 > 0,\ d_1 > 0$.
-Phase IV: $2a_2 + \eta^2 d_1 \ge 0,\ 2a_2 + \eta^2(b_1 + c_1) \ge 0$.
-Phase V: $d_1 \le 0,\ b_1 + c_1 - d_1 + \eta^2 e_1 \ge 0$.
-Phase VI: $a_2(\eta_1^2 - \eta_2^2) \ge 0$.

Fig.38a shows a two-dimensional cross-section (a_1, b_1) describing the sequence of transitions: I-III-VI-II-I, whereas Figs.38b-d represent cross-sections (b_1, c_1, d_1=const>0), (b_1, d_1, c_1=const>0) and (d_1, c_1, b_1=const>0) for negative values of a_1, which involve phases II, III, IV and V. Although the diagrams of Fig.38 are incomplete and, in some respect incorrect (since, for example, one finds that the first-order transition lines II-III, III-IV and III-V, in Figs.38b-d, coincide with their limit of stability lines) they provide a useful first approximation for the description of the location of the phases, and on the order of their mutual transitions.

3.3.6. Summary. Outline of a theory of phase diagrams. First-order phase transitions

From the examples of solutions of the radial problem discussed in this section, a number of rules can be deduced, and used as a guide in the choice of a phenomenological model. The inherent arbitrariness of this choice should be limited by the following procedure:

1) The order-parameter expansion associated with a given image-group has to take into account the full set of invariants (I_i), forming the integrity basis of the image-group. The minimal degree of the expansion, which corresponds to a "structurally stable" model, i.e. a model describing consistently the whole set of phases, is 2N, where N is the highest degree found among the invariants (I_i). For describing first-order transitions from the parent phase, instead of eventual second-order transitions, one has to consider (2N+2) degrees in the order-parameter components (η_j), and even higher degrees when isostructural transitions exist in the phase diagram. Structurally stable models do not require to keep in the order-parameter expansion the whole set of coupled invariants I_iI_j,

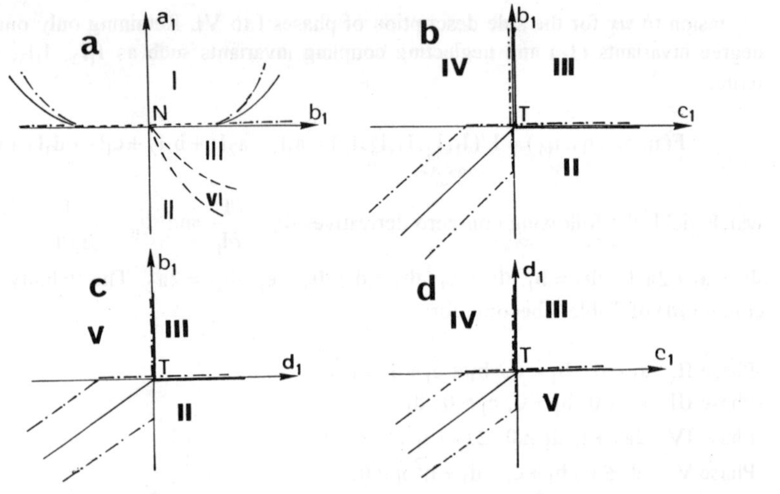

Fig. 38. Phase diagram corresponding to the four-component order-parameter expansion (1.178). The following conditions are assumed: (a) $a_2 > 0$; (b) $d_1 > 0$, $a_1 < 0$; (c) $c_1 > 0$, $a_1 < 0$; (d) $b_1 > 0$, $a_1 < 0$. Dashed, full and dashed-dotted lines are respectively second-order, first-order, and limit of stability lines.

$I_i I_j I_k$ ($i \neq j \neq k$) as their effect may be secondary for the essential topology of the phase diagram. On the other hand, one can reach a knowledge of the complete phase diagram by successive approximations, starting from the "zero approximation" (the fourth degree expansion) which allows to describe the transitions to the phases corresponding to the maximal isotropy subgroups of the parent phase.

2) One should start by the determination of the phase diagram in the orbit space (I_i), as it provides a useful preliminary description of the topology of the phases, i.e. their contact or neighboring in the real phase diagram, and is generally easier to work-out than the phase diagram in the space of the phenomenological coefficients. Note that when the dimension m of the basis of integrity is equal to the dimension n of the order-parameter space, the number of distinct phases is 2^n, while for m>n, the number of distinct phases is smaller than 2^n.

3) The minimization procedure of the order-parameter expansion $F(\eta_i)$ can be also performed using the form $F'(I_i)$ and the corresponding derivatives $\Phi_i = \frac{\partial F'}{\partial I_i}$, and $\Phi_{ij} = \frac{\partial^2 F'}{\partial I_i \partial I_j}$, since under this form the equations of state and stability conditions reveal more clearly the mutual stability of the phases, as well as the regions of the phase diagram forming independent "subregions".

4) The correspondence between the theoretical phase diagram, in the space of the phenomenological coefficients, and the phase diagram in the external variables space (X_e=T, P, x) imposes to assume a dependence on the X_e of *all* the coefficients a_1, b_1, c_1 ... of the invariants I_i, while the coefficients of the highest-power invariants I_i^2, I_i^3, I_iI_j ... must be assumed to be constant (see Sec.5). Generally, a *linear* dependence is assumed for $a_1(X_e)$, $b_1(X_e)$, $c_1(X_e)$,... in such a way that the real phase diagram (T-P, T-x, etc.) can be deduced by a linear transformation from the theoretical phase diagram, and thus displays the same configuration. However, close to singularities (e.g. N-phase points) or when strongly curved transition lines take place, a non linear dependence of $a_1(X_e)$, $b_1(X_e)$, $c_1(X_e)$,... may be necessary.

The preceding remarks outline roughly a formal theory of phase diagrams, not explicitely formulated up to now, which has also to make use of the results of the theory of singularities, presented in Sec.5. In this framework, the examples discussed in the present section provide also implicitely a qualitative classification of first-order transitions. Two main categories of first-order transitions can be distinguished. A first category is constituted by the transitions whose discontinuous character depends on the values of the coefficients in the order-parameter expansion, i.e. on the values of the external parameters and correlatively on the highest necessary degree at which the order-parameter expansion needs to be truncated. A second category contains transitions which can be predicted to be always discontinuous from symmetry considerations. This is for example the case of phase transitions between low symmetry phases which possess no group-subgroup relationship between their symmetries. It is also the case for isostructural transitions and for transitions which possess a cubic invariant in their order-parameter expansion. The next section focuses in more details on the classes of symmetry predicted first-order transitions.

4. Symmetry predicted first-order phase transitions

4.1. Introduction

Different classes of transitions can be predicted to be discontinuous on the sole basis of symmetry considerations, i.e. by considering the symmetry of the phases surrounding the transition.

1) The most obvious class of symmetry predicted first-order transitions are the transitions between phases which possess no group-subgroup relationship between their symmetries. One should distinguish in this respect two subclasses corresponding to very different phenomena:

a) Transition between two phases the structures of which are associated with different subgroups of a common parent structure, from which they can be deduced by slight distorsions. In other words, the two phases, although they have no mutual group-subgroup relationship, correspond to different equilibrium values of the same set of order-parameter components which transform as a single multidimensional irreducible representation of the symmetry group of the parent phase. Such type of transitions have been met in the preceding section for phase diagrams associated with image-groups witn $n \geq 2$ such as the

C_{nv}, O_h or T_d phase diagram, and correspond to a large number of experimental examples of ferroelectric and ferroelastic transitions [9a-d]. A textbook example is the orthorhombic-to-rhombohedral transition which occurs at 183 K in $BaTiO_3$ [39].

b) Reconstructive-type phase transitions for which the loss of group-subgroup relationship between the phases results entirely from the drastic character of the transition mechanism, which involves large atomic displacements, specific ordering processes etc.. Here the two phases do not possess, in general, a common parent structure within their phase diagram. The description of this category of transitions constitutes the subject of the remaining chapters of this book, and will not be discussed further in the present section.

2) Another important class of symmetry predicted discontinuous transitions, can be found in phase diagrams associated with *more than one irreducible (symmetry breaking) order-parameters*, i.e. in which some phases are induced by a *reducible* representations of the parent phase. This property is reflected by the low symmetry of the structure of these phases which can not be induced by a single irreducible representation of the parent phase, by their domain structure, and by the form of the transition anomalies. Examples of phase diagrams involving coupled order-parameters will be given in Sec.4.4.

3) Isostructural and antiisostructural phase transitions represent also a clearly recognizable class of first-order transitions, as the space groups of the phases possess the same symmetry (space-groups). A number of examples of isostructural transitions have been given in Sec.3.3. It has been noted that their distinctive property is to display a critical point of the liquid-gas type. Antiisostructural transitions were also found in the description of the C_{3v} phase diagram. We will summarize in Sec.4.5 the intrinsic properties of these two types of first-order transitions, and give additional illustrative examples of their occurence.

4) When a cubic invariant is allowed by symmetry in the order-parameter expansion, the transitions from the parent phase are necessarily first-order, except for an isolated point in the phase diagram (at which the coefficient of the cubic term vanishes). This is the formulation *a contrario* of the so-called Landau condition [1], and it has been illustrated in Sec.3.3 for the phase diagrams associated with the image C_1, C_{3v}, T_d and T. For the image-group C_3, to which correspond *two* cubic invariants a second-order transition has been shown to be completely forbidden, even for an isolated point of the phase diagram. In Sec.4.2 we will give the group-theoretical expression of the Landau condition and will discuss its applicability, which has been questioned by some authors [40, 41].

5) *Lock-in phases* occur generally in structurally modulated crystals, below the range of stability of the incommensurate structure [42]. They may also occur directly from the parent phase, across a first-order transition. Such situations have been described on a phenomenological basis by the competing influence of two types of invariants in the order-parameter expansion: i) Invariants including spatial derivatives of the order-parameter components, which express the tendency of the system to stabilize an incommensurately modulated structure. The existence of these so-called "Lifshitz invariants" can be predicted by symmetry on the basis of the criterion introduced by Lifshitz [42]. ii) High-degree invariants which favour the onset of lock-in phases.

According to the preceding scheme, one can identify (and predict the existence of) first-order "lock-in" transitions when the irreducible representation involved in the transition allows the existence of a Lifshitz invariant. In Sec.4.3 we discuss the actuality of this class of first-order transitions.

6) At last, one should mention the sufficient condition for a transition to be first-order, which has been proposed in the framework of the renormalization-group approach [43]. It has been hypothesized that transitions which do not possess a *stable fixed point* in their Landau-Wilson's ε-expansion would be necessarily first-order [44-46]. In Sec.4.6 we briefly discuss the reality of this conjecture with regard to the prediction of the order of experimentally observed structural transitions.

4.2. The Landau condition

Let us first come back on the elementary reason which forbids a cubic term in the order-parameter expansion of a second-order transition, and consider again the general form (1.16) of an expansion transforming as a single n-dimensional irreducible representation Γ of a space-group G_0. In (1.16) f_3, f_4,... are the successive terms of the Taylor expansion of F in function of the order-parameter components (η_j). For $T=T_c$ the minimum of F corresponds to $\eta_j=0$ for all j. Hence, for any infinitesimal values of η_j, the expansion $F-F_0$ must be positive. For such infinitesimal values, it is sufficient to consider the term of lowest degree, which at T_c is $f_3(\eta_j)$. Clearly, this term can not be positive for any set of infinitesimal η_j values, as if f_3 is positive for a given direction in the order-parameter space, the set ($-\eta_j$) corresponding to the opposite direction makes f_3 negative. Thus, in the presence of a third-degree term, the value $\eta_j=0$ does not correspond to a minimum of the order-parameter expansion but to a saddle point of this function (Fig.39a). Consequently, a second-order transition, for which the state $\eta_j=0$ is necessarily an equilibrium state, requires that no third-degree invariant exist in the expansion F.

The absence of $f_3(\eta_j)$ can be due to symmetry reasons, namely the impossibility to find a third degree polynomial invariant by the symmetry group G_0. Another "accidental" reason can be the vanishing at T_c of the temperature dependent coefficients in $f_3(\eta_j)$. Writing $f_3(\eta_j)=b_1(T,P)P_1(\eta_j)+b_2(T,P)P_2(\eta_j)+...$ where P_1, P_2,... are linearly independent polynomials of degree 3 separately invariant by G_0, one can see that the occurence of a continuous transition at (T_c,P_c) requires in addition to $\alpha(T_c,P_c)=0$ the fulfilment of the conditions $b_1(T_c,P_c)=b_2(T_c,P_c)=...=0$, the transition point being at the intersection of the lines $\alpha(T_c,P_c)=0$ and $b_i(T,P)=0$ (Fig.39b). If there are several independent $P_i(\eta_j)$ the simultaneous vanishing of the b_i at (T_c,P_c) will not be possible, and no continuous transition exists. If $f_3(\eta_j)$ includes a single $P_i(\eta_j)$ with the coefficient b(T,P), a second-order transition will be possible at the isolated point $\alpha(T_c,P_c)=b(T_c,P_c)=0$. If one excludes this singular type of transition, *the existence of a continuous transition requires that no third-degree polynomials of the order-parameter*

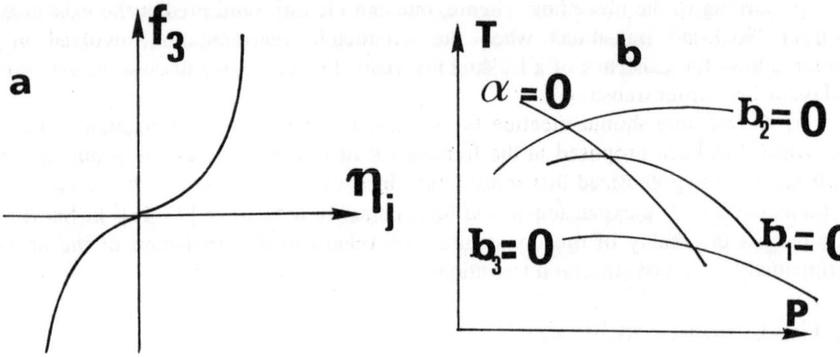

Fig. 39. Figures illustrating the facts: (a) that at η_j=0 a third-degree term $f_3(\eta_j)$ corresponds to a saddle point, and not to a minimum of the order-parameter expansion; (b) that when there are more than one independent cubic invariant, the simultaneous vanishing of the corresponding coefficients b_i(T,P) with α(T,P) is impossible.

components is invariant by G_0 (Landau condition [1]).

The $P_i(\eta_j)$ polynomials of degree 3 in the order-parameter components are the distinct invariant polynomials contained in the symmetrized third power of the irreducible representation Γ of G_0 (see Appendix I). Hence, the number of independent P_i is equal to the number of totally symmetric irreducible representations Γ_1 contained in $[\Gamma^3]$. Accordingly, a necessary and sufficient condition for the absence of third-degree term in the order-parameter expansion can be expressed as:

$$\Gamma_1 \not\subset [\Gamma^3] \qquad (1.179)$$

which is the group-theoretical form of the Landau condition.

The results of Sec.3.3 show that third-degree invariants are found for the one-dimensional image-group C_1, for the two-dimensional images C_3 and C_{3v}, and for the three-dimensional images T and T_d. For irreducible representations fulfilling the Lifshitz condition (see Sec.4.3), it has been shown [47] that cubic terms are found exclusively for rhombohedral, hexagonal or cubic space groups G_0, and for two, three and six-dimensional irreducible representations. From the fact that the Landau condition is fulfilled for irreducible representations corresponding to the image-group C_i, one can deduce [48] that when G is a subgroup of index 2 of G_0, the Landau condition is satisfied. In contrast, subgroup of index 3 of G_0 imply necessarily that condition (1.180) is violated [49].

A systematic investigation of the experimental examples of transitions associated with irreducible representations which do not fulfil the Landau condition [50] shows that this condition is a very reliable criterion to predict the first-order character of a transition, ruling out the argument raised by some authors [40, 41] who suggested that under the effect of thermal fluctuations, the coefficient of the cubic term could vanish. All the obsereved transitions are indeed first-order, sometimes very weakly denoting the vicinity of an isolated "Landau" critical point of second-order transition, which has been evidenced in some structural systems such as the mixed chromite compounds $Cr_{1-x}Ni_xCrO_4$ and $Fe_{1-x}Ni_xCrO_4$ [51]. In ferroelastic phase transitions in which the order-parameter identifies to spontaneous strain components, the coefficient of the cubic term corresponds to a combination of third-order elastic constants [52]. On the other hand for certain cubic-to tetragonal ferroelastic transitions the ratio c/a-1, which expresses the tetragonality of the low-symmetry phase, coincides with the third-degree invariant [53].

4.3. "Lock-in" transitions and the Lifshitz criterion

In Sec.2.3.3 we have shown that the irreducible order-parameter associated with a transition, can be distinguished from the other irreducible degrees of freedom by the fact that the coefficient α_{γ_0} multiplying the quadratic polynomial invariant $\sum_j \eta_{\gamma j}^2$ of the order-parameter components, vanishes at T_c, while the similar coefficients $\alpha_{\gamma \neq \gamma_0}$ corresponding to other degrees of freedom are strictly positive at T_c. Hence, in the neighborhood of T_c, α_{γ_0} is the smallest of the coefficients α_γ in Eq.(1.13). In the case of a structural transition Eq.(1.13) must be expressed differently since the high symmetry group G_0 is a crystallographic space group, i.e. the set of its irreducible representations $\Gamma_{k\gamma}$ *is a continuous set,* since k is any vector of the Brillouin-zone of G_0. Consequently the discrete sum (1.13) over the irreducible representations of G_0 must be replaced by a continuous sum over the volume V* of the Brillouin-zone, and also by a discrete sum over the small representations $\hat{\tau}_\gamma(\vec{k})$ of the "little" group $\bar{G}(\vec{k})$. One has:

$$F = F_0 + \int_{V^*} \left\{ \sum_\gamma \alpha_\gamma \left[\vec{k}, T, P, \rho_0(\vec{r}), \phi_{k'j'}(\vec{r})\right] \sum_j \eta_{kj}^2(\hat{\tau}_\gamma) \right\} d\vec{k} \tag{1.180}$$

where the normalized functions $\phi_{k\gamma}(\vec{r})$ carry the n-dimensional irreducible representation $\Gamma_{k\gamma}$ of G_0. The coefficients $\alpha_\gamma(\vec{k})$ constitute "branches", each branch being associated with a given index γ and a continuously varying vector $\vec{k}$.

Within the preceding continuous description, the order-parameter is the irreducible set η_{kj} corresponding to the minimum with respect to k of the lowest branch $\alpha_\gamma(\vec{k})$ in the vicinity of T_c. This minimum is equal to zero at T_c (Fig.40). Symmetry considerations impose to the various branches $\alpha_\gamma(\vec{k})$ the occurence of extrema at certain points of the corresponding Brillouin-zone. If the minimum of $\alpha_\gamma(\vec{k})$ coincides with one of the extrema, the order-parameter will possess a symmetry determined $\vec{k}$ vector whose location in the Brillouin-zone will not be submitted to variations when temperature or pressure are varied: the entire line of transitions $T_c(P)$ will be described by the same irreducible order-

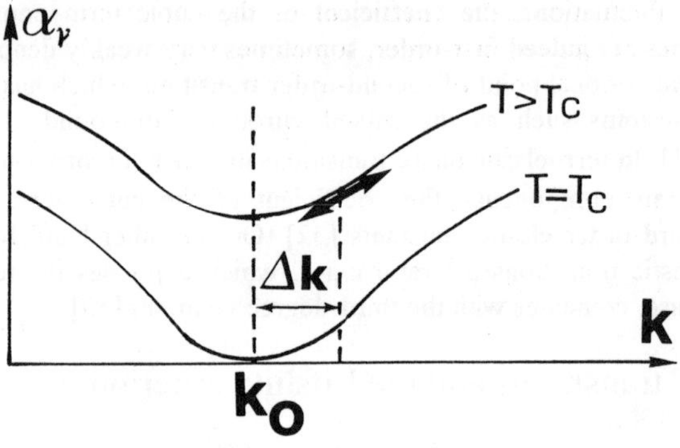

Fig. 40. Minimum at k_0 for the lowest branch $\alpha_\gamma(\vec{k}(T))$. At $\vec{k}=\vec{k}_0+\Delta\vec{k}$ $\alpha_\gamma(\vec{k})$ possesses an oblique tangent which corresponds to a linear term in the expansion (1.181).

parameter. By contrast, if the minimum of $\alpha_\gamma(\vec{k})$ does not coincide with a symmetry extremum, one can have a continuous variation of the order-parameter symmetry with T and P.

The symmetry requirements imposed to the order-parameter which imply the occurence of a symmetry determined extremum for the relevant coefficient α_{γ_0}, are expressed by a criterion initially proposed by Lifshitz [42] and clarified by Dzialoshinskii [54]. The detailed derivation of this criterion can be found in Refs.[55, 56]. We will restrict ourselves to stress its origin: Let $\Gamma_{k_0\gamma_0}$ in Eq.(1.180) be the irreducible representation associated with the coefficient $\alpha_{\gamma_0}(\vec{k}_0)$. Eq.(1.180) contains degrees of freedom corresponding to $\vec{k}=\vec{k}_0+\Delta\vec{k}$ whose coefficients $\alpha_\gamma(\vec{k})$ tend towards $\alpha_{\gamma_0}(\vec{k}_0)$ when $\Delta\vec{k}\to 0$. Depending if $\vec{k}_0$ is a general vector of the Brillouin-zone, or a vector ending at a point of high-symmetry, there can be one or several branches meeting at $\vec{k}_0$ for a given direction $\Delta\vec{k}$. For each branch i and direction $\Delta\vec{k}$ one can write the expansion:

$$\alpha_\gamma^i(\vec{k})=\alpha_{\gamma_0}(\vec{k}_0)+\sum_m\left[\frac{\partial\alpha_\gamma^i}{\partial(\Delta k_m)}\right]_{\Delta k=0}\cdot\Delta k_m+... \qquad (1.181)$$

where the Δk_m are the component of $\Delta\vec{k}$ in the Brillouin-zone. If $\alpha_{\gamma_0}(\vec{k}_0)$ is an extremum, the term in the second member of (1.181) which is linear in the Δk_m must vanish. In other words: *The presence of a linear term in the expansion (1.181) is incompatible with the symmetry of the order-parameter*. Reciprocally, *a necessary condition for a second-order transition to take place towards a (periodic) structure corresponding to* $\vec{k}_0$, *is the absence of linear terms in the* $\alpha_\gamma(\vec{k})$ *expansion*. This is the

Lifshitz criterion [42, 54] which can be formulated in the group-theoretical vocabulary [55, 56]: the antisymmetric square $\left\{\Gamma^2_{k_0\gamma_0}\right\}$ of the irreducible representation spanning the order-parameter has no representation in common with the vector representation of the symmetry group G_0 of the parent phase. One can also show that the absence of linear terms in (1.181) is equivalent to the absence in the order-parameter expansion, which transforms as $\Gamma_{k_0\gamma_0}$, of antisymmetric invariants of the form $\eta_i \frac{\partial \eta_j}{\partial u} - \eta_j \frac{\partial \eta_i}{\partial u}$ (so called Lifshitz invariants), which couple bilinearly the order-parameter components with their spatial derivatives, other than those which are full differentials [56].

The main consequence of the Lifshitz criterion is that it selects in the Brillouin zones associated with the fourteen Bravais lattices a small number of high-symmetry points, listed in Refs.[42, 19], which allow the existence of a second-order transition between crystalline (strictly periodic) structures. Reciprocally, the non-fulfilment of the criterion corresponds to two different possible situations [54, 57]:

1) A second-order (or slightly first-order) transition to an "inhomogeneous" phase corresponding to a non-strictly periodic structure associated with a temperature dependent $\vec{k}$-vector varying closely to $\vec{k}_0$. The experimental examples [58] reveal that such *incommensurate* phases are stable within a limited interval of temperature, and are generally followed, at lower temperature, by a first-order "lock-in" transition, at which the $\vec{k}$-vector "locks" at a value $\vec{k}_L$ (lock-in vector) corresponding to definite multiple of the basic periods of the parent structure.

2) A direct first-order transition to the lock-in phase, characterized by the wave-vector $\vec{k}_L$.

The two preceding situations are described within the same unifying phenomenological scheme, by the competing influence of the Lifshitz invariant $F_1\left(\eta_j, \frac{\partial \eta_j}{\partial u}\right)$ and of higher-order invariants $F_2(\eta_j)$ of the order-parameter. For small values of the η_j F_1, which is proportional to the square of the η_j, predominates giving rise to the incommensurate structure, whereas for larger η_j values the lock-in term F_2 becomes more significant and favours the stabilization of the lock-in phase. In a direct lock-in transition from the parent phase the underlying incommensurate structure is "squeezed" by the jump of the order-parameter at the first-order transition. A comprehensive description of this phenomenological picture can be found in Ref.[59].

In summary, the violation of the Lifshitz criterion leads to first-order lock-in transitions, which appear either below an intermediate incommensurate phase, or directly below the parent phase. In the first case the lock-in transition is easily predictible. In the latter case, one may distinguish lock-in transitions from standard first-order transitions by the value of the lock-in vectors which often display unusual fractional coordinates associated with "non-Lifshitz" points of the Brillouin-zone. Examples of materials undergoing direct or indirect lock-in transitions are given in Refs.[58, 59]. These transitions always possess a first-order character.

4.4. Phase transitions associated with reducible order-parameters

4.4.1. General features of the phase diagrams associated with a reducible order-parameter

The transitions examined up to now have been assumed to be related to a single irreducible order-parameter, in agreement with the essential statement of the Landau theory which associates to a second-order transition a single irreducible degree of freedom. This critical degree of freedom, which determines the symmetry of the low temperature phase, has been shown (Sec.2.4) to be possibly coupled to secondary (non-critical) degrees of freedom, which modify in a non-essential manner the equilibrium and non-equilibrium properties of the system. There exist however a number of structural systems reported experimentally, the phase diagram of which must be partly or entirely connected to more than one order-parameter. In particular, we will show in Chapters III and IV that a large fraction of the observed reconstructive transitions require two, and sometimes three symmetry breaking order-parameters for their description.

When more than one order-parameter has to be used for the description of a phase transition, it means that the number of phenomenological variables which have to be introduced in the model is not minimal, as for one order-parameter, and consequently that the theory is less predictive regarding the properties of the transition. On the other hand the linear or non-linear coupling between the order-parameters has the result that the modifications accompanying a transition (change of symmetry, thermal, dielectric and elastic anomalies, etc.) are more complex making more difficult a phenomenological description. Such a description is nevertheless realizable at least when a reduced number of order-parameters are involved, and we will see that specific consequences can be deduced from the corresponding model, regarding the topology of the phase diagrams and some physical properties of the systems.

Let us first examine the general symmetry aspects related to the existence of more than one order-parameter. If Γ_1, Γ_2,..., Γ_j is a set of irreducible representations of a space-group G_0, which corresponds to the symmetry of the parent phase, the sum $\Gamma_1 \oplus \Gamma_2 \oplus .. \oplus \Gamma_j$ forms a reducible representation of G_0. If G_k^i are the space groups respectively induced by the Γ_i (where k denotes the various low symmetry groups associated with the same irreducible representation of index i), one may expect in the phase diagram, the stabilization of low-symmetry phases corresponding to the intersecting symmetries $G_k^i \cap G_{k'}^i$ under the condition that a sufficiently strong coupling exists between the order-parameters transforming as the irreducible representations Γ_i and Γ_i'. Let us note that the first-order character of the transition from the parent phase to a phase resulting from a strong coupling of two order-parameters is unquestionable, since it implies for the coefficients α_i and α_i' of the corresponding quadratic invariants to change sign quasi-simultaneously, in a narrow interval of temperature. As a consequence, part of the order-parameter components will exhibit a discontinuous jump at the transition. The fact that more than one order-parameter is involved for a given low symmetry phase can be revealed by a variety of structural properties such as i) the existence of a superstructure which has to be related to more than one wave-vector, such as in Ag_2HgI_4 [60], $ZnSnAs_2$ [61] or N_3AlF_6 [62], ii) an unusual lowering of the rotational symmetry as compared to the parent phase symmetry, such as in ZrO_2 [63], in $Bi_4Ti_3O_{12}$ [64], or in a number of perovskites [65]. Other properties, such as the domain texture, the structure of the regions of coexistence with neighboring phases associated with a single order-parameter etc., should also be indicative of the contribution of more than one "primary" order-parameter to the transition mechanism. In the following subsections we will discuss the most typical phase diagrams associated with different types of couplings between two order-parameters, which have been mainly worked out by Gufan and his coworkers [66-72].

4.4.2. Biquadratic coupling between two order-parameters

The thermodynamic potential associated with a reducible representation $\Gamma_1 \oplus \Gamma_2$ contains invariants of the order-parameter components (η_i) and (ξ_j), corresponding respectively to Γ_1 and Γ_2, but also (for interacting order-parameters) mixed invariants which couple the two order-parameters. We will first discuss the most studied case of two one-component order-parameters η and ξ involving a biquadratic coupling $\eta^2\xi^2$, which transforms as the image-group C_{2h}, i.e. the one-dimensional irreducible representations Γ_1 and Γ_2 span the reducible vector representation of the point-group C_{2h}. The form of the order-parameter expansion is:

$$F(\eta,\xi) = F_0 + a_1\eta^2 + a_2\eta^4 + ... + b_1\xi^2 + b_2\xi^4 + ... + \delta\eta^2\xi^2. \qquad (1.182)$$

The integrity basis of $F(\eta,\xi)$ is formed by the two square invariants: $I_1 = \eta^2$, $I_2 = \xi^2$, and thus the minimal degree to which expansion (1.182) has to be considered, in order to describe the full set of low-symmetry phases, is the fourth degree.

a) The model corresponding to Eq.(1.182) with *two* fourth degree invariants in η *and* ξ, was first discussed by Lifshitz [73]. It contains five phenomenological parameters a_1, a_2, b_1, b_2 and δ, which are assumed to satisfy the conditions $a_2>0$, $b_2>0$, $\delta > -2\sqrt{a_2 b_2}$ insuring the positive definiteness of $F(\eta,\xi)$ for large values of η and ξ. The corresponding equations of state are:

$$\eta(a_1 + 2a_2\eta^2 + 2\delta\xi^2) = 0$$

$$\xi(b_1 + 2b_2\xi^2 + 2\delta\eta^2) = 0. \qquad (1.183)$$

In addition to the high-symmetry phase I, Eqs.(1.183) describe three low-symmetry phases with the following equilibrium values for η and ξ:

Phase II : $\eta_{II}^2 = -\dfrac{a_1}{2a_2}, \quad \xi_{II} = 0$

$$\text{Phase III}: \ \eta_{III}=0,\ \xi_{III}^2=-\frac{b_1}{2b_2}$$

$$\text{Phase IV}: \ \eta_{IV}^2=\frac{\delta b_1-2a_1b_2}{\Delta},\ \xi_{IV}=\frac{\delta a_1=2a_2b_1}{\Delta}, \text{with } \Delta=4a_2b_2-\delta^2. \quad (1.184)$$

Assuming the usual linear dependence on temperature and pressure for a_1 and b_1: $a_1(P,T)=a_T(T-T_{c1})+a_P(P-P_{c1})$, $b_1(P,T)=b_T(T-T_{c2})+b_P(P-P_{c2})$, one can easily draw, from Eqs.(1.184), the following conclusions: which are illustrated in Figs.41a-c: i) The transitions from the high symmetry phase I to phases II and III are always second order. ii) A direct transition from phase I to phase IV can only take place at a single point of the phase diagram defined by $a_1(P,T)=b_1(P,T)=0$. iii) The transitions between phases II and III, occur either as a first-order transition (Fig.41a) or across two second-order transition lines enframing phase IV (Fig.41b).

Two additional predictions reveal, however, the incompleteness of a fourth-degree expansion model, namely: 1) the stabilization of phase IV implies a negative interaction ($\delta<-2\sqrt{a_2b_2}$) between the order-parameters, which contradicts the positive definiteness condition for $F(\eta,\xi)$. 2) The model predicts the existence of a three-phase point N, shown in Fig.41c, of coordinates $a_1=b_1(a_2/b_2)^{1/2}$, $\delta=2(a_2b_2)^{1/2}$ which corresponds to a *first-order transition point located at the meeting of two second-order transition lines*, i.e. at N meet the second-order transition lines II-IV and III-IV, and the first-order transition line II-III.

The preceding shortcomings show that a more realistic phase diagram would require to consider a sixth-degree term in $F(\eta,\xi)$, for at least one of the order-parameters.

b) Let us take the sixth degree asymmetric expansion, first discussed by Holakovsky [74] and Gufan and Larin [67]:

$$F(\eta,\xi)=F_0+a_1\eta^2+a_2\eta^4+a_3\eta^6+b_1\xi^2+b_2\xi^4+\delta\eta^2\xi^2 \quad (1.185)$$

which yields the following equilibrium values for the low symmetry phases:

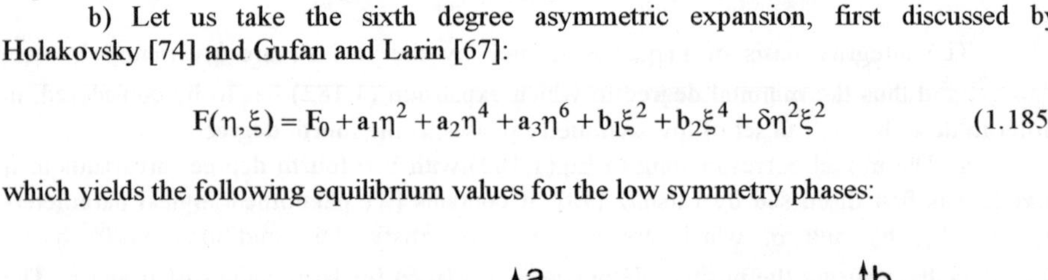

Fig. 41. Cross sections of the phase diagram corresponding to a fourth degree expansion (1.182) in η and ξ, for (a) $\Delta=4a_2b_2-\delta^2<0$, (b) $\Delta>0$, (c) Δ varying from positive to negative values. Full, dashed, and dashed-dotted lines are respectively first-order, second-order, and limit of stability lines.

$$\text{Phase II}: \ \eta_{II}^2 = \frac{-2a_1 + (4a_2 - 12a_1a_3)^{1/2}}{6a_3}, \quad \xi_{II} = 0$$

$$\text{Phase III}: \ \eta_{III} = 0, \quad \xi_{III}^2 = -\frac{b_1}{2b_2}$$

$$\text{Phase IV}: \begin{cases} a_1 + 2a_2\eta_{IV}^2 + 3a_2\eta_{IV}^4 + \delta\xi_{IV}^2 = 0 \\ b_1 + 2b_2\xi_{IV}^2 + \delta\eta_{IV}^2 = 0 \end{cases}. \tag{1.186}$$

Different phase diagrams are obtained from Eqs.(1.186), and from the corresponding stability conditions, depending on the value of $\Delta = 4a_2b_2 - \delta^2$. For $\Delta>0$ one gets phase diagrams which are similar to the ones represented in Fig.41. New topologies appear for $\Delta<0$, i.e. for the value $\eta^2 = -\frac{\Delta}{12}a_3b_2$ which corresponds to the appearence of a new transition line. The results of the minimization of $F(\eta,\xi)$ can be summarized as follows.
1) For $2\sqrt{a_2b_2}>\delta>0$ and $a_2>0$, the III-IV transition is always first order, and takes place on the line of equation:

$$a_1 = \frac{\delta}{2b_2}b_1 + \frac{\Delta^2}{64a_3b_2^2} \tag{1.187}$$

which is surrounded by the loss of stability line of phases IV and III, respectively given by:

$$a_1 = \frac{\delta}{2b_2}b_1 + \frac{\Delta^2}{48a_3b_2^2} \quad \text{and} \quad a_1 = \frac{\delta}{2b_2}b_1.$$

The II-IV second-order transition occurs on the line:

$$a_1 = \frac{2a_2}{\delta}b_1 - \frac{3a_3}{\delta^2}b_1^2. \tag{1.188}$$

At the three-phase point $N_2(a_1^N = \frac{\Delta\delta}{6a_3b_2},\ b_1^N = \frac{\Delta\delta^2}{12a_3b_2^2})$ the line (1.187) transforms into the line of first-order transition II-III. Figs.42a and 42b represent two different situations corresponding respectively to $a_2>0$ (the I-II transition is second-order) and $a_2<0$. For such negative values of a_2, the I-II transition becomes first-order and the corresponding line of transitions goes over continuously to the first-order transition line II-III, at $a_1 = a_2^2/4a_3$.
2) For $\gamma < -2\sqrt{|a_2|b_2}$, which expresses a sufficiently large negative interaction between the two order-parameters, one has a different first-order transition line between phases I and IV, following the "triggering" mechanism suggested by Holakovsky [74]. One has also two transition regimes between phases II and IV, which are separated by a tricritical point T_{tc} (Fig.42c). Let us also note the second three-phase point N_2 which lies either on the b_1 axis (Fig.42c) or above this axis (Fig.42d).

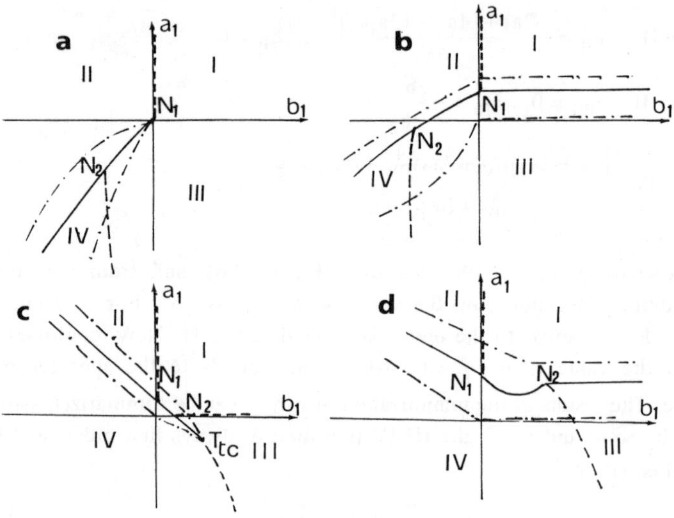

Fig.42. Phase diagrams corresponding to the asymmetric sixth degree expansion (1.185) for: (a) $\delta>2\sqrt{a_2b_2}$, $a_2>0$, (b) $\delta>2\sqrt{|a_2|b_2}$, $a_2<0$, (c) $\delta<-2\sqrt{a_2b_2}$, $a_2>0$, (d) $\delta<-2\sqrt{|a_2|b_2}$, $a_2<0$. Full, dashed and dashed-dotted lines are respectively first-order, second-order, and limit of stability lines.

c) At last, let us consider the symmetric sixth degree expansion:

$$F(\eta,\xi) = F_0 + a_1\eta^2 + a_2\eta^4 + a_3\eta^6 + b_1\xi^2 + b_2\xi^4 + b_3\xi^6 + \delta\eta^2\xi^2 \qquad (1.189)$$

for which the conditions $a_3>0$, $b_3>0$ are assumed. Figs.43a-h show various cross-sections (a_1,b_1) corresponding to different values of the phenomenological coefficients assuming either a positive (Figs.43a-d) or negative (Figs.43e-h) interaction between the two order-parameters. One can see that for $\Delta>0$, $a_2>0$ one finds again phase diagrams containing exclusively second-order transitions (as in Fig.41b). However for $\Delta>0$ and $a_2<0$, $b_2<0$ there is a region of coexistence of the phases II, III and IV and, in contrast to the asymmetric model, a first-order transition may occur directly between phases I and IV, and between phases II and IV. Accordingly, tricritical points may appear both on the II-IV and III-IV transition lines. Furthermore, for $\Delta<0$ one may have the "reentrant" sequences of phases II-IV-II and III-IV-III. Let us also note that triple points may replace the three-phase points, and that second-order phase transitions may disappear completely from the phase diagram.

d) Models based on the expansion (1.182) can also be used to construct the phase diagram corresponding to biquadratically coupled multi-dimensional order-parameters, as this expansion represents the *effective form* of the thermodynamic potential in which there is only one equilibrium value for each of the moduli $|\eta_i|$ and $|\xi_j|$ (see Chapters II-V). By

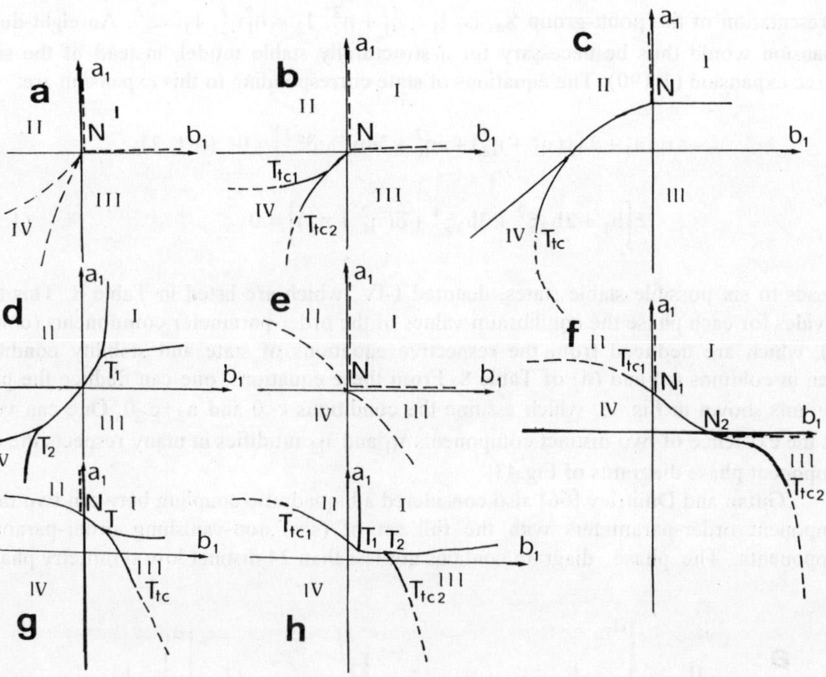

Fig.43. Phase diagrams corresponding to the symmetric sixth degree expansion (1.189) for: (a) $\delta>0$, $a_2>0$, $b_2>0$, $\Delta>0$; (b) $\delta>0$, $a_2<0$, $b_2<0$, $\Delta>0$; (c) $\delta>0$, $a_2<0$, $b_2>0$, $\Delta>0$; (d) $\delta>0$, $a_2<0$, $b_2<0$, $\Delta<0$; (e) $\delta<0$, $a_2>0$, $b_2>0$, $\Delta>0$; (f) $\delta<0$, $a_2>0$, $b_2>0$, $\Delta<0$; (g) $\delta<0$, $a_2<0$, $b_2>0$, $\Delta<0$; (h) $\delta<0$, $a_2<0$, $b_2<0$, $\Delta>0$. Full and dashed lines are respectively first and second-order transition lines. The T_{tc}, N and T points are tricritical, N-phase and triple points respectively.

contrast, when different non-zero equilibrium values are found for the order-parameter components, the resulting phase diagram may become more complex. This is illustrated by the model considered by Larin [69], of a biquadratic coupling between two three-component order-parameters (η_1, η_2, η_3) and (ξ_1, ξ_2, ξ_3) with $\eta_3 = \xi_2 = \xi_3 = 0$ and $\xi_1 \neq 0$, $\eta_1 \neq \eta_2 \neq 0$ for some of the phases. The effective expansion discussed by Larin is:

$$F(\eta_1, \eta_2, \xi) = a_1(\eta_1^2 + \eta_2^2) + a_2(\eta_1^2 + \eta_2^2)^2 + c(\eta_1^4 + \eta_2^4) + d(\eta_1^6 + \eta_2^6) +$$

$$+ b_1\xi^2 + b_2\xi^4 + b_3\xi^6 + \delta(\eta_1^2 + \eta_2^2)\xi^2 . \qquad (1.190)$$

Let us note that the integrity basis for the image-group associated with the expansion (1.190), which has the symmetry of the three-dimensional reducible vector representation of the point-group S_4, is: $I_1 = \eta_1^2 + \eta_2^2$, $I_2 = \eta_1^2\eta_2^2$, $I_3 = \xi^2$. An eight-degree expansion would thus be necessary for a structurally stable model, instead of the sixth-degree expansion (1.190). The equations of state corresponding to this expansion are:

$$\eta_i\left[a_1 + 2a_2(\eta_1^2 + \eta_2^2) + c\eta_i^2 + 3d\eta_i^4 + \delta\xi^2\right] = 0 \quad (i=1, 2)$$

$$\xi\left[b_1 + 2b_2\xi^2 + 3b_3\xi^4 + \delta(\eta_1^2 + \eta_2^2)\right] = 0. \qquad (1.191)$$

It leads to six possible stable states, denoted I-IV, which are listed in Table 8. This table provides for each phase the equilibrium values of the order-parameter components (column (b)), which are deduced from the respective equations of state and stability conditions given in columns (c) and (d) of Table 8. From these equations, one can deduce the phase diagrams shown in Fig.44, which assume the conditions $c<0$ and $a_2+c>0$. One can verify that the existence of two distinct components η_1 and η_2 modifies in many respects the one-component phase diagrams of Fig.43.

Gufan and Dmitriev [66] also considered a biquadratic coupling between two three-component order-parameters with the full set of (six) non-vanishing order-parameter components. The phase diagram contains not less than 24 distinct low symmetry phases,

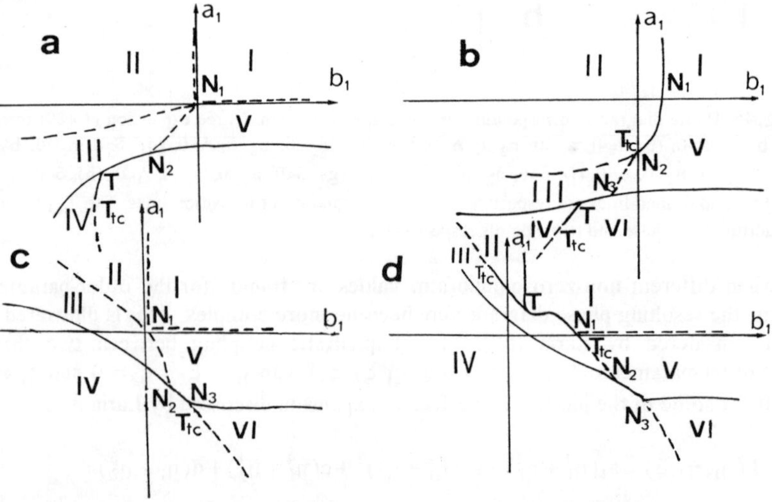

Fig. 44. Phase diagrams corresponding to the expansion (1.190) for $c<0$, $a_2+c>0$, $d>0$, $b_3>0$ and (a) $\Delta_0 = 4a_2b_2 - \delta^2 > 0$, (b) $\delta>0$, $\Delta_0<0$, (c) $\delta<0$, $\Delta_0<0$. Full and dashed lines are respectively first and second-order transition lines. N_1, N_2, N_3 are N-phase points. T is a triple point. T_{tc}, T_{tc1} and T_{tc2} are tricritical points.

Table 8. Stable phases obtained by minimization of the order-parameter exapnsion (1.190). Column (b): Equilibrium values of the order-parameter components. (c) Equations of State. (d) Stability conditions and equations of the limit of stability lines.

(a)	(b)	(c)	(d)
I	$\eta_1=\eta_2=\xi=0$	-	$a_1 \geq 0, b_1 \geq 0$
II	$\eta_1=\eta_2=0,\ \xi\neq 0$	$b_1+2b_2\xi^2+3b_3\xi^4=0$	$\begin{cases} b_1=0, b_2 \geq 0 \\ b_1=\frac{b_2^2}{3b_3}, b_2<0, b_1=\frac{2b_2}{\delta}a_1-\frac{3b_2}{\delta^2}a_1^2 \end{cases}$
III	$\eta_1\neq 0, \eta_2=0,\ \xi\neq 0$	$\begin{cases} a_1+2(a_2+c)\eta^2+3d\eta^4+\delta\xi^2=0 \\ b_1+2b_2\xi^2+3b_3\xi^4+\delta\eta^2=0 \end{cases}$	$\begin{cases} a_1\frac{\delta b_1}{2b_2}, \Delta_1=4(a_2+c)b_2-\delta^2 \geq 0 \\ a_1=\frac{\delta b_1}{2b_2}+\frac{\Delta_1}{48b_2^2 d}, \Delta_1<0 \\ a_1=\frac{\delta}{2b_2}b_1+\frac{\Delta_0}{3b_2 d}, c \leq 0\ (\Delta_0=4a_2b_2-\delta^2) \end{cases}$
IV	$\eta_1=\eta_2\neq 0,\ \xi\neq 0$	$\begin{cases} a_1+2(a_2+c)\eta^2+3d\eta^4+\delta\xi^2=0 \\ b_1+2b_2\xi^2+3b_3\xi^4+2\delta\eta^2=0 \end{cases}$	$\begin{cases} a_1=\frac{\delta b_1}{2b_2}, \Delta_2=4(2a_2+c)b_2-\delta^2 \geq 0 \\ a_1=\frac{\delta b_1}{2b_2}+\frac{\Delta_2^2}{12b_2 d}, \Delta_2<0, a_1=\frac{\delta b_1}{2b_2}+\frac{\Delta_0 c}{3b_2 d}+\frac{c^2}{3d} \end{cases}$
V	$\eta_1\neq 0, \eta_2=\xi=0$	$a_1+2(a_2+c)\eta^2+3d\eta^4=0$	$\begin{cases} a_1=0, a_3+c \geq 0, a_1=\frac{4a_2c}{3d} \\ a_1=\frac{(a_2+c)^2}{3d}, a_2+c<0, a_1=\frac{2(a_2+c)b_1}{\delta}-\frac{3b_1^2}{\delta^2}, b_1\delta<0 \end{cases}$
VI	$\eta_1=\eta_2\neq 0,\ \xi=0$	$a_1+2(2a_2+c)\eta^2+3d\eta^4=0$	$\begin{cases} a_1=0, 2a_2+c \geq 0, a_1=\frac{(4a_2c-c^2)}{3d} \\ a_1=\frac{(2a_3+c)^2}{3d}, 2a_2+c<0 \\ a_1=\frac{(2a_2+c)b_1}{d}-\frac{3b_1^2}{\delta^2}, (b_1\delta<0) \end{cases}$

among which 14 phases display non-vanishing components pertaining to both order-parameters. The integrity basis contains fourteen invariants. The complexity of the situation forbids to work-out the complete theoretical phase diagrams. Thus only a truncated fourth-degree expansion is used by the authors [66]:

$$F(\eta_1,\eta_2,\eta_3,\xi_1,\xi_2,\xi_3)=a_1I_1+a_2I_1^2+b_1I_2+b_2I_2^2+cI_3+dI_4+\delta I_5 \qquad (1.192)$$

with: $I_1=\eta_1^2+\eta_2^2+\eta_3^2$, $I_2=\xi_1^2+\xi_2^2+\xi_3^2$, $I_3=\eta_1^4+\eta_2^4+\eta_3^4$, $I_4=\xi_1^4+\xi_2^4+\xi_3^4$, $I_5=\eta_1^2\xi_1^2+\eta_2^2\xi_2^2+\eta_3^2\xi_3^2$. The phase diagram shown in Fig.45 gives a schematic view of a

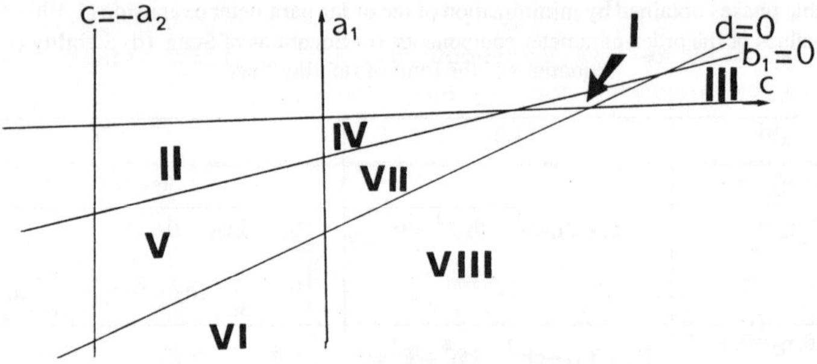

Fig. 45. Phase diagram corresponding to the fourth degree expansion (1.192). The notation of the phases corresponds to the following equilibrium values of the six-component order-parameter ($\eta_1, \eta_2, \eta_3, \ \xi_1, \xi_2, \xi_3$): I(000ξ00); II(η00000); III(000ξξξ); IV(ηηη000); V(η0000ξ); VI(η00ξξξ); VII($\eta_1\eta_2\eta_3$00ξ); VIII(ηηηξξξ).

small region of the full theoretical phase diagram, since it contains 8 among 24 low symmetry phases, and only second-order transitions due to the fourth-degree assumed in the expansion (1.192). It provides however an illustration of the intricate type of phase diagram, which can be obtained when considering coupled multicomponent order-parameter models.

A different biquadratic coupled one-component order-parameter model, containing a cubic invariant in the corresponding expansion:

$$F(\eta,\xi) = a_1\xi^2 + a_2\xi^3 + a_3\xi^4 + b_1\eta^2 + b_2\eta^4 + \delta\xi^2\eta^2$$

has been studied by Dmitriev and Rochal [75, 76]. It will be discussed in Chapter IV.

4.4.3. "Improper" coupling between two order-parameters

Another typical coupling between two order-parameters is of the form $\delta\eta^n\xi$ with $n \geq 2$. This type of "improper" coupling, which has been discussed in Sec.2.4 will be shown to result here in spontaneous values of ξ even in the absence of non-zero values for η. The lowest degree coupling (n=2) has been considered by Gufan and Torgashev [70, 71], Kut'in et al. [72, 77] and Lorman [33].

Let us first discuss the order-parameter expansion:

$$F(\eta,\xi) = F_0 + a_1\eta^2 + a_2\eta^4 + b_1\xi^2 + b_2\xi^4 + \delta\eta^2\xi \qquad (1.193)$$

assuming $a_2 > 0$, $b_2 > 0$. In the absence of interaction between the two order-parameters (δ=0), one gets the same four stable phases found in the biquadratically coupled order-parameter model (1.182). The existence of the non linear interaction (δ≠0) destabilizes phase II (η≠0, ξ=0) in the preceding model, and one has only three stable phases:

Phase I: $\eta=\xi=0$; Phase II: $\eta=0, \xi\neq 0$; Phase III: $\eta\neq 0, \xi\neq 0$.
In phases II and III the non-zero equilibrium value of ξ is induced by the δ-coupling, even in the absence of non-zero equilibrium values of η in phase II. This can be seen from the equations of state corresponding to (1.193):

$$\eta(a_1+2a_2\eta^2+\delta\xi=0, \quad 2b_1\xi+4b_2\xi^3+\delta\eta^2=0$$

from which the condition $\eta=0$ implies either $\xi=0$ or $\xi^2=-b_1/2b_2$. In this respect, let us stress that at variance with the improper coupling between "primary" and "secondary" order-parameters discussed in Sec.2.4, here ξ does not play the role of a secondary order-parameter, since its equilibrium values influence the symmetry of phases II and III.

Fig.46a shows the phase diagram associated with expansion (1.193). It contains a second-order transition between phases I and III, along the line $a_1=0$ for $b_1>\delta^2/4$. At the tricritical point T_{tc1} of coordinates $(\delta^2/4a_2,0)$ the preceding line becomes a first-order transition line, which transforms again, for negative values of b_1 into a second-order transition line, at a second tricritical point T_{tc2}. Between T_{tc1} and T_{tc2} one has a three-phase point of coordinates $(0, \delta^2/2^3(a_2b_2)^{1/2})$.

A consistent model actually requires to take into account in (1.193) a biquadratic coupling $\gamma\eta^2\xi^2$, which is always allowed by symmetry. As one can see in Fig.46b this adjunction does not modify the essential features of Fig.46a. By contrast the topology of the phase diagram is drastically changed when a cubic invariant ξ^3 exists in the order-parameter expansion:

$$F(\eta,\xi)=F_0+a_1\eta^2+a_2\eta^4+b_1\xi^2+b_2\xi^4+b_3\xi^3+\delta_1\eta^2\xi. \tag{1.194}$$

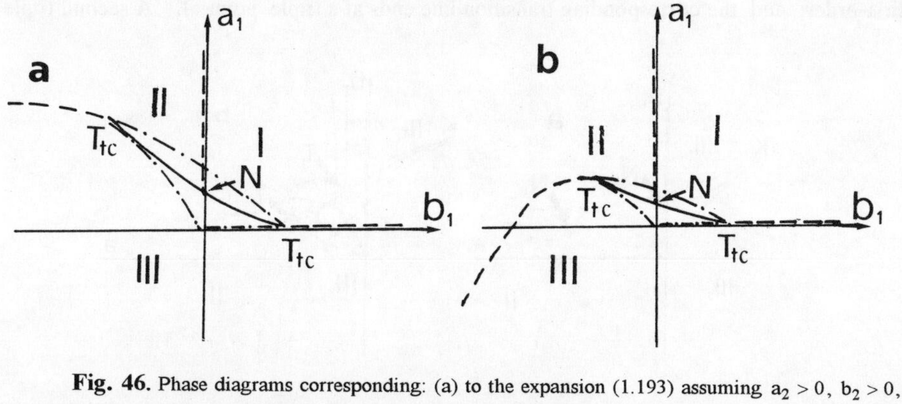

Fig. 46. Phase diagrams corresponding: (a) to the expansion (1.193) assuming $a_2>0$, $b_2>0$, $\Delta>0$; (b) to the expansion (1.193) with the additional term $\gamma\eta^2\xi^2$ with $\gamma>0$. Full, dashed and dashed-dotted lines are respectively first-order, second-order and limit of stability lines.

The same three phases I, II, III, are found, but with a different configuration for their regions of stability. Thus Fig.47a shows that the first-order transition line between phases I and II, given by:

$$b_1 = \frac{b_3^2}{4b_2} \qquad (1.195)$$

is prolongated for negative values of a_1, by a first-order transition line, separately two isostructural phases III_1 and III_2, of equation:

$$\tilde{a}_1 = -\frac{\delta a_1}{2a_2} - \frac{b_3}{2b_2}\tilde{b}_1 - \frac{b_3^2}{16b_2^2} = 0 \qquad (1.196)$$

with $\tilde{b}_1 = b_1 - \frac{\delta^2}{4a_2} - \frac{3b_3^2}{8b_2}$. This line terminates at a critical point K of coordinates ($\tilde{a}_1 = 0$, $\tilde{b}_1 = 0$). The diagram of Fig.47a also contains two three-phase points: N_1, at which meet the isostructural transition line (1.196) and the second-order transition line I-III_1 ($a_1 = 0$), and N_2 at which meet the first-order transition line I-II (1.195), and the second-order transition line II-III_2 of equation $b_1 = b_3^2 / 4b_2$.

Because of the cubic invariant ξ^3, which due to the non-linear interaction $\eta^2\xi$ is proportional to η^6, one has to include a sixth -degree invariant in (1.193) for a better consistency of the model:

$$F(\eta,\xi) = F_0 + a_1\eta^2 + a_2\eta^4 + a_3\eta^6 + b_1\xi^2 + b_2\xi^4 + b_3\xi^3 + \delta\eta^2\xi. \qquad (1.197)$$

It yields to the phase diagram represented in Fig.47b. One can see that although the stability region of phases I and II remains almost unchanged, the I-III, transition becomes first-order, and the corresponding transition line ends at a triple point T_1. A second triple

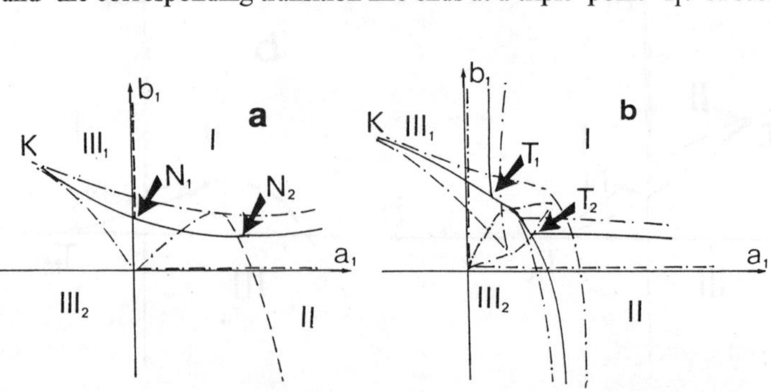

Fig. 47. Phase diagrams corresponding to the expansions (1.194) (a) and (1.197) (b). Full, dashed and dashed-dotted lines are respectively first-order, second-order and limit of stability lines.

point T_2 takes place at the convergence point of the first-order transition lines I-II and I-III.

4.4.4. "Pseudo-proper" coupling between two order-parameters

Another type of coupling may occur when two order-parameters representing different physical quantities transform as the same irreducible representation. Their "pseudo-proper" coupling, mentiond in Sec.2.4.4, is expressed by a bilinear invariant of the form $\eta_i\xi_j$, in the order-parameter expansion. Lorman [33] and Kut'in et al. [14] have discussed a model involving one-component order-parameters η, ξ which transform as the reducible two-dimensional image-group C_2. The corresponding basis of integrity is formed by three invariants:

$$I_1 = \eta^2, \quad I_2 = \xi^2 I_2 = \xi^2, \quad I_3 = \eta\xi. \tag{1.198}$$

The order-parameter expansion has the form:

$$F(\eta,\xi) = a_1\eta^2 + a_2\eta^4 + b_1\xi^2 + b_2\xi^4 + \delta\eta\xi + \gamma\eta^2\xi^2 \tag{1.199}$$

where the invariant I_3 expresses the interaction between the two order-parameters. The equations of state minimizing F are:

$$\begin{aligned} 2\eta\left(a_1 + 2a_2\eta^2 + \gamma\xi^2\right) + \delta\xi &= 0 \\ 2\xi\left(b_1 + 2b_2\xi^2 + \gamma\eta^2\right) + \delta\eta &= 0 \end{aligned} \tag{1.200}$$

which gives rise to *two* stable phases: Phase I: $\eta=\xi=0$, and Phase II: $\eta=\xi\neq0$. Fig.48a represents the configuration of the two phases in the orbit-space, whereas Figs.48c-d show the phase diagram in the phenomenological coefficients space. Thus, the transition between phase I and phase II is always second-order, and one has an isostructural transition between two variants of phase II. The metastability region in which those two variants coexist ends at a critical point of the liquid-gas type. In order to have a first-order regime between phases I and II, one needs to add a sixth-degree term in η or ξ to (1.199).

4.5. Isostructural and antiisostructural phase transitions

The liquid-gaz phase transition is the most commonly encountered example of isostructural transition. The discontinuous character of this transition derives from the fact that the two phases adjacent to the transition line are "qualitatively" identical but "quantitatively" distinct, following the formulation of Landau [1]. In other words, the two phases can not be distinguished by their symmetry, which is isotropic, but by the fluid's density, or the specific volume, which varies by several orders of magnitude between the two phases. This is true far from the *critical point*, which is the end-point of the line of

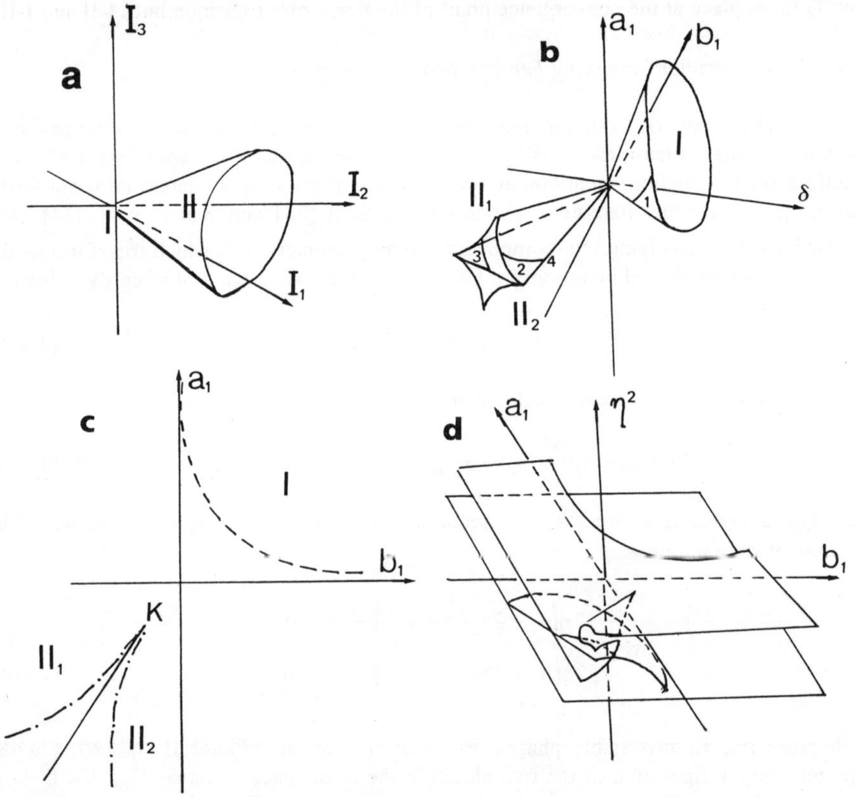

Fig. 48. Phase diagrams corresponding to the pseudoproper coupling model, based on expansion (1.199). (a) Configuration of the phases in the orbit-space. (b) Phase diagram in the (a_1, b_1, ξ) space. The numbers denote the second-order transition surface I-II (1), the surface limiting the region of coexistence of the isostructural phases II_1 and II_2 (2), and the isostructural transition plane (3). The edges limiting this plane are critical lines (4) which become a critical point in the cross-section (a_1, b_1). (c) Phase diagram in the (a_1, b_1) plane. Dashed, full and dashed-dotted lines are second-order, first-order and limit of stability lines. (d) Surfaces representing the square of the order-parameter η in the (a_1, b_1, η^2) space.

discontinuous transition in the pressure-temperature plane. Beyond the critical point the two phases are essentially indistiguishable.

It was first pointed-out by I.M.Lifshitz [78] that qualitative changes in the physical properties of a crystalline solid, could also take place without change in the structural symmetry (space-group). In the example given by Lifshitz of a "topological $2\frac{1}{2}$ transition" [78] the qualitative change consists in an electronic transition between closed and open Fermi surfaces. Another known examples is the metal-insulator transition, where the change concerns the number of electrons localized near a metallic ion [79]. Gufan and Lorman [80] have proposed in this case a one-dimensional model of isostructural transition assuming two types of ions with different radii. Lyotropic liquid crystals also provide examples of isostructural transitions related to the shape of the micellar agregates [81].

Among crystalline solids, there exist a small number of well established isostructural transitions, such as the transition between the FCC phases found in cerium [82], cesium [83] and chalcogenide lanthanides [84], or the transition in $(V_{1-x}Cr_x)_2O_3$ [85]. The isostructural transition is revealed by discontinuities in the cell parameter, and by the existence of a critical point of the liquid-gas type, which is found for example at 1.75 GPa and 550 K in Ce [86] or for x=0.17 and 580 K in $Sm_{1-x}Gd_xS$ [87]. Smooth anomalies of the dielectric and elastic constants as well as in the conductivity are generally observed at isostructural transitions.

The symmetry of the order-parameter (i.e. the equilibrium relationship between the order-parameter components) remains unchanged between two isostructural phases. In contrast, for *antiisostuctural transitions*, though the space groups of the phases are the same, the *sign* of the order-parameter components changes from one phase to another. Thus, the two phases can be distinguished qualitatively and no critical point will be found on a first-order transition line of antiisostructural transition. For instance, in the cubic-tetragonal transformations found in the solid solutions $Cu_{1-x}Ni_xCr_2O_4$ [88] and $Fe_{1-x}Ni_xCr_2O_4$ [89], the two tetragonal low-symmetry phases correspond to opposite signs of the quantity (c/a-1), where a and c are the tetragonal parameters. In terms of the spontaneous strain tensor components one has in one phase $e_{zz} > e_{xx} = e_{yy}$, whereas in the other phase $e_{zz} < e_{xx} = e_{yy}$. Accordingly, the two antiisostructural phases are associated with opposite values of the order-parameter components $(\eta_1\ 0)$ and $(-\eta_1\ 0)$ in the framework of the C_{3v}-type model assumed for the phase diagram [12, 53], the *cubic* invariant changing sign from one phase to the other.

A number of theoretical examples of isostructural and antiisostructural transitions have been disclosed in the phase diagrams reported in Secs.3.3 and 4.4. In the phase diagrams associated with a one-component order-parameter, a region of stability has been found for two isostructural phases (Fig. 14b) when assuming an eighth degree expansion (1.94) with $a_3 < 0$ and $a_4 > 0$. Increasing further the degree of the expansion offers the possibility of successive additional isostructural phases (Fig.13). In the case of a one-component order-parameter transforming as the image-group C_1, it has been shown that the "high" and "low" symmetry phases are isostructural, i.e. it represents the trivial case of isostructural transition induced by the identity representation.

In the case of a one-component order-parameter, the existence of an isostructural transition results from the strong non linearity of the interactions, which requires to consider a high-degree expansion. For multicomponent order-parameters, or reducible order-parameters another class of symmetry induced isostructural transitions is found, i.e. the existance of which is due to the symmetry of the order-parameter. Two different situations have to be distinguished in this respect, as pointed out by Lorman [33].

1) For image-groups not generated by reflections (C_n, O, T,...) the order-parameter expansion always possesses at least two orbits of maxima belonging to the same family of image-group, *which is a necessary condition for an isostructural transition to take place*. As shown in Figs.22 and 29, in the case of the C_4 and C_3 groups, the region of coexistence of the isostructural phases, takes place in the lowest symmetry phase, without any contact with the parent phase, i.e. the hysteresis region corresponding to the isostructural transition will decrease when approaching the parent phase.

2) For image-groups generated by reflections (C_{nv}, O_h, T_d,...) the preceding necessary condition is not always fulfilled. Besides, the phase diagrams found for C_{4v} (Fig.21), C_{3v} (Fig.27) and C_{6v} (Fig.30) show that isostructural transitions do not take place in the lowest symmetry phase, but within phases which can be reached across a line of transitions from the parent phase. In the case of the O_h image-group, Fig.34b shows that each of the low-symmetry phases undergoes an isostructural transition when a_1 is decreased.

The above mentioned condition applies also to image-groups associated with reducible order-parameters. This was shown in Sec.4.4 for an *improper* type of coupling between the order-parameters (Fig.47) or for a *pseudo-proper* type of coupling (Fig.48), this latest case corresponding to an image-group not generated by reflection (C_2). Phase diagrams associated with biquadratically coupled order-parameters have been shown to possess isostructural transitions [90] when high degree invariants are taken into account.

Antiisostructural phases have been found among the low symmetry phases for the image groups T_d (phases IV and VII) and T (phases III and IV). One can verify in Figs.35 and 37 that the corresponding antiisostructural transitions take place either across a first-order transition, or through an intermediate lower symmetry phase. This is also true for the antiisostructural phases II and III found for the C_{3v} image (Figs.26a and 26b).

4.6. First-order phase transitions and stable fixed points

In Secs.4.2 to 4.5 the Landau theory has been used as a starting point for the phenomenological characterization of the various types of first-order phase transitions. It is well known that in the vicinity of the transition point, this theory does not account correctly, in general, for the behaviour of physical quantities since it has the same range of validity as the mean-field approximation in microscopic theories. Early works had been performed in the Sixties [91, 92] in order to extend the Landau theory to include spatial inhomogeneous fluctuations of the order-parameter in the region of the transition. Later, the advent of the Renormalization Group methods [43, 93] has provided powerful techniques to study the influence of fluctuations on the qualitative and quantitative behavior of systems undergoing a phase transition. Thus, critical fluctuations have been shown to modify appreciably some predictions of the mean-field theory and eventually *the order of the transitions*. On that head, it has been established theoretically that a second-order transition can be transformed into a first-order one for certain relationships between the parameters of the system involving a growth of the fluctuations. For some transitions, predicted to be possibly continous by the Landau theory, the fluctuations can be so strong that the transitions are shown to be necessarily first-order for any values of the parameters of the system.

The preceding results are based on one of the crucial concepts of the Renormalization-Group theory, which is the association of *a stable fixed point* in the flow of the so-called Landau-Wilson Hamiltonian, with a second-order transition. Thus, when a transition is first-order it is thus interpreted as either because no appropriate stable fixed point exists, or because the existing stable fixed points are not physically accessible. More precisely, a number of authors [44-46, 94, 95] have suggested *that the lack of stable fixed points for a given system was a sufficient condition for the occurence of a fluctuation induced first-order transition.*

Along this line, Michel and Toledano [96] have specified the group theoretical conditions defining the Hamiltonians lacking a stable fixed point. The formulation of these conditions makes use of the concepts of centralizer and normalizer of a subgroup of O(n), where n is the dimension of the order-parameter: The centralizer G_C is the largest subgroup of O(n) which leaves invariant every vector in the n-dimensional order-parameter space, and the normalizer G_N is the largest subgroup of O(n) preserving the preceding space as a whole. The criterion established in Ref.[96] states that: 1) A stable fixed point is necessarily characterized by the coincidence of the centralizer and the normalizer associated to it: $G_C = G_N$, 2) If G_N is not a subgroup of at least one of the set of subgroups G_i of O(n), then the considered flow in the order-parameter space has no stable fixed point.

The preceding results have motivated a number of works focusing on the question of whether real systems do or do not have stable fixed points [97-100]. Unfortunately a number of restrictions have appeared to limit largely the bearing of such a question regarding to the prediction of the order of transitions. Thus, Brezin et al. [101] have proved that for n≤3, there is always a stable fixed point, which is the isotropic fixed point, i.e. corresponding to the isotropic part of the fourth degree invariants in the order-parameter expansion. Since most of the observed structural transitions fall into this case, it follows that for n≤3 the Renormalization-Group theory is not more predictive than the Landau theory. For n≥4 the critical behaviour is not known in the general case as the anisotropic terms of the expansion become relevant. The nature and stability of the fixed points are thus expected to depend closely of the type of Landau-Wilson Hamiltonian of the considered system. Accordingly, a particular study of each Hamiltonian with n≥4 is necessary, that was partly performed [18, 102, 103].

In Table 9 the results obtained for real systems with n≥4 have been collected. They correspond to structural *or* magnetic transitions induced by "active" irreducible representations, i.e. by representations which fulfil the Landau and Lifshitz criteria (upper

Table 9. Materials (column (a)) in which a transition induced by a single irreducible representation takes place. (b) High symmetry space-group. (c) and (d) Brillouin zone point and corresponding irreducible representation, in reference to the Table of Zak [17]. (e) Order-parameter dimension. (f) Existence (Yes) or absence (No) of a stable fixed point. (g) Order of the transition. (h) References.

(a)	(b)	(c)	(d)	(e)	(f)	(g)	(h)
VO_2	D_{4h}^{14}	R	τ_1 or τ_2	4	Yes	1	[104]
$CdSnAs_2$	T_d^2	W	τ_1 or τ_2	6	Yes	1	[61]
$NdSn_3$	O_h^1	X	τ_5 or τ_{10}	6	No	2	[105]
CeS	O_h^5	L	τ_1 or τ_4	4	No	2	[106]
UO_2	O_h^5	X	τ_{10}	6	No	1	[98]
ErSb	O_h^5	L	τ_3 or τ_6	8	No	2	[107]
MnO	O_h^5	L	τ_3 or τ_6	8	No	1	[98]
NbO_2	D_{4h}^{14}	$\frac{1}{4}\frac{1}{4}\frac{1}{4}$	τ_1	4	Yes	2	[97]
$Cd(NO_3)_2$	T_h^6	X	τ_1 or τ_2	6	Yes	1	[108]
MnS_2	T_h^6	X	τ_1 or τ_2	6	Yes	1	[94]
$CeAl_2$	O_h^7	$\frac{1}{2}\frac{1}{2}\frac{1}{2}$	τ_1	24	No	2	[109]
Eu, Cr	O_h^9	k 0 0	τ_1	12	No	1	[99]

part of the Table), or violating the Lifshitz criterion (bottom of the Table). A comparision of columns (e) and (g) of the Table shows that *no definite conclusions can be inferred from the existence or absence of stable fixed points with respect to the order of transition.* Thus, concerning "active" representations, no stable fixed point have been found for $NdSn_3$, CeS and ErSb though the transitions in these substances are second-order. Conversely a stable fixed point exists for VO_2 and $CdSnAs_2$ which both possess a first-order transition. Furthermore, VO_2 whose transition has a strongly first-order character and NbO_2, in which a second-order transition take place, are described by the same Hamiltonian. Only UO_2 and MnO which have no stable fixed points display effectively a first-order transition, in agreement with the above mentioned conjecture.

The transitions concerning "inactive" irreducible representations in Table 9 illustrate likewise the inconsistency of the conclusions which can be drawn from the existence or absence of stable fixed point for a given system about the order of its transition. Thus $CeAl_2$ has a second-order transition in the absence of stable fixed point, while $Cd(NO_3)_2$ and MnS_2 undergo first-order transitions, but possess stable fixed points.

Therefore, as the case stands at present, comparison to experimental data shows that even for $n \geq 4$ the absence of stable fixed point does not exclude transitions of second order, nor their presence exclude first-order transitions. So that the search for systems with no stable fixed point, does not seem conclusive for any value of n, regarding to the order of transitions.

5. Theory of singularities and phase transitions

5.1. Introduction

While Group Theory, and more especially the theory of Group Representations, underlies the "angular" aspect of the Landau theory of phase transitions, the "radial" part of the Landau theory grounds on the *theory of Singularities*. This theory, which is part of the Catastrophe theory, offers the possibility to place on a rigorous basis some empirical rules used in Sec.3.3 concerning for example the minimal degree to which the Landau polynomial has to be expanded in a "structurally stable" model, the role played by leading terms in the order-parameter expansion etc.. It systematizes also some of the topological properties of phase diagrams: classification of the singular points, form of the phase boundaries in their vicinity, multiplication of the singularities when additional degrees of freedom are taken into account etc..

Although some aspects of the theory of Singularities have been used a longwhile ago by physicists, the theory was formulated in an explicit form in the Seventies [110-114] and almost immediatly applied to the description of physical systems [115-117]. In the field of phase transitions the simplest form of the theory was first used to illustrate the applicability of its concepts to problems which had been already solved by the Landau theory, or by statistical physics, such as the theory of the critical point of a Van der Waals gas [118], the thermodynamic behaviour of a uniaxial ferromagnet under an external magnetic field [116], the standard properties of ferroelectric materials [119], or the elementary description of phase diagrams [120]. These works established a correspondence with the terminology used in the Landau approach: order-parameter (internal parameter), phenomenological coefficients (control parameters), equation of state (catastrophe manifold), Landau potential (miniversal deformation of the normal form of the singularity) etc.. Recently a number of more elaborated works have attempted to include symmetry considerations in the theory of singularities [14, 18, 32, 33, 121] and have shown that this theory could give a powerful stimulus to the Landau approach to complex phase transition problems. This is the case in particular, of the work of Kut'in et al. [14], to which we will often refer to in the present section.

5.2. Structural stability

5.2.1. Some definitions and concepts of the theory of singularities

Consider a smooth (infinitely differentiable) function F depending on n variables (x_i), and l parameters (α_j). The set of all F for fixed (α_j) is called the l-parameter family of functions. *Critical points* (x_i^0) of F are defined by dF=0. A critical point (x_i^0) is *degenerate* if the Hessian $\left[\det\left(\frac{\partial^2 F}{\partial x_i \partial x_j}\right)\right]_{x_i = x_i^0}$ is zero. On the contrary the critical point is non-degenerate. If the rank of the n×n matrix $\left(\frac{\partial^2 F}{\partial x_i \partial x_j}\right)$ at a degenerate critical point is r, the *co-rank* of the critical point is (n-r). The *multiplicity of the degeneracy* of a point M, is the maximum number of non-degenerate critical points into which it can be decomposed.

The most general arrangement of two objects in a given space is called the *transversality* of these objects. For instance, two intersecting lines in a plane are transversal, while parallel lines are non-transversal. Small variations of transversal objects do not change their mutual arrangement. When experiments performed on a given system under almost identical conditions yield, with some degree of accuracy, the same result (i.e. when they have a sufficient degree of reproducibility), then the system is said to be *structurally stable*. The concept of structural stability (which is analogous to the concept of robustness characterizing the insensivity of a system to initial conditions [122]) extends to the physical quantities describing the system. For example, if the physical phenomena taking place at a phase transition are structurally stable, the Landau potential describing these phenomena will be also said to possess structural stability.

The measure of the structural stability of a system is its transversality. One can speak for example of the transversality in the arrangement of the critical points associated with a given Landau expansion. Although transversal systems, or objects, are structurally stable, a systematic modification in their arrangement may lead ultimately to a qualitatively different (degenerate) arrangement. Figs.15 and 17 in Sec.3.3.2 illustrate this property. Thus the Landau potential F is transversal (i.e. it has non degenerate critical points) within each region of stability of phases I, II and III, when the coefficients (a_1, a_2), or equivalently the external parameters, vary within given limited intervals. However, when these coefficients reach critical values corresponding to the limit of stability lines, which are formed by degenerate critical points, then F becomes non-transversal. Furthermore changes in a_1 and a_2 again lead to transversality into another region of phase stability.

According to the preceding considerations a non-transversal situation can be observed when external variables are changed only close to a degenerate critical point. However, such point may arise for two fundamentally different reasons: as an "accidental" degeneracy or as the result of the symmetry of the system. Figs.49a and 49b illustrate the two types of situations. Fig.49a represents the phase diagram associated with the potential function: $F = \alpha_1 x + \alpha_2 x^2 + \alpha_3 x^4$ for $\alpha_2 < 0$ and $\alpha_3 > 0$. From the equations: $\frac{\partial^2 F}{\partial x^2} = \frac{\partial F}{\partial x} = 0$ one can deduce the equation of the limit of stability lines:

$$27\alpha_1^2 \alpha_3 + 8\alpha_2^3 = 0 \tag{1.201}$$

which are the dashed-dotted lines in Fig.49a. The critical points on these lines possess an accidental degeneracy which does not result from the symmetry of the system, i.e. small changes in the external parameters will destroy the relationship between the coefficients given by (1-201) and change the symmetry of the structurally unstable system. A different situation is in Fig.49b, which corresponds to the phase diagram associated with the family of functions $F=\alpha_1 x^2+\alpha_2 x^4$ for $\alpha_2>0$. Here the line of degenerate critical points $\alpha_1=0$ is due to symmetry requirement, i.e. suitable small changes in the external variables will preserve this line, as well as the symmetry of the system.

5.2.2 Splitting lemma. Diffeomorphisms

Since the Hessian of any smooth function $F(x_1,\ldots,x_n)$ is non-zero in the vicinity of a non-degenerate critical point, there exists a substitution

$$x_i \to \varepsilon_i(y_1,\ldots,y_n) \qquad (1.202)$$

which transforms F into a nondegenerate quadratic form:

$$F=\sum_{i=1}^{n}\alpha_i y_i^2 \quad \text{with} \quad \alpha_i=\pm 1. \qquad (1.203)$$

At a degenerate critical point some of the eigenvalues of the Hessian matrix of F are zero. The subspace spanned by the corresponding eigen vectors $(x_{s+1},\ldots,x_n)$ is called a *critical subspace*. Its dimension (n-s) is equal to the co-rank of the critical point. The remaining variables $(x_1,\ldots,x_s)$ are called *Morse variables*. The substitution (1.202) gives the form of F:

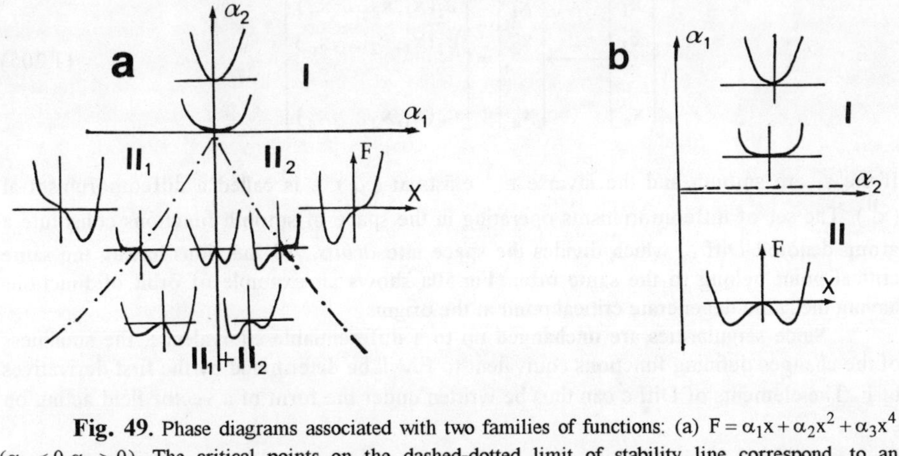

Fig. 49. Phase diagrams associated with two families of functions: (a) $F=\alpha_1 x+\alpha_2 x^2+\alpha_3 x^4$ $(\alpha_2<0,\alpha_3>0)$. The critical points on the dashed-dotted limit of stability line correspond, to an accidental degeneracy. (b) $F=\alpha_1 x^2+\alpha_2 x^4$ $(\alpha_2>0)$. The critical points on the second-order transition (dashed) line possess a symmetry induced degeneracy.

$$F = \sum_{i=1}^{s} \alpha_i y_i^2 + g(y_{s+1},\dots,y_n) \quad (\alpha_i = \pm 1) \tag{1.204}$$

where the function g obeys the conditions: $dg=d^2g=0$. Thus, the splitting lemma express the fact that: *at a degenerate critical point a smooth function F depending on n variables can split up into two part: a function of s Morse variables and a function of (n-s) non-Morse variables*. The singularity of F depends only on the form of g, and it remains unchanged by the addition of Morse variables. All functions F with the same $g(y_{s+1},\dots,y_n)$ are said to be *stably equivalent*.

The concept of structural stability has been defined hereabove with an accuracy up to *small changes* in the external variables. Thus, in the neighbourhood of a degenerate critical point, one can consider as equivalent all functions which can be deduced from a function F by small changes. Such changes can be obtained by substituting the variables (x_i) defining F by other variables (x_i') differing only slightly from the (x_i). The class of transformations $x_i \to x_i'$ corresponding to the same function F up to small changes are limited by the constraint imposed by the requirement of conservation of the singularity associated with the degenerate critical point. Indeed, *it is the singularity that determines all the properties of F*, and in turn *it is the decomposition of the singularity which determines the features of the phase diagram corresponding to F.*

The class of transformations preserving a singularity are called *diffeomorphisms*. In the general sense, a diffeomorphism is a smooth mapping of variables having a smooth inverse. More precisely, consider a mapping ε between two n-dimensional space $R^n \xrightarrow{\varepsilon} R^n$. It is represented by n functions ε_i as:

$$\begin{bmatrix} x_1 \\ x_2 \\ \vdots \\ x_n \end{bmatrix} \xrightarrow{\varepsilon} \begin{bmatrix} x_1' \\ x_2' \\ \vdots \\ x_n' \end{bmatrix} = \begin{bmatrix} \varepsilon_1(x_1,x_2,\dots,x_n) \\ \varepsilon_2(x_1,x_2,\dots,x_n) \\ \vdots \\ \varepsilon_n(x_1,x_2,\dots,x_n) \end{bmatrix}. \tag{1.205}$$

If the ε_i are smooth and the inverse ε^{-1} exists at (x_i^0), ε is called a diffeomorphism at (x_i^0). The set of diffeomorphisms operating in the space of smooth functions constitute a group denoted Diff ε, which divides the space into *orbits*. All functions having the same critical point belong to the same orbit. Fig.50a shows an example of orbit of functions having the same degenerate critical point at the origin.

Since singularities are unchanged up to a differentiable equivalence, the smallness of the changes defining functions equivalent to F will be determined by the first derivatives of F. The elements of Diff ε can thus be written under the form of a vector field acting on ε:

$$v = \sum_i V_i(x_1,\dots,x_n)\frac{\partial}{\partial x_i} \tag{1.206}$$

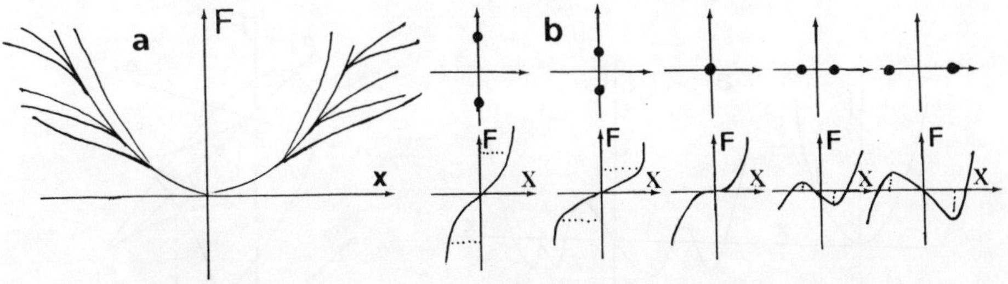

Fig. 50. (a) Orbit of functions F(x) having the same degenerate critical point at the origin. (b) Example of *abutment* of an orbit with a degenerate critical point by orbits with Morse critical points $F(x)=x^3$. $G(x)=x^3+\alpha_1 x$ ($\alpha_1>0$). The Morse critical points for G(x) are shown in the complex plane (upper part of the figure).

where V_i are smooth functions. If a diffeomorphism v shifts F along its orbit to a nearby point corresponding to the function G, then v takes the form: $v=\hat{1}+u$, where $\hat{1}$ is the identity operator. Accordingly, one has:

$$G = v(F) = (\hat{1}+u)(F) = F + u(F) = F + \sum_{i=1}^{n} \mu_i(x_1,\ldots,x_n)\frac{\partial F}{\partial x_i} \qquad (1.207)$$

where the μ_i are smooth functions. At the critical point $\partial F/\partial x_i = 0$, and Eq.(1.207) shows that G and F are equivalent up to a diffeomorphism. In other words, functions belonging to the same orbit differs only by terms which vanish at their common minima, i.e. *they possess the same equation of state*. Consequently, up to a diffeomorphism one can chose any representative of the same orbit. The most convenient representative is generally the one having the most compact form. It is called *the normal form of the singularity*.

5.3. Normal form of a singularity

5.3.1. Additional terminology

Let us consider the function $F(x)=x^3$. It belongs to the orbit with a degenerate critical point at x=0. On the other hand the function $G(x)=x^3+\alpha_1\, x$ ($\alpha_1>0$) possess two non-degenerate (Morse) critical points for $x=\pm\sqrt{-\alpha_1/3}$. Fig.50b shows that the orbits of G(x) *abut* the orbit of F(x) on both sides of the degenerate critical point. Generalizing, one can say that the orbits on which the Diff ε group acts are mutually arranged in such a way that the orbit of functions having a degenerate critical point *abut* on other orbit with Morse critical points corresponding to the *decomposition* of the degenerate critical point.

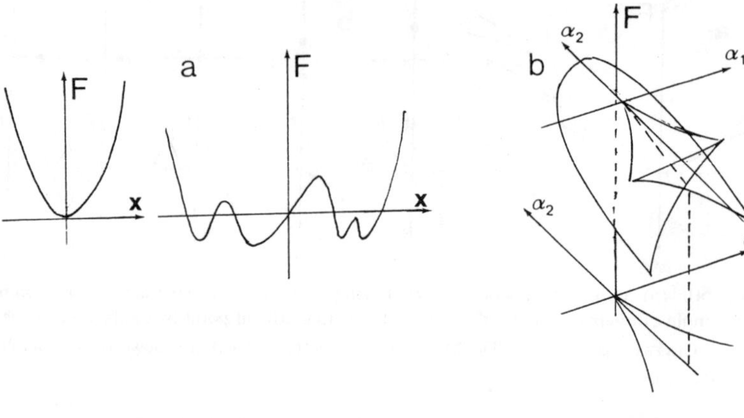

Fig. 51. (a) Miniversal deformation of a singularity of co-dimension 6: $F(x)=x^8+\alpha_6x^6+\alpha_5x^5+$ $+\alpha_4x^4+\alpha_3x^3+\alpha_2x^2+\alpha_1x$. (b) Relation between the bifurcation diagram of a singularity (up) and the phase diagram (down) for a miniversal deformation of a singularity of co-dimension 2: $F(x)=x^4+\alpha_2x^2+\alpha_1x$.

The family of all functions that includes the functions possessing a singularity as well as those with all of its decompositions, is called the *versal deformation of the singularity*. The parameters of this family form a space called the *deformation base*. The *co-dimension of the singularity* is the minimum number of parameters necessary to obtain all the decompositions of the singularity. If the co-dimension coincides with the dimension of the deformation base, then this deformation is called *miniversal*. Fig.50b shows an example of miniversal deformation with co-dimension 1 for a singularity at x=0. Fig.51a gives another example of miniversal deformation of a singularity (at x=0) corresponding to the function: $F(x)=x^8+\alpha_6x^6+\alpha_5x^5+\alpha_4x^4+\alpha_3x^3+\alpha_2x^2+\alpha_1x$. When the dimension of the deformation base is larger then the co-dimension the singularity evolves along extra parameters (see Sec.5.4).

The *bifurcation diagram of the singularity* of a function F is the dependence of the equilibrium values of F on the coefficients of the deformation base. The *phase diagram* is a projection of the bifurcation diagram of the singularity on the space of coefficients. Fig.51b shows the relation between the bifurcation diagram and the phase diagram for the function $F(x)=\alpha_1x+\alpha_2x^2+x^4$, in the neighborhood of a singularity (of co-dimension 2) which is a critical point of the liquid-vapor type (at x=0). Fig.51b also illustrates the following general property: At the phase stability boundary ($dF=d^2F=0$) the non-degenerate critical points (corresponding in this case to the solutions of the equation of state $\alpha_1+2\alpha_2x+4x^3=0$) merge, and thus the phase diagram of the system is preserved in the bifurcation diagram of the singularity.

5.3.2. Normal forms: in search of the Landau potential

From the preceding considerations one can deduce the successive steps which are necessary to obtain the suitable "structurally stable" Landau potential corresponding to a given phase diagram. At first one has to determine the degenerate critical point of maximal multiplicity (i.e. decomposing in the maximum number of Morse critical points) which is involved in the phase diagram. Then, from the potentials having such a singularity it is necessary to chose the most compact potential, namely to reduce the singularity to its *normal form.* A *miniversal deformation of the normal form of the singularity will be the desired form of the landau potential.*

Several methods have been proposed to reduce a singularity - a Landau potential - to normal form [114, 116, 117, 123]. Their description is beyond the scope of this book. The most general and systematic method (the so-called spectral sequence method [114]) is based on an algorithm which allows an exact determination of a finite number of normal forms of a singularity, and of their miniversal deformations, one of which being the more adapted to describe the properties of a given system near a degenerate critical point. The justification of this algorithm (see Ref.[14]) leads to distinguish two types of coefficients in the Landau potential corresponding to the miniversal deformation of the singularity: those depending on the external conditions (the parameters of the deformation base) and *the moduli* which do not depend on the external conditions. The bifurcation diagram in the space of deformation parameters determines the phase diagram and the other thermodynamic characteristics of the system, while the moduli determine the type of bifurcation diagram. Variation of the moduli in a phenomenological model is not permitted, since it may lead to a degeneracy of the normal form.

Let us illustrate the concept of normal form of the Landau potential by discussing its applicability to some of the image-groups considered in Sec.3.3.

a) C_{3v}.

The more general form of the Landau potential for the C_{3v} image-group is:

$$F = F_0 + a_1 I_1 + a_2 I_1^2 + \ldots + b_1 I_2 + b_2 I_2^2 + \ldots + c_{12} I_1 I_2 + \ldots \qquad (1.208)$$

where I_1 and I_2 are given by Eq.(1.137). If the experiment allows varying four external conditions (e.g. temperature, pressure, and the concentration of two different admixtures), then by an appropriate choice of parameters it is possible to set to zero four coefficients of the terms of lowest degrees in (1.208), namely a_1, b_1, c_{12} and a_2. Consequently, one can show [14] that a miniversal deformation of a normal form describing a phse diagram with variation of four external conditions is:

$$F = F_0 + a_1 I_1 + b_1 I_2 + a_2 I_1^2 + c_{12} I_1 I_2 + I_2^2 + I_2^2 + I_1^3 + I_1 I_2^2 + a_4 I_1^4 \qquad (1.209)$$

where a_1, b_1, c_{12} and a_2 are *variable parameters*, and a_4 is a *modulus* which does not depend on the external conditions. When *two* external conditions are varied, one finds analogously thet it is possible to restrict to the sixth degree form:

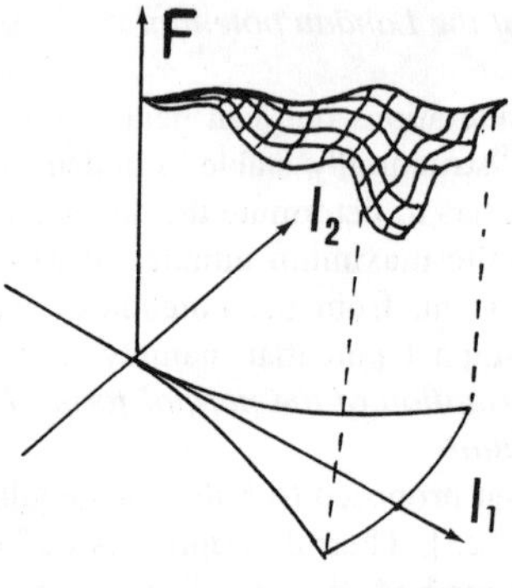

Fig. 52. Graph of the Landau potential corresponding to the image-group C_{3v}.

$$F = F_0 + a_1 I_1 + b_1 I_2 + I_1^2 + b_2 I_2^2 \qquad (1.210)$$

where a_1 and b_1 are variable parameters and b_2 is a modulus. It was already shown in Sec.3.3.3b that a degree lower than six in (1.208) corresponds to a *structurally unstable* model: In this case, any small change in the class of functions which are invariants by C_{3v}, results in a qualitative change in the result.

The phase diagram corresponding to (1.210) has been given in Fig.25a. In the terminology of the theory of singularities Figs.25b and 24 represent respectively the *manifold of states* and the *manifold of orbits* of the group C_{3v}. Fig.52 shows the *graph* of the Landau potential associated with the group C_{3v}.

b) C_4.

The image-group C_4 possesses three invariants in its integrity basis (see Sec.3.3.3a) and the fourth degree expansion (1.136) coincides with the deformation of the normal form of a singularity of co-dimension 1, where a_1 is a variable parameter and b_1 and c_1 are moduli. As stressed in Sec.3.3.3a variation of the moduli leads to unphysical results [33], and a miniversal deformation of a normal form requires not less than an eighth degree potential:

$$F = F_0 + a_1 I_1 + b_1 I_2 + c_1 I_3 + I_1^2 + b_2 I_2^2 \qquad (1.211)$$

where a_1, b_1 and c_1 are variable parameters, and b_2 is a modulus. Fig.22b shows the corresponding three-dimensional phase-diagram.

c) O_h.

The sixth degree expansion (1.169) has a form similar to the deformation of the normal form of co-dimension *two*:

$$F = F_0 + a_1 I_1 + b_1 I_2 + I_1^2 + I_3 + I_1 I_2 + a_3 I_1^3$$

where a_1 and b_1 are variable parameters, and a_3 is a modulus. One can thus construct phase diagrams of at the utmost *two* dimensions. An attempt to vary the parameter d_1 in (1.169) would lead to structurally unstable results. The phase diagram of Fig.34a which corresponds to three variable parameters, or equivalently three varying external conditions (e.g. temperature, pressure, and concentration) requires at least an eighth degree expansion, that allows to have a normal form of the singularity with a co-dimension 3.

In summary, we have seen that the determination of a structurally stable form of thermodynamic potential implies a number of constraints: The degree to which the potential can be truncated, the leading terms that must be kept in the expansion, and the constancy that one can assume for some of the coefficients, depend not only on the dimension of the order-parameter (the "co-rank of the singularity") and on its symmetry (which determines the integrity basis), but also on the number of varying external variables assumed in the experiment. Other obvious constraints play also a role in the form of the bifurcation diagram and on the limitation of the class of functions to which the Landau potential belong, such as the real character of the order-parameter, or the global minimality of the Landau potential at finite values of the order-parameter. This latter condition, which is imposed by the thermodynamic stability, requires in particular that all functions which belong to an orbit of Diff ε must be globally minimal. In the next section the importance of another requirement is stressed, which concerns the "locality" of the observable singularities.

5.3.3. Local nature of the observable singularities

In contrast to the definition of the Landau potential given in Sec.2.3.3 as a converging Taylor expansion of a small order-parameter, from the point of view of the theory of singularities there is no limitation on the magnitude of the variables on which depend the miniversal deformation of the normal form of singularity. Along the same line, using the mathematical apparatus of the theory, smooth non-linear interchanges of the variables may transform a converging series into a polynomial of finite degree.

By contrast, the theory of singularities imposes a limit on *the localization of the singularities*. Thus, when varying the external conditions, the Landau potential must allow a merging of the observable singularities in a single point. For systems presenting multisingularities, the requirement of *locality* must be satisfied near each of the isolated singularities. This property allows for example to understand why unimportant terms of the potential can be neglected for the description of an isolated second-order phase transition corresponding to the canonical form:

$$F = F_0 + \alpha\eta^2 + \beta\eta^4. \tag{1.212}$$

This expansion actually coincides with the normal form having a single variable parameter α, and a modulus β. Therefore it is sufficient for a *local* description of a second-order transition.

5.4. Restructuring of phase diagrams. Topological metamorphosis

5.4.1. External variables and observable singularities

The type of singularities exhibited by a system depends on the *internal* nature of the system, namely on its symmetry. However, as stressed in Sec.5.3.2 the possibility of describing a singularity by means of a Landau potential is also determined by the external conditions which fix the number of variable parameters in the potential. Which singularities of the Landau potential associated with a given system can be observed experimentally when changing the number of variable parameters (external conditions)? The answer to this question depends on the respective values of the co-dimension c of the singularity (s), and of the number l of variable parameters (α_i). Three qualitatively different situations can be distinguished.

1) l<c: At some point of the space of the variable parameters α_i one will find singularities of co-dimension l, entering into the decomposition of the singularity s. Thus, a l-dimensional phase diagram will contain in some l-dimensional cross-section the complete bifurcation diagram of the singularity s.

2) l=c: At some point of the space of α_i, one should find the singularity s. The phase diagram contains this singularity only in the complete bifurcation diagram.

3) l>c: The singularity s will be seen not at a single point, but on some (l-c)-dimensional manifold. The bifurcation diagram will be in this case the result of the evolution of the bifurcation diagram of the singularity s along the manifold. The observation of s requires to construct a phase diagram with a number of external parameters not less than c.

Another question arises in the cases with l<c or l>c, namely: what sorts of cross-sections of the bifurcation diagrams give the best account of the singularity s? The answer to this question is contained in the theory of the Restructuring of Phase Diagrams, developed in Refs.[111, 114], which will be briefly summarized.

5.4.2. Theory of the restructuring of phase diagrams

Let us first define some necessary terms of the theory of singularities, considering the canonical form of the Landau potential (1.212) in which we fix the value of the modulus β=1. Fig.53 represents the general form of the family of curves:

$$F(\eta) = \alpha\eta^2 + \eta^4 = \lambda \qquad (1.213)$$

with $\alpha^2/4 \le \lambda \le 0$. In the *non-degenerate* case $F(\eta)$ intersects the η axis at four points corresponding to the values $(\eta_1, \eta_2, \eta_3, \eta_4)$ of η. In the *degenerate* case, corresponding to the limits $\lambda = -\alpha^2/4$ and $\lambda = 0$, Eq.(1.213) has respectively 2 and 3 solutions among which one finds the roots of the equation of state:

$$\alpha\eta + 2\eta^3 = 0 \qquad (2.214)$$

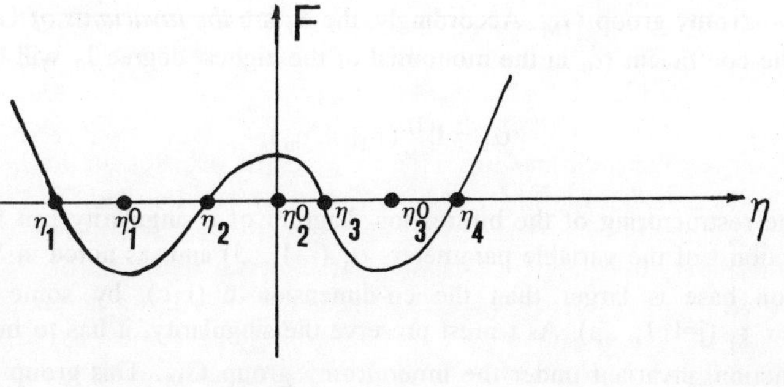

Fig. 53. Form of the curves (1.213) for $-\alpha^2/4 \le \lambda \le 0$. The quantities η_1-η_2, η_2-η_3, and η_3-η_4 are non-zero for a non-singular (non-degenerate) level of the Landau potential F(η), and go to zero for a singular (degenerate) level, at which $\eta_1=\eta_2=\eta_1^0$ *and* $\eta_3=\eta_4=\eta_3^0$ *or* $\eta_2=\eta_3=\eta_2^0$.

which are, by increasing order along the η axis (see Fig.53):

$$\eta_1^0 = -\left(-\frac{\alpha}{2}\right)^{1/2}, \quad \eta_2^0 = 0, \quad \eta_3^0 = \left(-\frac{\alpha}{2}\right)^{1/2}. \qquad (1.215)$$

The positions of the η_i (i=1-4) with respect to the η_j^0 (j-3) are defined by the *monodromy group* G_M *of the singularity* [14] which is the group of permutations g_m exchanging the η_i located on each side of *the same* η_j^0. For example: $g_M^1(\eta_1,\eta_2,\eta_3,\eta_4)\to(\eta_1,\eta_3,\eta_2,\eta_4)$ or $g_M^2(\eta_1,\eta_2,\eta_3,\eta_4)\to(\eta_2,\eta_1,\eta_4,\eta_3)$ etc.

Generalizing the preceding example, one can say that the non-singular levels of a Landau potential with a completely decomposed singularity, are described by the monodromy group G_M of the singularity. The group G_M is a generalization of the permutation group of the roots of the polynomials in one variable (the Galois group):

$$x^n + a_1x^{n-1} + a_2x^{n-2} + ... + a_n = 0 \qquad (1.216)$$

the coefficients of which can be written at the center of gravity of the roots $a_1 = 0$ (Vieta theorem):

$$a_1 = -\sum_{i=1}^{n} x_i, \quad a_2 = \sum_{i\neq j}^{n} x_i x_j, \quad ..., \quad a_n = \prod_{i=1}^{n} x_i. \qquad (1.217)$$

It is clear from (1.217) that the a_i are invariant with respect to permutations of the x_i. The same is true of the coefficient of the Landau potential (α_i with i=1,...,l) which are invariant with respect to permutations of the solutions of the equations of state, under the operations of the monodromy group G_M. Accordingly, the α_i are *the invariants of* G_M and, as a_2 in (1.217), the coefficient α_0 in the monomial of the highest degree l_0 will be the quadratic invariant:

$$\alpha_0 = I^{(2)}_{G_M}(\eta_1, \dots, \eta_n). \tag{1.218}$$

The restructuring of the bifurcation diagram of a singularity can be described by some function t of the variable parameters α_i (i=1,...,l) and, as noted in Sec.5.3.1, if the deformation base is larger than the co-dimension c (l>c), by some *extra variable parameters* τ_j (j=l+1,...,p). As t must preserve the singularity, it has to be a non-singular Morse function, invariant under the monodromy group G_M. This group leaves invariant the α_i as well as the τ_j but operates in a different way on the two sets of variables. Consequently the non-degenerate quadratic form is the sum of two terms:

$$t(\alpha_i, \tau_j) = \alpha_0 + \sum_{j=l+1}^{p} \sigma_j \tau_j^2 \quad \text{with } \sigma_j = \pm 1. \tag{1.219}$$

Therefore, one can come back on the three cases considered in the preceding section. If l<c, in order to obtain a significant cross-section of the bifurcation diagram of the singularity s, it is necessary to project it (i.e. to "reduce its transversality") in a (c-l)-dimensional space of the coefficients α_i. This procedure insures the structural stability of the obtained phase diagram.

If l=c, there are no "extra" variable parameters τ_j and $t=\alpha_0$. The bifurcation diagrams and the phase diagrams are reconstructed with t equal to a constant.

If l>c, in a c-dimensional cross-section one can find points with singularity s. Cross-section of dimension larger then c, e.g. c+1, c+2,...,c+l-1, will contain respectively 1, 2,..., (l-1)-dimensional spheres, hyperboloids, paraboloids,... filled with points with singularity s.

5.4.3. Topological metamorphosis

We will illustrate the preceding results by considering some concrete examples of restructuring of phase diagrams.

a) Reconstructing of the phase diagrams associated with the image group C_{3v}.

Let us consider again the image group C_{3v}, the phase diagrams of which have been discussed in Sec.3.3.3b. The sixth-degree expansion (1.141) can be written fixing the value of the moduli $a_2=b_2=1$:

$$F(I_1, I_2) = a_1 I_1 + b_1 I_2 + I_1^2 + I_2^2. \tag{1.220}$$

It corresponds to the phase diagram of Fig.25a the singularity of which (transition lines) are reproduced in Fig.54a. Changing the values of the moduli to $a_2=-1$, $b_2=1$ give other

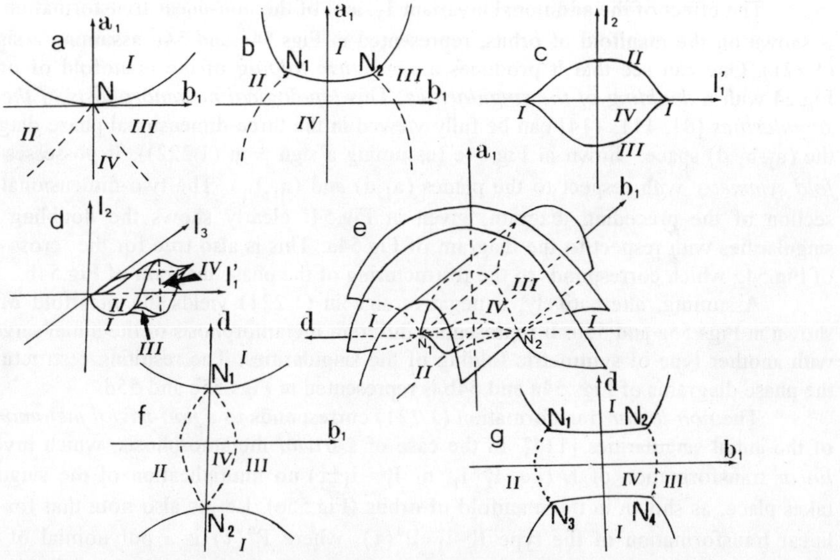

Fig. 54. Restructuring of the phase diagrams associated with the C_{3v} image-group assuming $t = a_1 + \tau^2$. (a) and (b): Initial C_{3v} phase diagrams corresponding to the Landau potential (1.222) for respectively $+I_1'^2$ and $-I_1'^2$. (c) and (d): Two and three-dimensional manifold of orbits after the restructuring process. (e) Three-dimensional restructured phase diagram in the external-variable space (a_1, b_1, d). (f) and (g): Cross-sections of the phase diagram corresponding to a restructuring of the phase diagrams shown in (a) and (b).

singularities (Fig.26c) which are represented schematically in Fig.54b. The co-dimension of the Landau potential (1.220) is c=2 and is thus adapted to the variation of two external conditions (e.g. pressure and temperature) on which a_1 and b_1 depend linearly. Assuming that a third external variable influences the experimental behaviour of the system described by the landau potential (1.220), will give a dimension l=3 to the deformation base, and requires to introduce an extra variable τ for obtaining a structurally stable phase diagram. Eq.(1.219) takes the explicit form:

$$t = a_1 \pm \tau^2. \tag{1.221}$$

Equivalently, one can say that the extravariable τ induces a non-linear transformation of the invariant $I_1 = \eta_1^2 + \eta_2^2$, which becomes $I_1' = \eta_1^2 + \eta_2^2 \pm \tau^2$, the invariant I_2 being unaffected by the transformation (1.221). By contrast an additional linear invariant $I_3 = \tau$ has to be taken in the order-parameter expansion which becomes (with $a_2 = \pm 1$):

$$F'(I_1', I_2, I_3) = a_1 I_1' + b_1 I_2 \pm I_1'^2 + I_2^2 + d\tau. \tag{1.222}$$

The effect of the additional invariant I_3, and of the non-linear transformation $I_1 \rightarrow I_1'$ is shown on the manifold of orbits, represented in Figs.54c and 54d assuming a sign + in (1.221). One can see that it produces a *symmetric folding* of the manofold of orbits of Fig.24 with a *doubling of the singularities*. This *topological metamorphosis of the initial singularities* [81, 111, 114] can be fully viewed in the three-dimensional phase diagram in the (a_1, b_1, d) space, shown in Fig.54e (assuming a sign + in (1.222)). It possesses a *two-fold symmetry* with respect to the planes (a_1,d) and (a_1,b_1). The two-dimensional cross-section of the preceding diagram, given in Fig.54f clearly shows the doubling of the singularities with respect to the diagram of Fig.54a. This is also true for the cross-section of Fig.54g which corresponds to the restructuring of the phase diagram of Fig.54b.

Assuming, alternatively, a negative sign in (1.221) yields the manifold of orbits shown in Figs.55a and 55b. It represents a different metamorphosis of the initial singularity, with another type of symmetric folding of the singularities. The resulting restructuring of the phase diagrams of Figs.54a and 54b is represented in Figs.55c and 55d.

The *non-linear* transformation (1.221) corresponds to a *non-trivial metamorphosis* of the initial singularities [114]. In the case of a *trivial* metamorphosis, which involves a *linear* transformation of I_1 (i.e. $I_1'=I_1$, or $I_1' = I_1 \pm \tau$) no multiplication of the singularities takes place, as shown in the manifold of orbits (Fig.55e). Let us also note that for a non-linear transformation of the type $I_1'=I_1+P^n(\tau)$, where $P^n(\tau)$ is a polynomial of degree n=3, 4, ... one should observe respectively a triple, quadrupling,... of the singularities which is however very improbable from the point of view of the theory of singularities. The topological metamorphosis illustrated by Fig.54 was shown to occur in a liquid of biaxial micellar aggregates [81, 124] where the N-phase point and the high-symmetry (isotropic) phase *reappear* when the temperature when the temperature is lowered, in a temperature-concentration phase diagram. The extra-variable τ in this experimental example, reflects a variation of the shape of the micellar aggregates.

b) Restructuring of phase diagrams associated with the image-group C_1.

Another experimental example of topological metamorphosis of singularity has been discussed by Kut'in et al. [14] and Lorman [33] for the solid solution $Sm_{1-x}R_xS$ in which the variations of temperature, pressure, and concentration of rare earth impurities R_x (R=La, Gd, Y) induce the doubling of a critical point of isostructural transition [87]. The Landau potential corresponds here to the trivial model of isostructural transition associated with the identity representation (see Sec.4.5) having the image-group C_1:

$$F(\eta) = a_1\eta + a_2\eta^2 + a_3\eta^3 + \eta^4 \tag{1.223}$$

which is transformed into the normal form:

$$F'(\xi) = \alpha_1\xi + \alpha_2\xi^2 + \xi^4 \tag{1.224}$$

via the change in variables: $\xi = \eta + a_3/4$, $\alpha_1 = a_1 - (a_3a_2)/2 + a_3^3/8$, $\alpha_2 = a_2 - 3a_2^2/8$. It corresponds to the phase diagram shown in Fig.49a. The co-dimension of the normal form

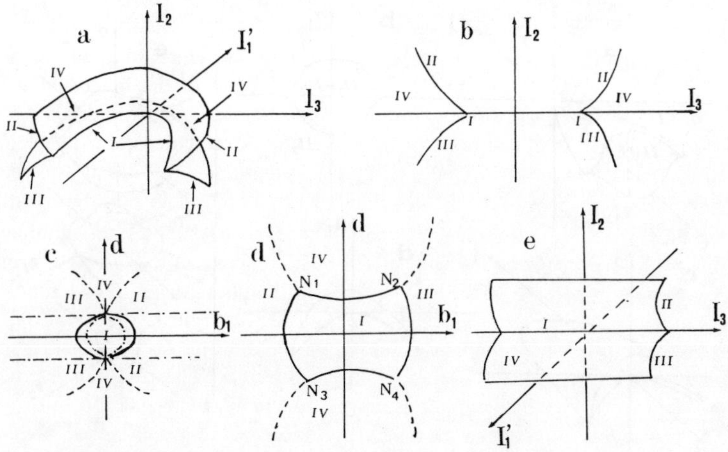

Fig. 55. Restructuring of the C_{3v} phase diagrams assuming $t = a_1 - \tau^2$. (a) and (b): 3-dimensional and 2-dimensional views of the manifold of orbits. (c) and (d): Cross-sections showing the restructuring of the phase diagrams of Figs.54a and 54b respectively. (e) Manifold of orbit for a trivial metamorphosis of the singularity corresponding to the Landau potential (1.220): $I_1' = I_1$.

of the Landau potential (1.224) is thus c=2, whereas the number of external variables is l=3. Introducing the extra variable τ, Eq.(1.219) is expressed by:

$$t = \alpha_2 - \tau^2. \qquad (1.225)$$

From Eq.(1.201) one can deduce the limit of stability lines:

$$27\alpha_1^2 + 8(t + \tau^2) = 0, \text{ and } \alpha_1 = 0$$

which yield the three-dimensional phase diagram shown in Fig.56a, in the space (α_1, τ, t). It represents a restructuring of the phase diagram of Fig.49a, with a topological metamorphosis of the singular point located at the origin (α_1=α_2=0). The two-dimensional cross-sections (α_1, τ) and (t, τ) (Fig.56b) show that the preceding singular point is splitted symmetrically with respect to the α_1 axis. Fig.56c gives the phase diagram corresponding to the "deformed" Landau potential form (1.223), in the space (a_1, τ, t) after the non-linear transformation $t = a_2 - \tau^2$. One can verify that the effect of the cubic invariant is to give an asymmetrical shape to the hypersurface of Fig.56a. The moon-quarter shape of the surface represented in the (a_1, τ) cross-section (Fig.56d), which terminates by two symmetric singular points, resembles closely to the surface limited by the two isostructural transition lines in $Sm_{1-x}R_xS$ [87].

Lorman has also considered the case where the concentration of two admixtures (x_1, x_2) are varied in addition to temperature and pressure [33]. In this case the number of

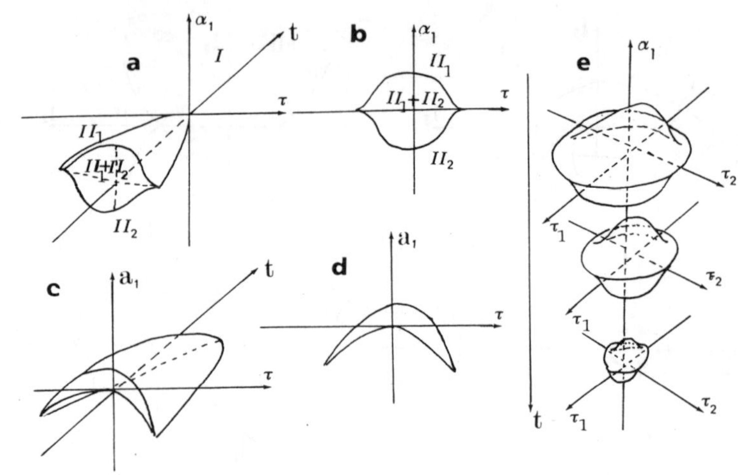

Fig. 56. Restructuring of the phase diagrams associated with the image-group C_1 (Fig.49a) assuming one extra variable τ (a-b-c-d) or two extra variables τ_1, τ_2 (e). (a) and (b): Phase diagram (a) and cross-section (b) corresponding to the potential (1.224). (c) and (d): Phase diagram (c) and cross-section (d) corresponding to the potential (1.223). (e) Phase diagram associated with the potential (1.223) assuming the non-linear transformation (1.226).

variable parameters is l=4, and one has to introduce two extra variables τ_1 and τ_2 in the restructuring procedure. Eq.(1.225) becomes:

$$t = \alpha_2 - \tau_1^2 - \tau_2^2 . \tag{1.226}$$

The corresponding phase digram is shown (Fig.56e) in the three-dimensional space $(\alpha_1, \tau_1, \tau_2)$ as well as the influence of the t-parameter.

c) Restructuring of the phase diagrams associated with the image-group C_i.

As a last example, let us discuss the restructuring of the phase diagrams associated with the canonical fourth and sixth-degree forms of the Landau potential, which transform as the image-group C_i. In the fourth-degree form (with the modulus β=1):

$$F(\eta) = \alpha\eta + \eta^4 \tag{1.227}$$

the co-dimension of the singularity is 1, and accordingly only a one-dimensional phase diagram (as in Fig.3) can be drawn. If two external variables are assumed, a two-dimensional phase diagram will require a non-linear transformation: $t = \alpha \pm \tau^2$, where τ is an extra-variable which depends linearly, as α, on the external variables. $F(\eta)$ becomes:

$$F(\eta) = (t \pm \tau^2)\eta^2 + \eta^4 . \tag{1.228}$$

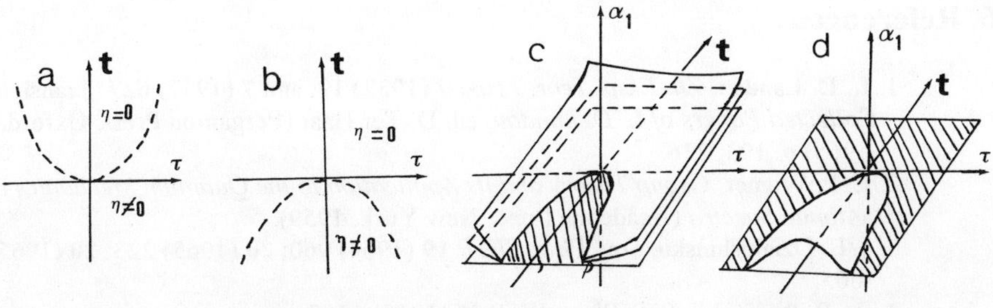

Fig. 57. (a) and (b): Restructuring of the phase diagram associated with the canonical Landau potential (1.227) assuming the non-linear transformations (a): $t = \alpha + \tau^2$, (b) $t = \alpha - \tau^2$. Restructuring of the phase diagram associated with the Landau potential (1.229) assuming the non-linear transformations: (c) $t = \alpha_2 + \tau^2$, (d) $t = \alpha_2 - \tau^2$. In (c) and (d) hatched planes are second-order transition surfaces, whereas first-order transition surfaces are curved. The planar parabola is a tricritical curve.

Minimization of $F(\eta)$ yields the phase diagrams shown in Figs.57a and 57b in the plane of the variables (t, τ). The topological metamorphosis consists here in a symmetric folding of the line of second-order transitions ($t=\pm\tau^2$) with respect to the t axis.

The sixth degree normal form of the Landau potential:

$$F(\eta) = \alpha_1\eta^2 + \alpha_2\eta^4 + \eta^6 \qquad (1.229)$$

possesses a singularity of co-dimension 2. One can thus construct a two-dimensional phase diagram with two varying external parameters as shown in Fig.12. If a third external variable is assumed, the introduction of an extra variable τ will transform nonlinearly the *fourth-degree coefficient* α_2:

$$t = \alpha_2 \pm \tau^2 \, . \qquad (1.230)$$

Figs.57c and 57d show the corresponding restructuring of the phase diagram of Fig.12, in the space (a_1, t, τ) for a sign + and - in (1.230) respectively. The topological metamorphosis of the singularities, transforms the tricritical point T_{tc} of Fig.12 in the tricritical parabola $t = \mp\tau^2$ whereas the lines of second and first-order transitions in Fig.12 become surfaces, possessing a symmetrical configuration with respect to the tricritical parabola.

6. References

1. L. D. Landau, *Zh. Eksp. Teor. Fiziki* **7** (1937) 19; and **7** (1937) 627. Transl. in *Collected Papers of L. D. Landau*, ed. D. Ter Haar (Pergamon Press, Oxford, 1965) p. 193; 216.
2. E. P. Wigner, *Group Theory and its Application to the Quantum Mechanics of Atomic Spectra* (Academic Press, New York, 1959).
3. I. E. Dzialoshinskii, *Sov. Phys. JETP* **19** (1964) 960; **20** (1965) 223; **20** (1965) 665.
4. L. P. Pitaevskii, *Sov. Phys. JETP* **37** (1960) 1267.
5. V. L. Indenbom, S. A. Pikin and E. B. Loginov, *Sov. Phys. Crystallogr.* **21** (1976) 632.
6. O. V. Kovalev, *Irreducible Representations of the Space Groups* (Gordon and Breach, New York, 1965).
7. J. Zak, A. Casher, H. Glück and Y. Gur, *The Irreducible Representations of Space Groups* (Benjamin, New York, 1969).
8. S. C. Miller and N. V. Love, *Tables of Irreducible Representations of Space Groups and Magnetic Corepresentations of Magnetic Space Groups* (Pruett, Boulder, 1964).
9. (a) P. Toledano and J.C.Toledano, *Phys. Rev.* **B14** (1976) 3097.
   (b) P. Toledano and J.C.Toledano, *Phys. Rev.* **B16** (1977) 386.
   (c) J. C. Toledano and P.Toledano, *Phys. Rev.* **B21** (1980) 1139.
   (d) P. Toledano and J.C.Toledano, *Phys. Rev.* **B25** (1982) 1946.
10. (a) Yu. M. Gufan, *Sov. Phys. Solid State* **13** (1971) 175.
    (b) Yu. M. Gufan and V. P. Sakhnenko, *Sov. Phys. JETP* **36** (1973) 1009.
    (c) Yu. M. Gufan and V. P. Sakhnenko, *Sov. Phys. Solid State* **16** (1974) 1034.
    (d) E. B. Vinberg, Yu. M. Gufan, V. P. Sakhnenko and Yu. I. Sirotin, *Sov. Phys. Crystallogr.* **19** (1974) 101.
    (e) Yu. M. Gufan and G. M. Chechin, *Sov. Phys. Crystallogr.* **25** (1980) 261.
    (f) Yu. M. Gufan and V. P. Popov, *Sov. Phys. Crystallogr.* **25** (1981) 527.
    (g) Yu. M. Gufan, V. P. Dmitriev, V. P. Popov and G. M. Chechin, *Phys. Met. Metall.* **46** (1980) 6.
    (h) Yu. M. Gufan, V. P. Dmitriev, V. P. Popov and G. M. Chechin, *Sov. Phys. Solid State* **21** (1979) 327.
    (i) Yu. M. Gufan and V. P. Dmitriev, *Phys. Met. Metall.* **53** (1982) 28.
    (j) Yu. M. Gufan and V. P. Dmitriev, *Phys. Met. Metall.* **53** (1982) 16.
11. H. T. Stokes and D. M. Hatch, *Isotropy Subgroups of the 230 Crystallographic Space Groups* (World Scientific, Singapore, 1988).
12. Yu. M. Gufan, *Structural Phase Transitions* (Nauka, Moscow, 1982).
13. L. Michel, *Reviews of Modern Physics* **52** (1980) 617.
14. E. I. Kut'in, V. L. Lorman and S. V. Pavlov, *Sov. Phys. Usp.* **34** (1991) 497.

15. L. Michel, J. C. Toledano and P. Toledano, in *Symmetries and Broken Symmetries in Condensed Matter Physics*, ed. N. Boccara (IDSET, Paris, 1981), p. 263.
16. L. Michel, in *Group Theoretical Methods in Physics*, ed. R. Sharp and B. Kolman (Academic Press, New York, 1977), p. 75.
17. J. C. Toledano and P. Toledano, *The Landau Theory of Phase Transitions* (World Scientific, Singapore, 1987).
18. J. C. Toledano, L. Michel, P. Toledano and E. Brezin, *Phys. Rev.* **B31** (1985) 7171.
19. P. Toledano, *Thesis* (Univ. Amiens, 1979).
20. P. Toledano, *Jap. Jour. Appl. Phys.* **24** (1985) 121.
21. Yu. M. Gufan, G.M. Vereshkov, P. Toledano, B. Mettout, R. Bouzerar and V. Lorman, *Phys. Rev.* **B51** (1995) 9228.
22. J. L. Birman, *Phys. Rev. Lett.* **17** (1966) 1216.
23. L. Michel, in *Regards sur la Physique Contemporaine* (CNRS, Paris, 1980), p. 157.
24. (a) J. C. Toledano and P. Toledano, *J. de Physique (France)* **41** (1980) 189.
(b) D. Mukamel and M. V. Jaric, *Phys. Rev.* **B29** (1984) 1465.
(c) M. V. Jaric, *Phys. Rev. Lett.* **51** (1983) 2073.
25. L. Michel and J. Mozrzymas, *Lecture Notes in Physics* **79** (1978) 447.
26. M. V. Jaric, *Phys Rev. Lett.* **48** (1982) 164.
27. J. S. Kim, *Phys. Rev.* **B31** (1985) 1433.
28. L. D. Landau and I. M. Khalatnikov, *Dokl. Akad. Nauk SSSR* **96** (1954) 469. Transl. in *Collected Papers of L. D. Landau*, ed. D. Ter Haar (Pergamon Press, Oxford, 1965) p. 101.
30. Yu. M. Gufan and E. S. Larin, *Sov. Phys. Dokl.* **23** (1978) 754.
31. Yu. M. Gufan and E. S. Larin, *Izv. Akad. Nauk SSSR, Ser. Fiz.* **43** (1979) 14.
32. E. I. Kut'in, *Candidate Dissertation* (Univ. of Rostov on the Don, 1988).
33. V. L. Lorman, *Candidate Dissertation* (Univ. of Rostov on the Don, 1988).
34. S. Galam and D. M. Hatch, *Phys. Rev.* **B34** (1986) 7813.
35. Yu. M. Gufan and V. P. Sakhnenko, *Sov. Phys. JETP* **42** (1976) 728.
36. Yu. M. Gufan amd E. S. Larin, *Sov. Phys. Solid State* **20** (1978) 998.
37. Yu. M. Gufan , V. P. Dmitriev and A. N. Sadkov, *Izv. Akad. Nauk SSSR, Ser, Fiz.* **47** (1983) 692.
38. A. F. Devonshire, *Philos. Mag.* **40** (1949) 1040.
39. *Landolt-Börnstein, Ferro- and Antiferroelectric Substances*, Band 3 (Springer Verlag, Berlin, 1969).
40. S. Alexander, *Solid State Commun.* **14** (1974) 1069.
41. S. Alexander and D. J. Amit, *J. Phys.* **A8** (1975) 1988.
42. E. M. Lifshitz, *Zh. Eksp. Teor. Fiz.* **11** (1941) 255.
43. K. G. Wilson and M. E. Fisher, *Phys. Rev. Lett.* **28** (1972) 240.
44. S. A. Brasovskii and I. E. Dzialoshinskii, *Sov. Phys. JETP Lett.* **21** (1975) 164.
45. P. Bak, S. Krinsky and D. Mukamel, *Phys. Rev. Lett.* **36** (1976) 52.
46. A. L. Korzhenevskii, *Sov. Phys. JETP* **44** (1976) 751.

47. P. Toledano and G. Pascoli, *Ferroelectrics* **25** (1980) 427.
48. Ref.[17], p. 53.
49. E. Brown and L. V. Meisel, *Phys. Rev.* **B13** (1976) 213.
50. P. Toledano and G. Pascoli, in *Symmetries and Broken Symmetries in Condensed Matter Physics*, ed. N. Boccara (IDSET, Paris, 1981), p. 291
51. M. Kataoka and J. Kanomori, *J. Phys. Soc. Japan* **32** (1972) 113.
52. P. Toledano, M. M. Fejer and B. A. Auld, *Phys. Rev.* **B27** (1983) 5717.
53. V. P. Sakhnenko and V. M. Talanov, *Sov. Phys. Solid State* **21** (1980) 1401.
54. I. E. Dzialoshinskii, *Sov. Phys. JETP* **19** (1964) 960.
55. Y. Onodera and Y. Tanabe, *J. Phys. Soc. Japan* **45** (1978) 1111.
56. S. Goshen, D. Mukamel and S. Shtrikman, *Int. J. Magnetism* **6** (1974) 221.
57. A. P. Levanyuk and D. G. Sannikov, *Sov. Phys. Solid State* **18** (1976) 1122.
58. *Incommensurate Phases in Dielectrics*, eds. R. Blinc and A. P. Levanyuk (North Holland, Amsterdam, 1986).
59. Ref. [17], p. 215-306.
60. K. W. Browal, J. S. Kasper and H. Wiedemeier, *J. Solid State Chem.* **10** (1974) 20.
61. J. Jerphagnon, in the *Proceedings of Ternary Compaunds Conference* (Institute of Physics, London, 1976).
62. E. G. Stewart and H. P. Rooksby, *Acta Cryst.* **6** (1953) 49.
63. G. M. Wolton, *Acta Cryst.* **17** (1964) 763.
64. J. F. Dorian, R. E. Newnham and D. K. Smith, *Ferroelectrics* **3** (1971) 17.
65. S. C. Abrahams, J. L. Bernstein and J. P. Remeika, *Mater. Res. Bull.* **9** (1974) 1613.
66. Yu. M. Gufan and V. P. Dmitriev, *Sov. Phys. Crystallogr.* **25** (1980) 6.
67. Yu. M. Gufan and E. S. Larin, *Sov. Phys. Solid State* **22** (1980) 270.
68. V. P. Dmitriev, *Sov. Phys. Crystallogr.* **25** (1981) 683.
69. E. S. Larin, *Sov. Phys. Solid State* **26** (1984) 1820.
70. Yu. M. Gufan and V. I. Torgashev, *Sov. Phys. Solid State* **22** (1980) 951.
71. Yu. M. Gufan and V. I. Torgashev, *Sov. Phys. Solid State* **23** (1981) 1129.
72. E. I. Kut'in, V. L. Lorman and Ya. E. Cherner, *Sov. Phys. Solid State* **29** (1988) 1819.
73. E. M. Lifshitz, *Zh. Eksp. Teor. Fiz.* **14** (1944) 353.
74. J. Holakovsky, *Phys. Stat. Sol. (b)* **56** (1973) 615.
75. V. P. Dmitriev and S. B. Rochal, *J. Low Temp. Phys.* **14** (1988) 456.
76. S. B. Rochal, *Candidate Dissertation* (Univ. of Rostov on the Don, 1989).
77. Yu. M. Gufan, E. I. Kut'in, V. L. Lorman and E. N. Sidorenko, *Sov. Phys. Solid State* **29** (1987) 432.
78. I. M. Lifshitz, *Sov. Phys. JETP* **11** (1960) 30.
79. N. F. Mott and E. A. Davis, *Electronic Processes in Non-Crystalline Materials* (Claredon, Oxford, 1971).
80. Yu. M. Gufan and V. L. Lorman, *Sov. Phys. Solid State* **25** (1983) 198.
81. P. Toledano and A. M. Figueiredo Neto, *Phys. Rev. Lett.* **73** (1994) 2216.
82. A. K. Singh, *High Temp.-High Press.* **12** (1980) 47.

83. H. T. Hall, L. Merrill and J. D. Barnett, *Science* **146** (1964) 1297.
84. A. Chatterjee, A. K. Sing and A. Jayaraman, *Phys. Rev.* **B6** (1972) 2275.
85. A. A. Bugaev, B. P. Zakharchenya and F. A. Chudnovskii, *Metal-Insulator Phase Transition and Its Applications* (Nauka, Leningrad, 1979).
86. A. Jayaraman, *Phys. Rev.* **137** (1965 179.
87. I. L. Aptekar' and E. Yu. Tonkov, *Sov. Phys. Solid State* **21** (1979) 110.
88. Y. Kyno and S. Miyahara, *J. Phys. Soc. Japan* **21** (1966) 2737.
89. H. J. Levinstein, M. Robbins and C. Capio, *MRS Bull.* **7** (1972) 27.
90. Yu. M. Gufan and E. S. Larin, *Sov. Phys. Solid State* **29** (1987) 4.
91. V. L. Ginzburg, *Sov. Phys. Solid State* **2** (1960) 1824.
92. A. P. Levanyuk, *Sov. Phys. Solid State* **8** (1964) 1294.
93. K. G. Wilson and J. Kogut, *Phys. Rep.* **12C** (1974) 75.
94. I. E. Dzialoshinskii, *Sov. Phys. JETP* **45** (1977) 1014.
95. G. Grinstein and D. Mukamel, *J. Phys.* **A15** (1982) 233.
96. L. Michel and J. C. Toledano, *Phys. Rev. Lett.* **54** (1985) 1832.
97. D. Mukamel, *Phys. Rev. Lett.* **34** (1975) 481.
98. D. Mukamel and S. Krinsky, *Phys. Rev.* **B13** (1976) 5078.
99. P. Bak and D. Mukamel, *Phys. Rev.* **B13** (1976) 5086.
100. I. E. Dzialoshinskii, in *Symmetries and Broken Symmetries in Condensed Matter Physics*, ed. N. Boccara (IDSET, Paris, 1981), p. 29.
101. E. Brezin, J. C. Le Guillou and J. Zinn-Justin, *Phys. Rev.* **B10** (1974) 892.
102. E. Meimarakis and P. Toledano, *Jap. J. Appl. Phys.* **S24-2** (1985) 380.
103. D. M. Hatch and J. W. Felix, *Jap. J. Appl. Phys.* **S24-2** (1985) 176.
104. J. R. Brews, *Phys. Rev.* **B1** (1970) 2557.
105. J. M. Hastings, L. M. Corliss, W. Kunnmann, R. Thomas, R. J. Begum and P. Bak, *Phys. Rev.* **B22** (1980) 1327.
106. D. Mukamel and D. J. Wallace, *J. Phys.* **C13** (1979) L851.
107. P. Bak, S. Krinsky and D. Mukamel, *Phys. Rev. Lett.* **36** (1976) 52.
108. P. Toledano and G. Pascoli, *Ferroelectrics* **25** (1980) 427.
109. M. Ma and J. Solyom, *Phys. Rev.* **B21** (1980) 5262.
110. R. Thom, *Structural Stability and Morphogenesis* (Benjamin, Reading, Mass., 1975).
111. V. I. Arnol'd, *Commun. Pure Appl. Math.* 29 (1976) 557.
112. R. Gilmore, *J. Math. Phys.* **A18** (1977) 17.
113. I. N. Stewart, *Math. Proc. Cambridge Phylos. Soc.* **88** (1980) 429.
114. V. I. Arnol'd, A. N. Varchenko and S. M. Gusein-Zade, *Singularities of Differentiable Maps*, Vol. 1 and 2 (Burkhaüser, Boston, 1985 and 1988).
115. T. Poston and I. N. Stewart, *Catastrophe Theory and Its Applications* (Pitman, London, 1978).
116. R. Gilmore, *Catastrophe Theory for Scientists and Engineers* (Wiley, N. Y., 1981).
117. J. M. T. Tompson, *Instabilities and Catastrophes in Science and Engineering* (Wiley, N. Y., 1982).
118. D. H. Fowler, *Forwars Theor. Biol.* **4** (1972) 1.

119. K. Okada, *Catastrophe Theory and Phase Transitions* (Trans Tech Publ. Zürich, 1993).
120. M. C. M. Vendrik, *Physica* **A99** (1979) 103 and 513.
121. V. I. Arnol'd, V. A. Vasil'ev, V. V. Goryunov and O. V. Lyashko, *Singularities II: Classification and Applications*, in *Modern Problems in Mathematics. Fundamental Directions*, Vol. 39 (VINITI, Moscow, 1989).
122. A. A. Andronov and I. S. Pontryagin, *Dokl. Akad. Nauk SSSR* **14** (1937) 247.
123. M. Golubitsky and V. Guillemin, *Stable Maps and Their Singularities* (Springer-Verlag, N. Y., 1974).
124. P. Toledano, A. M. Figueiredo Neto, V. Lorman, B. Mettout and V. Dmitriev, *Phys. Rev.* **E52** (1995) 5040.

CHAPTER II

DENSITY-WAVE THEORY OF RECONSTRUCTIVE PHASE TRANSITIONS

1. Introduction

In the preceding chapter (Sec.2.3.3), we have seen that the formulation of the Landau theory of structural transitions uses an expansion of the increment $\delta\rho(\vec{r})$ of the probability density, expressing the difference between the initial density in the high-symmetry phase $(\rho_0(\vec{r}))$ and the final low-symmetry phase $(\rho(\vec{r}))$, expanded as function of the basic functions of a representation $\Gamma_{k\gamma}$ of the space-group G_0 of the initial phase. This expansion has the form:

$$\delta\rho(\vec{r}) = \rho(\vec{r}) - \rho_0(\vec{r}) = \sum_{k\gamma i} \eta^i_{k\gamma} \varphi^i_{k\gamma}(\vec{r}). \qquad (2.1)$$

The wave-vector $\vec{k}$, located in the first Brillouin zone, characterizes the translational symmetry of the basic functions $\varphi^i_{k\gamma}$, which are linear combinations of atomic functions associated with the crystalline structure. The index γ labels the representations $\Gamma_{k\gamma}$ of G_0, and the index i runs over the distinct functions spanning the representation space ε. For a given γ, the set of coefficients $\eta_{k\gamma}$ define the order-parameter, which describes the total distortion (or change in atomic distribution) of the initial structure at the transition. This distortion results from the contribution of each $\varphi^i_{k\gamma}$, and its magnitude is not a priori limited.

The linear coupling between the $\eta^i_{k\gamma}$ and $\varphi^i_{k\gamma}$ in (2.1) means that either of these two quantities can be chosen as forming the basis of the relevant (symmetry breaking) representation. As a consequence the non-equilibrium thermodynamic potential associated with the transition: $F(T, P, \delta\rho)$ can be considered as a function of the $\eta^i_{k\gamma}$ instead of the $\varphi^i_{k\gamma}$. The substitution in F of the *functions* $\varphi^i_{k\gamma}$, by the *scalar* order-parameter components $\eta^i_{k\gamma}$, makes the problem of the description of a transition, more general and easy to handle, as it does not require specific informations about the concrete structural modifications taking place at the transition. One needs only to know the symmetry of

$\Gamma_{k\gamma}$, or equivalently the corresponding space group change $G_0 \rightarrow G$, for a description of the essential features of the transition.

This is one of the main achievements of the Landau theory of phase transitions, as in such an abstract form, it allows to describe whole classes of transitions in a unified way, exclusively on the basis of symmetry considerations, and requires a minimal number of experimental informations about the transition. It leads also to establish universal behaviors depending exclusively on the symmetry of the order-parameter. *However, it is precisely this universal aspect which forbids the applicability of the Landau theory to transitions in which the group-subgroup relationship, between the structures of the phases, is lost.*

Another important limitation of the conventional formulation of the Landau theory of structural phase transitions for its application to group-subgroup unrelated structures, is *the space-group description of the structures.* A space-group is defined as a set of symmetry elements corresponding to the spatial transformations exchanging the coordinates of the sites in a given structure. The success and usefulness of the space-group concept, is that it allows to define classes of structures, without considering the specific configuration of atoms. By contrast, it forbids to account for *the transformation of a symmetry element into another*, or for *the onset of new symmetry elements*, which characterize reconstructive transformations, as it will be illustrated in the next section. A more adapted approach to describe the structures in the case of reconstructive transitions, is to consider them as *specific configuration of sites.* Such a representation of a structure leaves the possibility to transform a symmetry element into another (e.g. six fold axis of a structure transforms into the four fold axis of another structure across the transition), by a continuous deformation of the initial sites, and thus to expand this deformation in symmetric functions, *without imposing a group-subgroup relationship between the structures of the phases.*

In this chapter we will show that a phenomenological theory can be formulated for reconstructive phase transitions, by generalizing and modifying a number of fundamental aspects of the conventional Landau scheme [1-5]. In the view of constructing this theory, we will use the formalism developed for X-ray analysis, in which a structure is expressed in terms of the interference of an infinite number of density waves. This density-wave approach will be first introduced through simple working examples, corresponding to the main classes of reconstructive transitions.

2. Introductory examples to the density-wave approach

2.1. A first example of displacive reconstructive phase transition

2.1.1. Crystallographic description of the transition mechanism

Let us consider a crystal of hexagonal D_{6h}^4 symmetry in which only the two-fold positions 2(c):$(\frac{1}{3}\ \frac{2}{3}\ \frac{1}{4};\ \frac{2}{3}\ \frac{1}{3}\ \frac{3}{4})$ are occupied. We assume that the degree of freedom associated with the transition mechanism, is a set of antiparallel displacements of the atoms, denoted ξ_i with i=1, 2, from their initial 2(c) positions, along the direction **Z** of

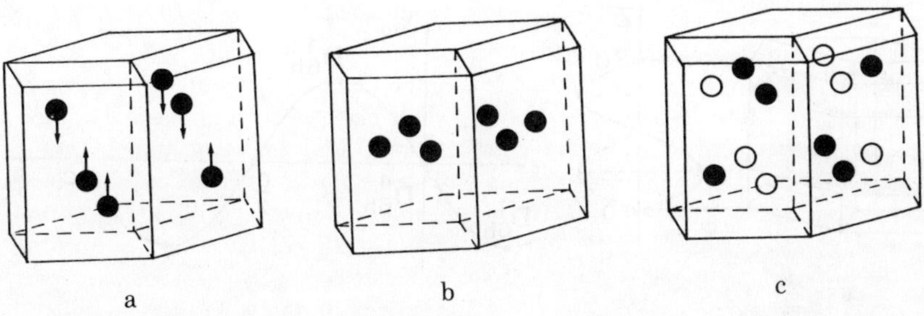

Fig.1. Antiparallel shifting mechanism from the D_{6h}^4 structure (a) which leads to D_{6h}^1 (b). Fig.(c) represents the two orientational domains of the D_{6h}^4 structure. Full and open circles pertain to two different domains.

the 6-fold axis (Fig.1a). For small arbitrary displacements of equal magnitudes $\xi_1 = -\xi_2 = \xi$, the displaced atoms do not lie in a single plane, so that the screw axes are lost, and the six-fold symmetry of the crystal is lowered to the rhombohedral symmorphic space group D_{3d}^3, which is a subgroup of D_{6h}^4. If ξ increases, the rhombohedral symmetry is preserved until ξ reaches the specific value $\xi_0=\frac{c}{4}$, where *c* is the hexagonal lattice parameter. For this value, the z coordinate of the two atoms becomes identical (z=0, or $z=\frac{1}{2}$) and, as they lie in the same plane, the symmetry of the crystal *increases* to the simple hexagonal structure of symmetry D_{6h}^1, as illustrated in Fig.1b. This hexagonal structure possesses 2 atoms per unit cell, and is equitranslational to the initial hexagonal structure. Accordingly D_{6h}^1 *is not a subgroup of* D_{6h}^4, i.e. *the group-subgroup relationship has been lost in the preceding displacive mechanism for the specific displacement of magnitude* ξ_0.

Let us note that when displacing the atoms in opposite direction (Fig.1c), one gets the same sequence of structures: rhombohedral D_{3d}^3 for $|\xi| < \frac{c}{4}$, and hexagonal D_{6h}^1 for $|\xi| = \frac{c}{4}$, but the corresponding crystal lattice is turned by 60° around the six-fold axis, with respect to the lattice obtained for opposite displacements. In other words, one gets *differently oriented domains* of the rhombohedral and simple hexagonal structures. Along the same line, for antiparallel displacements of magnitude $|\xi| = \frac{c}{2}$, the initial symmetry D_{6h}^4 is restored, but the corresponding lattice is rotated by 60° with respect to the original hexagonal lattice (Fig.1c).

2.1.2. Density-wave description of the phase transition

Let us express the distribution of atoms in the considered structures by a *lattice function* $\psi(\vec{r})$, defined as the sum of the first harmonics of a Fourier series:

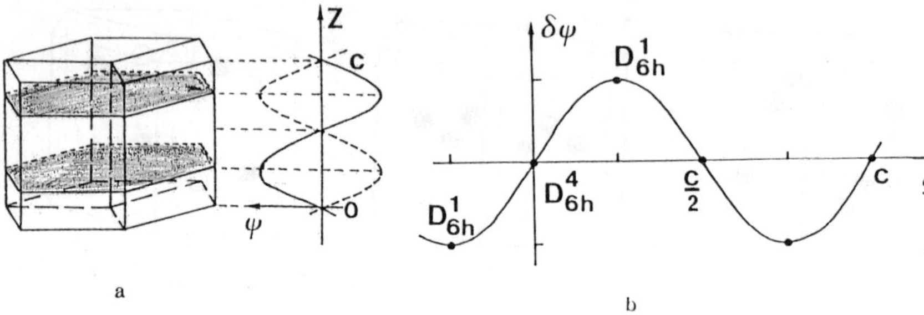

Fig.2. (a) Connection between density waves and atomic planes for the D_{6h}^4 structure. (b) Periodic dependence of the lattice function increment $\delta\psi(\xi)$.

$$\psi(\vec{r}) = \frac{1}{N}\sum_{j=1}^{N} A_j e^{i\vec{k}(\vec{r}-\vec{R}_j)} + \text{c.c.} \tag{2.2}$$

where $\vec{r}$ is the position vector, N is the number of atoms in the unit cell, the A_j are normalized amplitudes, $\vec{k}$ is the wave-vector associated with the first harmonics and $\vec{R}_j$ denote the position of the j-th atom. As shown in Fig.2a, Eq.(2.2) can be interpreted as a sum of *stationary density waves*, which propagate in the direction $\vec{k}$ perpendicular to the atomic planes forming the structure, i.e. the maxima of the waves coincide with the average atomic positions.

In the initial (D_{6h}^4) structure, one has only two atoms per unit-cell, located in planes orthogonal to the Z axis at the coordinates z_0^1 and z_0^2=-z_0^1 , which are distant by $\frac{c}{2}$. The corresponding lattice function can thus be written in the direction Z of the displacements, (neglecting the “noncritical” density waves along X and Y) as:

$$\psi_0(z) = \frac{A}{2}\left[\cos c^*(z_0^1 - z) + \cos c^*(z_0^2 - z)\right] \tag{2.3}$$

where $A=A_1=A_2$, and k=c* is the primitive reciprocal lattice translation along Z. When shifting the atoms by the antiparallel displacements of magnitude $\xi_1 = -\xi_2$, the lattice function becomes:

$$\psi(z) = \frac{A}{2}\left[\cos c^*(z_0^1 + \xi_1 - z) + \cos c^*(z_0^2 + \xi_2 - z)\right]. \tag{2.4}$$

When $z_0^1 = c/4 = -z_0^2$ and $\xi_2 = -\xi_1 = \xi$, the increment $\delta\psi(z,\xi) = \psi_0(z,\xi) - \psi(z,\xi)$ takes the form:

$$\delta\psi(z,\xi) = \frac{A}{2}\left[\cos c^*\left(\frac{c}{4} + \xi - z\right) + \cos c^*\left(-\frac{c}{4} - \xi - z\right) - \cos c^*\left(-\frac{c}{4} - z\right) - \cos c^*\left(\frac{c}{4} - z\right)\right]. \quad (2.5)$$

Since $\frac{c^*c}{4} = \frac{\pi}{2}$ one gets:

$$\delta\psi(z,\xi) = -2A \cdot \sin(c^*\xi) \cdot \cos(c^*z). \quad (2.6)$$

Whereas for unequal magnitude of the shifts $\xi_1 \neq -\xi_2$ one has:

$$\delta\psi(z,\xi_1,\xi_2) = -A\left[\sin(c^*\xi_1) - \sin(c^*\xi_2)\right]\cos(c^*z). \quad (2.7)$$

Eqs.(2.6) and (2.7) express *the periodic deformation of the initial hexagonal lattice as a function of the magnitude of the displacements*. One can see in Fig.2b following Eq.(2.6) that the D_{6h}^4 structure is obtained for $\xi = 0, \frac{c}{2}, c, \frac{3c}{2}, \ldots$, whereas the D_{6h}^1 structure corresponds to the successive displacements $\xi = \frac{c}{4}, \frac{3c}{4}, \frac{5c}{4}, \ldots$. For all other displacements, the structure of the crystal is lowered to the rhombohedral D_{3d}^3 structure, which correspond to the maximal substructure common to the D_{6h}^4 and D_{6h}^1 structures. Let us note, that for the hexagonal structures, two successive displacements (e.g. $\xi = (0, \frac{c}{2})$ for D_{6h}^4, and $\xi = (\frac{c}{4}, \frac{3c}{4})$ for D_{6h}^1) coincide with the *pairs of differently oriented domains* mentioned in Sec.2.1.1. Let us also stress that Eq.(2.6) holds for *infinite* displacements.

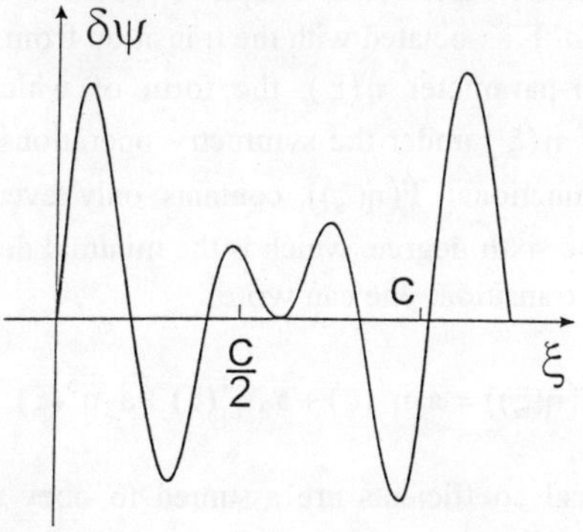

Fig.3. Dependence of $\delta\psi(\xi)$ as given by Eq.(2.7) for two unequal antiparallel shifts ξ_1 and ξ_2.

For unequal magnitudes of the displacements $\xi_1 \neq \xi_2$, one should use Eq.(2.7), which corresponds to a family of aperiodic curves determined by the respective values of ξ_1 and ξ_2. Fig.3 shows the curve corresponding to $\xi_1/\xi_2 = \sqrt{2}$, in which the initial hexagonal symmetry is lowered to C^1_{3v} for general displacements. At the extrema of the curve, the crystal structure increases to D^1_{6h}, whereas the initial symmetry is recovered for the points of intersection of the curve with the axis of displacements. At last, let us mention that the situation in which the displacements are not parallel to the Z direction, requires a more elaborated approach, but will not lead, in general, to a periodic dependence of the structural deformation as a function of displacements.

2.1.3. Thermodynamic potential and phase diagram

In the preceding density-wave approach to the D^4_{6h}-D^1_{6h} reconstructive transition, the *order-parameter* corresponding to the transition mechanism (the deformation of the structure) is expressed in function of the atomic displacements, which constitute the *variational parameters*. Eq.(2.6) provides the explicit dependence of the structural deformation, in function of the displacements of magnitude ξ. Hence one can define the order-parameter η as the variational part of $\delta\psi(z,\xi)$ namely:

$$\eta(\xi) = \eta_0 \cdot \mathrm{Sin}(c * \xi) \tag{2.8}$$

where η_0 is a normalizing constant. η fulfills the standard conditions imposed to the order-parameter in the Landau theory, namely η=0 for the initial phase, and $\eta \neq 0$ for the rhombohedral phase as well as for the "limit" hexagonal phase (D^1_{6h}) which satisfies the additional conditions to coincide with extrema of $\eta(\xi)$: $\partial\eta/\partial\xi$=0.

Following the procedure described in Chapter I (Secs.2.2 and 2.3), one can express the *thermodynamic potential* F associated with the transition from the initial D^4_{6h} phase, by an expansion of the order-parameter $\eta(\xi)$, the form of which is determined by the transformation property of $\eta(\xi)$ under the symmetry operations of D^4_{6h}. As $\eta(\xi)$ is an odd function of ξ, the functional $F(\eta(\xi))$ contains only even powers of the order-parameter. Restricting to the sixth degree, which is the minimal degree necessary to have a direct first-order D^4_{6h}-D^1_{6h} transition, one can write:

$$F(\eta(\xi)) = a_1\eta^2(\xi) + a_2\eta^4(\xi) + a_3\eta^6(\xi) \tag{2.9}$$

where the phenomenological coefficients are assumed to obey the standard conditions: $a_1 = a_{10}(T - T_c)$, $a_{10} > 0$, a_2=const, a_3>0.

The minimization of F has to be performed with respect to the variational parameter ξ, so that the equation of state can be written under the general form:

$$\frac{\partial F}{\partial \eta}\cdot\frac{\partial \eta}{\partial \xi}=0 \tag{2.10}$$

which gives explicitly:

$$2\eta\frac{\partial \eta}{\partial \xi}\left(a_1+2a_2\eta^2+3a_3\eta^4\right)=0\ . \tag{2.11}$$

The stability condition has the form:

$$\frac{\partial^2 F}{\partial \eta^2}\cdot\left(\frac{\partial \eta}{\partial \xi}\right)^2+\frac{\partial F}{\partial \eta}\cdot\frac{\partial^2 \eta}{\partial \xi^2}\geq 0 \tag{2.12}$$

which is expressed by:

$$2\left(a_1+2a_2\eta^2+3a_3\eta^4\right)\left\{\left(\frac{\partial \eta}{\partial \xi}\right)^2+\eta\frac{\partial^2 \eta}{\partial \xi^2}\right\}+8\eta^2\left(\frac{\partial \eta}{\partial \xi}\right)^2\left(a_2+3a_3\eta^2\right)\geq 0 . \tag{2.13}$$

Equation (2.11) and (2.13) yield, using the form (2.8) of $\eta(\xi)$, three possible stable states:

1) the initial D_{6h}^4 phase (I) for Sin ($c^*\xi$)=0, namely for $\xi=0, \frac{c}{2}$, c, ... , and $a_1 \geq 0$;

2) the hexagonal D_{6h}^1 phase (phase II) for Cos ($c^*\xi$)=0, i.e. for $\xi=\frac{c}{4}, \frac{3c}{4}, \frac{5c}{4}, \ldots$, and $a_1 \leq \left(-2a_2-3a_3\eta_0^2\right)\eta_0^2$;

3) the rhombohedral D_{3d}^3 phase (phase III) for the equilibrium value of the order-parameter:

$$\eta=\pm\left[\frac{-2a_2+\left(4a_2^2-12a_1a_3\right)^{1/2}}{6a_3}\right]^{1/2} \tag{2.14}$$

which is stabilized for $a_1 \leq a_2^2/3a_3$.

The corresponding phase diagram is shown in Fig.4, in the plane of the phenomenological ratios (a_1/a_3 ; a_2/a_3). Thus, the reconstructive transition I-II is always first-order, whereas the I-III transition can be either first- or second-order, the two regimes being separated by a *tricritical point* . The fact that the II-III transition is always second-order in Fig.4, results from the sixth power assumed for $F(\eta(\xi))$, and also from the condition $a_3>0$. As it will be shown in Sec.2.1.5, a higher degree expansion, or a negative value for a_3 , would give a possible first-order transition regime between phases II and III.

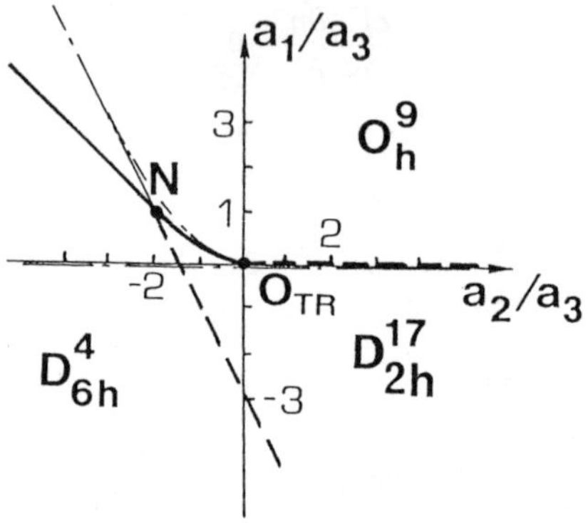

Fig.4. Phase diagram representing the order-parameter expansion (2.9) in the plane $(a_1/a_3, a_2/a_3)$. Solid and dashed lines are first-order and second-order transition lines respectively. Dashed-dotted lines are limit of stability lines. I, II and III denote the phases of symmetry D^4_{6h}, D^1_{6h} and D^3_{3d}, respectively. O is a tricritical point and N is a three phase point.

The phase diagram contains a *three-phase point* , at which merge three transition lines: i) the first-order transition line between phases I and II. Its equation is $a_1 = -a_2\eta_0^2 - a_3\eta_0^4$, which corresponds *to a straight line* , ii) the first-order I-III transition line, which is a parabolic branch: $a_1 = a_2^2/4a_3$, iii) the line of second-order order transitions II-III, which is also a *straight line* , of equation $a_1 = -2a_2\eta_0^2 - 3a_3\eta_0^4$. Let us note that this line prolongates the limit of stability line between phases I and II. The coordinates of the three-phase point are: $a_1 = a_3\eta_0^4$, $a_2 = -2a_3\eta_0^2$.

The form of the $F(\xi_e)$ curves, where ξ_e denotes the equilibrium value of the displacements, is represented in each of the relevant region in Fig.4, neglecting the thermal expansion contribution $F_0(T)$ to the thermodynamic potential. The main distinctive feature of the limit D^1_{6h} phase, is that the value of $F(\xi_e)$ corresponding to the minima of $F(\xi)$ remains constant across the phase, as ξ_e corresponds to *fixed* displacements.

2.1.4. Temperature dependence of the order-parameter, and of the critical displacements

In phases I and II, the order-parameter has temperature-independent values, $\eta=0$ and $\eta = \eta_0$ respectively, whereas in phase III, η varies with temperature as expressed by Eq.(2.14). the shape of $\eta(T-T_c)$ across different sequences of phases, indicated in Fig.5a, is given in Fig.5b. Thus, one has a typical step-like shape for the reconstructive I-II transition (Fig.5b-1,2) and the usual second- or first-order transition temperature dependence of $\eta(T-T_c)$ for the I-II transition (Fig.5b-3,5). The second-order II-III

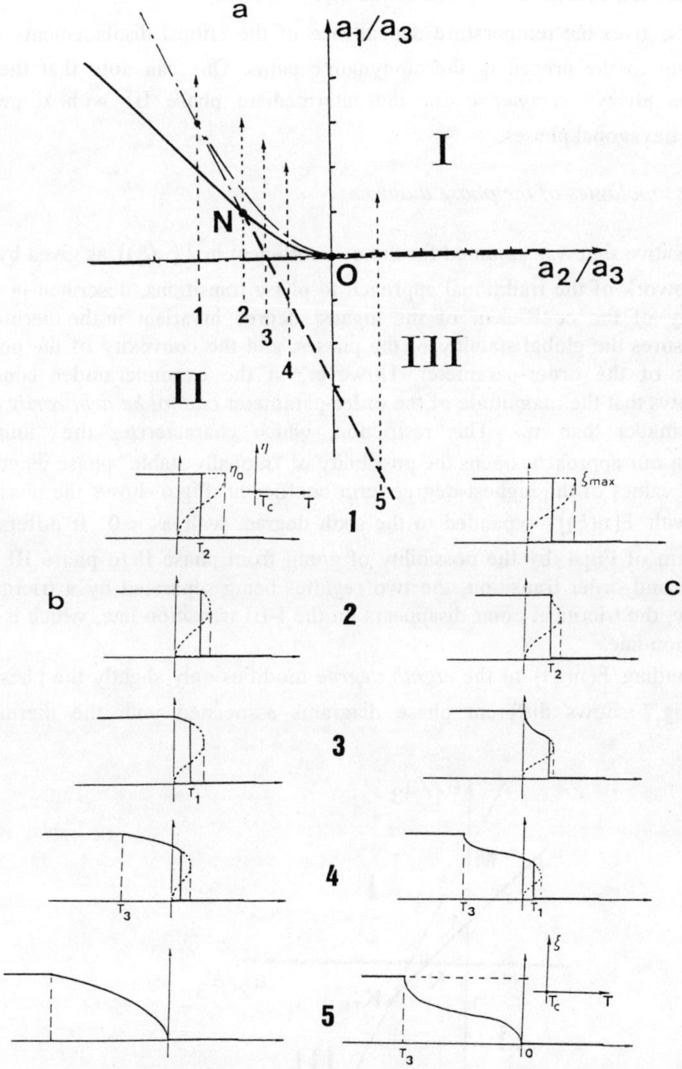

Fig.5. (a) Same phase diagram as in Fig.4 on which are indicated the thermodynamic paths, denoted 1 to 5, corresponding to different sequences of phases. (b) Temperature dependence of the order-parameter for the various paths shown in (a). (c) Temperature dependence of the critical displacements for the paths indicated in (a).

transition exhibit a change in the slope of the $\eta(T-T_c)$ curve.

Fig.5c gives the temperature dependence of the critical displacements $\xi(T-T_c)$, corresponding to the preceding thermodynamic paths. One can note that the shape of $\xi(T-T_c)$ is always *symmetric* for the intermediate phase III with respect to the neighboring hexagonal phases.

2.1.5. Other topologies of the phase diagram

A positive sign was assumed for the a_3 coefficient in $F(\eta(\xi))$ as given by Eq.(2.9). In the framework of the traditional approach to phase transitions, described in Chapter I, the positivity of the coefficient of the highest degree invariant in the thermodynamic potential, insures the global stability of the phases, and the convexity of the potential for large values of the order-parameter. However, in the example under consideration, Eq.(2.8) shows that the magnitude of the order-parameter *cannot be arbitrarily large*, and is always smaller than η_0. This restriction, which characterizes the "limit" phases introduced in our approach, opens the possibility of "globally stable" phase diagrams, even for negative values of the highest-degree term coefficient. Fig.6 shows the phase diagram associated with $F(\eta(\xi))$, expanded to the sixth degree, with $a_3 < 0$. It differs from the phase diagram of Fig.4, by the possibility of going from phase II to phase III through a first *or* a second order transition, the two regimes being separated by a tricritical point. Correlatively, the tricritical point disappears on the I-III transition line, which is a second-order transition line.

Expanding $F(\eta(\xi))$ to the *eighth degree* modifies only slightly the phase diagram of Fig.6. Fig.7 shows different phase diagrams associated with the thermodynamic potential:

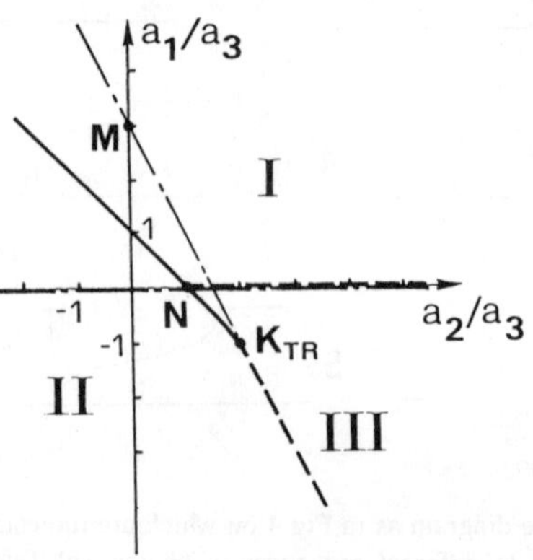

Fig.6. Phase diagram associated with the sixth degree expansion (2.9), assuming $a_3 < 0$. Same notation as in Fig.4. K_{TR} is a tricritical point.

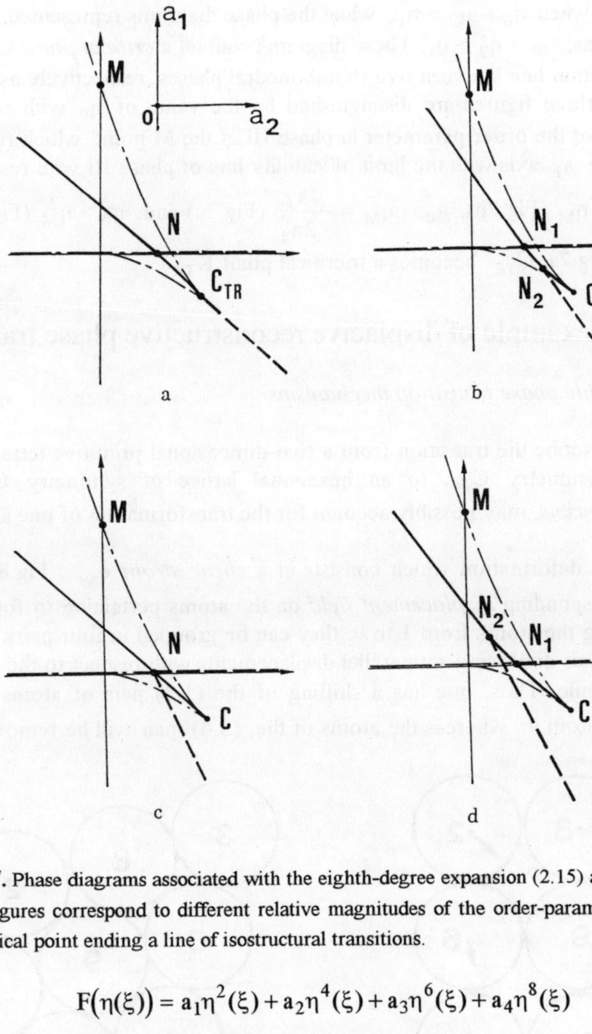

Fig.7. Phase diagrams associated with the eighth-degree expansion (2.15) assuming $a_3 < 0$ and $a_4 > 0$. The figures correspond to different relative magnitudes of the order-parameter indicated in the text. C is a critical point ending a line of isostructural transitions.

$$F(\eta(\xi)) = a_1\eta^2(\xi) + a_2\eta^4(\xi) + a_3\eta^6(\xi) + a_4\eta^8(\xi) \tag{2.15}$$

assuming $a_3 < 0$ and $a_4 > 0$. The diagrams are discriminated by the relative values of η_0 and (η_1, η_2), where η_1 and η_2 are the solutions of the equation:

$$a_1 + 2a_2\eta^2 + 3a_3\eta^4 + 4a_4\eta^6 = 0 \tag{2.16}$$

which are physically acceptable, i.e. which correspond to $\eta_1^2 > 0$, $\eta_2^2 > 0$, $\eta_1 \neq \eta_2$. Thus

Fig.7a is obtained when $\eta_0 = \eta_2 > \eta_1$, while the phase diagrams represented in Fig.7b,c,d, fulfill the conditions: $\eta_0 > \eta_2 > \eta_1$. These diagrams contain a *critical point* C_{TR} ending an isostructural transition line between two rhombohedral phases, respectively associated with η_1 and η_2. The three figures are distinguished by the value of η_0 with respect to the equilibrium value of the order-parameter in phase III at the M point, which represents the intersection of the a_1 axis with the limit of stability line of phase III with respect to phase II. One has: $\eta_0^2 < \eta_M^2$ (Fig.7b), $\eta_0^2 = \eta_M^2 = -\frac{a_3}{2a_4}$ (Fig.7c) and, $\eta_0^2 > \eta_M^2$ (Fig.7d). Let us also note that in Fig.7a, C_{TR} becomes a tricritical point K_{TR}.

2.2. A second example of displacive reconstructive phase transition

2.2.1. Three possible phase transition mechanisms

We will describe the transition from a two-dimensional primitive tetragonal lattice, of point-group symmetry C_{4v}, to an hexagonal lattice of symmetry C_{6v} . *Three alternative mechanisms*, may possibly account for the transformation of one structure into the other:

1) A macroscopic deformation which consists in a *shear strain* e_{xy} . Fig.8a shows the effect of the corresponding *displacement field* on the atoms pertaining to four tetragonal unit-cells. Denoting the atoms from 1 to 9, they can be grouped in four pairs, (1,2), (3,4), (5,6), (7,8), each pair displaying antiparallel displacements with respect to the central atom 9, which is immobile. Thus, one has a shifting of the (1,2) pair of atoms in the $[11]$ direction *towards* atom 9, whereas the atoms of the (3,4) pair will be removed from the

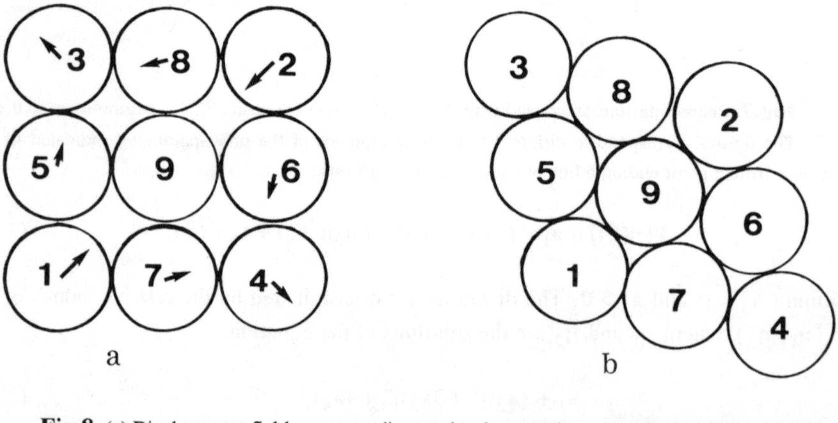

Fig.8. (a) Displacement field corresponding to the shear strain e_{xy} applied to the atoms of four tetragonal unit-cells of the C_{4v} structure. (b) Hexagonal atomic packing corresponding to the C_{6v} structure.

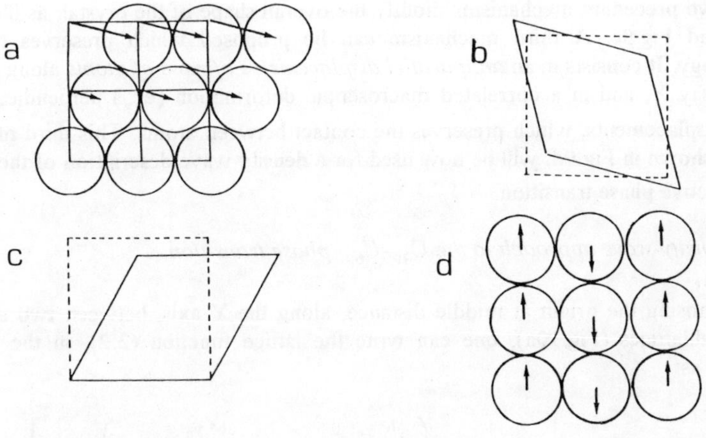

Fig.9. (a) Parallel shifts of atomic rows, transforming the C_{4v} into C_{6v}. (b) and (c): Modifications of the shape of the crystal (b): subsequent to the displacive field shown in Fig.8c, and (c): resulting from the parallel shifting of atomic rows. (d) Antiparallel displacement of atomic rows lowering the C_{4v} symmetry to C_{2v}.

central atom along $\left[\bar{1}\,1\right]$. On the other hand, the remaining pairs of atoms, (5,6) and (7,8), will get closer to the $\left[\bar{1}\,1\right]$ direction, as represented in Fig.8a and Fig.8b. If ξ is the magnitude of the displacements of one pair of atoms, let say (1,2), the tetragonal lattice will correspond to $\xi=0$, and for arbitrary small shifts the fourfold symmetry will decrease to orthorhombic (C_{2v}). For a given magnitude ξ^{max} of the displacements, the pair of atoms (1,2) will get in contact with the central atom, and the lattice will acquire the hexagonal symmetry C_{6v}. ξ^{max} corresponds to the maximal value of the shear strain e_{xy}^{max}.

2) In the framework of our approach to the transition mechanism, the preceding description is not satisfactory as the variational parameter e_{xy} has a (maximal) limit value, corresponding to an hexagonal close packing of the atoms. Another equivalent description can be given in terms of *parallel shifts of atomic rows*, as represented in Fig.9a. The transition is associated with two correlated (bilinearly coupled) mechanisms: a collective displacement of parallel rows of atoms along one axis (X), and a macroscopic deformation in the perpendicular direction (Y) which keeps the atoms in contact. The advantage of this second description of the transformation, is that it shows *the periodicity of the mechanism as a function of the displacements*. Thus for small displacements of parallel rows of atoms, the symmetry decreases to orthorhombic, whereas for displacements corresponding to $a/2$, where a is the tetragonal lattice parameter, the hexagonal lattice is formed. When increasing the displacements, one will get periodically the tetragonal lattice (for displacements of magnitude a) and again the orthorhombic and hexagonal lattices.

3) The two preceding mechanisms modify the overall shape of the crystal, as illustrated in Fig.9b and Fig.9c. A third mechanism can be proposed which preserves the crystal morphology. It consists in an *antiparallel displacements of rows of atoms* along one of the axis, let say X, and in a correlated macroscopic deformation (e_{yy}) perpendicular to the atomic displacements, which preserves the contact between atoms. This third mechanism, which is shown in Fig.9d, will be now used for a density wave description of the C_{4v}-C_{6v} reconstructive phase transition.

2.2.2. Density-wave approach to the C_{4v}-C_{6v} phase transition

Chosing the origin at middle distance, along the Y axis, between two antiparallel atomic sublattices (Fig.10a), one can write the lattice function (2.2), in the tetragonal structure, as:

$$\psi_0(x,y) = A \cdot \left[e^{-i(a^*x+b^*y)} \left(e^{i\frac{b^*}{2}\frac{y_0}{2}} \cdot e^{ia^*x_0} + e^{-i\frac{b^*}{2}\frac{y_0}{2}} \cdot e^{ia^*x_0} \right) + \text{c.c.} \right] \tag{2.17}$$

where a* and b* are the tetragonal reciprocal lattice vectors, and x_0=0, y_0=b, the initial atomic positions. After an antiparallel displacement ξ along the X axis, of two consecutive rows of atoms, the lattice function becomes:

$$\psi(x,y,\xi) = A \cdot \left[e^{-i(a^*x+b^*y)} \left(e^{i\frac{b^*}{2}\frac{y_0}{2}} \cdot e^{ia^*(x_0+\xi)} + e^{-i\frac{b^*}{2}\frac{y_0}{2}} \cdot e^{ia^*(x_0-\xi)} \right) + \text{c.c.} \right] \tag{2.18}$$

which yields the increment of the lattice function at the transition:

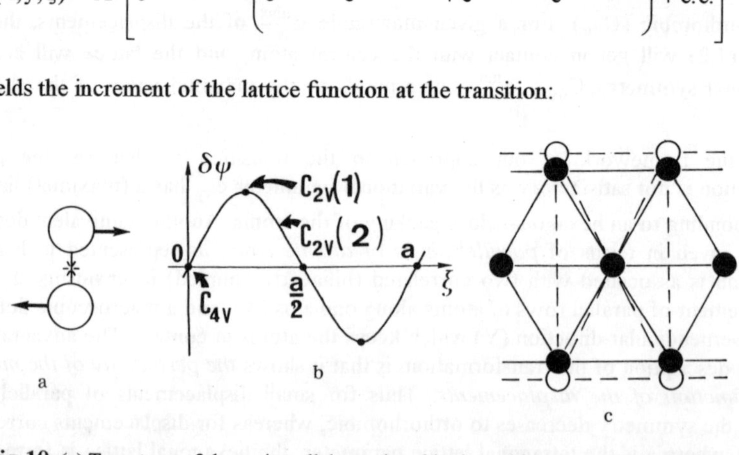

Fig.10. (a) Two atoms of the antiparallel atomic sublattice, with the coordinate of the origin at middle distance between them. (b) The $\delta\psi(\xi)$ curve follows Eq.(2.19). (c) Effect of the strain component e_{yy} on the $C_{4v} - C_{6v}$ mechanism. Open and full circles represent respectively the initial and final atomic positions.

$$\delta\psi(x,y,\xi) = 2A\cdot\cos(a^*x + b^*y)\cdot\text{Sin}(a^*\xi)\,. \tag{2.19}$$

It expresses the periodic deformation of the structure in function of the atomic displacements ξ , from the initial tetragonal structure. However, as these displacements constitute only part of the transition mechanism, the tetragonal lattice will not be transformed into the hexagonal lattice. This can be viewed on the $\delta\psi(\xi)$ curve, represented in Fig.10b, which shows that one gets periodically the tetragonal structure, for $\xi = 0, \frac{a}{2}, a, ...$, and *two distinct orthorhombic structures* of symmetry C_{2v}: a *primitive* orthorhombic structure with *two atoms per unit-cell* (i.e. the tetragonal cell is doubled along b by the "optical" antiparallel shifting mechanism), and a *centered* orthorhombic structure, with *one atom per unit-cell.* This latter structure is a "limit" phase, corresponding to the fixed displacements $\xi = \frac{a}{4}, \frac{3a}{4}, \frac{5a}{4}, ...$ associated with the extrema of $\delta\psi(\xi)$.

In order to obtain the hexagonal structure, one has, in addition, to take into account the macroscopic deformation along Y, which transforms the centered orthorhombic lattice onto the hexagonal lattice. Fig.10c shows the effect of the strain component e_{yy} which brings about the b/a ratio to its hexagonal value $\sqrt{3}$, or equivalently the angle between the diagonals to 60°, the transformation occurring continuously, without loss of contact between the atoms.

Let us emphasize that in the present example, the more simple situation has been assumed, in which the displacements occur along a single direction, and the deformation is in the orthogonal direction. As a result, the limit hexagonal phase takes place when the atoms get in contact. One may find more complex cases in which either the limit phase is obtained geometrically, i.e. the atoms are in contact without an increase in symmetry, or conversely the symmetry increases without contact between atoms. Such situations are found in more complex structures, than the one considered in this section, and will be discussed in Chapter III.

Eq.(2.19) shows that the effective order-parameter $\eta(\xi)$ of the transition has the same symmetry and form as the order-parameter of the D^4_{6h}-D^1_{6h} transition given by Eq.(2.8). Accordingly the same form (2.9) of the thermodynamic potential, and the same phase diagram topology, as represented in Fig.4, hold for the C_{4v}-C_{2v}-C_{6v} sequence of phases (in which the phases denoted I, II and III coincide respectively with the phases of symmetries C_{4v} , C_{6v} and C_{2v}). However, it should be stressed that the full symmetry of the order-parameter in the present example, may involve *two components*, only one of which has been considered for the C_{4v}-C_{6v} transition. Furthermore, in order to obtain the hexagonal structure, one has to include in Eq.(2.9) a *coupling* between the order-parameter $\eta(\xi)$ and the spontaneous strain component e_{yy}. The two first mechanisms, discussed in Sec.2.2.1 require a *bilinear* coupling $\eta\, e_{yy}$. In the third mechanism, assumed in our density-wave approach, one has a *non-linear* coupling of the $\eta^2 e_{yy}$ type.

2.3. An example of order-disorder reconstructive phase transition

2.3.1. Mechanism of the O_h^9-D_{6h}^1 ordering transition

Let us assume a transition from a disordered body centered cubic structure, in which one has a random distribution of atoms and vacancies each BCC position being occupied with a probability 1/3, to a fully ordered simple hexagonal structure in which the probability of occupancy is 1. As shown in Fig.11, the hexagonal unit-cell (space group D_{6h}^1, Z=1) corresponds to a *tripling* of the cubic BCC unit cell (space group O_h^9, Z=1/3), with the following relations between the corresponding basic translations:

$$\vec{a}_h = \vec{a}_c - \vec{b}_c\,,\quad \vec{b}_h = \vec{b}_c - \vec{c}_c\,,\quad \vec{c}_h = \vec{a}_c + \vec{b}_c + \vec{c}_c\,. \tag{2.20}$$

Accordingly, the $\vec{c}_h$ axis coincides with the diagonal of the primitive cubic unit-cell, whereas $\vec{a}_h$ and $\vec{b}_h$ are diagonals of the primitive cubic faces, and one has $|\vec{c}_h|/|\vec{a}_h| = \sqrt{3/8}$.

The preceding ordering mechanism involves exclusively a change in the distribution of atoms and vacancies among the positions, but does not displace the initial positions, and requires no deformation of the cubic lattice. In the final hexagonal positions, the atoms fill the full circles shown in Fig.12, instead of occupying the initial open circles, corresponding to the cubic positions. The current description of their *exchange* process [6], is formulated in terms of *uncorrelated diffusion* and *jump* of the atoms between *fixed* positions. It *virtually* follows the same scheme described in the preceding examples: The difference is that the incompletely ordered structures (with a probability of occupancy comprised between 1/3 and 1) possesses the same symmetry (D_{6h}^1) as the fully ordered structure,which represents the *limit state.* Table 1, which is extracted from the International

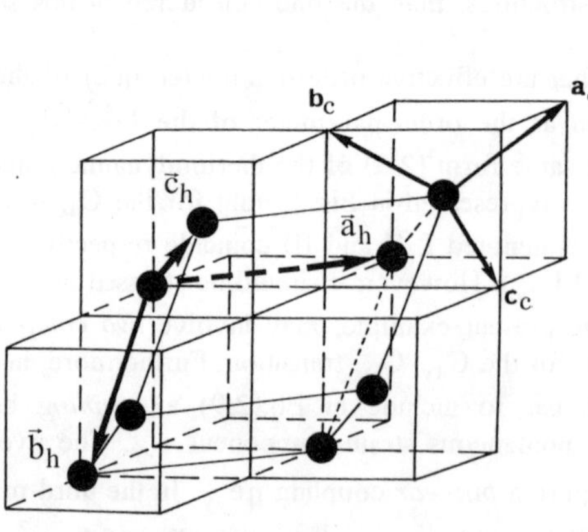

Fig.11. Connection between the cubic and hexagonal unit-cells following (2.20). The arrows are the basic lattice vectors of the BCC and hexagonal cells.

Tables for Crystallography [7] supports this description. It represents the symmetry of the D_{3d}^3 rhombohedral structure, for different occupancies of the positions. One can see in the table that for definite occupancies, the overall structural symmetry increases to the hexagonal space-group D_{6h}^1.

Table 1. Possible symmetry of the D_{3d}^3 space group (column (1)) for different occupancies of positions given in column (2).

(1)	(2)
D_{6h}^1	(a) in (a) or (b)
D_{6h}^1	(e) in (c)
D_{3d}^3	(d) in (d)
D_{6h}^1	(f) in (e) or (f)
D_{6h}^1	(j) in (g) or (h)
D_{3d}^3	(i) in (i)

The preceding increase in symmetry can also be understood in terms of probability distribution functions. In the initial cubic structure one has only one atomic distribution function of symmetry O_h^9. When the initial atomic position splits, the preceding function also splits into three independent functions the set of which is invariant by the D_{3d}^3 symmetry. When the fully ordered state is reached, the coefficients of two among these three functions vanish, and the distribution function corresponding to the non-vanishing coefficient acquires the symmetry of a special (isolated) atomic position, which is is D_{6h}^1.

Let us stress that in diffraction experiments, the partially ordered-fully ordered transition coincides with the disappearance of superstructure reflections corresponding to the vacant positions. In reconstructive phase transition, this is usually interpreted in terms of *stacking faults,* the number of which decreases when the occupancy varies from 1/3 to 1. In other words the process which brings about a full ordering state may occur continuously and *does not necessarily correspond to a phase transition.*

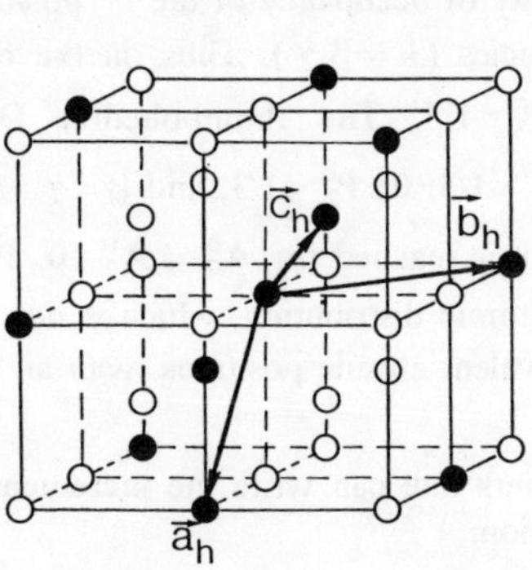

Fig.12. Cubic to hexagonal ordering mechanism. Full and open circles are, respectively, occupied and vacant positions in the hexagonal structure.

2.3.2. Density-wave description of the reconstructive O_h^9-D_{6h}^1 transition

Fig. 12 illustrates the cubic-to-hexagonal change in atomic probability distribution. Using Eq.(2.2) one can express the corresponding reconstructive transition mechanism in terms of density-waves, which gives the form of the lattice function, valid for the different structures:

$$\psi(z) = e^{-\frac{4\pi i z}{a\sqrt{3}}}\left\{A_1 e^{\frac{4\pi i}{a\sqrt{3}}z_1} + A_2 e^{\frac{4\pi i}{a\sqrt{3}}\left(z_1+\frac{a}{2\sqrt{3}}\right)} + A_3 e^{\frac{4\pi i}{a\sqrt{3}}\left(z_1+\frac{a}{\sqrt{3}}\right)}\right\} + \text{c.c.} \qquad (2.21)$$

where A_1, A_2 and A_3 are the amplitudes associated with the three, possibly inequivalent crystallographic positions (0 0 0), $(\frac{1}{3}\ \frac{2}{3}\ \frac{1}{3})$ and $(\frac{2}{3}\ \frac{1}{3}\ \frac{2}{3})$ written in hexagonal coordinates, and corresponding respectively to the z coordinates: $z_1=0$, $z_2 = z_1 + \frac{a}{2\sqrt{3}}$, $z_3 = z_1 + \frac{a}{\sqrt{3}}$, where a is the cubic lattice parameter.

The z-dependent factor before the bracket in (2.21), expresses the periodicity of the structures along the Z hexagonal direction. The exponential terms inside the bracket account for the geometry of the structure, and thus remain unchanged at the O_h^9-D_{6h}^1 transition. The ordering process modifies exclusively the occupancy of the atomic positions, which determines the values of the density wave amplitudes A_i (i=1-3). In this respect, one can write:

$$A_i^k = A \cdot \delta P_i^k \qquad (2.22)$$

where the P_i^k denote the probability of occupancy of the ith position corresponding either to an atom ($k=\alpha$) or to vacancies ($k=\beta,\gamma$). Thus, in the cubic structure, one has $P_1^\alpha=P_1^\beta=P_1^\gamma=1/3$ and $P_1^\alpha=P_2^\alpha=P_3^\alpha=1/3$. The rhombohedral D_{3d}^3 atomic distribution corresponds to $1/3<P_1^\alpha\leq 1$, $0\leq P_1^\beta<1/3$, $0\leq P_2^\alpha<1/3$, and $\beta=\gamma$, $A_2^k = A_3^k$ for all k. The limit hexagonal atomic distribution is reached for $A_2^\alpha = A_3^\alpha=0$, $P_1^\alpha=1$, $P_1^\beta=P_1^\gamma=0$. Let us also note that the more general atomic distribution, which is not stabilized in the present mechanism, involves three inequivalent atomic positions, with an overall symmetry C_{3v}^1, and $A_1^k \neq A_2^k \neq A_3^k$.

Using the preceding numbers one can write the increment of the lattice function (2.21) at the disorder-order transition:

$$\delta\psi(z) = e^{-\frac{4\pi i z}{a\sqrt{3}}}\left\{\eta A^\alpha\left(P_1^\alpha - \frac{1}{3}\right) + (1-\eta)A^\alpha\left[e^{\frac{4\pi i}{a\sqrt{3}}z_2} + e^{\frac{4\pi i}{a\sqrt{3}}z_3}\right]\right\} + \text{c.c.} \qquad (2.23)$$

where η *represents the order parameter of the ordering transition, defined as*:

$$\eta = \frac{3}{2}\delta P_1^\alpha . \tag{2.24}$$

δP_1^α is the increment of the atomic probability of occupancy P_1^α from the cubic to the ordered phases. As in the initial cubic structure $P_1^\alpha=1/3$, and in the ordered phase $1/3 < P_1^\alpha \leq 1$, one has thus $0 \leq \delta P_1^\alpha \leq 2/3$ consistently with the standard property of the order-parameter associated with an order-disorder transition, namely to vary from $\eta=0$ in the disordered cubic phase, to $\eta=1$ in the fully ordered hexagonal phase.

2.3.3. Symmetry of the order-parameter and phase diagram

The density waves associated with the three atomic sublattices, expressed in Eq.(2.21) transform as the C_{3v} point group, which is the group of permutation of three equivalent objects. The phase diagrams associated with the image-group C_{3v}, have been discussed in details in Chapter I (Sec.3.3.3). They correspond to a two-component order parameter (η_1, η_2). Although in the present example the actual dimensionality of the order-parameter is eight (i.e. it transforms as an eight-dimensional irreducible representation of the O_h^9 space group associated with the wave vector $\vec{k} = \frac{1}{3}\left(\vec{a}_c^* + \vec{b}_c^* + \vec{c}_c^*\right)$ of the BCC Brillouin-zone), one has two non-vanishing *effective* components, which can be written as:

$$\eta_1 = \delta P_1^\alpha + \varepsilon^* \cdot \delta P_2^\alpha + \varepsilon \cdot \delta P_3^\alpha \; ; \; \eta_2 = \delta P_1^\alpha + \varepsilon \cdot \delta P_2^\alpha + \varepsilon^* \cdot \delta P_3^\alpha \tag{2.25}$$

with $\varepsilon = e^{2\pi i/3}$.

Among the possible C_{3v} phase diagrams shown in Sec.3.3.3 of Chapter I the one which is the more adapted to our example is represented in Fig.13a, in the case of a *fully occupied* O_h^9 structure, and in Fig.13b, for the *partially filled* cubic structure assumed in our description. One can see that the two anti-isostructural low-symmetry structures of symmetry D_{3d}^3 in Fig.13a, which correspond to distinct binary alloys AB_2 and A_2B transforming one into the other by chemical decomposition across a line of first-order transition, become respectively in Fig.13b hexagonal (D_{6h}^1) and rhombohedral (D_{3d}^3) *binary mixtures* of atoms (A) and vacancies ($\square$). The occupancy of atoms and vacancies in the structures of the $A\square_2$ and $A_2\square$ mixture can be easily deduced from Fig.12.

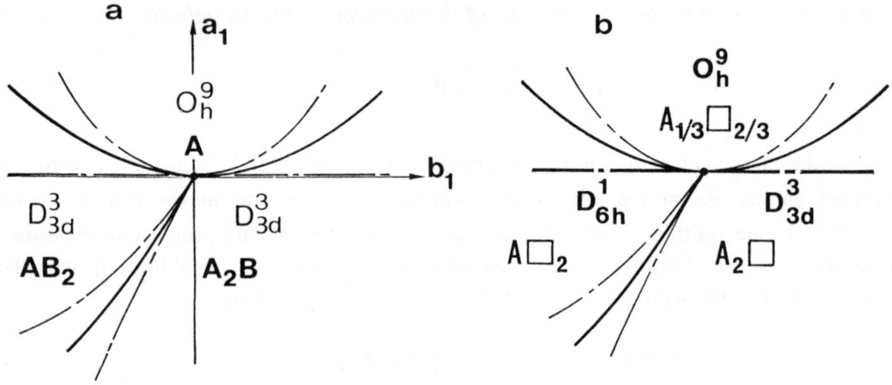

Fig.13. Phase diagram corresponding to the image-group C_{3v} associated with (a) the O_h^9-D_{3d}^3 transition (full occupancy of the lattice) and (b) the O_h^9-D_{6h}^1 transition (fractional occupancy of the lattice). Full lines and dashed-dotted lines are respectively first-order and limit-of-stability lines.

2.4. An example of periodic-aperiodic reconstructive phase transition

2.4.1. Crystallographic description of the tetragonal-octagonal transformation mechanism

As an illustrative example of the density-wave approach to a crystal-aperiodic quasicrystal reconstructive phase transition, we will describe the simple transformation from a planar tetragonal structure, of point-group symmetry C_{4v}, to an octagonal structure of symmetry C_{8v} . The unit-cell of the tetragonal structure contains 8 atoms in general positions (x,y), formed by two squares, rotated with respect to each other as represented in Fig.14a. If we apply to this structure a *displacement field*, corresponding to antiparallel shifts of atoms, which rotate the initial squares by opposite angles, the atoms pertaining to two squares will form a *regular octagon*. The obtained structure remains however tetragonal, although it possesses *local* octagonal motifs. In order to break the periodicity of the tetragonal lattice one has to displace sets of close-pairs of atoms, indicated in Fig.14b, in such a way that they will occupy regular octagonal positions on a given coordination circle. This additional mechanism corresponds to a *inhomogeneous deformation* of the structure, as two different coordination circles involve displacements of different magnitudes.

In summary the C_{4v}-C_{8v} transformation can be described as the result of two correlated mechanism: a homogeneous deformation which increases the local symmetry of the atoms to C_{8v} , and an inhomogeneous deformation which achieves the overall quasiperiodic symmetry. Let us stress that the final quasiperiodic structure is *defective*, as

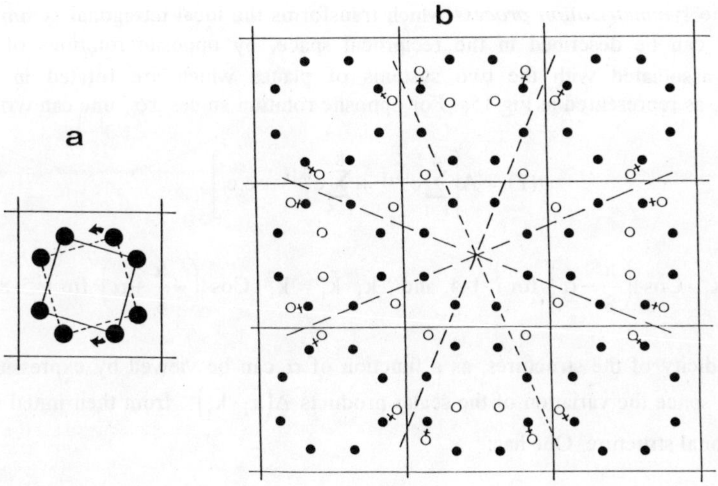

Fig.14. (a) Opposite rotation of the two squares leading to the octagonal basis. (b) Representation of the mechanism leading to the breaking of the translational symmetry. Full and open circles are occupied and vacant octagonal positions, respectively. The ideal octagonal quasilattice is obtained after the merging at crossed positions of close pairs of full and open circles.

shown in Fig. 14b, due to the existence of regularly ordered *vacant* octagonal sites which are distributed according to the initial tetragonal symmetry. One needs a further *disordering* of the sites, through a diffusion process, which will lead to the ideal octagonal quasiperiodic structure.

2.4.2. Density-wave description of the $C_{4v} - C_{8v}$ mechanism

Let us show that the preceding combined displacive mechanism can be expressed in terms of *inhomogeneous rotations* of the system of density waves associated with the initial tetragonal structure. The lattice functional corresponding to the tetragonal unit-cells is:

$$\psi_t(\vec{r}) = A\left[\sum_{i=1}^{8} e^{i\vec{k}_i\vec{r}} + c.c.\right] \tag{2.26}$$

where the $\vec{k}_i$ are the eight branches of the star $\vec{k}_1^*$, $\vec{k}_1$ being a wave-vector located in general position in the tetragonal Brillouin-zone. Eq.(2.26) expresses, in terms of density waves, the family of eight planes forming the tetragonal structure, and transforming one into another, under the symmetry operations of C_{4v}.

The *symmetrization process* which transforms the local tetragonal symmetry into octagonal can be described in the reciprocal space, by opposite rotations of four $\vec{k}_i$ branches associated with the two systems of planes which are rotated in opposite directions, as represented in Fig. 15a. For opposite rotation angles $\pm\alpha$, one can write:

$$\psi(\vec{r}) = A\left[\sum_{i=1}^{4} e^{i\vec{k}_i\vec{r}} + \sum_{i=5}^{8} e^{i\vec{k}'_i\vec{r}} + \text{c.c.}\right] \tag{2.27}$$

where:

$$\vec{k}_i \cdot \vec{k}'_i = k_i^2 \cdot \text{Cos}4\left(\frac{\pi}{8} - \alpha\right) \text{ for i=1-4, and } \vec{k}_i \cdot \vec{k}'_i = k_i^2 \cdot \text{Cos}4\left(-\frac{\pi}{8} + \alpha\right) \text{ for i=5-8.} \tag{2.28}$$

The periodicity of the structures, as a function of α can be viewed by expressing in the reciprocal space the variation of the scalar products $\Delta\left(\vec{k}_i \cdot \vec{k}'_i\right)$, from their initial values in the tetragonal structure. One has:

$$\Delta\left(\vec{k}_i \cdot \vec{k}'_i\right) = k_i^2\left[\text{Cos}4\left(\frac{\pi}{8} - \alpha\right) + \text{Cos}4\left(-\frac{\pi}{8} + \alpha\right)\right] = 2k_i^2 \cdot \text{Sin}4\alpha\,. \tag{2.29}$$

Fig. 15b indicates the structures corresponding to different values of α. Thus, an octagonal local structure is formed for the successive angles α=0, $\frac{\pi}{4}$, $\frac{\pi}{2}$, $\frac{3\pi}{4}$, ..., whereas the extrema, given by α=$\frac{\pi}{8}$, $\frac{3\pi}{8}$, $\frac{5\pi}{8}$,... coincide with tetragonal local structures formed with four atoms per unit-cell, and turned alternately by an angle of $\pi/8$. For general values of α, one has the initial local tetragonal basis, with 8 atoms per unit-cell.

The preceding description concerns the homogeneous deformation which transforms the initial tetragonal structure in a *rational approximant* of the octagonal

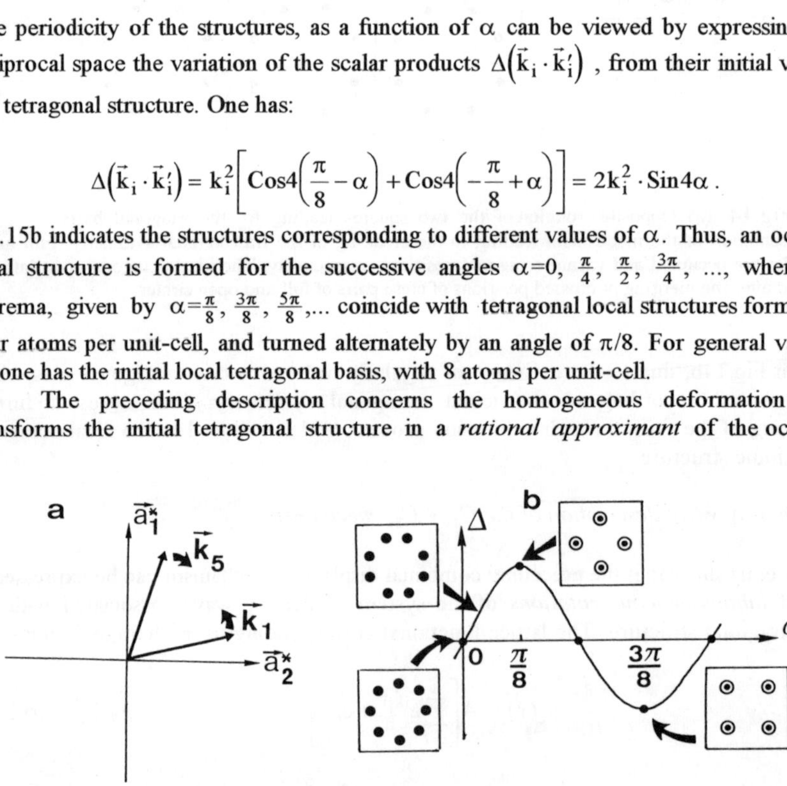

Fig.15. (a) Opposite rotation of a pair of branches of the star $\vec{k}_1$ associated with the symmetrization process, which transforms the local tetragonal basis into octagonal. (b) $\Delta(\alpha)$ dependence following (2.29). The squares indicate the local basis for relevant values of α.

quasicrystal, i.e. the formation of an octagonal basis *within* the tetragonal unit-cell. In order to break the tetragonal translational symmetry and to obtain a long-range octagonal quasilattice, one has to perform an *inhomogeneous deformation* of the structure which brings a number of atoms to occupy vacant octagonal positions *created* by the homogeneous deformation, as shown in Fig.14b. The description of this symmetry breaking mechanism in terms of density waves, entails to take into account *the whole structure of the crystal*, in order to determine, numerically, the atomic positions in the quasicrystal state. The corresponding procedure requires a number of experimental data, such as the atomic positions in the crystal structure, and the orientation of the wave-vector in the quasicrystal structure and thus cannot be carried on in the tetragonal-octagonal working example, under consideration. However, the necessary steps for finding the atomic positions in the octagonal structure assuming that the tetragonal atomic positions are known, can be briefly depicted (and will be illustrated by concrete examples in Chapter V). The lattice function corresponding to the symmetrized tetragonal structure is:

$$\psi(\vec{r}, \vec{r}_k, \vec{k}_i') = \frac{1}{N} \sum_{k=1}^{N} \sum_{i=1}^{8} A \left[e^{i\vec{k}_i'(\vec{r}_k - \vec{r})} + \text{c.c.} \right] \tag{2.30}$$

where the sum runs over the *total number N of atoms contained in the sample*. In order to obtain equal and maximal intensities for the Bragg reflections in the octagonal quasiperiodic structure, one has to maximize the *amplitude* of ψ:

$$A\left(\vec{r}_k, \vec{k}_i'\right) = \frac{1}{N} \sum_{k=1}^{N} \sum_{i=1}^{8} A \cdot \mathrm{Cos}\left(\vec{k}_i' \vec{r}_k\right) \tag{2.31}$$

with respect to the $\vec{k}_i'$ *and* to the $\vec{r}_k$. The maximization process, which is performed in the vicinity of the initial tetragonal positions, provides the *critical wave-vector* $\vec{k}_i^{oct} = \vec{k}_i' + \Delta\vec{k}_i'$, and the *critical displacements* $\vec{\xi}_k^{oct} = \vec{r}_k^{oct} - \vec{r}_k^{tet}$, associated with an octagonal quasiperiodic structure. For non-critical wave vectors $\vec{k}_i'' \neq \vec{k}_i^{oct}$, or (and) non-critical displacements ξ_k, the resulting structure is incommensurate (non octagonal) with respect to the initial crystal structure. Accordingly, the octagonal structure appears as a *limit situation* to an incommensurate regime.

3. Generalization of the preceding examples: Density-wave theory of reconstructive phase transitions

3.1. Differences in the phenomenological descriptions of displacive and order-disorder reconstructive phase transitions

Let us summarize the main features characterizing the phenomenological approach to the preceding working examples of reconstructive phase transitions. In this approach the structures of the phases surrounding a transition are described in terms of *lattice functions,* which are sums of stationary plane waves, associated with atomic planes and corresponding to the first harmonics of the Fourier expansions of the respective lattice functions. The form of the lattice function *increment* $\delta\psi$, expresses the transition mechanism. In this respect, it has been shown through the considered examples, that *the transition order-parameter can be identified to the* ξ*-dependent part (phase) of* $\delta\psi$ *for displacive mechanisms, and to the amplitude of* $\delta\psi$ *for order-disorder mechanisms.* The fact that the two classes of mechanisms are related to *distinct physical quantities defining* $\delta\psi$, has the following consequences on their respective phenomenological approach:

1) Displacive mechanisms give rise to a periodic dependence of the order-parameter η as a function of the critical atomic displacements ξ, and consequently to a *periodic* thermodynamic potential $F(T,\eta(\xi))$. This potential has the meaning of a *truncated Fourier series*, since it is formed by powers of the increment of the first harmonics of a Fourier expansion. The loss of group-subgroup relationship between the symmetries of the structures, occurs for certain definite displacements at which the system acquires new symmetry elements. The minimization of $F(\eta(\xi))$ with respect to ξ , yields two types of stable states: "Landau" states which are group-subgroup related to the parent state symmetry, and "limit" states which have no group-subgroup relation with the parent structure and coincide with the extrema of $\eta(\xi)$. For these limit states, the temperature dependence of the corresponding reconstructive transitions displays a step-like behavior.

For displacive reconstructive transitions to aperiodic structures, one must take into account the additional variation (in length and orientation) of the transition wave-vector $\vec{k}$. In this case the stable states are obtained by a maximization of the lattice functional increment $\delta\psi(\vec{k},\xi)$ of the whole structure, both with respect to $\vec{k}$ and to the displacements ξ.

2) For order-disorder mechanisms, the order-parameter η is a function of the atomic occupancy of the structure and is proportional to the increment of probability of occupancy of definite atomic sites. The loss of group-subgroup relationship with respect to the disordered structure, is the result of a selective occupancy of the preceding sites, which may correspond either to partially ordered, or to fully ordered states. These latter states coincide with a limit situation, corresponding to a maximal probability of occupancy of the sites equal to unity. At variance with displacive-type transitions, the system does not exhibit an increase in symmetry when going from partially to fully ordered states, i.e. the order-parameter varies continuously (without transition) with increasing occupancy of given sites. In this framework, "Landau" states which are group-subgroup related to the disordered structure can be interpreted as *defective limit states.*

The thermodynamic potential $F(\xi)$ associated with order-disorder reconstructive transitions, is a standard Taylor expansion of the order-parameter, and the stable (Landau and limit) states are obtained by usual minimization of F with respect to η. As fully ordered states correspond to fixed (temperature independent) probabilities of occupancy of given sites, they will be distinguished from partially ordered state, by a step-like dependence of the order-parameter in function of temperature.

The preceding considerations refer to situations in which the transition mechanisms are either purely displacive or purely of the order-disorder type. There exist a number of concrete examples of reconstructive phase transitions, discussed in Chapter IV, in which the two sorts of mechanisms are simultaneously involved, and thus require a combined phenomenological treatment. In these intermediate cases both the amplitude and the phase of the lattice functional $\delta\psi$ are varying quantities which depend respectively on the atomic occupancy of the structure *and* on critical displacements.

In contrast to the content of this section, which underlines the differences in the approaches to the various categories of reconstructive transitions, the next section stresses the unifying character of the density wave theory of reconstructive transitions, by providing its general formulation.

3.2. Formulation of the density-wave theory, and its connection with the Landau theory

3.2.1. Lattice function, probability density, order-parameter

The electron density in a crystal $\rho(x,y,z)$ varies periodically along any direction through the lattice, going across a complicated cycle of peaks and minima, and is thus described in structure analysis, by Fourier series as can be done for any periodic function. The electron density at the point of coordinates (x,y,z) can be written for a unit-cell of volume V [8] by:

$$\rho(x,y,z,) = \frac{1}{V}\sum_{-\infty}^{+\infty}{}_{h,k,l}F(h,k,l)\cdot e^{2\pi i(hx+ky+lz)} \qquad (2.32)$$

where each Fourier term in the series, associated with the coefficient F(h,k,l), represents a stationary system of density waves, parallel to the (h,k,l) family of atomic planes. Each set of waves is capable of diffracting incident rays with a certain intensity $|F(h,k,l)|^2$, for the successive parallel sheets of electron density act as reflecting planes.

Although the sum (2.32) is infinite for a full determination of the atomic positions, a number of simplifications [8,9] using the symmetry of the crystal, allow to restrict to a finite number of harmonics, for each sublattice. Furthermore, when dealing with the sole description of the structural modifications occurring at a phase transition, one can restrict to the critical sublattice primarily involved in the transition mechanism, and to the first harmonics which contain the relevant information on the periodicity of the lattice, and on the average positions of the atoms. Summing up over the equivalent atoms contained in the unit-cell in real space, instead of the atomic planes in reciprocal space, the probability density given by Eq.(2.32) reduces to the lattice function, defined by Eq.(2.2).

At a structural transition, which may involve in the more general case a change in the wave vector *and* in the atomic positions, the increment $\delta\psi$ of the lattice function from the initial structure (with atomic positions $\vec{R}_j^0$ and k-vector $\vec{k}_0$) to the final structure ($\vec{R}_j = \vec{R}_j^0 + \Delta\vec{R}_j$, $\vec{k} = \vec{k}_0 + \Delta\vec{k}$), at a point defined by the position vector $\vec{r}$, is:

$$\delta\psi(\vec{r}) = \frac{1}{N}\sum_{j=1}^{N}\left[A'_j e^{i\vec{k}(\vec{R}_j-\vec{r})} - A_j e^{i\vec{k}_0(\vec{R}_j^0-\vec{r})} + \text{c.c.}\right] \tag{2.33}$$

where the sum runs over the N atoms of the unit-cell of the *final structure*, in which the normalizing density wave coefficients are A'_j. Two main specific forms can be used for Eq.(2.33), depending on the nature of the transition mechanism:

1) For order-disorder transitions between periodic structures, one has fixed positions $\vec{R}_j = \vec{R}_j^0$, and $\Delta\vec{k}$=0. Eq.(2.33) simplifies to:

$$\delta\psi(\vec{r}) = \frac{1}{N}\sum_{j=1}^{N}\left[\left(A'_j - A_j\right)\cdot e^{i\vec{k}_0(\vec{R}_j^0-\vec{r})} + \text{c.c.}\right]. \tag{2.34}$$

Besides, one can write: $A'_j - A_j = A^\alpha \cdot \Delta P_j^\alpha$ where the ΔP_j^α represents the variation at the transition of the probabilities of occupancy of an atom of type α in position j. Using the property that the ΔP_j^α, for different j, differ only by their sign, Eq.(2.34) becomes:

$$\delta\psi(\vec{r}) = A^\alpha\left|\Delta P^\alpha\right|\left[\frac{1}{N}\sum_{j=1}^{N} e^{i\vec{k}_0(\vec{R}_j^0-\vec{r})}\right]. \tag{2.35}$$

In (2.35), $\eta = A^\alpha\left|\Delta P^\alpha\right|$ defines the (scalar) transition order-parameter, and the multiplying factors $\frac{1}{N}\sum_{j=1}^{N} e^{i\vec{k}_0(\vec{R}_j^0-\vec{r})}$ represent the normalized basic functions spanning the order-parameter space ε. These functions possess the same symmetry as the δ -atomic functions, and transform as irreducible representations of the initial crystal structure.

2) For displacive reconstructive transitions between periodic crystal structures $A'_j = A_j$ for all j and $\Delta\vec{k}$=0. Eq.(2.33) can be written as:

$$\delta\psi(\vec{r}) = \frac{1}{N}\sum_{j=1}^{N} A_j\cdot\left\{e^{-i\vec{k}_0\vec{r}}\cdot\left[e^{i\vec{k}_0\left(\vec{R}_j^0+\Delta\vec{R}_j\right)} - e^{i\vec{k}_0\vec{R}_j^0}\right] + \text{c.c.}\right\}. \tag{2.36}$$

Assuming only one type of atoms (A_j=A for all j), the collective displacements of the atoms at the transition can be decomposed into parallel displacements, which do not modify the relative structural arrangement of atoms, and *antiparallel displacements of pairs of atoms*. The critical sublattice can thus be subdivided into two antiparallel sublattices, indexed j_1 and j_2, with $\vec{R}^0_{j_1} = -\vec{R}^0_{j_2}$, and $\Delta\vec{R}^0_{j_1} = -\Delta\vec{R}^0_{j_2}$. Eq.(2.36) is transformed into:

$$\delta\psi(\vec{r}) = \frac{A}{N}\cdot\cos(\vec{k}_0\vec{r})\cdot\sum_{j_1}^{N}\left\{\mathrm{Sin}\left[\vec{k}_0\left(\vec{R}^0_{j_1} + \Delta\vec{R}_{j_1}\right)\right] - \mathrm{Sin}(\vec{k}_0\vec{R}^0_{j_1})\right\} \tag{2.37}$$

In (2.37) the position of the atoms coincide with the maxima of the density waves, so for different indices j_1, the arguments of the corresponding functions differ only by periods of these functions. Accordingly one has the final expression of $\delta\psi$:

$$\delta\psi(\vec{r}) = \frac{A}{N}\cdot\cos(\vec{k}_0\vec{r})\cdot\left\{\mathrm{Sin}\left[\vec{k}_0\left(\vec{R}^0_1 + \Delta\vec{R}_1\right)\right] - \mathrm{Sin}(\vec{k}_0\vec{R}^0_1)\right\} \tag{2.38}$$

where $\vec{R}^0_1$ is an arbitrary atomic position in the critical sublattice. In (2.38) the set of $\cos(\vec{k}_0\vec{r})$ functions identifies to the basic atomic functions spanning the representation space ε of the initial structure. Thus, each of the transition order-parameter components has the general form:

$$\eta = A\left\{\mathrm{Sin}\left[\vec{k}_0\left(\vec{R}^0_1 + \Delta\vec{R}_1\right)\right] - \mathrm{Sin}(\vec{k}_0\vec{R}^0_1)\right\} \tag{2.39}$$

i.e. it is a scalar periodic function of the magnitude of the atomic antiparallel displacements which constitute the transition mechanism.

In the case of a purely displacive periodic-aperiodic transition, one has again $A'_j = A_j$ for all j but $\Delta\vec{k} \neq 0$. Analogous considerations lead to the increment of the lattice functional for the whole structure:

$$\delta\psi(\vec{r}) = \frac{1}{N}\sum_{j_1=1}^{N}\sum_{k} A_{j_1}\cdot\cos(\vec{k}_0 + \Delta\vec{k})\vec{r}\cdot\left\{\mathrm{Sin}\left(\vec{k}_0 + \Delta\vec{k}\right)\left(\vec{R}^0_{j_1} + \Delta\vec{R}_{j_1}\right) - \mathrm{Sin}\left(\vec{k}_0\vec{R}^0_{j_1}\right)\right\} \tag{2.40}$$

which generalizes Eq.(2.38) to an inhomogeneous (aperiodic) structure. The sinusoidal factor in Eq.(2.40) represents formally the order-parameter of the crystal-aperiodic quasicrystal reconstructive phase transition. In real space, the order-parameter is a quasi-periodic function with respect to the initial crystal lattice. *It becomes a periodic function in the multidimensional space spanned by the* $(\vec{k}_0 + \Delta\vec{k})$ *wave-vectors.*

The definition of the order-parameter given herebefore for order-disorder type reconstructive phase transitions, namely as a linear combination of increments of probability of occupancy for given atomic sites, is the standard definition assumed in the approaches to order-disorder transitions [10]. It coincides with the "degree of order" introduced in the original papers of Landau [11,12] for positional ordering in binary mixtures, and for the orientational ordering in NH_4Cl. By contrast, for displacive-type reconstructive phase transitions, our definition of the order-parameter, given by Eq.(2.39), as a periodic function of the critical displacements, represents obviously a *generalization for large displacements*, of the *linear relationship* between order-parameter and small displacements [13], usually assumed for second- or slightly first-order displacive transitions. For small magnitude of ΔR_1^0 in Eq.(2.39), one has as a first approximation:

$$|\eta| \approx A k_0 \left|\Delta R_1^0\right| . \tag{2.41}$$

The expressions (2.35) and (2.38) found for the increment of lattice functional $\delta\psi(\vec{r})$, appear as formally similar to the increment of probability density $\delta\rho(\vec{r})$, given by Eq.(2.1) in the Landau theory, as the two quantities are bilinear product of the order-parameter and of atomic functions, which span the representation space of the initial crystal structure. However, they are obtained from different premises. In the Landau theory [12], $\delta\rho(\vec{r})$ is assumed to be small. This has the correlated consequences that a group-subgroup relationship necessarily exists between the initial and final structures, and that the order-parameter components (the $\eta^i_{k\gamma}$ in Eq.(2.1)) are necessarily small quantities. In our definition no restriction is made on the magnitude of $\delta\psi(\vec{r})$, and as a consequence the symmetry groups which leave invariant $\psi(\vec{r})$ and $\psi_0(\vec{r})$, are not necessarily related by a group-subgroup relationship. This is illustrated in Fig.16, which shows that *for a large increment, a system may acquire new symmetry elements.*

Another consequence of the group-subgroup relationship assumed in the Landau

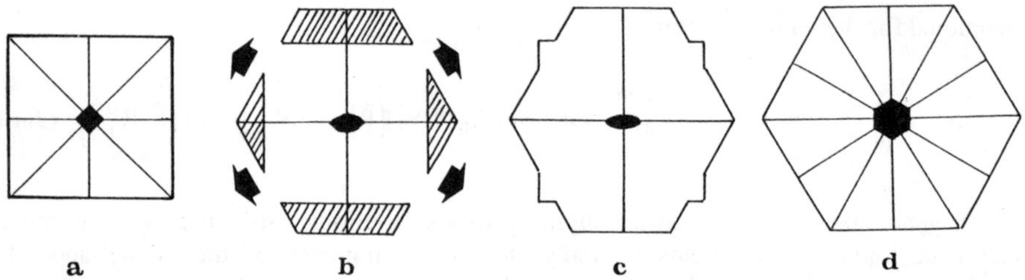

Fig.16. Illustrative example showing that for a *large and specific* increment $\delta\rho$, symmetries of the densities ρ and ρ_0 in Eq.(2.1) *may be not group-subgroup related.* Thus, for the large increment shown in (b), which has a orthorhombic (C_{2v}) symmetry, the initial figure (a), of tetragonal symmetry (C_{4v}) may either decrease in symmetry (Fig.(c)) or change to the hexagonal symmetry (C_{6v}) for a specific size of the increment (Fig.(d)).

theory between $\delta\rho(\vec{r})$ and $\rho_0(\vec{r})$, is that the bilinear form found for $\delta\rho(\vec{r})$ can be obtained directly from general group theoretical considerations, namely from the property that $\delta\rho(\vec{r})$ can be expanded into a sum of functions, the number of which is equal to the number of symmetry elements of $\rho_0(\vec{r})$. Accordingly, the Landau theory is formulated without considering any specific transition mechanism, and *without expressing* $\rho(\vec{r})$ and $\rho_0(\vec{r})$. In our formulation, the form of $\delta\psi(\vec{r})$ is *deduced* from the calculated forms of $\psi_0(\vec{r})$ and $\psi(\vec{r})$, *assuming a given type of mechanism.* If such formulation loses in generality, it has the advantage to *relate explicitly the order-parameter to the physical mechanism.*

In the two following sections, we will show that essential distinctive features of our approach, with respect to the Landau approach, are reflected in the structure of the order-parameter space, and in the mathematical nature of the thermodynamic potential associated with a reconstructive phase transition.

3.2.2. Order-parameter space for reconstructive phase transitions

Equations (2.35) and (2.38) reveal that the order-parameter space of reconstructive phase transitions, denoted hereafter Σ, differs in an essential manner from the order-parameter space ε used in the description of second-order transitions. In the standard Landau theory, the order-parameter is represented by a *fixed vector* $\vec{\eta}$ in the n-dimensional vectorial space ε. Every vector direction $\vec{\eta}$ in ε is left invariant by the isotropy group I_η of $\vec{\eta}$, and the symmetry group G of the phase stable below the transition, is related to the isotropy group I_η^{eq} of the vector $\vec{\eta}^{eq} = (\eta_1^{eq} ... \eta_n^{eq})$, which *fix the corresponding symmetry direction in* ε. For reconstructive transitions the magnitude of $\vec{\eta}^{eq}$ (and of the components η_i^{eq}) *is not determined* but *is a function of a variational parameter associated with the transition mechanism.* Two situations can be distinguished in this respect:

(1) For displacive reconstructive transitions $\left|\vec{\eta}^{eq}(x_i)\right|$ is a function of the critical displacements x_i. It is a *nonlinear, bounded, periodic* function, and depending on the magnitude of $\left|\vec{\eta}^{eq}(x_i)\right|$, the order-parameter exhibits different symmetries, which may correspond to a subgroup G of G_0, but also to a supergroup G' of G. The increase in symmetry from G to G' occurs for specific points of the distribution determined by $\left|\vec{\eta}(x_i)\right|$ *which correspond to the holohedry* (maximal symmetry of the Bravais lattice) *of G.* These special points of symmetry degeneracy of the structure, are located on the *edges* of the Σ-space, which coincide with the *maximal values* of the $\left|\vec{\eta}^{eq}(x_i)\right|$ function.

(2) For order-disorder reconstructive transition $\left|\eta(\delta P^\alpha)\right|$ is a linear function of the increment of probability of occupancy of selected atomic sites i. The function varies from 0 to 1, and its interval of definition is determined by the concentration (number of atoms

per unit-cell) of the initial structure. Its maximal values is realized for fully ordered states, and determines the *edge* of the Σ-space.

Thus, the essential difference in the description of reconstructive transition as compared to the standard description of structural transitions [11, 13] is due to the fact that *the magnitude* of the variational parameters associated with the transition mechanism, *influences the symmetry of the order-parameter*, and thus it has to be taken into account in the symmetry approach to the transition. By contrast in the standard Landau theory the dependence of $|\vec{\eta}|$ as a function of the variational parameters is a trivial (linear) function of the variational parameter, which has no effect on its symmetry, and accordingly, *the description of the transition mechanism* can be separated from the considerations which lead to the order-parameter symmetry.

Fig. 17a illustrates the mathematical difference existing between the ε and Σ order-parameter spaces. One can see that while ε is a n-dimensional vectorial space, Σ is a n-dimensional *functional space*, the structure of which *depends on the periodicity of the function of the variational parameter describing the transition mechanism.* The operator h transforming ε into Σ can be written:

$$h = g \circ f^{-1} \tag{2.42}$$

where f is a linear projection operator, transforming a set of displacements into the basis of an irreducible representation $\Gamma_{k\gamma}$, and g is a transcendental operator which associates the

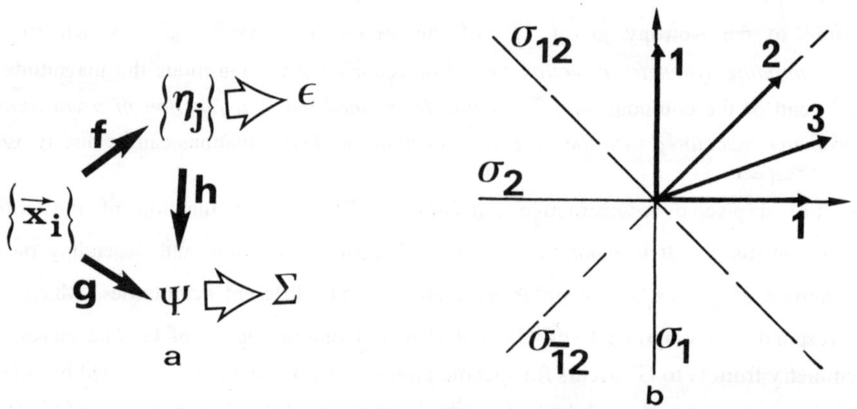

Fig.17. (a) Theoretical scheme showing the connection between ε and Σ, i.e. between the set of m displacements $\vec{x}_i$, the n scalar order-parameter components η_j, and the lattice functional ψ which depends on linear combinations of basic density waves $\exp\{i\vec{k}(\vec{R}_0 + \vec{x}_i)\}$. (b) Orientations of the "vector" order-parameter $\vec{\eta}$ in the ε_2 space associated with the C_{4v} - C_{6v} transition. 1, 2 and 3 are the orientations corresponding respectively to the symmetries C_{6v}, C_{4v} and C_{2v}.

set of displacements to the *variation* of the basic density waves (or simply to density waves for ordering transitions).

As an illustration of the preceding considerations, let us come back on the examples discussed in Sec.2. In the first example of displacive (D^4_{6h}-D^1_{6h}) transition, the subgroup $G=D^3_{3d}$ is induced by a one-dimensional irreducible representation (B_{1g}) associated with the center of the hexagonal Brillouin-zone (denoted $\tau_5(\vec{k}_{16})$ in Kovalev's notations [14]). The Σ-space is formed by the function (2.8) of critical displacements along Z, which has two symmetrical edges $\pm\eta_0$. The limit states have the symmetry $G'=D^1_{6h}$ corresponding to the holohedry of the D^3_{3d} lattice. They occur at the special points of the D^1_{6h} unit-cell shown in Fig.2.

The second case of displacive reconstructive transition described in Sec.2.2 shows explicitly the difference and connection existing *between the Σ-space and the critical displacement space*. Here, the ε-space is two-dimensional, as the group $G=C_{2v}$ is associated with a bidimensional irreducible representation ($\tau_3(\vec{k}_{15})$, with $\vec{k}_{15}=\frac{1}{2}\vec{a}^*$) of the initial C^1_{4v} planar space-group. In the ε-space three distinct *orientations* of the order-parameter $\vec{\eta}$ are found, as shown in Fig.17b.

1) For $\eta_1\neq 0$, $\eta_2=0$ (or for $\eta_1=0$, $\eta_2\neq 0$) the direction of $\vec{\eta}$ coincides with critical displacements (ξ_a *or* ξ_b) along X *or* Y, and its *magnitude* is $|\vec{\eta}(\xi)|=\eta_0\cdot \mathrm{Sin}(a^*\xi)$. The holohedry of the C_{2v} lattice increases to C_{6v}, when taking into account the tensile deformation (e_{xx} *or* e_{yy}), for the specific displacements of magnitude $\xi=\frac{a}{4}, \frac{3a}{4}, \frac{5a}{4}, \ldots$ along X *or* Y.

2) For $\eta_1=\eta_2=\eta\neq 0$, $\vec{\eta}$ is oriented along the diagonals of the (η_1, η_2) plane in the ε-space, and its magnitude is $|\vec{\eta}(\xi)|=\sqrt{2}\eta_0\cdot \mathrm{Sin}(a^*\xi)$, i.e. it corresponds to *equal* critical displacements $\xi_a=\xi_b=\xi$, along X *and* Y. For arbitrary values of ξ, one has $G=C_{4v}$ with *four* atoms per unit-cell, whereas for the special displacements $\xi_a=\xi_b=\frac{1}{2}a$ the preceding symmetry increases to $G'=C_{4v}$, *with one atom per unit-cell*, as shown in Fig.10b. The G'-lattice is turned by 45° with respect to the initial tetragonal lattice.

3) For $\eta_1\neq\eta_2\neq 0$, $\vec{\eta}$ has a general direction in the (η_1, η_2) plane and corresponds to *unequal* critical displacements $\xi_a\neq\xi_b$ along the X and Y axes. The magnitude of the $|\vec{\eta}(\xi)|$ function is: $\eta_0\left(\mathrm{Sin}^2(a^*\xi_a)+\mathrm{Sin}^2(b^*\xi_b)\right)^{1/2}$, and there are no special points at which the lowest C_{2v} symmetry may increase.

In the example of order-disorder O^9_h-D^1_{6h} transition discussed in Sec.2.3, the ε-space is eight-dimensional, although the orientation of $\vec{\eta}$ is invariant by a two dimensional subspace ε^G, where $G=D^3_{3d}$. Besides, G represents a distribution of positions never realized as a stable state, due to the existence of vacancies determined by the initial concentration 1/3. This specific value increases the symmetry of G, to D^1_{6h} which *is* the holohedry of the structure. The magnitude of $\vec{\eta}$ varies from 0 to 1, when $1/3\leq P^{\alpha}\leq 1$. All other fractional concentrations of the initial lattice (1/2, 2/3, ...) will *preserve* the rhombohedral symmetry D_{3d}^3, and the D_{6h}^1 symmetry will not take place in this case.

In the tetragonal-octagonal transition described in Sec.2.4, the ε-space is eight-dimensional, as the C_{4v}-C_{8v} symmetry change requires a star k* of the wave-vector with eight branches. However, while starting from the crystal structure in the description of the crystal-quasicrystal transition is physically justified (we usually *know* the structure and properties of the crystal, and try to work-out the structure and properties of the quasicrystal), it corresponds to an unusual scheme, with respect to the properties of the Σ-space. As shown in Fig.15b, the lowest symmetry G *coincides* with the crystal structure (C_{4v} symmetry with *8* atoms per unit-cell). The symmetry *increases* to C_{8v} or C_{4v} (with *4* atoms per unit-cell) for specific displacements ξ_{max} *and specific orientations of the transition wave-vector* $\Delta\vec{k}_{max}$, which are obtained by maximizing both the magnitude and orientation of the order-parameter with respect to ξ and $\vec{k}$.

3.2.3. Thermodynamic potential, equations of state, stable states

The non-equilibrium thermodynamic potential associated with a reconstructive transition has the general form:

$$F\left[T,P,\delta\psi(x_j)\right] = F_0(T,P) + F\left[\eta_{ki}(x_j)\right] \qquad (2.43)$$

where the n-components η_{ki} transform as an irreducible representation $\Gamma_{k\gamma}$ of the space group G_0 of the initial structure, and determine the possible stable directions of the order-parameter $\vec{\eta}$ in the ε-space. The magnitude of $\vec{\eta}$:

$$\left|\vec{\eta}(x_{j})\right| = \left[\sum_{i=1}^{n}\eta_{ki}^2(x_j)\right]^{1/2} \qquad (2.44)$$

depends on the m variational parameters x_j.

The *form* of the order-parameter expansion $F\left[\eta_{ki}(x_j)\right]$ is determined, as usually, by the symmetry of $\Gamma_{k\gamma}$ (see Chapter I, Sec.2.3). By contrast, *the minimization of the expansion has to be performed with respect to the variational parameters* x_j , *and not with respect to the* η_{ki} , as in the standard procedure. This leads to m equations of state:

$$\frac{\partial F}{\partial x_j} = \sum_i \frac{\partial F}{\partial \eta_{ki}} \cdot \frac{\partial \eta_{ki}}{\partial x_j} = 0 \qquad (j=1\text{-}m) \qquad (2.45)$$

and to stability conditions, which consist in finding the range of positiveness of the eigenvalues of the $(m\times m)$ matrix built from the second derivatives of $F\left[\eta_{ki}(x_j)\right]$ with respect to x_j:

$$\left\| \frac{\partial^2 F}{\partial x_j \partial x_k} \right\| \geq 0 . \tag{2.46}$$

Equation (2.45) clearly shows the existence of *two different sorts of stable states*: the "Landau" states, which correspond to the usual minima determined by the conditions $\frac{\partial F}{\partial \eta_{ki}}=0$ (i=1-n), and the limit states, associated with the conditions : $\frac{\partial \eta_{ki}}{\partial x_j}=0$ (j=1-m).

The minimization procedure corresponding to Eqs.(2.45) and (2.46) is the more general, and it allows to work out all the Landau *and* limit states which may take place as the result of a transition (reconstructive or not reconstructive) from a given structure, of symmetry group G_0, for a given irreducible representation $\Gamma_{k\gamma}$. However, the problem posed by the interpretation of reconstructive phase transitions, that we actually aim to investigate in this book, usually appears under a more restricted, but more complex, form: *Given a reconstructive phase transition between two known structures (associated with the unrelated group-subgroup symmetries* G_0 *and* G'*), 1) what is the microscopic (atomistic) mechanism of the transition? 2) which theoretical framework allows to describe the specific macroscopic features (phase diagram, physical properties) of the transition?* The practical procedure which allows to answer to these questions, can be decomposed as follows.

1) Determination of the maximal substructure, of space group $G=G_0 \cap G'$, common to the structures surrounding the transition, from which one can deduce, following the standard procedure indicated in Chapter I, the n-dimensional irreducible representation $\Gamma_{k\gamma}$ of the G_0 space-group associated with the virtual $G_0 \rightarrow G$ "Landau" transition. G corresponds to given relationships between the n order-parameter components, *which fix the orientation of* $\vec{\eta}$ *in the* ε*-space*.

2) Taking into account the symmetry of the degree of freedom $\Gamma_{k\gamma}$, one can find the *possible* atomic mechanisms transforming the intermediate virtual structure (G) into the final one (G'). The *nature* of the mechanism (displacive, order-disorder, or a combination of the two types of processes) and the selection of the *more probable* steps of reorganization from one structure to another, have to be chosen taking into account their consistency with the experimentally known properties of the transition. For example, in the case described in Sec.2.2.1, three possible mechanisms correspond, a priori, with the C_{4v}-C_{6v} structural change, but only one preserves the crystal shape.

3) Once the effective transition mechanism is singled out, one expresses the structures in terms of the lattice functions ψ_0 and ψ, and deduce the form of the increment $\delta\psi(x_j)$, which provides the magnitude of the order-parameter $\left|\vec{\eta}(x_j)\right|$ in function of the relevant variational parameters x_j. *This determines the structure and*

symmetry of the corresponding functional space Σ, and the critical variational parameters x_j^c at which the $G_0 \rightarrow G$ group-subgroup relationship is lost.

4)For a thermodynamic description of the transition, which gives the phase diagram in which the various structures are inserted, as well as the macroscopic properties of the phases, one can then restrict to the use of the effective order-parameter expansion:

$$F\left[\left|\vec{\eta}(x_j)\right|\right] = \frac{\alpha}{2}\left|\vec{\eta}(x_j)\right|^2 + \frac{\beta}{3}\left|\vec{\eta}(x_j)\right|^3 + \frac{\gamma}{4}\left|\vec{\eta}(x_j)\right|^4 + ... \qquad (2.47)$$

in which the allowed non-vanishing successive powers of $\left|\vec{\eta}(x_j)\right|$ are determined by the symmetry of $\Gamma_{k\gamma}$.

The various stages of the phenomenological approach to reconstructive phase transitions, may be simplified, or on the contrary complicated, depending on the nature of the mechanism.

i) For displacive transitions between strictly periodic structures, the wave-vector is fixed ($k=k_0$), and for a given direction $\vec{\eta}$ in the ε_n-space, the order-parameter is a sinusoidal function of the critical displacements given by Eq.(2.39). The extrema of this function, given by $\frac{\partial \eta_{ki}}{\partial x_j}=0$, provide directly the special points of the lattice at which the structure of space-group G′, may arise, whereas the intersections of the sinusoid with the displacement axis coincide with different orientational domains of the initial structure. The effective order-parameter expansion (2.47) is a sum of powers of sinusoidal functions, i.e. it represents a Fourier series, corresponding to the successive higher harmonics of the series. One can generally restrict to the lowest degrees (fourth or sixth) of $F\left[\left|\eta_{ki}(x_j)\right|\right]$, as the sinusoidal arc separating the initial structure from the first maximum (one fourth of a period) expresses a fifth-degree approximation of the sinus with respect to the x_j, as illustrated in Fig.18.

ii) For displacive reconstructive transitions to aperiodic structures, k is also a variable which determines the orientation of $\vec{\eta}$ in the ε-space, and the magnitude of $\left|\vec{\eta}(x_j)\right|$ in the Σ-space. Thus, the determination of the equilibrium, aperiodic, state requires a minimization procedure with respect to k, and not only with respect to the displacements. This can be performed either by minimizing the lattice functional increment, given by Eq.(2.40), or by a minimization of the non-equilibrium thermodynamic potential $F\left[\left|\vec{\eta}_{ki}(x_j)\right|\right]$ with respect to k and to the x_j. In this case, the dependence on k is taken into account by a renormalization in k-space (see Sec.4.3 of Chapter I) of the coefficient $\alpha(k)$ of the second-degree invariant $\sum_i \eta_{ki}^2$, as in the current approach to incommensurate structures [16].

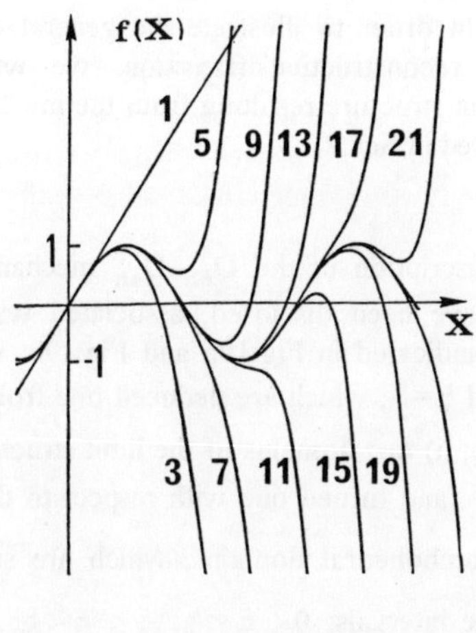

Fig.18. Approximation of the Sin(x) function by successive functions x^n of increasing power.

iii) For order-disorder transitions, the x_j variables identify to the increments δP_j^α of probability of occupancy of the j position by the α-atom. As the η_{ki} are *linear* functions of these variables, Eq.(2.45) reduces to the usual equation of state : $\frac{\partial F(\eta_{ki})}{\partial \eta_{ki}}=0$. The magnitude of $\left|\vec{\eta}\left(\delta P_j^\alpha\right)\right|$ determines the partial or full ordering or the states. As already noted in Sec.3.2.2, the distinction between "Landau" and limit states is here meaningless as G and G′ have the same symmetry. This stems from the fact that in our description, we have not considered the number of atoms in the unit-cell of the initial structure (the concentration) as a variable, but as a fixed number, which determines the lower limit of the δP_j^α in the ordered states. The loss of group-subgroup relationship for specific fractional concentrations, which is typical of morphotropic transitions [15], reveals that the temperature-concentration phase diagram of order-disorder transitions *would correspond to a non-linear dependence of the order-parameter,* with a more complex structure than the linear function $\left|\vec{\eta}\left(\delta P_j^\alpha\right)\right|$ found for *temperature dependent* reconstructive order-disorder mechanisms.

4. Specific properties of reconstructive phase transitions

4.1. Orientational and antiphase domains

The domain structure of the phases which may be stabilized in a reconstructive transition process, differs from the domain structure found in structural transitions with group-subgroup related phases. In order to illustrate the general rules followed by the domain texture induced in a reconstructive transition, we will first describe the orientational and antiphase domain structure resulting from the mechanisms corresponding to the three first examples discussed in Sec.2.

Example 1.

In the crystallographic description of the D_{6h}^4-D_{6h}^1 mechanism (Sec.2.1.1) *eight* distinct orientational domains have been disclosed, associated with critical antiparallel displacements along the Z axis, indicated in Fig.19a and Fig.19b: i) *two* domains of the initial D_{6h}^4 structure for $\xi=0$, and $\xi=\frac{c}{2}$, which are deduced one from another by a *simple* rotation of 60° around the Z axis; ii) *two* domains of the limit structure D_{6h}^1, obtained for the critical shifts $\xi=\frac{c}{4}$ and $\xi=\frac{3c}{4}$, and turned one with respect to the other by the screw rotation $\left\{C_6 \middle| 00\frac{c}{2}\right\}$; iii) *four* rhombohedral domains, which are stabilized for arbitrary shifts pertaining to the respective intervals: $0<\xi<\frac{c}{4}$, $\frac{c}{4}<\xi<\frac{c}{2}$, $\frac{c}{2}<\xi<\frac{3c}{4}$, and $\frac{3c}{4}<\xi<c$. They form two different *pairs*, corresponding to the two different orientations of the D_{6h}^4 structure, the transformation properties of which are summarized in Fig.19c. In one pair, one domain is deduced from the other by a simple six-fold rotation, whereas in the other pair, it is deduced by a six-fold screw rotation. As the hexagonal and rhombohedral structures are equitranslational, no antiphase domains are found.

Example 2.

The antiparallel displacements of rows of atoms along X, which take place at the C_{4v}-C_{6v} mechanism, shown in Fig.9d, involve a doubling of the b tetragonal lattice parameter. Accordingly, the resulting orthorhombic C_{2v} unit-cell possesses two possible antiphase domains, represented in Fig.20a, which are deduced one from another by the lost $\vec{b}$-translation. Each of these antiphase domains is associated with two orientational orthorhombic domains, shown in Fig.20b, which are deduced one from another by the four-fold rotation, lost at the transition. For the specific shifts $\xi=\frac{a}{4}$, $\frac{3a}{4}$, ... the two preceding domains become hexagonal domains, providing a suitable tensile strain e_{yy} is taken into account. Starting from these hexagonal domains, one gets for arbitrary shifts three orthorhombic orientational domains, turned by 60° with respect to each other, around the created six-fold axis. At last, for the specific shifts corresponding to the tetragonal C_{4v} structure ($\xi=0$, $\frac{a}{2}$, a, ...), one gets three distinct tetragonal domains, forming 60° angles between them.

Fig.20c summarizes the total number of orientational domains, which are realized in the transition process, namely 3 tetragonal, 2 hexagonal, and 6 orthorhombic domains. The preceding numbers are doubled when distinguishing antiphase domains. Let us note that *cycling again from one phase to the other, will not increase the number of domains, as fixed symmetry planes* (e.g. perpendicular to X and Y) *determine the absolute orientation of the domains.*

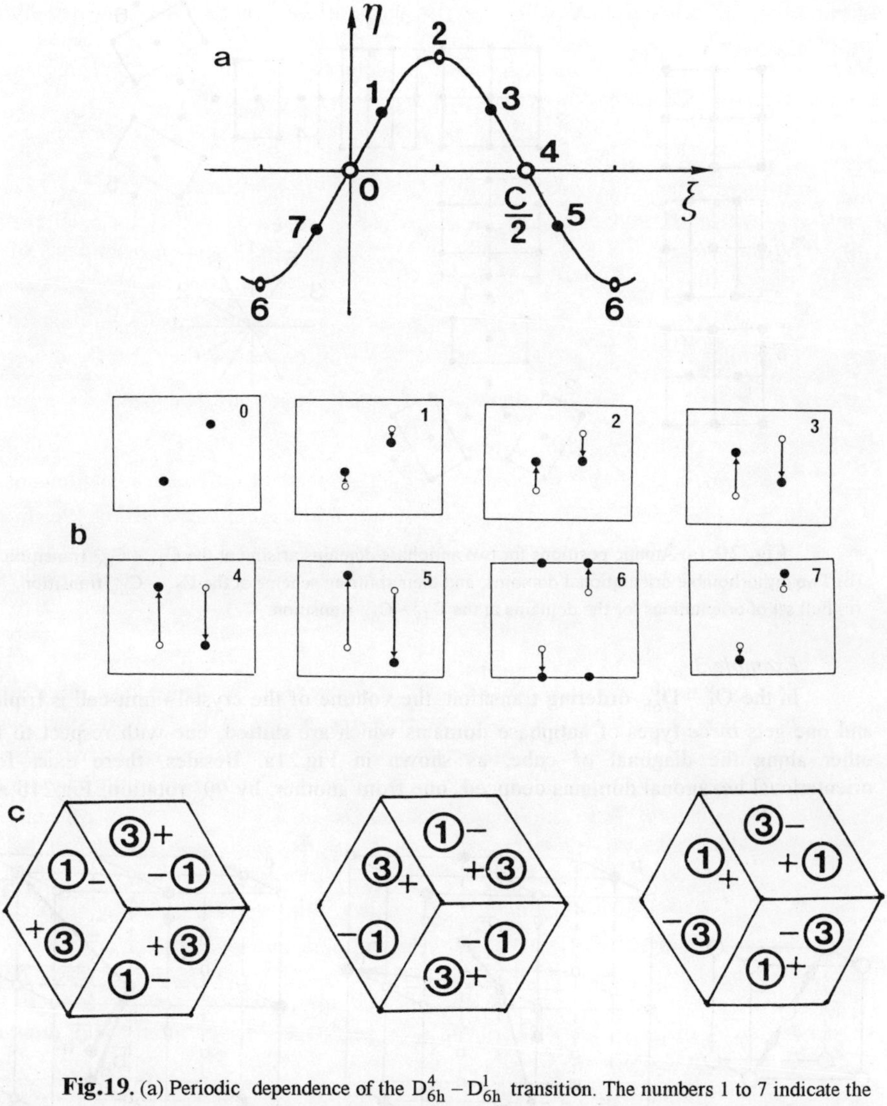

Fig.19. (a) Periodic dependence of the $D_{6h}^4 - D_{6h}^1$ transition. The numbers 1 to 7 indicate the critical displacements corresponding to the different domains represented in (b). (c) Orientational properties of two different pairs of rhombohedral domains. + and - indicate the directions of the shifts along the Z axis for the atoms located at c/4 and 3c/4 (denoted respectively by **1** and **3** on the figure).

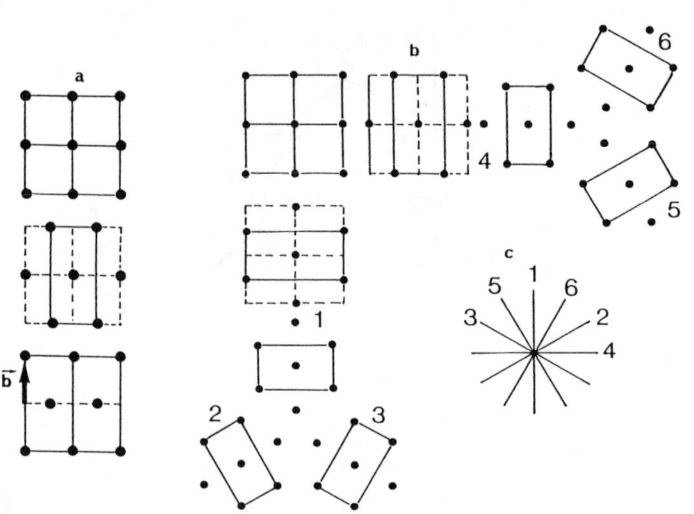

Fig. 20. (a) Atomic positions for two antiphase domains arising at the $C_{4v} - C_{6v}$ transition. (b) Two orthorhombic orientational domains, and their splitting scheme at the $C_{2v} - C_{6v}$ transition. (c) Full set of orientations for the domains at the $C_{4v} - C_{6v}$ transition.

Example 3.

In the $O_h^9 - D_{6h}^1$ ordering transition, the volume of the crystal's unit-cell is tripled, and one gets *three* types of antiphase domains which are shifted, one with respect to the other along the diagonal of cube, as shown in Fig.21a. Besides, there exist four orientational hexagonal domains deduced, one from another, by 90° rotation. Fig.21b and

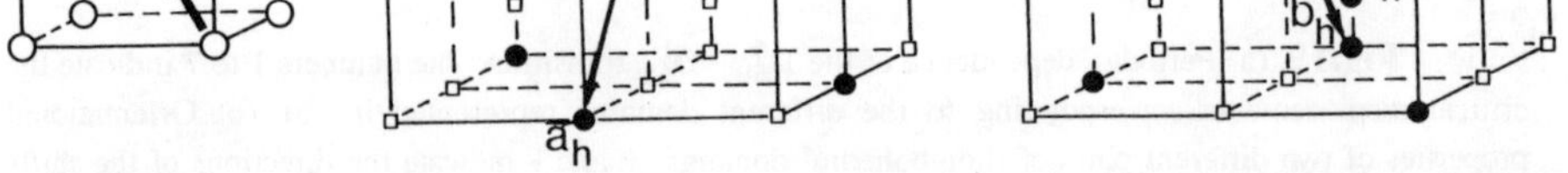

Fig. 21. (a) Shift between the antiphase domains arising at the $O_h^9 - D_{6h}^1$ transition. (b) and (c): Two different orientational hexagonal domains rotated one from another by the C_4^y rotation.

Fig.21c represent two different domains rotated around the C_4^y axis. These domains have the same distribution as in the virtual D_{3d}^3 ordered structure, which becomes hexagonal for a specific concentration 1/3 of the initial cubic unit cell, i.e. the domain structure of the ordered phase displays the symmetry of the distribution of the sites. The reverse disordering transition D_{6h}^1-O_h^9 does not create distinct cubic domains, as all the atomic positions remain fixed at the transition, and only the permutation group of the sites is modified.

Accordingly, the preceding examples show that, while for order-disorder reconstructive transitions, the domain textures of the phases, follows the standard scheme existing for group subgroup related phases, *this is not the case for displacive reconstructive transitions*. The domain structure induced by such transitions possess a number of distinctive features. Thus, one finds that *the two phases surrounding the transition possess orientational domains*. The orientations of these domains depend on the magnitude of the critical displacements, and in each phase the domains transform into one another by the symmetry operations belonging exclusively to the *other* phase. If, in the transformation process, the Landau state of symmetry group $G = G_0 \cap G'$ is stabilized, the number of its domains is given by the rule:

$$n = \frac{n_{G_0} \times n_{G'}}{n_G^2} \qquad (2.48)$$

where n_{G_0}, $n_{G'}$, and n_G are the respective orders of the groups G_0, G' and G. Concrete verification of the preceding properties are discussed in Chapters III and IV.

4.2. Temperature dependence of the order-parameter and related physical quantities

Since a displacive reconstructive transition to a limit state occurs for fixed critical displacements (given by the extrema $\partial\eta/\partial\xi = 0$) the corresponding temperature dependence of the order-parameter, displays a typical step-like behavior, shown in Fig.22a. Reciprocally, the constancy of $\eta(T)$ below the transition reflects only the existence of a limit state, but does not imply necessarily the absence of a group-subgroup relationship between the phases. For example, a step-like behavior is found for the liquid-solid transition [17], and also for a number of first-order transitions in ferroic materials (e.g. in langbeinites [18]), in which the complexity of the structure preserves the group-subgroup connection of the phases (see Sec.5). When a displacive reconstructive transition occurs via an intermediate Landau state, one may have different behaviors for $\eta(T)$, as shown in Fig.22b and Fig.22c.

A constant limit $\eta=1$ is also reached by the order-parameter in the fully ordered state resulting from a disorder-order transition. Thus, the same step-like shape represented in Fig.22a should be observed for reconstructive order-disorder transitions. However, due

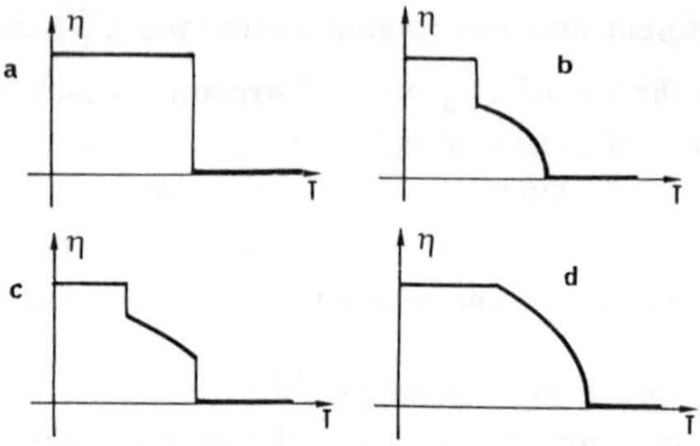

Fig.22. (a) Step-like dependence of η(T) at a direct transition to a limit state. (b) Temperature dependence of η(T) for a second-order transition to a Landau state followed by a transition to a limit state, (c) for a first-order transition to a Landau state, followed by a transition to a limit state, (d) for an ordering transition to a fully ordered state going through the partially ordered state.

to the fact that partially ordered states, with $0<\eta<1$, do not differ in their symmetry from the fully ordered state, the $\eta(T)$ curve is more liable to display the dependence shown in Fig.22d, in which the partially ordered-fully ordered limit, corresponds to a *break* between the "Landau" and limit regimes. A comparison with experiment is here uneasy (see Chapter IV), and to our knowledge, the more generally observed behavior coincides with the curve represented in Fig.22e, in which the limit regime is reached asymptotically, without break.

Another distinctive property of reconstructive transitions, is the temperature dependence of the specific heat $C_P = -T\frac{\partial^2 F}{\partial T^2}$. From the constant values of the order-parameter in the two phases surrounding a reconstructive transition, one can easily deduce that the corresponding values of their specific heats are *equal*, and thus *no jump* ΔC_P should take place at the transition. This is in contrast with the *upward* jump observed on cooling for first (or second)-order transition to Landau states (Fig.11b in Chapter I). However, from the finite discontinuity in entropy $S = -\frac{\partial F}{\partial T}$ at the transition, one can infer that $C_P(T)$ will display a *narrow peak* at the transition temperature, as shown in Fig.23a.

Table 2 summarizes the expressions, in function of the phenomenological coefficients of the thermodynamic potential (2.9), of the latent heats $L = \frac{\Delta S}{T_{tr}} = -\frac{1}{T_{tr}} \cdot \frac{\partial \Delta F_{eq}}{\partial T}$ and specific heat jumps ΔC_P, for the various transitions occurring in the phase diagram of Fig.4. Fig.23b and Fig.23c illustrate the corresponding behaviors of $C_P(T)$ for, respectively, a second- and first-order transition to a Landau state (e.g. between states I and III in Fig.4). Fig.23d shows the variation of $C_P(T)$ for the transition (II-III in Fig.4) between a Landau state and a limit state.

Table 2. Latent and specific heats for the phase transitions represented in Fig.4.

	Transition temperature	L	ΔC_p
I-III (2nd order)	T_c	0	$\frac{T_c a_{10}^2}{2a_2}$
I-III (1st order)	$T_1 = T_c + \frac{a_2^2}{4a_{10}a_3}$	$-\frac{a_{10}}{T_1} \cdot \frac{a_2}{6a_3}$	$-\frac{T_1 a_{10}^2}{a_2}$
I-II	$T_2 = T_c - \frac{a_2 + a_3}{a_{10}}$	$-\frac{a_{10}}{T_2}$	0
II-III	$T_3 = T_c - \frac{2a_2 + 3a_3}{a_{10}}$	0	$\frac{T_3 a_{10}^2}{2(2a_2 + 3a_3)}$

Using the forms (2.8) and (2.9) of the order-parameter η and thermodynamic potential $F(\eta)$, respectively, one can calculate the temperature dependence of the susceptibility associated with the order-parameter : $\chi = \lim_{\bar{\eta}\to 0}\left(\frac{\partial \eta}{\partial \bar{\eta}}\right)$, where $\bar{\eta}$ is the field conjugated to η. One finds for the phases I, II and III:

$$\chi_I = (2a_1)^{-1} \; ; \; \chi_{II} = \left(2a_1 + 12a_2\eta_0^2 + 30a_3\eta_0^4\right)^{-1} \; ; \; \chi_{III} = \left[-8\left(a_1 + a_2\eta_0^2\right)\right]^{-1} .$$

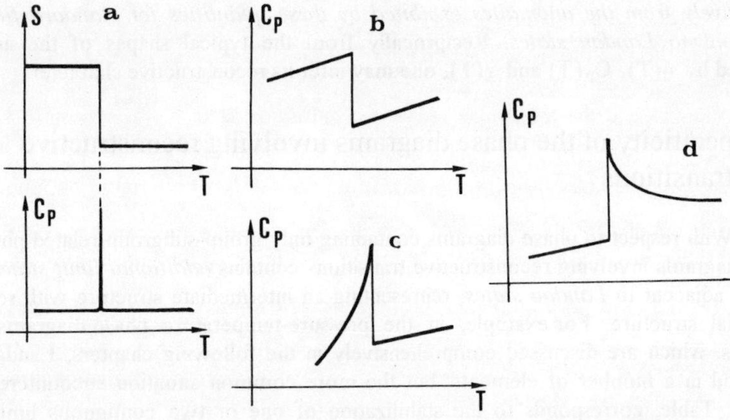

Fig. 23. (a) Dependence on temperature of the entropy S(T), and specific heat $C_P(T)$ at a reconstructive transition. (b) $C_P(T)$ for a second-order and (c) a first-order transition to a Landau state. (d) $C_P(T)$ for a second-order transition between a Landau state and a limit state.

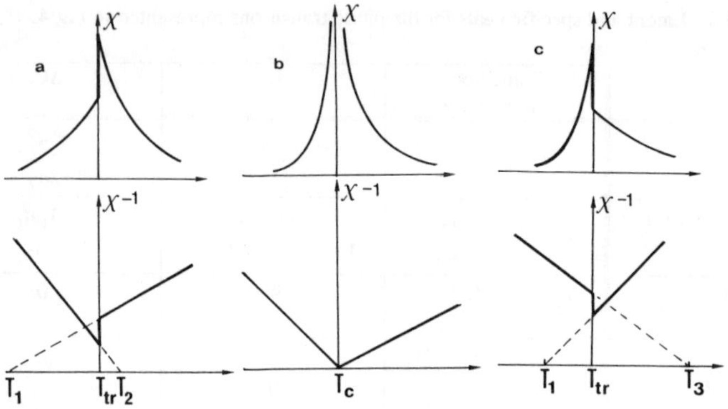

Fig. 24. Temperature dependences of the susceptibility $\chi(T)$ and its inverse $\chi^{-1}(T)$ for (a) a reconstructive transition, (b) a second-order transition to a Landau state, (c) a transition between a Landau state and a limit state.

The corresponding temperature dependences $\chi(T)$ across the I-II, I-III, and II-III transitions and the inverse susceptibilities $\chi^{-1}(T)$ are shown, respectively, in Figs.24a-c. One can see that the reconstructive I-II transition is characterized by a *downward* jump of $\chi(T)$, while an upward jump is usually found for the first-order transition I-II to a Landau state as shown in Chapter I (Sec.3.3.2).

In summary, we have shown that *the anomalies found at reconstructive transitions for the order-parameter* $\eta(T)$, *the specific heat* $C_P(T)$ *and the susceptibility* $\chi(T)$, *differ qualitatively from the anomalies exhibited by these quantities for standard first-order transitions to Landau states.* Reciprocally from the typical shapes of the anomalies displayed by $\eta(T)$, $C_P(T)$ and $\chi(T)$, one may infer its reconstructive character.

4.3. Specificity of the phase diagrams involving reconstructive transitions

With respect to phase diagrams containing only group-subgroup related phases, the phase diagrams involving reconstructive transitions contains *additional limit states*, which may be adjacent to *Landau states*, representing an intermediate structure with respect to the initial structure. For example, in the pressure-temperature phase diagrams of pure elements, which are discussed comprehensively in the following chapters, Landau states are found in a number of elements, but the more common situation encountered in the Periodic Table, corresponds to the stabilization of one or two contiguous limit phases below the initial structure.

The fact that the order-parameter has a constant value in a limit phase results in general properties of the phases boundaries, which in a two dimensional (Pressure-Temperature) phase diagram, can be expressed as follows.

PROPERTY 1: The lines of first-order transitions between limit phases as well as the lines limiting their stability regions are *straight lines*.
Proof: The effective order-parameter expansion (2.47) can be written under the general form:

$$F(x_j) = \sum_i a_i(P,T,...) \cdot f^{n_i}(x_j) \tag{2.49}$$

where the $f^{n_i}(x_j)$ are n_i-th degree monomials of the x_j, and the a_i are phenomenological coefficients, which depend linearly on the external variables P, T, ... The equations of the limit of stability lines which are : $\left\| \frac{\partial^2 F}{\partial x_j \partial x_k} \right\| = 0$ (j, k=1,...,m), take the explicit form:

$$\sum_i n_i \cdot (n_i - 1) \cdot a_i(P,T,...) \cdot f^{n_i-2}(x_j) = 0. \tag{2.50}$$

As the (n_i-2)-th degree monomials $f^{n_i-2}(x_j)$ are constant in the limit phases, the *linear* equations of the a_i (2.50), define *hyperplanes* in the n-dimensional space A_n of the coefficients a_i. The hyperplanes are straight lines in two-dimensional *cross-sections* of the A_n-space. As the Pressure-Temperature plane can be obtained by a linear transformation of any of the preceding cross-sections, the same property holds in the (P, T) plane, in which the limit of stability lines of the limit phases appear as straight lines.

First-order transition lines are lines of equal energies. If $F(\xi_j^I)$ and $F(\xi_j^{II})$ are the equilibrium values of the order-parameter expansions corresponding to the limit phases denoted I and II, the equality of the expansions on the first-order transition line I-II, gives:

$$\sum_i a_i(P,T,...) \cdot \left[f^{n_i}(x_j^I) - f^{n_i}(x_j^{II}) \right] = 0.$$

As the term between brackets is constant, one has again a linear equation relating the a_i. The same arguments given here above leads to the conclusion that the first-order transition line I-II is a straight line.
PROPERTY 2. The limit of stability and transition lines of a limit phase, with respect to the initial phase, are tangent, respectively, to the limit of stability and transition lines of the adjacent Landau state, with respect to the initial phase. The point of tangency of the transition lines coincides with the three-phase point, at which merge the initial, limit and Landau phases.
Proof: In the A_n-space of the phenomenological coefficients a_i, the equation of the hyperspace $\Omega(a_i)$=0 limiting a Landau phase, can be approximated by a polynomial:

$$\Omega(a_i) = \sum_i a_i \cdot f^{n_i}(x_j) = 0 \tag{2.51}$$

where f is a non linear function of the variational parameters x_j, which are functions of the a_i . The three-phase point N at which the three phases merge corresponds to constant values of the x_j , and thus to constant $f(x_j)$. Thus, at the N-point, Eq.(2.51) becomes the equation of a hyperplane $\sum_i a_i=0$, which is necessarily tangent to the hypersurface $\Omega(a_i)=0$. Besides, as the preceding hyperplane contains the N-point, it identifies to the hyperplane separating, in the A_n-space, the limit and initial phases. In projection in a two-dimensional section of A_n , the hyperplane and hypersurface become transition *lines*. The proof for the limit of stability lines, follows an identical scheme.

Properties 1 and 2 are clearly illustrated in the phase diagram of Fig.4, in the (a_1,a_2) plane of the phenomenological coefficients pertaining to the thermodynamic potential (2.9). Thus, the first-order transition line separating the initial (I) and limit (II) phases, has the straight line equation: $a_1 = -a_2\eta_0^2 - a_3\eta_0^4$. It prolongates the (parabolic) line of transitions between phases I and III at the three-phase point of coordinates: $a_1 = a_3\eta_0^4$, $a_2 = -2a_3\eta_0^2$. On the other hand, the limit of stability line of phase II with respect to phase I, also obeys the equation of a straight line ($a_1 = -2a_2\eta_0^2 - 3a_3\eta_0^4$). It is tangent to the line of stability limit of phase III with respect to phase I, at the point of coordinates ($a_1 = 3a_3\eta_0^4$, $a_2 = -3a_3\eta_0^2$).

4.4. Phonon spectra and reconstructive phase transitions

A temperature dependent softening of phonon branches corresponding to the critical modes is expected, and generally observed, for second-order or weakly first-order displacive structural transitions. For more pronounced first-order displacive transitions, between group-subgroup related structures, the softening decreases with the decrease in amplitude of the critical oscillations. At reconstructive transitions, for which the critical displacements between the initial and limit phases are *fixed*, there is a priori *no critical fluctuations*, and thus *no softening should be observed.* However, this property should be verified only far from the domain of stability of the intermediate Landau state, and provided a secondary, temperature dependent, displacive mechanism should not be involved in the transition.

No softening is observed in the phonon spectra of second-order order-disorder transitions, as the average atomic positions remain fixed across such transitions, and the jumps of atoms between the sites, are supposed to be uncorrelated. This picture is still reinforced for discontinuous order-disorder transitions between group-subgroup related and unrelated structures. The absence of mode softening holds, however, for purely ordering mechanisms. In Chapter IV, we show that ordering processes often take place with correlated displacive mechanisms, which may result in slight softening of some phonon branches.

The absence of temperature dependence of the phonon spectrum for reconstructive transitions, constitutes however only a negative test. A more positive confirmation of the mechanisms proposed in our theoretical description of these transitions can be found in the frequency dependence on the reduced k-vector of the phonon spectra, as it can be used for verifying the coincidence of the structures between the initial and limit phase, in the directions of space *preserved* by the reconstruction of the lattice, i.e. one expects to find a coincidence of the phonon branches for the two structures in these directions and a softening of $\omega(k)$ at special points corresponding to the translational connection of the structures. The phonon spectrum may also serve as a test to check the influence of macroscopic deformations, which are often involved in reconstructive transition mechanisms, as such deformations modify the distance between equivalent sites in the initial and final structure (the *metric* of the structures). The *optical* phonon curves will thus display *shifts* with respect to their location in the undeformed phase, whereas *acoustic* phonons branches may show changes in their *slopes*.

5. Applicability of the density-wave theory of reconstructive phase transitions

5.1. Summary: Specificity of reconstructive phase transitions

The phenomenological description of reconstructive phase transitions presented in this chapter, has disclosed the distinctive properties of this class of transitions, on a theoretical ground. Thus, when starting from an initial structure, a reconstructive transition to a limit phase occurs for definite *fixed* values of the variational parameters, for which a simple (group-subgroup) symmetry relationship with the initial structure is lost. The initial and limit phases *both* possess a domain structure reflecting the symmetry elements which are lost when going from one phase to the other. The physical anomalies which characterize the transition also display typical features, such as the step-like temperature (or pressure) dependence of the order-parameter, the absence of jump at the transition for the specific heat, and the corresponding drop of the susceptibility. The particularities of the phase diagrams, and of the dynamical properties (phonon spectra) of reconstructive phase transitions, have also been stressed.

The specificity of reconstructive transitions is also reflected in the mathematical structure of the quantities which are used in the approach. Thus, the order-parameter space *includes* the periodicity of the transition mechanism, as this mechanism also determines the symmetry of the order-parameter. For a given *direction* of the order-parameter in the representation space, the transition mechanism is expressed by the magnitude of the order-parameter, which is either a periodic function of the critical displacements, or a scalar function of the increment of occupancy of probability of definite atomic sites. Consequently the *effective* order-parameter expansion depends exclusively on the magnitude of the order-parameter, and is thus a Fourier series for displacive transitions. The various stable states arising from the initial state, are obtained by minimizing the order-parameter expansion, with respect to the variational parameters. This is made possible by the fact that the *explicit form* of the magnitude of the order-parameter in function of these parameters has been *deduced* from the increment of the lattice functional between the initial and limit structures, which are calculated using a density-wave formalism.

For the interpretation of reconstructive transitions in concrete systems a number of points in our approach still require to be clarified, as for example *the choice of the initial structure*.

5.2. Parent phase and initial structure

In the usual Landau theory of structural phase transitions the parent phase is generally chosen as the crystal phase of *maximal symmetry*, which is also, with a few well understood exceptions, the phase which is the more stable at high temperature, below the melt. The lower symmetry phases, appearing at lower temperatures, correspond to subgroups (G_i) of the parent phase space-group G_0 , and the G_i transform as irreducible (or reducible) representations of G_0 . The consistency of this simple scheme is attested by the observed properties of the low symmetry phases (domain structure, spontaneous macroscopic tensors), which can be *deduced* from the $G_0 \rightarrow G_i$ broken symmetry.

For reconstructive transitions the situation is less comfortable and the choice of the parent phase is not obvious, and *may be arbitrary*. In most cases the experimental observations *cannot* discriminate between parent and "daughter" phase: absence of group-subgroup relation between the phases, existence of orientational domains in both phases, each of which may be considered as the *undeformed* state, for the macroscopic strains, which are generally the sole spontaneous tensor involved in reconstructive transitions.

Arbitrariness in the choice of the *initial* structure mainly occurs for displacive reconstructive transitions. For instance, in the two examples of transitions discussed in Sections 2.1 and 2.2, one may *chose* as initial structures indifferently the D_{6h}^1 and C_{6v} phases *instead* of the D_{6h}^4 and C_{4v} phases, respectively, as taken in our description. In this case, there is no "natural" parent phase, and one may better speak of *initial structure*. The choice of the "best" initial structure in concrete cases, must be then determined by physical considerations such as the location of the phases in the phase diagram (the highest temperature phase) or by specific geometrical considerations (as the closest packing of atoms in one phase).

In the framework of our approach the indetermination in the existence of an *actual* parent phase can be foreseen in the *dependence of the order-parameter in function of the critical displacements*, and in the *form* of the effective order-parameter expansion. Thus, in the case of the $D_{6h}^4 \rightarrow D_{6h}^1$ transition, Eq.(2.8) and Fig.3 show that the sinusoidal dependence of $\eta(\xi)$ is *symmetric* with respect to the displacement axis, namely that the maxima *and* minima $\pm\eta_{max}(\xi)$ both correspond to *the same* D_{6h}^1 structure. In other words, one gets periodically the same sequences of phases for opposite critical shifts. As a result the effective thermodynamic potential (2.9) contains only *even* powers of $|\vec{\eta}(\xi)|$. A similar situation is found for the $C_{4v} \rightarrow C_{6v}$ order-parameter dependence on ξ (Eq.(2.19) and Fig.10b) and order-parameter expansion.

By contrast, when a displacive reconstructive transition, is associated with an *asymmetric* periodic dependence of the order-parameters in function of the critical displacements, or in other words when the maxima and minima of $\eta(\xi)$ correspond to *different* structures, *the choice of the initial structure is not free, and only one of the two structures surrounding the transition may play this role*. In this case, one finds that $\eta(\xi)$ has a more complex form, and the effective order-parameter expansion necessarily contains a *cubic invariant*. In such a situation, which is found for the $\beta-\omega$ transition discussed in Chapter III, the initial structure identifies unambiguously with the parent phase.

By contrast, no ambiguity exists in the choice of the parent phase for order-disorder reconstructive transitions as the diffraction patterns clearly distinguish the disordered phase from the ordered phases. Even when the transition takes place between two differently ordered structures, our description (see Chapter IV) assumes a *latent disordered parent phase*, which is the *maximal substructure* common to the two ordered structures. The choice of the parent phase is also well determined for crystal-quasicrystal transitions, as our approach to these transitions, given in Chapter V, reveals that the quasicrystal structure *is the parent structure*, although for practical reasons (i.e. the fact that we *investigate* the quasicrystal state starting from a known crystal state), we chose in this case the crystal structure as the *initial* structure.

5.3. Limit (non-Landau) states and limit (Landau) states

In phase diagrams involving reconstructive transitions two types of phases are distinguished: 1) *Landau phases*, which are group-subgroup related to the initial structure. They display a usual temperature dependence of the order-parameter below the transition, and are obtained by a minimization of the thermodynamic potential with respect to the order-parameter. 2) *Limit (non-Landau) phases* which possess no group-subgroup relationship, with respect to the initial structure. They are characterized by a *constant* magnitude of the order-parameter (temperature independent), and correspond to the *extrema* of $|\vec{\eta}(\xi)|$ with respect to the variational parameter ξ. Note that these phases fulfil the condition $\frac{\partial^2 F}{\partial \xi^2} \geq 0$ and thus *coincide with the boundaries of the stability regions of Landau phases*.

While the absence of group-subgroup relationship between two structures on either sides of a transition is a clear-cut signature of its reconstructive character, the constancy of the order-parameter below a transition does not imply *a contrario* that the transition is of the reconstructive type. For example, in first-order transitions to Landau phases, the transition mechanism may reach its limit below the transition discontinuity. This is the case for the liquid-solid transition [17] and for some ferroelectric, or ferroelastic, transitions [18] where the spontaneous polarization, or strain, appear as saturated below the transition. This class of "Limit Landau phases" which fulfil the simultaneous conditions:

$$\frac{\partial \eta}{\partial \xi}=0 \quad \text{and} \quad \frac{\partial^2 F}{\partial \xi^2}=0 \qquad (2.52)$$

below the transition point also contains a number of "unachieved" reconstructive transitions, for which the preservation of the group-subgroup relationship, with respect to the initial structure, is related to specific crystallogeometrical restrictions. This may occur, for example, when a structure reaches a geometrical limit imposed by a close packing of the atoms. Another illustration is found in some complex structures where "non-critical" sublattices preserve the overall symmetry of the structure, while the displacements of "critical" sublattices reach a limit giving rise to new symmetry elements for the corresponding substructure. In Chapter III we describe concrete examples of "limit-Landau phases" among actinide elements, in which it is the non-spherical shape of the atoms that preserves the group-subgroup relationship with the parent (BCC) structure. Let us stress that one can find situations in which a limit Landau phase transforms into a non-Landau phase, as it will be shown in the description of the BCC-HCP transition (Chapter III, Sec.3.3).

5.4. Intermediate nature of the mechanism in reconstructive transitions

In this chapter, in order to disclose the theoretical behaviors of reconstructive transitions, we have insisted upon the differences existing between displacive and order-disorder mechanisms. However, in the interpretation of concrete examples of reconstructive transitions the two mechanisms often appear as correlated. This is mainly true for the ordering processes which are almost always accompanied by collective displacements (see Chapter IV). The connection between ordering and displacive mechanism in reconstructive transition stems from the fact that these transitions realize in some respect an intermediate situation between ordering and displacive processes. In order to make this point more clear, we will first briefly recall the current description of ordering and displacive mechanisms.

A system undergoing a purely ordering type of transition, can be considered as a set of positions *fixed in space*, and filled by heterogeneous objects: atoms and vacancies, or atoms and ions of different elements, etc. . The ordering process causes a change in the distribution of objects among the positions, but does not displace the positions. Strictly speaking, the symmetry of the probability distribution function of the objects, and the corresponding symmetry group, obtained from experiment, is the result of averaging of the real atomic distribution. The coincidence between the distribution functions of the different types of particles and the actual distribution, is obtained for the fully ordered state. In systems undergoing a purely displacive transformation, the crystallographic positions are fully occupied by identical particles, which are shifted in the transition process. This process induces a *crystallographic nonequivalence* in the atomic sites, which were initially equivalent this being realized even among chemically identical atoms or ions. In all cases the atoms do not exchange diffusionally. There is also an important difference in the dynamics of the two types of transitions. The diffusional motion of atoms in ordering processes is not correlated, whilst the atomic shifts in a displacive mechanism is a collective process.

Within the preceding classical scheme, reconstructive transitions correspond to an *intermediate situation* between the two types of mechanisms. On the one hand, a displacive reconstructive transition, directly from the initial structure to a non-Landau phase, transfers atoms from one fixed site to another fixed site, without any intermediate position, but the corresponding atomic displacements occur collectively. On the other hand in ordering reconstructive transitions, the uncorrelated character of the atomic diffusion is only partially verified. This appears for example in the reordering mechanism between close packed structures. In this mechanism the change in position for any individual atom from position A to position B requires creation of high-energy defects connected with the absence of vacant interstices. It is thus energetically more advantageous to shift simultaneously (collectively) all the atoms of the layer A. Besides, a relative displacement of layers appears as uncorrelated, but actually has correlations: If the original stacking of layers is random, e.g. ABCBA..., and the second layer has to be shifted from position B to final position C, it causes a displacement of the successive layers, which cannot occupy position C in the new structure due to geometrical constraints.

For example, in the reordering FCC-HCP transition mechanism, described in Chapter IV, the reordering of the layers between the two structures appears as clearly correlated. The three-layer FCC periodic stacking ABCABC... must be replaced by the two-layer HCP ABABAB... stacking. This can be realized by periodically repeated shifts of each AB unshifted pair: CA moves in the direction C→A (Fig.25) and becomes AB, while BC displaces from B to A, and also becomes AB. The next AB pair is unshifted etc. Thus, *the reordering process becomes periodically correlated.* Other concrete examples, described in the following chapters, illustrate the property that reconstructive transition mechanisms allow a continuous *crossover* from ordering to displacive mechanisms. This is made possible by the *invariability* of the order-parameter value, which is *constant and maximal*, when going from a limit fully ordered phase, to a limit displacive phase.

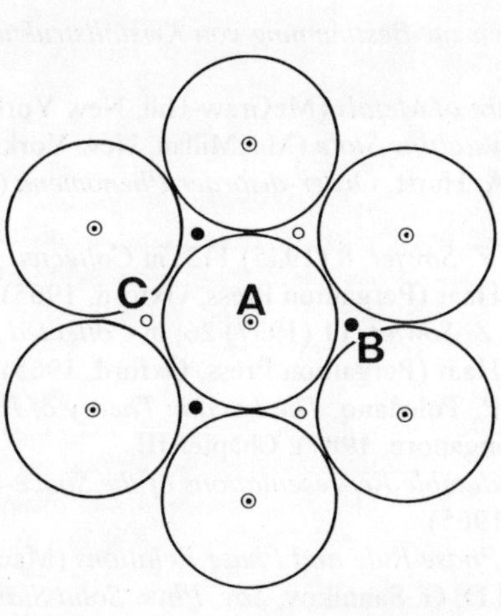

Fig. 25. Close atomic packing. A, B and C represent three atomoc positions, within the atomic layer (A) and in the nearest adjacent layers (B and C).

5.5. Degree of discontinuity of first-order phase transition

As a final remark, let us stress that the approach described in this chapter provides a general *measure* of the discontinuous character of first-order transitions, which consists in the *magnitude of the order-parameter* $|\vec{\eta}(\xi)|$ as function of the variational parameter. This degree of discontinuity of first-order transitions can be calculated explicitly from the initial and final structures of the phases. It varies nonlinearly between zero (for second-order transitions) and 1 (for reconstructive transitions), and usefully complete the numbers which are usually quoted as a measure of the discontinuity of a phase transition (latent heat, discontinuity of the lattice parameters, magnitude of the shifts, jump of entropy etc....). In the following chapters, some quantitative applications of this criterion are given.

6. References

1. V. P. Dmitriev, S. B. Rochal, Yu. M. Gufan and P. Toledano, *Phys. Rev. Lett.* **60** (1988) 1958.
2. V. P. Dmitriev, S. B. Rochal, Yu. M. Gufan and P. Toledano, *Phys. Rev. Lett.* **62** (1989) 844.
3. V. P. Dmitriev, S. B. Rochal, Yu. M. Gufan and P. Toledano, *Phys. Rev. Lett.* **62** (1989) 2495.
4. Yu. M. Gufan, V. P. Dmitriev, and P. Toledano, *Sov. Phys. Solid State* **30** (1988) 613.
5. V. P. Dmitriev, Yu. M. Gufan and P. Toledano, *Phys. Rev.* **B44** (1991) 7248.
6. N. G. Parsonage and L. A. K. Staveley, *Disorder in Crystals* (Clarendon Press, Oxford, 1978).
7. *International Tabellen zur Bestimmung von Kristallstrukturen* (Borntrager, Berlin, 1935).
8. C. S. Barrett, *Structure of Metals* (McGraw-Hill, New York, 1952), p. 622.
9. W. L. Bragg, *The Crystalline State* (MacMillan, New York, 1934).
10. H. S. Green and C. A. Hurst, *Order-disorder Phenomena* (Interscience, New York, 1964).
11. L. D. Landau, *Phys. Z. Sowjet.* **8** (1935) 113; in *Collected Papers of L. D. Landau*, ed. D. Ter Haar (Pergamon Press, Oxford, 1965), p. 96.
12. L. D. Landau, *Phys. Z. Sowjet.* **11** (1937) 26; in *Collected Papers of L. D. Landau*, ed. D. Ter Haar (Pergamon Press, Oxford, 1965), p. 193
13. J. C. Toledano and P. Toledano, *The Landau Theory of Phase Transitions* (World Scientific, Singapore, 1987), Chapter III.
14. O. V. Kovalev, *Irreducible Representations of the Space-Groups* (Gordon and Breach, New York, 1965).
15. S. T. Bowden, *The Phase Rule and Phase Relations* (MacMillan, London, 1945).
16. A. P. Levanyuk and D. G. Sannikov, *Sov. Phys. Solid State* **18** (1976) 1122.
17. H. N. V. Temperley, *Changes of State* (Interscience, New York, 1956).
18. *Landolt-Bornstein Tables for Ferro- and Antiferroelectric Substances* (Springer-Verlag, Berlin, 1969), vol.3, group 3.

CHAPTER III

DISPLACIVE RECONSTRUCTIVE PHASE TRANSITIONS

1. Introduction

In our theoretical description, displacive transitions reflect the most strikingly the specificity of the reconstructive transition mechanism. In the same way, it is from the experimental results reported on typical examples of displacive mechanisms, that our understanding of the particular features of reconstructive transitions was made possible: stabilization of "Landau" and limit phases in the phase diagrams of uranium, existence of two orientational domains of the BCC parent structure in iron, X-ray and electron microscopy diffraction data showing the direction of the atomic shifting mechanism and the special positions corresponding to an increase of the rhombohedral to hexagonal ω symmetry in titanium, absence of phonon softening for the β - ω transition or incomplete softening in the vicinity of specific points of the BCC Brillouin zone for the martensitic transitions in lithium and sodium.

Displacive reconstructive transitions display a large variety of complex behaviors, and although our approach offers a unifying framework for their properties, we do not pretend in this chapter to discuss in details the set of different mechanisms evidenced in the field. However, in the study of particular examples we try to show how the main current problems are inserted in our phenomenological description: important role of elastic strains, of defects, and of some lattice instabilities, thermodynamics of martensitic transformations, premartensitic behaviors etc... .

In the first place, we have chosen to illustrate the applicability of our approach, to allotropic transitions of elemental crystals, because the simplicity of their structures allow more clear considerations on the transitions mechanisms. Some examples of alloys and dielectrics are also discussed. The succession of the treated examples goes from the more simple situation (the β - ω transitions in Ti and Zr) to the more complicated one (the martensitic transformations in Li and Na).

2. The β – ω transition

2.1. The transition mechanism

The β - ω transition takes place in titanium, zirconium, hafnium, and in some of their alloys (TiCr, TiV, TiMo, TiNb, ZrNb...) [1]. Ti and Zr display a very similar phase diagram, shown schematically in Fig.1a [2]. Thus, the β - ω transition occurs directly at high pressures and high temperatures (above about 9 GPa and 940K in Ti, and above 6 GPa and 970K in Zr), whereas the α (HCP) phase appears as an intermediate structure at room temperature and atmospheric pressure. The transition is found to be completely reversible, with a small latent heat and a slight hysteresis [3].

The β-phase is body centered cubic, and the ω-phase can be found in two slightly different structures: an hexagonal form (space group D_{6h}^1, Z=3) [4], and a rhombohedral modification (space group D_{3d}^3, Z=3) [5], which is obtained from the "ideal" hexagonal ω-phase with a low concentration of impurities (1.2% of oxygen in Ti_3V) or in more concentrated alloys (TiCr, ZrNb) [1].

In Ti, the average shifting of the Ti atoms from the β to the ω positions can be deduced from the superstructures observed in the electron diffraction and X-ray patterns [6,7]. It takes place along the $[111]$ cubic direction which is parallel to the c axis of the hexagonal unit-cell (Fig.1b). The antiparallel displacements of the atoms along the preceding direction, and the formation of the hexagonal ω-layers is shown in Fig.1c. Fig.1d represents the $(0\bar{1}1)$ cubic plane, in which one can see, that the corner atoms of the unit-cell do not move at the transition, but only the two atoms located within the unit-cell, which are displaced along $[111]_c$. Depending on the magnitude of the displacements, five phases may arise:

1) The β-phase, corresponding to unshifted atoms. It has the parent O_h^9 symmetry, with one atom in the elementary unit-cell (Z=1).

2) The hexagonal ω-phase (denoted ω_h hereafter), which appears as the result of a shift of $a\sqrt{3}/12$, where a is the cubic (β) lattice parameter. It possesses the symmetry D_{6h}^1 with Z=3.

3) The β'-phase, isostructural to β, which is induced by a shift of $a\sqrt{3}/6$, and can be deduced from β by a rotation of 60° around $[111]$, i.e. it constitute a different orientational domain of the β structure.

4) The h-phase arising for a shift of $a\sqrt{3}/3$. It has the same hexagonal space group as the ω_h-phase (D_{6h}^1), but with one atom per unit-cell.

5) The rhombohedral ω-phase (ω_{rh}) associated with any general shift along $[111]$. It has the space group D_{3d}^3 with Z=3.

In the ω_{rh}-phase the initial one fold cubic position (000) splits into two positions (000) and ($\frac{1}{3}$ $\frac{2}{3}$ z). Characterizing by the parameter $\xi = z - \frac{1}{3}$ the magnitude of the displacements from the β structure along $[111]$, one can see from Fig.1d, which is also a section of the hexagonal D_{3d}^3 cell parallel to the three fold axis, that it is the specific atomic shift $\xi = a\sqrt{3}/12$ (1/6 in $a\sqrt{3}/2$ unit) which leads to the ω_h-phase. It corresponds to z=1/6+1/3=1/2, which is a symmetric position *giving rise to a set of new symmetry elements of the resulting structure, which pertain to the holohedry of the lattice*. In particular, we obtain a mirror plane, perpendicular to the three-fold axis, and also a six-fold axis oriented along $[111]_c$. As a consequence, the group-subgroup relationship is lost

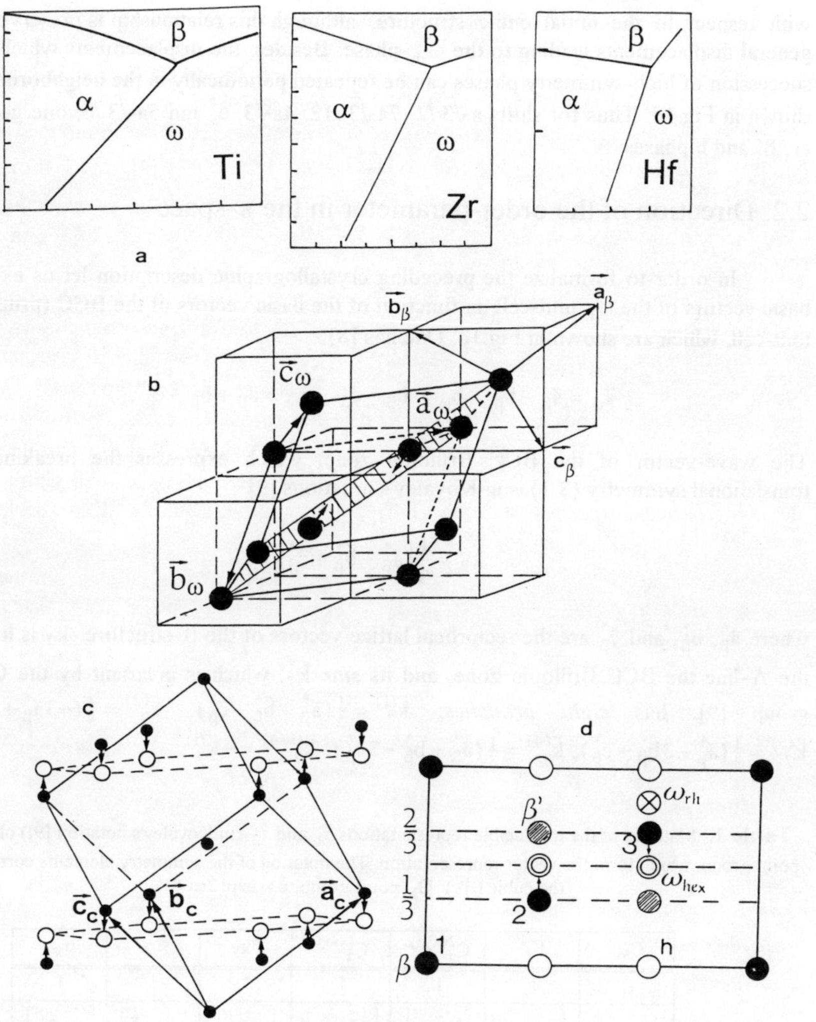

Fig.1. (a) Phase diagrams of Ti, Zr, and Hf. (b) Connection between the cubic β and hexagonal ω structures. The hatched plane is the $(0\bar{1}1)$ cubic plane. (c) Antiparallel displacements from the β structure leading to the formation of the hexagonal ω layers. (d) Critical shifts along the [111] cubic direction, leading to the formation of the ω phase. Full, double, hatched and open circles represent the atomic positions in the β, ω_{hex}, β' and h phases, respectively. The crossed circle indicates the general position in the $(0\bar{1}1)$ plane, corresponding to the atomic positions in the ω_{rh} phase.

with respect to the initial cubic structure, although this relationship is preserved for the general displacements leading to the ω_{rh}-phase. Besides, the displacements which give the succession of high-symmetry phases can be repeated periodically in the neighboring cells as shown in Fig. 1d. Thus for shifts $a\sqrt{3}/2$, $7a\sqrt{3}/12$, $4a\sqrt{3}/6$, and $5a\sqrt{3}/6$, one gets the β, ω, β' and h phases.

2.2. Direction of the order-parameter in the ε-space

In order to formalize the preceding crystallographic description let us express the basic vectors of the ω_h unit-cell, in function of the basic vectors of the BCC (primitive) β-unit-cell, which are shown in Fig. 1b. One has [8]:

$$\vec{a}_\omega = \vec{a}_\beta - \vec{b}_\beta \ , \ \vec{b}_\omega = \vec{b}_\beta - \vec{c}_\beta \ , \ \vec{c}_\omega = \vec{a}_\beta + \vec{b}_\beta + \vec{c}_\beta \ . \tag{3.1}$$

The wave-vector of the BCC Brillouin zone, which expresses the breaking of the translational symmetry (3.1) is in Kovalev's notations [9]

$$\vec{k}_7 = \frac{1}{3}(\vec{a}_\beta^* + \vec{b}_\beta^* + \vec{c}_\beta^*) \tag{3.2}$$

where $\vec{a}_\beta^*$, $\vec{b}_\beta^*$, and $\vec{c}_\beta^*$ are the reciprocal lattice vectors of the β-structure. $\vec{k}_7$ is located on the Λ-line the BCC Brillouin zone, and its *star* $\vec{k}_7^*$, which is invariant by the C_{3v} point group [9], has *eight branches*: $\vec{k}_7^{(1)} = \frac{1}{3}(\vec{a}_\beta^* + \vec{b}_\beta^* + \vec{c}_\beta^*)$, $\vec{k}_7^{(2)} = \frac{1}{3}(-3\vec{a}_\beta^* + \vec{b}_\beta^* + \vec{c}_\beta^*)$, $\vec{k}_7^{(3)} = \frac{1}{3}(\vec{a}_\beta^* - 3\vec{b}_\beta^* + \vec{c}_\beta^*)$, $\vec{k}_7^{(4)} = \frac{1}{3}(\vec{a}_\beta^* + \vec{b}_\beta^* - 3\vec{c}_\beta^*)$, $\vec{k}_7^{(i+4)} = -\vec{k}_7^{(i)}$.

Table 1. Matrices of the irreducible representations $\hat{\tau}_1$ and $\hat{\tau}_3$ (in Kovalev's notation [9]) of the C_{3v} point group, which form the vector representation. The notation of the symmetry elements correspond to the cubic (β) O_h point group. $\varepsilon = \exp(2\pi i/3)$.

C_{3v}	C_1	$C_3^{2(xyz)}$	$C_3^{(xyz)}$	$\sigma_{\bar{x}y}$	$\sigma_{\bar{y}z}$	$\sigma_{\bar{x}z}$
$\hat{\tau}_1$	1	1	1	1	1	1
$\hat{\tau}_3$	$\begin{bmatrix}1 & \\ & 1\end{bmatrix}$	$\begin{bmatrix}\varepsilon & \\ & \varepsilon^2\end{bmatrix}$	$\begin{bmatrix}\varepsilon^2 & \\ & \varepsilon\end{bmatrix}$	$\begin{bmatrix} & 1\\ 1 & \end{bmatrix}$	$\begin{bmatrix} & \varepsilon\\ \varepsilon^2 & \end{bmatrix}$	$\begin{bmatrix} & \varepsilon^2\\ \varepsilon & \end{bmatrix}$

The irreducible representations corresponding to the shifts of the atoms of the β-structure, which involve by all the branches of $\vec{k}_7^*$, are constructed from the two "little" representations (see Appendix I) $\hat{\tau}_1(A_{1z})$ and $\hat{\tau}_3(E_{xy})$ of the vector representation of the C_{3v} point group, the matrices of which are given in Table 1. The corresponding eight- and sixteen-dimensional irreducible representations, denoted $\tau_1(\vec{k}_7^*)$ and $\tau_3(\vec{k}_7^*)$ account respectively for the atomic shifts along $[111]$ (longitudinal phonons) and within the cubic plane(111) (transverse phonons). $\tau_3(\vec{k}_7^*)$ *is not* associated with the β-ω transition mechanism, and one has to consider only $\tau_1(\vec{k}_7^*)$. From the matrices generating this representation, which are listed in Table 2 and constructed following the method described in Appendix I, one can determine (using the geometrical method of Chapter I, Section 3.3.2) the possible low symmetry groups, subgroups of O_h^9, which may arise below a transition from the β-structure, for any shifts of the atoms parallel to $[111]$ and to the other corresponding diagonals. Among the 21 possible equilibrium states, given in Table 3, one can see that only two states possess a symmetry involving the three-fold multiplication of the BCC unit-cell, following Eq.(3.1). These states possess the symmetry groups C_{3v}^1 and D_{3d}^3 and column (b) of Table 3 reveals that the corresponding basic vectors φ_{ki} (i=1-8) have zero coefficients η_i except φ_{k1} and φ_{k5}.

The basis functions φ_{k1} and φ_{k5} can be written in terms of the shifts ξ_i (i=1-27) of the 27 atoms, numbered from 1 to 27 in Fig. 1b, which pertain to the 27-fold unit-cell involving all the $k_7^{(i)}$ components. One has:

Table 2. Matrices generating the 8-dimensional irreducible representation $\hat{\tau}_1(\vec{k}_7^*)$. Columns represent diagonal matrices. The zero elements are not indicated in non-diagonal matrices. $\varepsilon = \exp(2\pi i/3)$.

$\vec{a}_\beta$	$\vec{b}_\beta$	$\vec{c}_\beta$	$\{U_x \vert \vec{0}\}$	$\{U_y \vert \vec{0}\}$
$\begin{bmatrix} \varepsilon \\ 1 \\ \varepsilon \\ \varepsilon \\ \varepsilon^* \\ 1 \\ \varepsilon^* \\ \varepsilon^* \end{bmatrix}$	$\begin{bmatrix} \varepsilon \\ \varepsilon \\ 1 \\ \varepsilon \\ \varepsilon^* \\ \varepsilon^* \\ 1 \\ \varepsilon^* \end{bmatrix}$	$\begin{bmatrix} \varepsilon \\ \varepsilon \\ \varepsilon \\ 1 \\ \varepsilon^* \\ \varepsilon^* \\ \varepsilon^* \\ 1 \end{bmatrix}$	$\begin{bmatrix} &1&&&&&& \\ 1&&&&&&& \\ &&&1&&&& \\ &&1&&&&& \\ &&&&&1&& \\ &&&&1&&& \\ &&&&&&&1 \\ &&&&&&1& \end{bmatrix}$	$\begin{bmatrix} &&1&&&&& \\ &&&1&&&& \\ 1&&&&&&& \\ &1&&&&&& \\ &&&&&&1& \\ &&&&&&&1 \\ &&&&1&&& \\ &&&&&1&& \end{bmatrix}$

$\{C_3^{2(xyz)} \vert \vec{0}\}$	$\{U_{\bar{x}y} \vert \vec{0}\}$	$\{I \vert \vec{0}\}$
$\begin{bmatrix} 1&&&&&&& \\ &&&1&&&& \\ &1&&&&&& \\ &&1&&&&& \\ &&&&1&&& \\ &&&&&&&1 \\ &&&&&1&& \\ &&&&&&1& \end{bmatrix}$	$\begin{bmatrix} &&&&1&&& \\ &&&&&&1& \\ &&&&&1&& \\ &&&&&&&1 \\ 1&&&&&&& \\ &&1&&&&& \\ &1&&&&&& \\ &&&1&&&& \end{bmatrix}$	$\begin{bmatrix} &&&&1&&& \\ &&&&&1&& \\ &&&&&&1& \\ &&&&&&&1 \\ 1&&&&&&& \\ &1&&&&&& \\ &&1&&&&& \\ &&&1&&&& \end{bmatrix}$

$$\vec{\varphi}_{k1}=\{(\xi_1+\xi_8+\xi_{12}+\xi_{13}+\xi_{14}+\xi_{15}+\xi_{17}+\xi_{20}+\xi_{21})+$$
$$+\varepsilon(\xi_2+\xi_3+\xi_4+\xi_{16}+\xi_{18}+\xi_{19}+\xi_{22}+\xi_{23}+\xi_{25})+$$
$$+\varepsilon^*(\xi_5+\xi_6+\xi_7+\xi_9+\xi_{10}+\xi_{11}+\xi_{24}+\xi_{26}+\xi_{27})\}\cdot\vec{c}_\omega,$$

$$\vec{\varphi}_{k5}=\{-(\xi_1+\xi_8+\xi_{12}+\xi_{13}+\xi_{14}+\xi_{15}+\xi_{17}+\xi_{20}+\xi_{21})-$$
$$-\varepsilon(\xi_5+\xi_6+\xi_7+\xi_9+\xi_{10}+\xi_{11}+\xi_{24}+\xi_{26}+\xi_{27})-$$
$$-\varepsilon^*(\xi_2+\xi_3+\xi_4+\xi_{16}+\xi_{18}+\xi_{19}+\xi_{22}+\xi_{23}+\xi_{25})\}\cdot\vec{c}_\omega. \qquad (3.3)$$

From Table 3, one can see that the equilibrium state of symmetry C_{3v}^1 corresponds to the relationship: $\eta_1 \neq \eta_5 \neq 0$, $\eta_{i\neq 1,5} = 0$. Thus, it is associated with *non-symmetric* antiparallel shifts of atoms denoted 2 and 3 in Fig. 1d. The state of symmetry D_{3d}^3, *which identifies to the ω_{rh}-phase*, is associated with *symmetric shifts*, determined by the equilibrium conditions:

$$\eta_1 = \eta_5 \neq 0, \quad \eta_{i\neq 1,5} = 0 \qquad (3.4)$$

which can also be written, taking into account in (3.3) the translationally equivalent atoms:

$$\vec{\varphi}_{k1} = \vec{\varphi}_{k5} = (\xi_{19} - \xi_{27})\cdot\vec{c}_\omega. \qquad (3.5)$$

In other words, *the direction of the 8-dimensional order-parameter $\vec{\eta}$, associated with the $\beta \rightarrow \omega_{rh}$ transition, in the ε_8 representation space of $\tau_1(\vec{k}_7^*)$, is fixed by the equilibrium conditions (3.4).* In the next section, we will use this result to find the dependence of the order-parameter modulus $|\vec{\eta}|$ in function of the magnitude ξ of the antiparallel shifts.

2.3. Periodicity of the order-parameter magnitude

The lattice function describing the extended unit-cell involved by the variable displacements $\vec{r}_j$ along $[111]$, from the β structure, can be written:

$$\psi(\vec{r}_j) = \frac{1}{27}\sum_{j=1}^{27}\sum_{m=1}^{8} A_j \cdot \exp\left\{i\vec{k}_7^{(m)}(\vec{R}_j^0 - \vec{r}_j)\right\} + \text{c.c.} \qquad (3.6)$$

where the sum runs over the 27 atoms contained in the extended cell, the positions of which are characterized by the vectors $\vec{R}_j^0$, and over the 8 branches of $\vec{k}_7^*$. Restricting to the substructure determined by the two "active" branches $\vec{k}_7^{(1)} = -\vec{k}_7^{(5)}$, corresponding to

Table 3. The 21 stable low-symmetry subgroups induced by $\hat{\tau}_1(\vec{k}_7^*)$. Columns (b), (c) and (d) indicate respectively the equilibrium relationships between the order-parameter components, the corresponding space groups, and the multiplication of the initial BCC unit-cell.

(a)	(b)								(c)	(d)
1	η	0	0	0	η	0	0	0	D_{3d}^3	3
2	0	η	η	0	0	η	η	0	D_{2h}^{23}	9
3	η	η	η	η	η	η	η	η	O_h^9	27
4	η_1	0	η_1	0	η_2	0	η_2	0	C_{2v}^{18}	9
5	η_1	0	0	0	η_2	0	0	0	C_{3v}^1	3
6	η_1	η_1	η_1	η_1	η_2	η_2	η_2	η_2	T_d^3	27
7	η_1	η_2	0	0	η_2	η_1	0	0	C_{2v}^{18}	9
8	η_1	η_1	η_2	η_2	η_2	η_2	η_1	η_1	C_{4v}^9	27
9	η_1	η_2	η_2	η_1	η_1	η_2	η_2	η_1	D_{2h}^{25}	27
10	η_1	η_2	η_1	η_1	η_1	η_2	η_1	η_1	D_{3d}^5	27
11	η_1	η_2	0	0	η_1	η_2	0	0	C_{2h}^3	9
12	η_1	η_2	η_3	η_2	η_1	η_2	η_3	η_2	C_{2h}^3	27
13	η_1	η_2	η_3	η_2	η_3	η_2	η_1	η_2	C_{2v}^{20}	27
14	η_1	η_2	η_1	η_2	η_3	η_4	η_3	η_4	C_{2v}^{20}	27
15	η_1	η_2	η_3	η_4	η_1	η_2	η_3	η_4	C_i^1	27
16	η_1	η_2	η_3	η_4	η_2	η_1	η_3	η_4	C_s^3	27
17	η_1	0	η_2	0	η_3	0	η_4	0	C_s^3	27
18	η_1	η_2	η_3	η_4	η_1	η_3	η_2	η_4	C_2^3	27
19	η_1	η_2	η_1	η_1	η_3	η_4	η_3	η_3	C_3^4	27
20	η_1	η_2	η_3	η_2	η_4	η_5	η_6	η_5	C_s^3	27
21	η_1	η_2	η_3	η_4	η_5	η_6	η_7	η_8	C_1^1	27

the antiparallel shifts ξ_1 and ξ_2 associated with the $\beta \rightarrow \omega_{rh}$ transition, one needs to consider only the 3 atoms contained in the ω_{rh} unit-cell. (3.6) takes the simplified form:

$$\psi(\xi_1,\xi_2) = \frac{A}{3}\left[1+\exp\frac{4\pi i}{a\sqrt{3}}\left(\frac{a}{2\sqrt{3}}+\xi_1\right)+\exp\frac{4\pi i}{a\sqrt{3}}\left(\frac{a}{\sqrt{3}}-\xi_2\right)\right]+\text{c.c.} \tag{3.7}$$

where the first term into the bracket corresponds to the atoms denoted 1 in Fig.1d, which are unshifted after the transition, while the second and third terms in (3.7) relate to atoms of type 2 and 3, which are shifted in opposite directions. In the ω_{rh}-phase, one has $\xi_1=\xi_2=\xi$, whereas in the β-phase $\xi_1=\xi_2=0$. Accordingly the corresponding lattice functions are:

$$\psi_{\omega_{rh}}(\xi) = A\left[\operatorname{Sin}\left(\frac{4\pi\xi}{a\sqrt{3}}+\frac{\pi}{6}\right)-\frac{1}{2}\right] \tag{3.8}$$

and ψ_β=0. Thus the magnitude of the order-parameter $|\vec{\eta}(\xi)|$ in function of the critical displacements ξ, which is determined by the increment $\delta\psi(\xi) = \psi_{\omega_{rh}} - \psi_\beta = \psi_{\omega_{rh}}$ has, up to a normalizing constant η_0, the same form (3.8):

$$|\vec{\eta}(\xi)| = \eta_0\left\{\operatorname{Sin}[f(\xi)]-\frac{1}{2}\right\} \tag{3.9}$$

with $f(\xi) = \left(\frac{4\pi\xi}{a\sqrt{3}}+\frac{\pi}{6}\right)$.

Fig.2 represents the dependence of $|\vec{\eta}(\xi)|/\eta_0$. It shows that for successive specific critical shifts ξ, one gets periodically the sequence of β, ω_h, β' and h phases. More precisely, for ξ=0, 1, 2, 3,... (in $a\sqrt{3}/2$ units) one has η=0 for the parent β-phase; for ξ=1/6, 7/6, 13/6,... η/η_0=1/3 and ω_h-phase is obtained; for ξ=1/3, 4/3, 7/3,... one gets again η=0 (β'-phase) and for ξ=2/3, 5/3, 8/3,.. η/η_0=-1, which corresponds to the h-phase. For all other arbitrary displacements ξ, the ω_{rh}-phase is stabilized.

2.4. Theoretical phase diagram for the $\beta-\omega$ transition

The dependence of $|\vec{\eta}(\xi)|$, as expressed by (3.9) and shown in Fig.2, *is not symmetric* with respect to the β and ω_h phases. As underlined in Chapter II (Sec.3.5.2) this asymmetry is indicative of the fact that β represents the actual parent structure, and that ω_h cannot be used as the initial structure for describing the β-ω transition. This must also be reflected in the form of the effective order-parameter expansion, which will be shown to contain a *cubic* term, and in the asymmetry of the corresponding phase diagram.

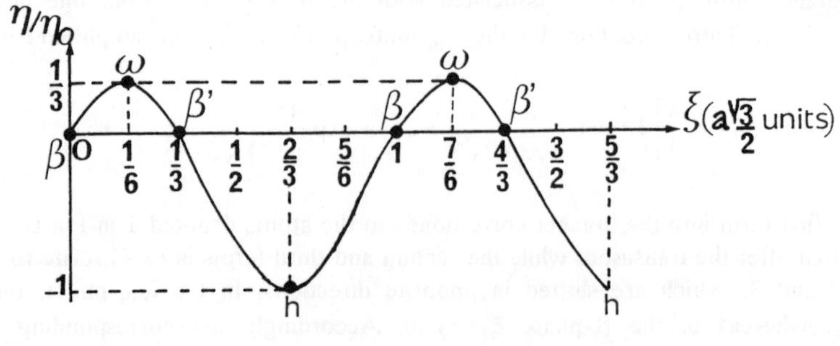

Fig. 2. $|\vec{\eta}(\xi)|$ dependence at the β–ω transition following Eq.(3.9).

When restricting to *fourth degree terms*, one can construct, using the transformation properties of the matrices given in Table 2, four independent invariants of the order-parameter components:

$$I_0=\sum_{i=1}^{8}\eta_i^2,\quad I_1=\sum_{i=1}^{8}\eta_i^3,\quad I_2=\eta_1\eta_2\eta_3\eta_4+\eta_5\eta_6\eta_7\eta_8,\quad I_3=\sum_{i\neq j\neq k\neq l}\eta_i\eta_j\eta_k\eta_l-I_2. \quad (3.10)$$

Taking into account the equilibrium conditions (3.4), one has I_2=I_3=0, $I_0=\eta^2$, $I_1=\eta^3$. The effective thermodynamic potential can be written:

$$F=F_0(P,T)+a_1|\vec{\eta}(\xi)|^2+b_1|\vec{\eta}(\xi)|^3+a_2|\vec{\eta}(\xi)|^4. \quad (3.11)$$

Minimization with respect to ξ, using (3.9), yields the equation of states:

$$\left[\mathrm{Sin}(f(\xi))-\frac{1}{2}\right]\cdot\mathrm{Cos}(f(\xi))\cdot\left\{2a_1+3b_1\cdot\left[\mathrm{Sin}(f(\xi))-\frac{1}{2}\right]+4a_2\cdot\left[\mathrm{Sin}(f(\xi))-\frac{1}{2}\right]^2\right\}=0. \quad (3.12)$$

When using also the stability condition $\frac{\partial^2 F}{\partial\xi^2}\geq 0$, one has the following equilibrium states:

(1) the parent β (or β') phase corresponding to η=0, Sin(f(ξ))=1/2 and $a_1\geq 0$;

(2) the ω_h phase given by $\frac{\partial\eta}{\partial\xi}$=0 and η=1/3, namely for Cos(f(ξ))=0 and Sin(f(ξ))=+1. Its domain of stability is determined by $a_1\leq\frac{1}{4}(-3b_1-2a_2)$;

(3) the h phase which obeys the conditions $\frac{\partial\eta}{\partial\xi}$=0, η=-1, expressed by Cos(f(ξ))=0 and Sin(f(ξ))=-1. It is stabilized for $a_1\leq\frac{9}{4}(b_1-2a_2)$;

(4) the ω_{rh} phase coinciding with the Landau solution $\frac{\partial F}{\partial\eta}$=0, and -1<$\eta$<1/3, $\eta\neq 0$. It gives two (temperature dependent) solutions: Sin(f(ξ))= $=\frac{1}{2}+\frac{1}{8a_2}\left\{-3b_1\pm(9b_1^2-32a_1a_2)^{1/2}\right\}$, associated with two ($\omega_{rh}^+$ and ω_{rh}^-) *antiisostructural* phases determined by opposite shifts in Fig.1d. The stability region in which ω_{rh}^+ and ω_{rh}^- are located, is bounded by three limit of stability lines: with respect to the β-phase ($a_1\leq\frac{9b_1^2}{32a_2}$), to the ω_h-phase ($a_1\geq\frac{1}{4}(-3b_1-2a_2)$) and to the h-phase ($a_1\geq\frac{1}{8}(6b_1-a_2)$).

The corresponding phase diagram (Fig.3) is represented in the plane (a_1,b_1). One can verify that the β-phase is separated from the *limit states* ω_h and h, by the first-order transition *straight lines*: $a_1 = \frac{1}{4}(-2b_1 - a_2)$, and $a_1 = \frac{3}{4}(2b_1 - 3a_2)$, respectively, whereas the β-ω_{rh} first-order transition line is the parabola: $a_1 = \frac{b_1^2}{4a_2}$. Between the group-subgroup unrelated phases ω_{rh}^+ and ω_{rh}^- one has also a first-order transition line (b_1=0), whereas the lines ω_h-ω_{rh}^+ and ω_h-ω_{rh}^- correspond to second-order transitions. Their equations, $a_1 = \frac{1}{4}(-3b_1 - 2a_2)$ and $a_1 = \frac{1}{8}(6b_1 - a_2)$, are straight-line equations, as the result of the fourth-degree power assumed for $F(\vec{\eta}(\xi))$. Finally, there are four singular points in the phase diagrams: (i) a triple point, located on the β-ω_{rh} first-order transition line, and ending the ω_{rh}^+-ω_{rh}^- transition line; (ii) three *three-phase* points N_1, N_2 and N_3 at which merge three lines of first- *and* second-order transitions.

2.5. Discussion

The displacive mechanism which has been used in our density-wave approach to the β-ω transition [10] is based on clear-cut electron and X-ray diffraction data [6,7] and confirms previous theoretical descriptions of this transition [11, 12]. The mechanism is also attested by the $\omega(\vec{k})$ phonon dispersion curves reported for Ti, Zr and Hf [13-15] which reveals a strong dip close to $\vec{k} = \frac{2}{3}(111)$ (the longitudinal phonon mode), in agreement with the lattice instability expressed by Eq.(3.1). A temperature dependence of this mode, when approaching the transition from the β-phase, has been vainly searched [13, 15],

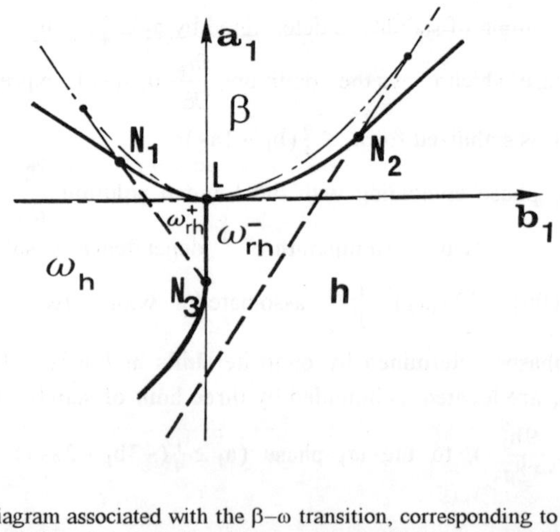

Fig.3. Phase diagram associated with the β–ω transition, corresponding to the thermodynamic potential (3.11). Full, dashed and dashed-dotted lines represent first-order, second-order and limit of stability lines, respectively. N_1, N_2, N_3 are three-phase points, whereas L is a triple point. ω_{rh}^+ and ω_{rh}^- are antiisostructural phases.

confirming the absence of softening with temperature (or pressure) as predicted in our model for direct transitions to limit-states. In this respect, it can be noted that the reconstructive character of the β-ω transition is more reliably reflected in the magnitude of the critical displacements (and of the order-parameter) reported experimentally [6, 7] than by the latent heat measurement [3] which suggests a weakly first-order discontinuity for the transition. The observation of four orientational domains of the ω_h phase, corresponding to the four (111) cubic planes, in titanium alloys [4, 5] brings another confirmation of the assumed connection between the β and ω phases.

The fact that the ω_{rh} phase replaces the ω_h phase when adding a small concentration of oxygen, or when doping with a small amount of metal (Cr, Nb) [5, 16] suggests that the corresponding experimental thermodynamic paths are close to the region of the three-phases point N_1, in the theoretical phase diagram of Fig.3. This may also be confirmed by the fact that the h-phase, which is located in a different region of the phase diagram, was not found experimentally. Besides, when considering a model of hard sphere atoms, instead of points as in Fig.1d, the atoms would realize in the h-phase a more unstable configuration than the close-packing structure of the α-phase. Thus the presence of this latter phase in the phase diagram of Ti, Zr and Hf may have the result of destroying the stability of the h-phase.

3. The BCC-HCP transition: The role of macroscopic deformation

3.1. Burgers mechanism

The BCC-HCP transition is, together with the BCC-FCC transition, the most commonly encountered type of reconstructive transition in crystals of elements. It is found in about twenty elements (seeTable 5). The BCC phase is, with some exceptions the more stable at high temperatures and high pressures. Figures 4a-f represents schematically some typical phase diagrams containing a direct BCC-HCP transition [17]. The transition is always strongly first-order and corresponds to large atomic displacements, of the order of 1 A° [18].

The mechanism generally assumed for the BCC-HCP transition is the so-called Burgers mechanism [19], proposed for the $\beta-\alpha$ transition in Zr. It can be divided into three stages which are represented in Figures 5a-c. In the first stage the $(110)_{BCC}$ planes transform into the $(001)_{HCP}$ planes via a shearing mechanism (Fig.5a). Second, a homogeneous shifting of the atoms in the (110) planes leads to a FCC structure (Fig.5b). Third, the displacement of one out of two planes in the $[110]_{BCC}$ direction leads to the formation of the HCP structure (Fig.5c). In other words, Burgers mechanism involves three different and simultaneous instabilities associated with distinct primary (symmetry breaking) order-parameters. We will first reformulate this mechanism, in a phenomenologically more simple manner [20].

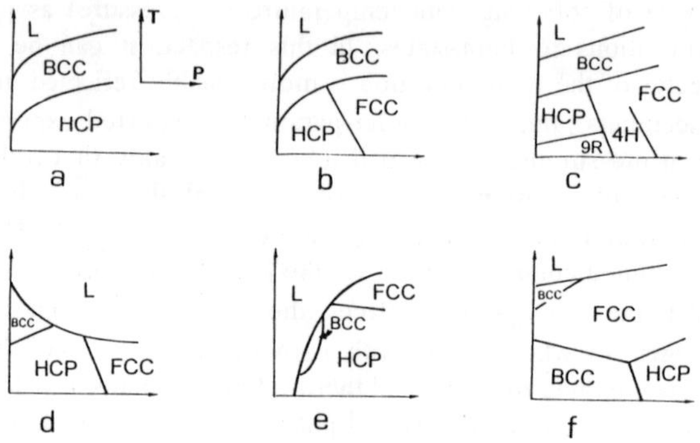

Fig.4. Topology of the BCC-HCP transition line in elemental crystals. The represented phase diagrams are found, for example, in Be (a), heavy lanthanides (b), Sm (c), Ba (d), He (e) and Fe (f).

3.2. Burgers mechanism reformulated

In contrast with the preceding scheme we propose a BCC-HCP transition mechanism based on a single *primary* order-parameter. It consists in an *antiparallel shifting* of the atoms lying in the (110) BCC plane, along the $\pm[1\bar{1}0]$ directions, which *induces* as a *secondary* effect, a *compression* of the cubic unit-cell along the fourfold axis $[001]$, and a simultaneous decompression along the two other ($[100]$ and $[010]$) fourfold axes. Fig.6a shows the average shifting of the atoms from the BCC to the HCP structure whereas Fig.6b represents the projection of the HCP unit-cell on the (110) BCC plane.

A more concrete view of the assumed correlated mechanisms at the atomistic level is shown in Figs.7a and 7b. Each (110) BCC layer is organized in such a manner that, in any triangle formed by neighboring atoms, two sides are formed by atoms which touch each other, and the third side is formed by separated atoms. It is on the third side that the

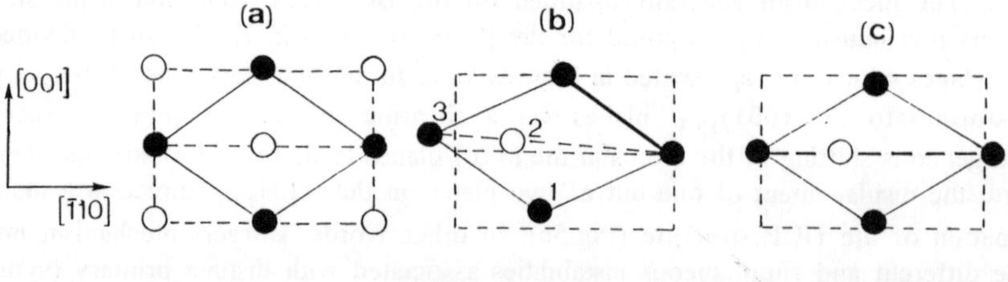

Fig.5. Projection of the BCC and HCP unit-cells on the (110) BCC plane. Solid and open circles are atoms in positions 1/2 and 0 respectively. (a) Initial BCC structure. (b) HCP structure: the atoms denoted 1, 2, 3 are shifted along the same $[1\bar{1}2]$ BCC direction. (c) HCP structure, following our reformulation of Burgers mechanism: the displacements are along the $[1\bar{1}0]$ BCC direction. The deformation along [001] BCC is induced by the shifting of the open sublattice.

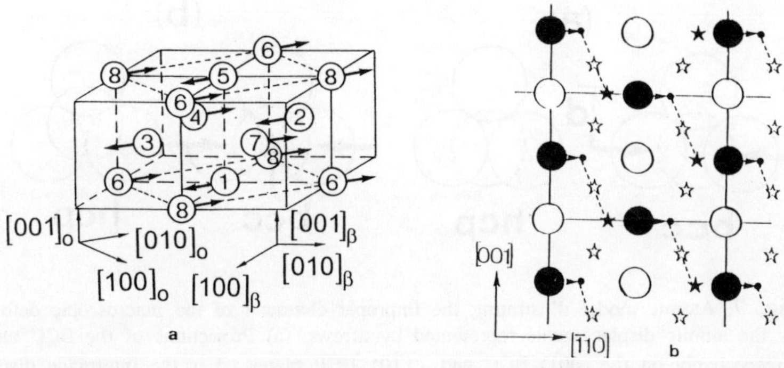

Fig. 6. Average shifting of the atoms from the BCC to the HCP structures. Solid and open circles are atoms located in adjacent (0 and 1/2) layers. (a) The small cube (thick lines) is the BCC unit-cell. The dotted parallelepiped is the unit-cell of the orthorhombic (D_{2h}^{17}) structure. (b) Projection on the (110) BCC plane. The shifting from the BCC (large circles) to the HCP structure (small full dots) is $a\sqrt{2}/12$. The open and solid stars correspond to virtual positions at which the symmetry would increase to O_h^9 and D_{6h}^4, respectively.

atoms of next layer are located. The displacement of one layer with respect to the following one in the $\left[1\bar{1}0\right]$ BCC direction (i.e. the shifting of each atom of the upper layer towards the center of the corresponding triangle, which is formed by the atoms of the lower layer) allows the separated atoms to come in contact, and thus to form a regular double-layered hexagonal close-packed structure. The *reverse* sliding of one close-packed layer with respect to the next layer, at *fixed interlayer distance*, breaks one of each of the three pairs of atoms touching each other, again forming a BCC lattice.

In order to formalize the preceding crystallographic description, let us first write the relationships between the basic vectors of the HCP (space group D_{6h}^4, Z=2) and BCC (O_h^9, Z=1) unit-cells, which are shown in Fig.6a:

$$\vec{a}_h = \vec{a}_c + \vec{b}_c + \vec{c}_c, \quad \vec{b}_h = -\vec{c}_c, \quad \vec{c}_h = \vec{a}_c - \vec{b}_c. \tag{3.13}$$

The *doubling* of the BCC unit-cell involved by (3.13) corresponds to *one branch*, $\vec{k}_9^{(2)}$, of the star of the wave-vector $\vec{k}_9$ [9], of the BCC Brillouin-zone, ending at the zone-boundary N-point. The star $\vec{k}_9^*$ has *six* branches, the coordinates of which are:

$$\vec{k}_9^{(1)} = \left(\frac{\pi}{a} \ \frac{\pi}{a} \ 0\right), \quad \vec{k}_9^{(2)} = \left(\frac{\pi}{a} \ -\frac{\pi}{a} \ 0\right), \quad \vec{k}_9^{(3)} = \left(\frac{\pi}{a} \ 0 \ \frac{\pi}{a}\right),$$

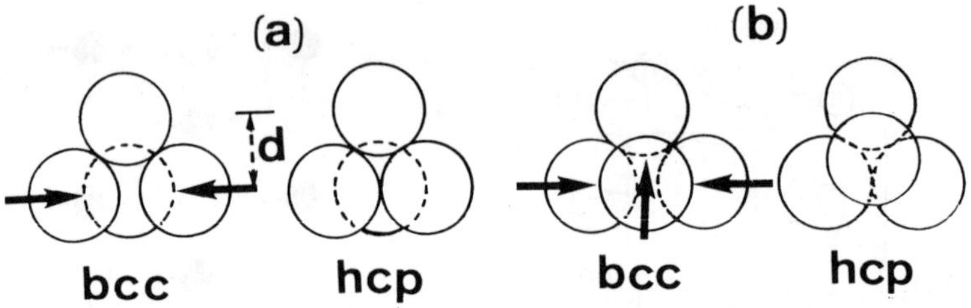

Fig. 7. Atomic model illustrating the improper character of the macroscopic deformation induced by the atomic displacements represented by arrows. (a) Projections of the BCC and HCP structures respectively on the (001) BCC and $(\bar{1}10)$ HCP planes. d is the interplane distance. (b) Projections of the BCC and HCP structures, respectively, on the (110) BCC and (001) HCP planes.

$$\vec{k}_9^{(4)} = \begin{pmatrix} \frac{\pi}{a} & 0 & -\frac{\pi}{a} \end{pmatrix}, \quad \vec{k}_9^{(5)} = \begin{pmatrix} 0 & \frac{\pi}{a} & \frac{\pi}{a} \end{pmatrix}, \quad \vec{k}_9^{(6)} = \begin{pmatrix} 0 & \frac{\pi}{a} & -\frac{\pi}{a} \end{pmatrix}. \tag{3.14}$$

On the other hand the antiparallel displacement field, represented in Fig.6a, transforms as the basic function:

$$\varphi(x,y) = (x+y)(\Delta_1 - \Delta_2 - \Delta_3 + \Delta_4 + \Delta_5 - \Delta_6 + \Delta_7 - \Delta_8) \tag{3.15}$$

where the Δ_i are the shifts associated with the atoms numbered 1 to 8 in Fig.6a. $\varphi(x,y)$ transforms as the one-dimensional little irreducible representation, labeled $\hat{\tau}_4$ in Kovalev's tables [9], the matrices of which are, for the invariance group D_{2h}^{xy} of $\vec{k}_9^{(1)}$:

	C_1	C_2	$U_{\bar{x}y}$	U_{xy}	I	σ_z	$\sigma_{\bar{x}y}$	σ_{xy}
$\hat{\tau}_4$	1	−1	1	−1	−1	1	1	−1

(3.16)

Using (3.14) and (3.16), one can construct the generating matrices of the *six-dimensional* irreducible representation $\tau_4(\vec{k}_9^*)$, of the (symmorphic) space group O_h^9:

$$\{U_x|\vec{0}\} \begin{bmatrix} & 1 & & & & \\ 1 & & & & & \\ & & & 1 & & \\ & & 1 & & & \\ & & & & 1 & \\ & & & & & 1 \end{bmatrix} \quad \{U_y|\vec{0}\} \begin{bmatrix} & 1 & & & & \\ 1 & & & & & \\ & & 1 & & & \\ & & & 1 & & \\ & & & & & 1 \\ & & & & 1 & \end{bmatrix} \quad \{C_3^{2(xyz)}|\vec{0}\} \begin{bmatrix} & & & & 1 & \\ & & & & & 1 \\ 1 & & & & & \\ & 1 & & & & \\ & & 1 & & & \\ & & & 1 & & \end{bmatrix} \quad \{U_{xy}|\vec{0}\} \begin{bmatrix} 1 & & & & & \\ & 1 & & & & \\ & & & & 1 & \\ & & & & & 1 \\ & & 1 & & & \\ & & & 1 & & \end{bmatrix} \quad \{I|\vec{0}\} \begin{bmatrix} 1 & & & & & \\ & 1 & & & & \\ & & 1 & & & \\ & & & 1 & & \\ & & & & 1 & \\ & & & & & 1 \end{bmatrix}$$

(3.17)

The geometrical method described in Chapter I (Sec.3.3.2) provides, using (3.17) the set of 33 low-symmetry groups, subgroups of O_h^9, which are induced by $\tau_4(\vec{k}_9^*)$. They are listed in Table 4. One can verify in this Table that only one low-symmetry state, of space-group D_{2h}^{17}, involves a *doubling* of the O_h^9 unit-cell. It coincides with the branch $\vec{k}_9^{(2)}$ and to the equilibrium solution $(0\eta 0000)$ of the *six-dimensional order-parameter*, transforming as $\tau_4(\vec{k}_9^*)$.

In summary, the direction in the ε_6-space of the six-dimensional order-parameter $\vec{\eta}$, which corresponds to the primary mechanism proposed for the BCC-HCP transition (the antiparallel atomic shifts along [110]) is given by:

$$\eta_2 = \eta \neq 0, \quad \eta_1 = \eta_3 = \eta_4 = \eta_5 = \eta_6 = 0. \tag{3.18}$$

However this primary order-parameter leads to a phase of orthorhombic symmetry D_{2h}^{17}. The additional transformation which brings about an increase of the D_{2h}^{17} symmetry to D_{6h}^4 (the HCP structure) corresponds to the following requirements:

(1) A compression of the cubic unit-cell along one of the fourfold axis. This *secondary deformation mechanism*, has the symmetry of the two-dimensional irreducible representation of the O_h^9 space-group at the center (Γ point) of the BCC Brillouin zone ($\tau_3(\vec{k}_{11}^*) \equiv E_g$ in Kovalev's notation [9]), which is spanned by the two combinations of the strain tensor components:

$$\zeta_1 = \frac{1}{\sqrt{6}}(e_{xx} + e_{yy} - 2e_{zz}), \quad \zeta_2 = \frac{1}{\sqrt{2}}(e_{xx} - e_{yy}). \tag{3.19}$$

(2) A *specific magnitude* for the antiparallel shifts along $[110]$:

$$|\xi_{HCP}| = \frac{a\sqrt{2}}{12} \tag{3.20}$$

where a is the cubic lattice parameter.

When the two preceding requirements are fulfilled, the D_{2h}^{17} symmetry *enlarges* to D_{6h}^4. This enlargement of symmetry mechanism is illustrated in Fig.8. This figure, which is a projection of the BCC structure in the $(1\bar{1}0)$ BCC plane, shows that the angle between the three fold axes, corresponding to arcCos1/3=70°32' in the cubic cell, *becomes* 60° in the hexagonal cell.

A striking property, which can be foreseen from Fig.6b, is the *periodic connection* that can be found between the overall symmetry undertaken by the structure and the virtual displacements of the atoms along the $[110]$ direction. Thus, for general arbitrary

Table 4. The 33 stable low-symmetry subgroups of O_h^9 induced by $\tau_4(\vec{k}_9^*)$. Columns (b), (c) and (d) indicate, respectively, the equilibrium relationships between the order-parameter components, the corresponding space groups, and the multiplication of the initial BCC unit-cell.

(a)	(b)						(c)	(d)
1	η	$-\eta$	η	$-\eta$	$-\eta$	η	T_d^6	8
2	η	$-\eta$	0	0	0	0	D_{4h}^{14}	4
3	0	0	η	$-\eta$	$-\eta$	η	D_{4h}^{19}	8
4	0	0	η	0	0	$-\eta$	C_{2h}^3	4
5	η	0	$-\eta$	0	$-\eta$	0	D_{3d}^6	8
6	0	0	$-\eta$	η	η	η	D_{4h}^{17}	8
7	0	η	0	η	0	$-\eta$	C_{3v}^5	4
8	0	η	0	0	0	0	D_{2h}^{17}	2
9	0	η_1	0	η_2	η_2	0	C_{2h}^6	8
10	η_1	0	η_2	η_2	$-\eta_2$	η_2	C_{2v}^{18}	8
11	0	0	η_1	η_2	$-\eta_2$	η_1	C_{2h}^3	8
12	0	0	η_1	0	0	η_2	C_i^1	4
13	η_1	0	η_2	0	0	η_2	C_s^3	4
14	η_1	η_1	η_2	η_2	$-\eta_2$	$-\eta_2$	C_{4v}^{11}	8
15	η_1	η_2	$-\eta_1$	$-\eta_2$	η_1	0	C_{3v}^6	8
16	η_1	η_1	$-\eta_2$	η_2	η_2	0	D_{2h}^{28}	8
17	η_1	η_2	0	0	0	0	D_{2h}^{12}	4
18	η_1	$-\eta_1$	η_2	η_2	$-\eta_2$	$-\eta_2$	D_{2d}^{12}	8
19	0	0	η_1	η_2	$-\eta_1$	η_2	C_{2h}^6	8
20	η_1	$-\eta_1$	η_2	η_2	0	0	D_{2h}^{25}	8
21	η_1	0	η_2	0	0	η_3	C_1^1	4
22	η_1	0	η_2	η_3	$-\eta_3$	η_2	C_s^3	8
23	0	η_1	0	η_2	η_3	0	C_i^1	8
24	0	η_1	η_2	η_3	η_3	$-\eta_2$	C_2^3	8
25	η_1	η_2	η_3	$-\eta_3$	η_4	$-\eta_4$	C_2^3	8
26	η_1	η_2	η_3	η_4	$-\eta_3$	$-\eta_4$	C_s^3	8
27	0	0	η_1	η_2	η_3	η_4	C_i^1	8
28	η_1	$-\eta_1$	η_2	η_2	η_3	η_4	C_s^3	8
29	η_1	η_2	η_3	η_4	η_5	η_6	C_1^1	8
30	η_1	η_2	η_3	$-\eta_3$	$-\eta_3$	η_3	C_{2v}^{19}	8
31	η_1	$-\eta_1$	η_2	η_2	η_3	$-\eta_3$	D_2^8	8
32	η_1	η_2	0	0	η_3	$-\eta_3$	C_{2h}^3	8
33	η_1	η_1	$-\eta_2$	η_2	$-\eta_3$	η_3	C_{2v}^{20}	8

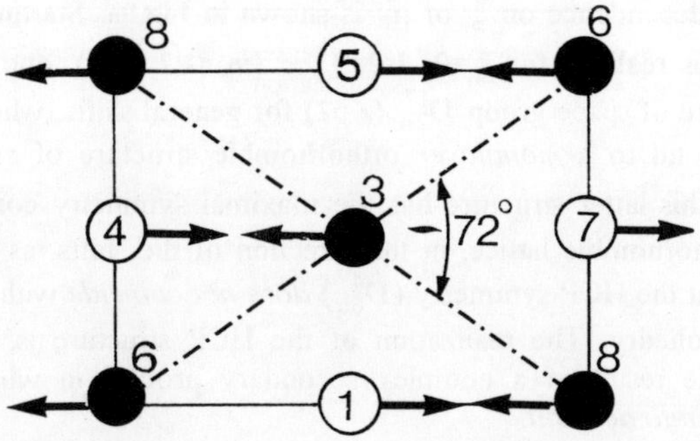

Fig. 8. Shifting mechanism at the BCC-HCP transition in projection on the $(1\bar{1}0)$ plane. After the shifting, the angle between the diagonals becomes 60°. Solid and open circles are atoms located in two adjacent layers.

displacements $|\xi|$ along [110], the system exhibits the symmetry D_{2h}^{17}, while for the specific displacements $|\xi_{HCP}| = a\sqrt{2}/12$, $7a\sqrt{2}/12$, $13a\sqrt{2}/12$,... the crystal acquires the D_{6h}^{4} symmetry. For other specific displacements, namely $a\sqrt{2}/4$, $3a\sqrt{2}/4$, $5a\sqrt{2}/4$,... the symmetry increases to D_{6h}^{1} (Z=1). We will now express this periodicity of the magnitude of $|\vec{\eta}|$ as a function of ξ within a density-wave description of the BCC-HCP transition.

3.3. Periodic dependence of the order-parameter magnitude

The lattice function ψ of a unit-cell, associated with a general shift of magnitude $\pm\xi$ in the direction [110], can be written in the (x,y) plane as:

$$\psi(\xi) = A\left(e^{\frac{2\pi i}{a}\cdot\frac{a}{4}} \cdot e^{\frac{4\pi i}{a\sqrt{2}}\xi} + e^{-\frac{2\pi i}{a}\cdot\frac{a}{4}} \cdot e^{-\frac{4\pi i}{a\sqrt{2}}\xi} + \text{c.c.} \right) \qquad (3.21)$$

where the origin of the coordinates has been chosen at middle distance between two atoms shifted in antiparallel directions. In the BCC structure $\xi=0$ and $\psi_{BCC}(\xi)=0$. Thus, the increment $\delta\psi(\xi)$ corresponding to the variation of $\psi(\xi)$ from BCC to the HCP structure, has the same form (3.21), which can be simplified, giving the following dependence of the order-parameter magnitude, in function of the critical shifts:

$$|\vec{\eta}(\xi)| = \eta_0 \cdot \text{Sin}\frac{4\pi}{a\sqrt{2}}\xi. \qquad (3.22)$$

The periodic dependence on ξ of $|\vec{\eta}|$ is shown in Fig.9a. Starting from the initial O_h^9 structure, which is realized for ξ=0, 1/2, 1, ... (in $a\sqrt{2}$ unit), one gets a two-layer orthorhombic structure of space group D_{2h}^{17} (Z=2) for general shifts, whereas the extrema of the curve correspond to a *monolayer* orthorhombic structure of *symmorphic* space group D_{2h}^{19} (Z=1). This latter structure has the maximal symmetry compatible with the holohedry of the orthorhombic lattice, in the direction of the shifts, as shown in Fig.9b, which also reveals that the HCP symmetry (D_{6h}^4) *does not coincide* with any special point of the preceding holohedry. The realization of the HCP structure is, as shown in the preceding section, the result of a complex secondary process in which *two physical constraints are taken into account.*

The first constraint is *the preservation of the interlayer distance between the (110) cubic planes in the HCP structure*, where they becomes the $(001)_h$ planes. This is a general property of the crystals of elements undergoing a BCC-HCP transformation, which are listed in Table 5. This table shows that, except for Ba and ^{3}He (which will be discussed below) the variation of the interlayer distance is smaller than 3%, when going from one structure to the other. When this constraint is satisfied, the orthorhombic (D_{2h}^{19}) symmetry *increases* to hexagonal (D_{6h}^1, Z=1) as represented in Fig.9c.

In order to obtain the D_{6h}^4 structure, another constraint should be imposed, which consists in *taking into account the actual size of the atoms*, i.e. to exclude from the theoretical description of the transition process, shifts of atoms after they have entered into contact. Two types of contacts can be distinguished: the ones which occur, for example, between pairs of atoms denoted (3,4), (5,6), or (1,8) in Fig.8, which result from the primary (antiparallel shifting) mechanism, and the ones which take place in the

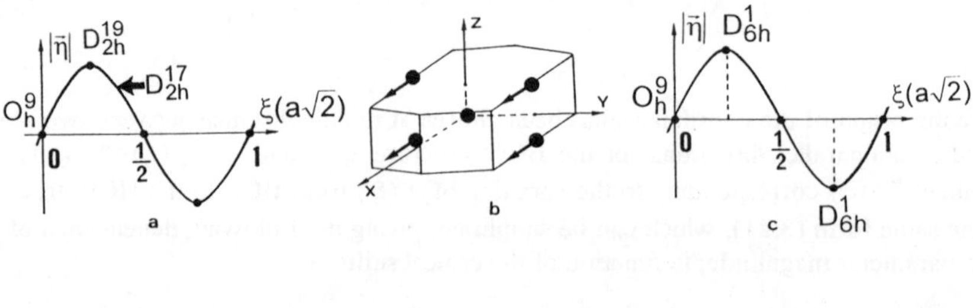

Fig. 9. (a) Periodic dependence of the order-parameter following Eq.(3.22). (b) Holohedry of the orthorhombic lattice, in the direction of the shifts. (c) Periodic dependence of the order-parameter at the BCC-HCP transition, when the preservation of the interlayer distance is taken into account.

Table 5. Interlayer distances d (in A°) between the cubic (110) planes (column 3) and hexagonal (001) planes (column 5) for the elements listed in the first column. $\frac{d_c - d_h}{d_c}$ is the relative variation of d when going from the BCC to the HCP structure.

Elements	HCP		BCC		$\frac{d_c - d_h}{d_c}$
	c_h	d_h	a_c	d_c	(%)
Ti	4.732	2.37	3.330	2.36	-0.42
Zr	5.150	2.58	3.620	2.56	-0.78
Hf	5.470	2.74	3.610	2.55	-7.45
Be	3.626	1.81	2.550	1.80	-0.56
Ca	6.440	3.22	4.380	3.10	-3.87
Ba	6.154	3.08	5.050	3.57	13.73
Fe	4.370	2.19	2.870	2.03	-7.88
Ti	5.530	2.77	3.882	2.75	-0.73
^{3}He	4.986	2.49	3.963	2.80	11.07
^{4}He	5.798	2.90	4.110	2.91	0.34
Gd	5.700	2.85	4.060	2.87	0.7
Tb	5.6936	2.85	4.020	2.84	-0.35
Dy	5.6475	2.82	3.980	2.81	-0.36
Yb	6.3859	3.19	4.440	3.14	-1.59
Y	5.7306	2.87	4.110	2.91	1.38

perpendicular direction, such as between the pairs (4,8), (1,3) or (6,7) in Fig.8. These latter contacts are the result of the secondary mechanism assumed in our model, namely of the compression (or decompression) along the fourfold axes, and are equivalent to collective atomic shifts, which are, for instance, along the Z axis: $\Delta z = \frac{1}{2}|\vec{a}_c - \vec{c}_h|$, or in terms of the atomic radius: $\Delta z = \frac{1}{2}(4R/\sqrt{3} - 2R)$.

Fig. 10a illustrates the role of the contacts between the atomic spheres, which is *to create a geometrical barrier*, acting as a *mirror plane*, perpendicular to the direction of the shifts. *This barrier modifies the " map" of the special points of the lattice at which the symmetry of the structure may increase*. Fig. 10b shows that it produces a reduction by 3/2 of the periodicity of the shifts, so that the unit length $a\sqrt{2}$ in Fig.9a becomes $2a\sqrt{2}/3$). Consequently the exponential terms $\exp(\pm\frac{4\pi i}{a\sqrt{2}}\xi)$ in (3.21) become $\exp(\pm\frac{6\pi i}{a\sqrt{2}}\xi)$, and the magnitude of the order-parameter $|\vec{\eta}(\xi)|$ takes the form:

$$|\vec{\eta}(\xi)| = \eta_0 \cdot \mathrm{Sin}\frac{6\pi}{a\sqrt{2}}\xi \qquad (3.23)$$

represented in Fig. 10c. In this figure one gets periodically the initial BCC structure (for ξ=0, 1/2, 1, ... in $a\sqrt{2}/3$ units) and the HCP structure for ξ=1/4, 3/4, 5/4, ... in agreement with our crystallographic description.

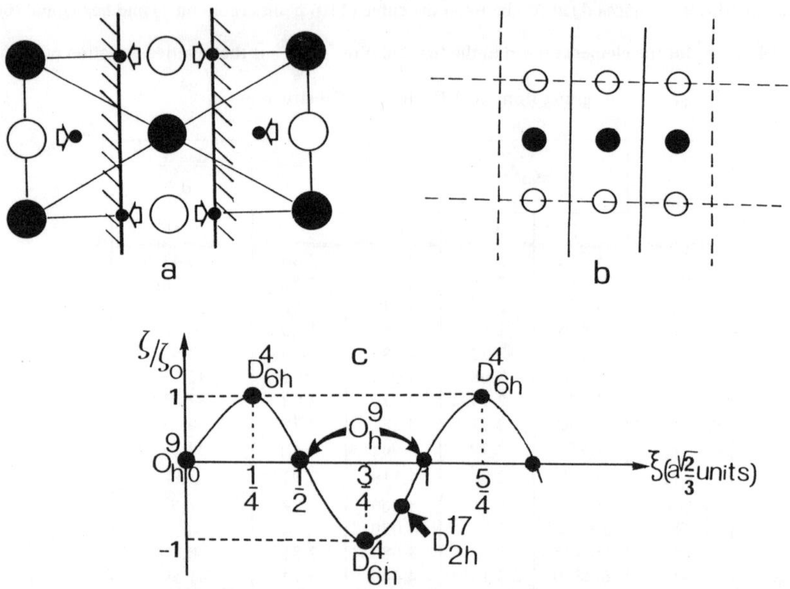

Fig. 10. (a) Creation of a geometrical barrier, due to contacts between atoms. (b) Subsequent reduction of the periodicity of the shifts. (c) Periodicity of the order-parameter at the BCC-HCP transformation, following Eq.(3.23).

3.4. Alternative description of the BCC-HCP transition: The HCP parent phase

The functional dependence $|\vec{\eta}(\xi)|$ represented in Fig.10c deserves two observations: (1) It has been obtained by modifying the periodicity associated with the "primary" antiparallel shifting mechanism, in order to account for necessary constraints: the preservation of the interlayer distance between definite families of planes, and the "hard sphere" character of the atoms. These constraints are taken into account by the secondary mechanism assumed in our reformulation of Burgers mechanism.

(2) The sinusoidal curve is *symmetric* with respect to the displacement axis *for the* D^4_{6h} *structure*. Following the general scheme given in Sec.5.2 of Chapter II, this means that *the HCP structure may also play the role of the initial structure in the BCC-HCP transition.*

In order to verify this point, let us write the lattice function ψ, starting from the HCP structure. The antiparallel shifts, which lead to the BCC structure take place along the hexagonal $[1\bar{1}0]$ directions. For an arbitrary shift $\pm\xi$, one has:

$$\psi(\xi) = A\left\{\exp\left[\frac{4\pi i}{a\sqrt{3}}\left(-\frac{a\sqrt{3}}{12}+\xi\right)\right]+\exp\left[\frac{4\pi i}{a\sqrt{3}}\left(\frac{a\sqrt{3}}{12}-\xi\right)\right]+\text{c.c.}\right\}. \tag{3.24}$$

In the initial (HCP) phase: $\xi=0$ and $\psi_{HCP}=1$. The increment $\delta\psi(\xi)=\psi_{HCP}-\psi(\xi)$, which is proportional to the order-parameter magnitude is thus:

$$\delta\psi = 2A\cdot\left[\operatorname{Sin}\left(\frac{4\pi}{a\sqrt{3}}\xi+\frac{\pi}{6}\right)-\frac{1}{2}\right] \tag{3.25}$$

The functional dependence (3.25), given in Fig.11a, shows an asymmetric behavior with respect to the ξ axis, as the maxima of the curve correspond to the tetragonal structure D_{4h}^{17} (Z=1) (after taking into account the macroscopic deformation), and the minima, to the simple hexagonal structure of symmetry D_{6h}^{1} (Z=1). The body centered tetragonal symmetry increases to BCC when applying a decompression along the [110] hexagonal direction, which coincides with the Z axis. This *secondary* mechanism increases the angle between the two fold axes of the hexagonal structure from 60° to 70°32'. Let us stress that the O_h^9 space-group *figures among the symmetries reached for special points* of the D_{2h}^{17} holohedry, as shown in Fig.11b, where one can see that it coincides with specific values of the shifts along the axes of coordinates.

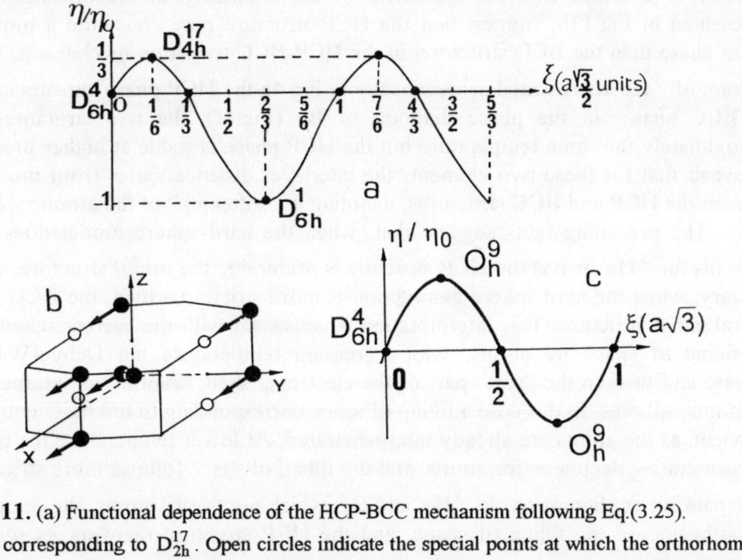

Fig. 11. (a) Functional dependence of the HCP-BCC mechanism following Eq.(3.25). (b) Holohedry corresponding to D_{2h}^{17}. Open circles indicate the special points at which the orthorhombic symmetry increases to cubic. (c) Periodic dependence of the HCP-BCC order-parameter, following Eq.(3.26).

Two clarifying remarks have also to be made on the curve of Fig.11a. (1) The curve has been obtained by considering the atoms as located at *points* of the lattice. This leads for the phase of D^1_{6h}, to a configuration in which the atoms of one layer are exactly over the atoms of the next layer, i.e. a situation which implies a considerable increase of the interlayer distance, or a noticeable deformation of the atoms. *As neither happen* in the close-packed model under consideration, one has again to introduce the geometrical constraints imposed by the size of the atoms and by the invariability of the interlayer distance when going from one structure to another. In particular, one can see in Fig.6b, that the realistic path followed by the atoms starting from the HCP positions is to be shifted along the directions $[120]_{HCP}$, $[210]_{HCP}$ or $[110]_{HCP}$, i.e. to pass through valleys between touching atoms. *This excludes the phases of symmetry* D^1_{6h} *to be formed*, and yields a simplified dependence of the order-parameter modulus, as compared to (3.25), which is:

$$\eta = \eta_0 \cdot \mathrm{Sin}\left(\frac{8\pi}{3a\sqrt{3}}\xi\right). \tag{3.26}$$

The corresponding curve given in Fig.11c, expresses a more realistic situation than the curve of Fig.11a. In particular it is *symmetric* with respect to the other possible (BCC) initial structure.

(2) The two curves shown in Figs.11a and 11c, and the better compatibility of an initial HCP structure with the symmetry of the holohedry of the orthorhombic lattice, represented in Fig.11b, suggest that the HCP structure may constitute a more "natural" parent phase than the BCC structure, in the HCP-BCC transition mechanism. In the phase diagram of ^{3}He represented schematically in Fig.4e the HCP phase appears indeed *above* the BCC phase. In the phase diagram of Ba (Fig.4f), the two structures appear at approximately the same temperature but the HCP phase is stable at higher pressure. Table 5 Reveals that for these two elements the interlayer distance varies from more than 10% between the HCP and BCC structures, denoting a " softening" of the atomic spheres.

The preceding facts suggest that, when the hard sphere model does not strictly apply (as for ^{3}He or Ba) the HCP structure is preferably the *initial* structure, while on the contrary, when the hard sphere assumption is more strictly verified, the BCC phase is the natural initial structure. This interpretation is consistent with the current scheme depicting the filling of space by atoms: with increasing temperature the Debye-Waller factors increase and thicken the "soft" part of the electronic shell, favoring an interpenetration of the atomic spheres. In this case a filling of space corresponding to the BCC configuration is sufficient, as the atoms are already interpenetrated. At lower temperature the possibility of interpenetrating decreases for atoms, and the filling of space follows more strictly the HCP close packing configuration. In ^{3}He and Ba, which are "soft" atoms, the temperature has less influence on the filling of space, and the HCP structure recovers its role of parent structure.

3.5. Thermodynamics of the BCC-HCP transition: Improper symmetry breaking strains

The dependences on critical shifts of the order-parameter magnitudes associated with the BCC-HCP and HCP-BCC transitions, as given by (3.23) and (3.26) respectively, are *symmetric* with respect to the displacement axis, and thus correspond both to an effective order-parameter expansion of the form:

$$F\left(\left|\vec{\eta}(\xi)\right|\right) = F_0(P,T) + \frac{\alpha}{2}\left|\vec{\eta}(\xi)\right|^2 + \frac{\beta}{4}\left|\vec{\eta}(\xi)\right|^4 + \frac{\gamma}{6}\left|\vec{\eta}(\xi)\right|^6 \qquad (3.27)$$

which has to be expanded at least up to the sixth degree, in order to account for a region of stability including a direct BCC-HCP transition line. (3.27) is identical to the expansion (2.9) discussed in Chapter II, and thus, the topology of the corresponding phase diagram, represented in Fig.12, is identical to the phase diagram of Fig.4 in Chapter II. The thermodynamic path followed for the elements listed in Table 5, is far from the region of stability of the orthorhombic phase, which among elements is found for example in uranium (see Sec.6 of the present Chapter).

Expansion (3.27) reflects the symmetry of the primary order-parameter associated with the BCC-HCP mechanism. For a full phenomenological description of the transition, one needs to add to (3.27) a coupling term which accounts for the macroscopic deformation involved in the transition mechanism. The strain components which couple to the primary order-parameter can be found by decomposing the symmetrized *second* power of $\tau_4(\vec{k}_9)$ into the irreducible representations of the point-group O_h. One finds:

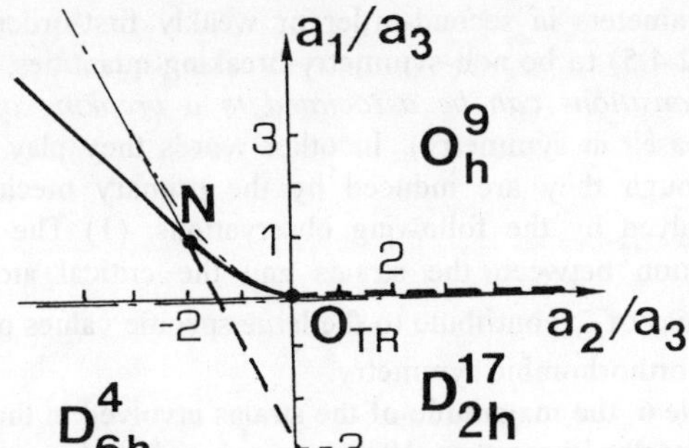

Fig. 12. Phase diagram containing the BCC and HCP phases, following Eq.(3.27) . Full, dashed, and dashed-dotted lines are respectively first, and second-order transitions, and limit of stability lines. N is a three-phase point. 0 is a tricritical point.

$$\left[\tau_4(\vec{k}_9)\right]^2 \cong A_{1g} + E_g + F_{2g} \tag{3.28}$$

where A_{1g}, E_g and F_{2g} are respectively the one, two, and three dimensional irreducible representations of O_h, spanned by the following combinations of the strain-tensor:
(i) $e_{xx} + e_{yy} + e_{zz}$, which transforms as A_{1g}, is the rigidity, expressing the change in volume of the crystal at the transition.
(ii) (ζ_1, ζ_2) defined by Eq.(3.19), which has the symmetry of E_g, account for the compression, or decompression, along the fourfold axes.
(iii) (e_{yz}, e_{xz}, e_{xy}) which span F_{2g}, are the shear strains describing an eventual modification of the interlayer distance.

In the BCC-HCP mechanism, one can neglect, in a first approximation, the change in volume, and the variation of the interlayer distance. Besides, the equilibrium value of the tetragonal deformation of the unit-cell is: $\zeta_1 = \zeta \neq 0$, $\zeta_2 = 0$ (see Section 4 of this Chapter). Accordingly the effective coupling term is:

$$F(\eta, \zeta) = \delta \cdot \left|\vec{\eta}(\vec{\xi})\right|^2 \cdot \zeta \tag{3.29}$$

where δ is a negative coupling constant.

Let us stress, that although the form of (3.29) reflects the "improper" character of the secondary strain ζ, the role of ζ, as described in our mechanism, results in an *increase* (to D_{6h}^4) of the orthorhombic symmetry D_{4h}^{17}, at least when considering an initial BCC structure, and in the stabilization of the HCP phase. Thus, in contrast to the property of secondary order-parameters in second-order or weakly first-order structural transitions (see Chapter I, Sec.2.4.5) to be non-symmetry-breaking quantities, *the secondary strains in reconstructive transitions can be associated to a breaking of symmetry* (or more exactly to an "increase" in symmetry). In other words they play the role of a distinct order-parameter, though they are induced by the primary mechanism. This apparent contradiction is resolved by the following observations: (1) The coupling term (3.29) expresses a connection between the strains and the critical atomic shifts ζ. Thus, sufficiently large values of ζ contribute to the large specific values of ζ which correspond to an increase of the orthorhombic symmetry.
(2) As shown in Table 6, the magnitude of the strains involved in the BCC-HCP transition in elements, are generally larger than 10%, i.e. one order of magnitude larger than the strains usually involved in ferroelastic transitions [21].

The primary instability, which transforms as $\tau_4(\vec{k}_9^*)$, coincides with an optical phonon mode at the N-point of the BCC Brillouin zone. Except for Li and Na (see Sec.8 of this Chapter), no detectable temperature dependent softening of this mode has been observed in crystals of the elements listed in Table 6, confirming, on the one hand, the

Table 6. Magnitude of the spontaneous macroscopic strain e_{zz}, involved in the BCC-HCP transition, for the elements listed in column (a). Note the large values of e_{zz} found for ^{3}He and Ba.

Elements	HCP a_h (A°)	BCC a_c (A°)	e_{zz}
Ti	2.976	3.33	0.11
Zr	3.24	3.62	0.11
Hf	3.33	3.61	0.08
Be	2.291	2.55	0.1
Ca	3.94	4.38	0.1
Ba	3.901	5.05	0.23
Fe	2.705	2.87	0.06
Ti	3.48	3.882	0.1
^{3}He	3.046	3.963	0.23
^{4}He	3.555	4.11	0.14
Gd	3.629	4.06	0.11
Tb	3.601	4.02	0.1
Dy	3.5903	3.98	0.1
Yb	3.8799	4.44	0.13
Y	3.6474	4.11	0.11

absence of intrinsic softening predicted in our approach for direct reconstructive transitions to limit state, and on the other hand, that the transition occurs far from the region of stability of the orthorhombic phase. By contrast a dip in the dispersion curve at, or close to, the N point was clearly observed in Ti, Zr, Hf, Na, Ba, and Fe [22], and also in the BCC phase of some other elements, such as Mo, Cr, Nb, W, K or Cs [22] in which the HCP phase has not been recognized up to now.

The secondary strain ζ corresponds to the acoustical transverse mode Σ_4. It may have two types of dynamical effects: a *direct one*, such as a change in the slope of the Σ_4 mode with temperature (the softening of the $c_{11} - c_{12}$ elastic constant), or an *indirect effect* connected with the coupling to the primary instability, that would result in a shift of the dip at the N-point observed in the dispersion phonon branch $\left[\frac{1}{2}\frac{1}{2}0\right]$, related to the change of the lattice parameters below the transition.

At last, among the other properties deriving from our approach to the BCC-HCP transition, let us briefly describe the corresponding domain structure. Starting from a BCC structure, one should observe *six orientational domains* in the HCP phase, transforming into each other by the three-fold cubic axes lost at the transition, and two types of antiphase domains, corresponding to the loss of the cubic a_c and b_c lattice parameters. In the reversed HCP-BCC transition, no antiphase domains should be found in the cubic phase, as the transition is induced by a zone-center instability, but three orientational domains should be created, which can be deduced one from another by the six fold rotations lost at the transition.

4. The BCC-FCC transition

4.1. The Bain deformation

The BCC-FCC transition is almost as frequent, among elements, as the BCC-HCP transition. It is found in about twenty elements, in the phase diagram of which the BCC phase appears generally at high temperatures, below the melt. The stability of the FCC structure, which is more close packed than the BCC structure, increases with pressure, and in the phase diagram of K, Cs and Rb, FCC stability region appears only at high pressures, at approximately the same temperature than the BCC phase. Figures 13a,b,c,... show the main types of phase diagrams of the elements in which the two phases are inserted. As the BCC-HCP transition, the BCC-FCC transition is always found to be strongly first-order, and associated with large atomic displacements, of the order of the unit-cell parameters.

One can visualize the way a BCC lattice transforms onto a FCC lattice, via a deformation which stretches the cubic unit-cell by the same extent along one of the fourfold axes (e.g. [001]) and compresses it by the same extent along the other fourfold axes. Such a macroscopic deformation, which is represented in Figs.14a and 14b, is called the *Bain deformation* [23]. Fig.14b shows the connection between the BCC and FCC (O_h^5, Z=1) unit-cell, which yields the following orientation relationship between the corresponding, ($\vec{a}_c, \vec{b}_c, \vec{c}_c$) and ($\vec{a}_1^F, \vec{a}_2^F, \vec{a}_3^F$), lattice parameters:

$$\vec{a}_1^F \| \vec{a}_c, \quad \vec{a}_2^F \| (\vec{a}_c + \vec{b}_c + \vec{c}_c), \quad \vec{a}_3^F \| (\vec{a}_c + \vec{c}_c). \tag{3.30}$$

This translational relationship is associated with a wave-vector located at the center (Γ-point) of the BCC Brillouin zone. The spontaneous strains which account for the Bain deformation, are the same combinations of the strain-tensor components, ζ_1 and ζ_2 given

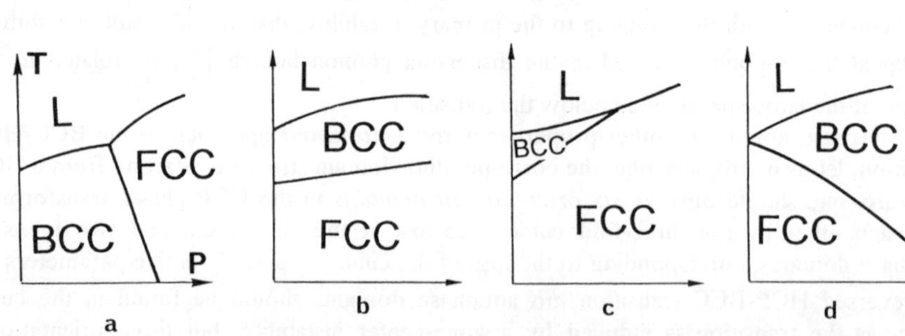

Fig. 13. Topology of the BCC-FCC transition line in some elemental crystals. The represented phase diagrams are found, for example, in Rb (a), Mn (b), La (c) and Yb (d).

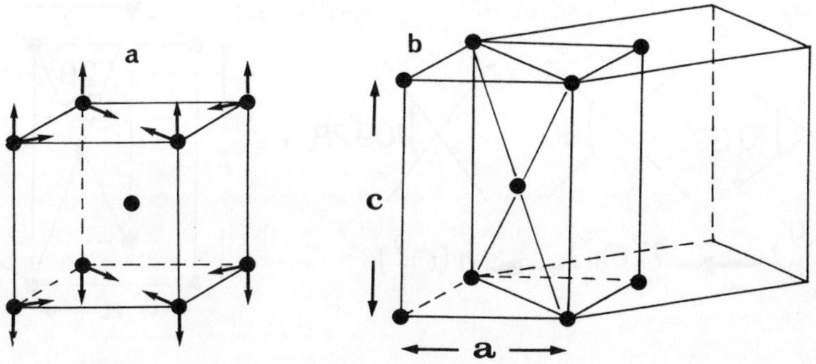

Fig. 14. Macroscopic deformation representing the Bain deformation: (a) corresponding atomic displacements; (b) connection between the BCC and FCC unit-cells.

by (3.19), which play the role of a secondary deformation in Burgers mechanism. The corresponding two-dimensional irreducible representation ($\tau_3(k^*_{11})$ in Kovalev's notations [9]) has the following generating matrices:

$\{U_x, U_y, I\|\vec{0}\}$	$\{U_{xy}\|\vec{0}\}$	$\{C_3^{xyz}\|\vec{0}\}$
$\begin{bmatrix}1 & 0\\0 & 1\end{bmatrix}$	$\begin{bmatrix}1 & 0\\0 & -1\end{bmatrix}$	$\frac{1}{2}\begin{bmatrix}-1 & -\sqrt{3}\\\sqrt{3} & -1\end{bmatrix}$

which lead to the following, possibly stable, low symmetry groups: D_{4h}^{17} (Z=1) for $\zeta_1 = \zeta \neq 0$, $\zeta_2 = 0$ (or $\zeta_2 = \zeta \neq 0$, $\zeta_1 = 0$), and D_{2h}^{25} (Z=1) for $\zeta_1 \neq \zeta_2 \neq 0$. When, for the first solution, one has the specific ratio c/a=$\sqrt{2}$, where a and c are the tetragonal lattice parameters, the tetragonal space-group D_{4h}^{17} *identifies as the face centered cubic space-group* O_h^5. The preceding ratio corresponds to the multiplication of the conventional BCC lattice parameters, by a factor $\sqrt{3}/2$ along X and Y, and by a factor $\sqrt{3/2}$ along Z. Accordingly the numerical values of the spontaneous strain components are: $e_{xx} = e_{yy} = \sqrt{3}/2$ -1, and $e_{zz} = \sqrt{3/2}$ -1, which give: ζ=1-1/$\sqrt{2} \cong 0.293$. Besides, in order to keep the atoms in contact at the transition, one has to consider as a secondary order-parameter, the rigidity: $\delta = \frac{1}{\sqrt{3}}(e_{xx} + e_{yy} + e_{zz})$, which measures the decrease in volume at the transition. One finds numerically: $\delta = 1 + 1/\sqrt{2} - \sqrt{3} \cong -0.025$.

As the BCC-HCP transition, the Bain deformation has a *periodic character*, which can be seen by considering the angle θ between the diagonals of the (110) cubic planes, shown in Fig.15a. Thus in the BCC lattice $\theta \cong 70°32'$. For θ =90°, the (110) BCC planes

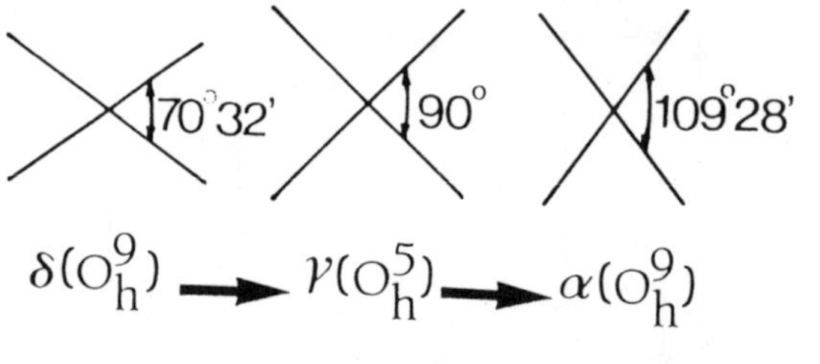

a

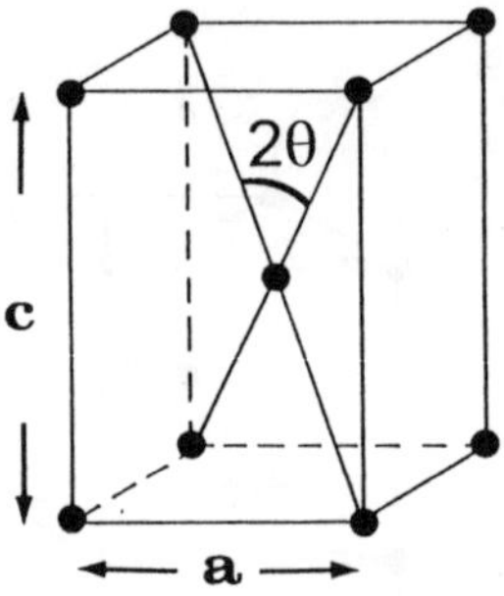

b

Fig. 15. (a) Periodic character of the Bain deformation. (b) Angle θ between the diagonals of the deformed cube, used as a variational parameter in Eq.(3.31).

transform into the (010) FCC planes. Further increase of the angle θ up to $\theta \cong 109°28'$ brings about the crystal again to the BCC symmetry, but with a different orientation, turned by 90° with respect to the initial BCC structure. For other arbitrary values of θ the lattice remains tetragonal (Fig.15b). One can calculate the dependence of $\zeta(\theta)=\frac{2}{\sqrt{6}}\left[e_{xx}(\theta)-e_{zz}(\theta)\right]$ by expressing e_{xx} and e_{zz} *independently* in function of the angle θ. One finds, using Fig.15b:

$$e_{xx}=\sqrt{\frac{3}{2}}\cdot \mathrm{Sin}2(\theta_0+\Delta\theta)-1,\quad e_{zz}=\sqrt{3}\cdot \mathrm{Cos}2(\theta_0+\Delta\theta)-1 \tag{3.31}$$

where $\theta_0=\frac{1}{2}\mathrm{arcSin}\sqrt{3/2}=\frac{1}{2}\mathrm{arcCos}\sqrt{1/3}\cong 27°22'$, is one fourth of the large angle (109°28') between the diagonals in the BCC unit-cell, and $\Delta\theta$ is the variational angle. Figs.15a and 15b show the variations of e_{xx} and e_{zz} in function of $\Delta\theta$. One can see that *three* structures appear periodically for definite values of $\Delta\theta$: the BCC structure for $e_{xx}=e_{zz}=0$, the FCC structure for maximal e_{xx} and e_{zz}, and a planar hexagonal (D^1_{6h}) structure, represented schematically in Fig.16c. This latter unphysical atomic configuration results from the fact that the variations laws (3.31) do not hold for any value of θ (or $\Delta\theta$), as this parameter represents an absolute measure of the angle between the diagonals, and is thus restricted by the following constraints: (1) the arguments of $e_{xx}(\Delta\theta)$ and $e_{zz}(\Delta\theta)$ should be limited as they represent a macroscopic deformation of the crystal; (2) After reaching the FCC close-packed structure, a further deformation of the unit-cell along the $[100]_{BCC}\parallel[110]_{FCC}$ directions is impossible, at least if one assumes a hard-sphere model; as in the preceding directions, close-packed chains of atoms are formed with atoms in contact with each other.

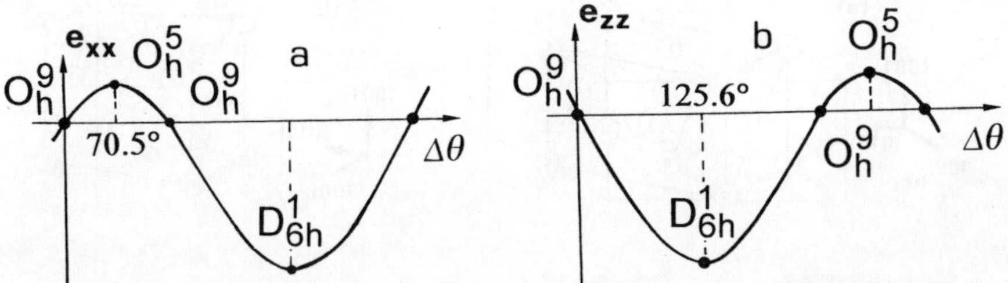

Fig. 16. Dependence of e_{xx} and e_{zz} on the variation of the diagonal angle $\Delta\theta$ at the BCC-FCC transition.

In order to remove the preceding restrictions, *one should express the order-parameter as a function of the average displacements of the atoms*. This brings us to analyze the atomistic mechanism which underlies the Bain deformation.

4.2. The Bain deformation revised

In the shearing deformation model, currently proposed for the BCC-FCC transition [24] as equivalent to the Bain deformation mechanism, the structural stability of the BCC structure is lowered with respect to an homogeneous shear strain of the (011) planes in the $\pm[0\bar{1}1]$ directions, which gives rise to a FCC structure. An experimental support to this interpretation is the decrease of the shear modulus $C_{11} - C_{12}$ observed in a number of BCC metals when approaching the FCC phase. We can propose the following path, in reference to Figs.17a and 17b, which materializes at the atomistic level the preceding macroscopic shearing mechanism. (1) The layer containing the atoms labeled 7-12 is shifted by a$\sqrt{2}$/6 (where a is the cubic lattice parameter) in the $[011]$ BCC direction, the atom in 0 position reaching the center of a triangle formed by the atoms denoted 8, 9 and 11; (2) the next layer, formed by atoms 1-6 will be subsequently shifted by a$\sqrt{2}$/3. This will place each atom initially in 0 position, at the center of the corresponding triangles of the following layer; (3) at last, a shifting of the third layer by a$\sqrt{2}$/2 will force the atoms located in this layer to occupy equivalent positions, i.e. the corner atoms (0,0) will lie on a straight line passing by the centers of the triangles (8, 9, 11) and (2, 4, 5) and orthogonal to the three successive layers.

Hence, the layers containing the atoms labeled 0 in Figs.17a and 17b will be equivalent crystallographically, i.e. the resulting structure will consist of a three-layered close-packed structure (a FCC structure) since the shears will have eliminated the repulsive effect of one layer with respect to the neighboring ones.

Let us stress that the hereabove microscopic description of the Bain deformation is closely analogous to Burgers mechanism, as both mechanisms preserve the interlayer distance, and thus the repulsive effect of one layer with respect to the others. The essential

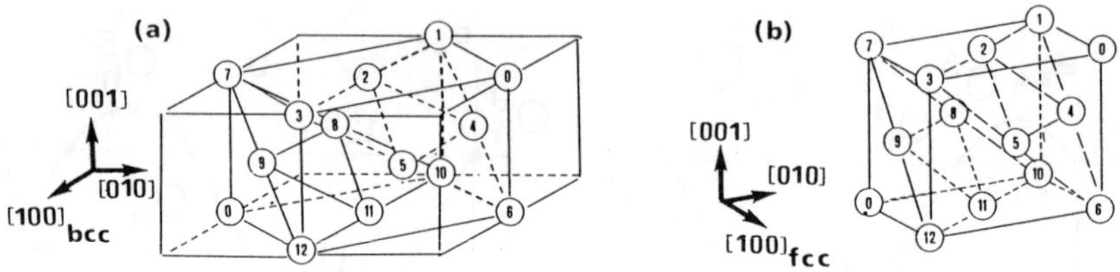

Fig. 17. Unit-cell of (a) the BCC and (b) the FCC structures. The numbering of the atoms is used in the description of the BCC-FCC transition mechanism given in the text.

difference is that, in the BCC-HCP transition, two successive layers are assumed to shift in opposite directions, i.e. the displacements are associated with a soft mode at the surface of the BCC Brillouin zone, whereas for the BCC-FCC transition, *two successive layers are sheared in the same direction*, and thus correspond to a zone-center soft mode.

The periodicity of the atomic displacements associated with the BCC-FCC transition, is shown in Fig. 18a. The mechanism can be depicted in terms of antiparallel shifts of the $(\bar{1}10)$ atomic planes in the $\pm[110]$ directions. A zero shift is chosen at middle distance between two arbitrary planes, and the successive planes correspond to increasing shifts ξ_n enumerated 1, 2, 3,..., n. The shifts of the atoms in the n-th plane are:

$$\xi_n = \frac{c}{4}(2n+1)\cdot \mathrm{tg}\alpha \tag{3.32}$$

where α is the angle between the Z axis and the direction formed by the shifted atoms (Fig. 18a), which characterizes the macroscopic deformation.

The lattice function expressing the structure for a shift ξ is:

$$\psi(\xi) = A\left[e^{\frac{2\pi i}{c}\frac{c}{4}}\cdot e^{\frac{4\pi i}{\tilde{a}}\xi} + e^{-\frac{2\pi i}{c}\frac{c}{4}}\cdot e^{-\frac{4\pi i}{\tilde{a}}\xi} + \mathrm{c.c.}\right] \tag{3.33}$$

with $\tilde{a} = \frac{2}{3}\, a\sqrt{2}$. For $\xi = 0$ (the BCC structure) $\psi(0) = 0$, and the increment $\delta\psi(\xi)$ associated with the BCC-FCC transition can be simply expressed as:

$$\delta\psi(\xi) = A \cdot \mathrm{Sin}\frac{3\pi}{a\sqrt{2}}\xi\,. \tag{3.34}$$

The corresponding dependence of the order-parameter is shown in Fig. 18b. It expresses a physically realistic scheme, as compared to Figs. 16a or 16b, and holds for non-

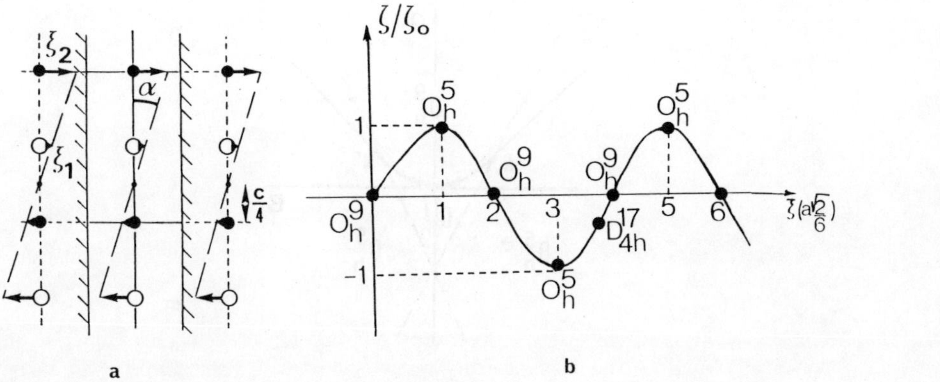

Fig. 18. (a) Periodicity of the antiparallel shifting mechanism at the BCC-FCC transition. (b) Periodic dependence of the order-parameter following Eq.(3.34).

limited magnitude of ξ, as large antiparallel shifts are compatible with a limited macroscopic deformation and respect the constraints imposed by the close packing of atoms. In particular, the "layer" D^1_{6h} phase found in the preceding curves is absent from Fig. 18b.

4.3. Phase diagram associated with the BCC-FCC transition

In the sinusoidal curve shown in Fig. 18b, the FCC phase is symmetric with respect to the ξ-axis. In our approach, this fact is interpreted as the possibility for the FCC structure to be the initial structure in the BCC-FCC transition. In the phase diagrams of elements [17], one effectively finds the FCC phase at higher temperatures than the BCC phase in ^{3}He (Fig. 13c). In iron (Fig. 13a) the FCC structure is located between two BCC (δ and α) phases [17], providing a striking confirmation of the interplay between the BCC and FCC structures shown in Fig. 18b. In K, Cs, and Rb the BCC and FCC phases appear at approximately the same temperature, but the region of stability of the FCC phase extends at higher pressures (Fig. 13d).

The form of the order-parameter expansion associated with the BCC-FCC transition, is determined by the symmetry of the order-parameter (ξ_1,ξ_2) and by the equilibrium condition ($\xi_1 \neq 0$ *or* $\xi_2 \neq 0$) which gives the orientation of the vector $\vec{\eta}$ in the ε_2-space, showing that the FCC phase may correspond to two possible antiisostructural configurations. It yields:

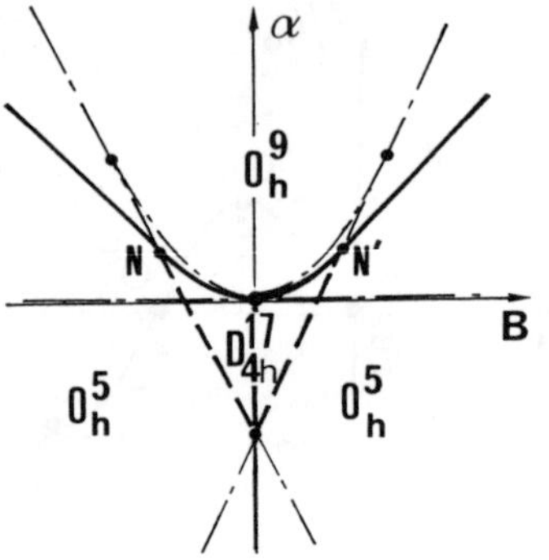

Fig. 19. Phase diagram associated with the BCC-FCC transition and with the potential (3.35). The symmetric character of the phase diagram results from a folding with respect to the a_1 axis of the right-hand part of the phase diagram of Fig.3. Full, dashed and dashed-dotted lines, are respectively, first and second-order transition lines, and limit of stability lines. N and N' are three-phase points. The two symmetric O_h^5 regions correspond to antiisostructural phases.

$$F(T,P,|\vec{\zeta}(\xi)|) = F_0(T,P) + \frac{\alpha}{2}|\vec{\zeta}(\xi)|^2 + \frac{B}{3}|\vec{\zeta}(\xi)|^3 + \frac{\beta}{4}|\vec{\zeta}(\xi)|^4 \qquad (3.35)$$

with $|\vec{\zeta}(\xi)| = \zeta_0 \cdot \mathrm{Sin}\frac{3\pi i}{a\sqrt{2}}\xi$. This *symmetric form* of ζ, does not contradict the existence of a *cubic invariant* in (3.35), since it results from the limitations imposed to the dissymmetrical functions, represented in Figs.16a and 16b, in order to take into account the physical constraints inherent to the Bain deformation. Thus, the phase diagram of Fig.19, which corresponds to (3.35) is obtained from the phase diagram of Fig.3, by keeping only the right-hand side of the phase diagram of Fig.3 (the physically acceptable part) and by a subsequent reflection with respect to the a_1 "plane". It finally gives a *symmetric* topology with respect to the coefficient B in (3.35).

Our description of the BCC-FCC transition, illustrates the property of the order-parameter periodicity, to be fully expressed only in function of the critical atomic shifts, and not in function of a macroscopic variational parameter, consistently with the microscopic nature of the density-wave approach. Verifications of this description can be found: 1) in the existence of the tetragonal Landau intermediate state, which is found in a number of alloys (D_{4h}^{17} *is* the symmetry of "martensite" in carbon steels [18]) and in some elements. For example, one has in plutonium the full sequence of structures: $\varepsilon(O_h^9)$ - $\delta'(D_{4h}^{17})$ - $\delta(O_h^5)$. The same sequence is also found in protactinium [17]. Three tetragonal domains are usually observed in the tetragonal structure, which should be also found in the FCC structure or in the BCC structure when coming for the FCC phase.

4.4. The BCC-FCC phase diagram

Our description of the BCC-HCP and BCC-FCC transitions shows that the HCP and BCC structures may both constitute the *initial* structure in the framework of our approach. Such a possibility disappears when the FCC and HCP phases coexist in the same phase diagram, far from the eventual region of stability of the BCC phase. This situation is found, for example, for some elements, such as He, Co, Pb, Si, La, Ce, Pr, Nd, or in compositions such as ZnS or SiC [2, 17]. In this case, the FCC-HCP transition loses its displacive nature, in relation with the close packing of the two structures, and acquires an *ordering* character, the mechanism of which is described in Chapter IV.

The situation is different when the FCC and HCP structures coexist with the BCC phase. This is for example the case in Fe and Tl, in the phase diagram of which the three phases merge at a triple point. Here, the transition mechanisms and corresponding phase diagram can be simply described by combining the Burgers and Bain deformation mechanisms, taking the BCC structure as the initial structure [20]. In order to investigate the region at which the three phases coexist, one should use the thermodynamic potential:

$$F(T,P,|\vec{\eta}|,|\vec{\zeta}|) = F_0(T,P) + F_1(|\vec{\eta}(\xi_1)|) + F_2(|\vec{\zeta}(\xi_2)|) + F_3(|\vec{\eta}|,|\vec{\zeta}|) \qquad (3.36)$$

where $F_1(|\vec{\eta}(\xi_1)|)$ and $F_2(|\vec{\zeta}(\xi_2)|)$ coincide, respectively, with (3.27) and (3.35), and can be rewritten for the sake of homogeneity:

$$F_1(\eta) = \frac{a_1}{2}\eta^2 + \frac{a_2}{4}\eta^4 + \frac{a_3}{6}\eta^6, \text{ and } F_2(\zeta) = \frac{b_1}{2}\zeta^2 + \frac{b_2}{3}\zeta^3 + \frac{b_3}{4}\zeta^4.$$

ξ_1 and ξ_2 denote the independent shifts associated with the BCC-HCP and BCC-FCC transitions, respectively. F_3 expresses the coupling between the two order-parameters. From the symmetries of the order-parameters, one finds:

$$F_3(\eta,\zeta) = \delta\zeta\eta^2 + \frac{\gamma}{2}\zeta^2\eta^2. \qquad (3.37)$$

Minimization of F with respect to ξ_1 and ξ_2, yields the equations of state:

$$\eta\left(a_1 + a_2\eta^2 + a_3\eta^4 + 2\delta\zeta + \zeta^2\right)\frac{\partial\eta}{\partial\xi_1} = 0, \qquad (3.38)$$

$$\left(b_1\zeta + b_2\zeta^2 + b_3\zeta^3 + \delta\eta^2 + \gamma\zeta\eta^2\right)\frac{\partial\zeta}{\partial\xi_2} = 0. \qquad (3.39)$$

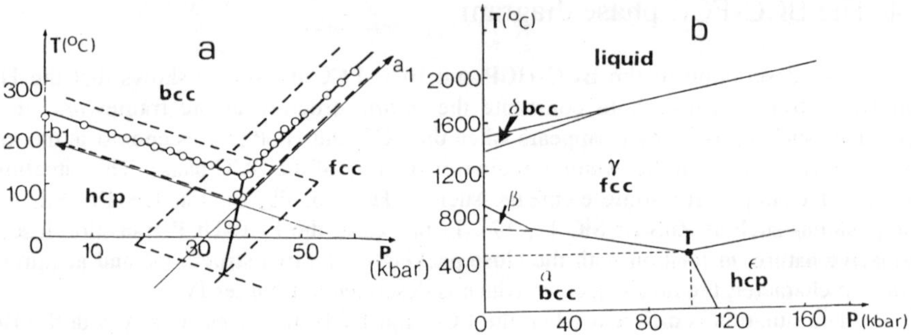

Fig. 20. (a) Phase diagram, close to the triple point, corresponding to the thermodynamic potential (3.36). The lines correspond to the best fit with the experimental points found by Jayaraman et al. [27] for Thallium. (b) Pressure-temperature phase diagram of Iron from ref. [17].

Restricting the description of these equations to the region of the phase diagram in which the three phases are close to the triple point, one obtains the phase diagram of Fig.20a. A striking feature of this diagram, which contains the two *limit*, HCP and FCC, phases, is that all the transition lines and limit of stability lines are *straight lines*, in agreement with the general properties established in Chapter II (Sec.4.3). The equations corresponding to the lines of first-order transitions are:

$$a_1 + a_2 + a_3 = 0 \ \ \text{(BCC-HCP)}, \quad b_1 + b_2 + b_3 = 0 \ \ \text{(BCC-FCC)},$$
$$a_1 + a_2 + a_3 = b_1 + b_2 + b_3 \neq 0 \ \ \text{(FCC-HCP)}.$$

The preceding lines intersect at the triple point T, whose coordinates are in the (a_1, b_1) plane $(-a_2 - a_3; -b_2 - b_3)$. The equations giving the limit of stability lines are :

(1) $a_1 = -\gamma - \delta$ (for the BCC phase into the HCP phase),
(2) $b_1 = -(3b_2 + b_3)$ (for the FCC into the HCP phase),
(3) $b_1 = -\gamma$ (for the BCC into the FCC phase),
(4) $a_1 = (2a_2 + a_3)$ (for the HCP into the BCC phase).

The pressure-temperature phase diagrams of iron, schematized in Fig.20b, provides a good verification of the preceding theoretical diagram. At atmospheric pressure iron crystallizes from the melt into the BCC structure (δ-Fe) and at lower temperature transforms into the austenite FCC phase (γ-Fe) and again into a BCC structure (α-Fe). With increasing pressure, a triple point is found [25] at $P_T \cong 97$ kbar and $T_T \cong 450°C$, at which the γ and α phases meet the HCP ε phase. The metastable regions of the BCC phase into the FCC and HCP phases, shown in Fig.20a, have been verified by Poniatovskii [26], who observed an overcooling of the BCC phase into FCC and HCP iron. The lines separating the phases in Fig.20b are shown from a insufficient number of experimental points [2], in order to confirm conclusively their predicted straight line character. In contrast, the phase diagram of Thallium, measured by Jayaraman et al. [27] in the vicinity of the triple point, clearly confirms this property. In Fig.20a the experimental points given by these authors are superposed to the theoretical curves calculated for the transition lines and limit of stability lines. A linear transformation of the (a_1, b_1) plane into the pressure-temperature plane has been performed for comparison with experiment. One can verify in Fig.20a that a very good fit is obtained between the theoretical lines and the experimental points.

It has already been pointed out, that the δ and α BCC phases of iron can be interpreted as two differently oriented structures, predicted from the periodic dependence of the order-parameter shown in Fig.15a. No available data allows one to compare the different orientations assumed in our model for the δ and α structures. A diffraction spectrum of the two phases would provide such a verification since cycling across the transition should progressively lead to a replacement of twofold positions (e.g. in the cubic direction $[110]$) by a fourfold diffraction pattern.

In conclusion, for iron and thallium the FCC-HCP transitions is interpreted as a displacive transition between two low-symmetry phases, which are both derived from a common parent BCC structure. This interpretation does not hold when the BCC phase is absent from the phase diagram, or does not have a region of coexistence with the FCC and HCP phases, as it will be shown in Chapter IV.

5. Landau and limit phases in the phase diagrams of actinides

5.1. Polymorphism of the actinide group

In the preceding sections we have considered the more widespread displacive mechanisms involved in reconstructive transitions. In some groups of structures (such as actinides or lanthanides), or in the case of crystals displaying complex phase diagrams, with a large number of different phases, reconstructive transitions to limit structures, and first- or second-order transitions to Landau phase, may occur conjunctively. This is the case of the actinide group of elements which will be described in this section.

The polymorphism found in crystals of the actinide group [28, 29] is unquestionably the richest among the elements. In addition to the BCC, FCC and HCP structures, which are found all over the Periodic Table, actinides display a number of low-symmetry structures (e.g. tetragonal, orthorhombic, monoclinic) that are prominent features of this family of elements. Thus, the α phase of neptunium, and the β phase of plutonium constitute model examples of *idiosyncratic structures* [30], i.e. they are unique among all known structures, and the understanding of their formation has been for a long time an unsolved problem. Besides, the incommensurate phase observed in α-uranium [31, 32] is the sole case of structural incommensurability reported for crystals of elements.

Although the pressure-temperature phase diagrams of a large number of actinide elements were thoroughly investigated [2, 17], there have been only recently [33-35] attempts to describe theoretically the corresponding phase transformations mechanisms. We will show, in this section, that a unified description of the structures of actinide crystals can be given in the framework of our approach. More precisely we will show that *almost*

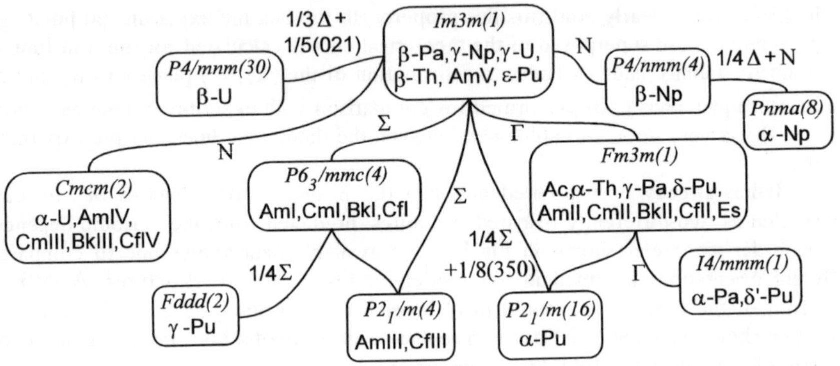

Fig. 21. Symmetry relationship between the structures, in actinide elements, following our description. The letters on the lines connecting the various symmetries, correspond to the points and lines of the BCC Brillouin zone associated with the proposed mechanisms.

all the structures, found in crystals of actinide elements can be derived from the same parent BCC structure, via a small number of displacive mechanisms. Furthermore, we will illustrate on the examples of uranium and neptunium, the property that *the structures can be understood on the basis of the shape of the electronic shell of the individual atoms* [36].

Anticipating on the results obtained in this section, we summarize on Fig.21, the connections proposed for the structures found in crystals of actinide elements with the assumed "parent" BCC structure. Three main situations can be distinguished, which in the Periodic Table correspond to the lighter, heavier and intermediate actinides, respectively. They will be described in separated subsections.

5.2. Lighter actinides: Full realization of the Bain deformation

In the lighter actinides, namely actinium, thorium and protactinium, the more complete sequence of structures: BCC- centered tetragonal (D_{4h}^{17})- FCC, which is fully realized in Pa, and only partially in Th and Ac (Fig.22), corresponds to the complete sequence of phases involved in the Bain deformation mechanism, as described in Sec.4. In Pa, the β-BCC structure transforms into tetragonal α-Pa, at about 1450 K [37,38]. The values of the lattice constants in the two phases [37] give the following numbers at room temperature: $\zeta = \frac{1}{\sqrt{6}}(e_{xx} + e_{yy} - 2e_{zz}) \cong 0.15$, and a/c=0.825. The FCC phase of Pa is observed by quenching the high temperature form [39] and corresponds to the value $\zeta \cong 0.31$. In Th one observes at 1678 K [40] a strongly first-order transition between the BCC (β-Th) and FCC (α-Th) phases. At room temperature, the deformation is $\zeta \cong 0.02$.

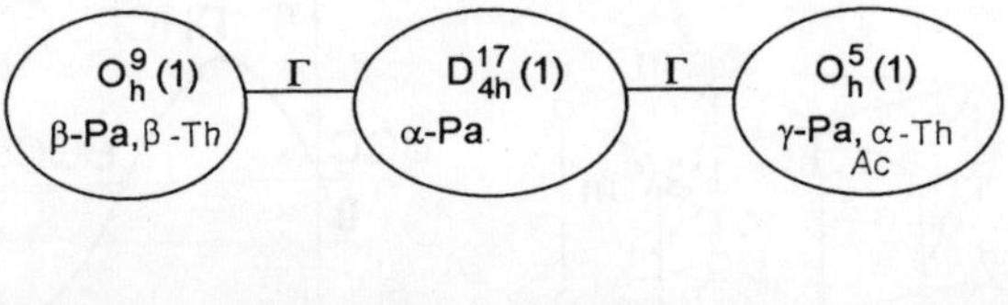

Fig. 22. Symmetry relationship between the phases found in light actinides.

The pressure-temperature phase diagram of Ac, was not investigated, and only the FCC structure was observed in this element at room temperature and atmospheric pressure.

5.3. Heavier actinides: A variant of Burgers mechanism

A large variety of structures are found in the phase diagrams of the heavier actinide elements (Fig.23) (americium, curium, berkelium, californium, einsteinium,...), the more complete set of five distinct structures having been identified in Am [41], which are denoted AmI (D_{6h}^4, Z=4) AmII (O_h^5, Z=1), AmIII (C_{2h}^2, Z=4), AmIV (D_{2h}^{17}, Z=2) and AmV (O_h^9, Z=1). The AmV (BCC) $\rightarrow$AmII (FCC) transition, which takes place [42] with increasing temperature and atmospheric pressure at T_c=1352 K corresponds again to the (direct) Bain deformation mechanism.

The sequence of phases AmI$\rightarrow$AmIII$\rightarrow$AmIV, which is found below 800 K, with increasing pressure [43], can be understood in the framework of a *variant* of Burgers mechanism described in Sec.3.2. Thus, the three corresponding structures are obtained from the BCC structure as shown in Fig.24a, by an antiparallel shifting of the atoms in the (110) BCC planes, along the $\pm[\bar{1}10]$ BCC directions. This primary mechanism induces,

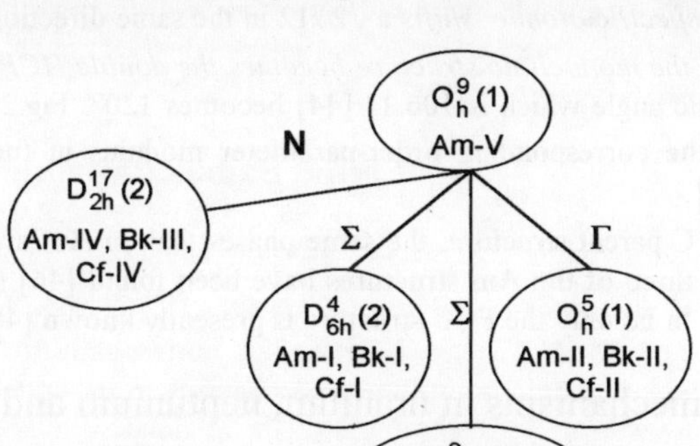

Fig. 23. Symmetry relationship between the phases found in heavy actinides.

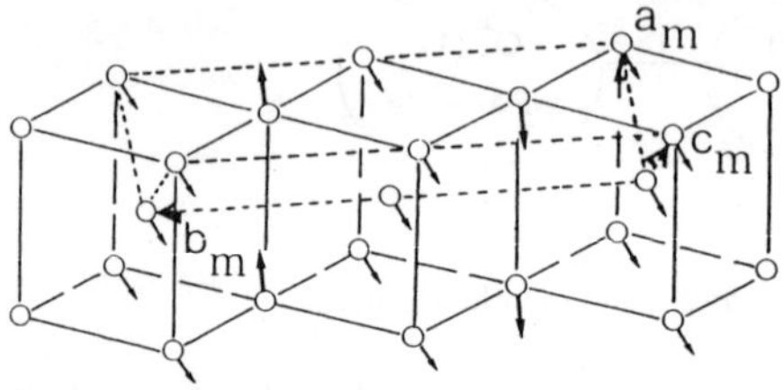

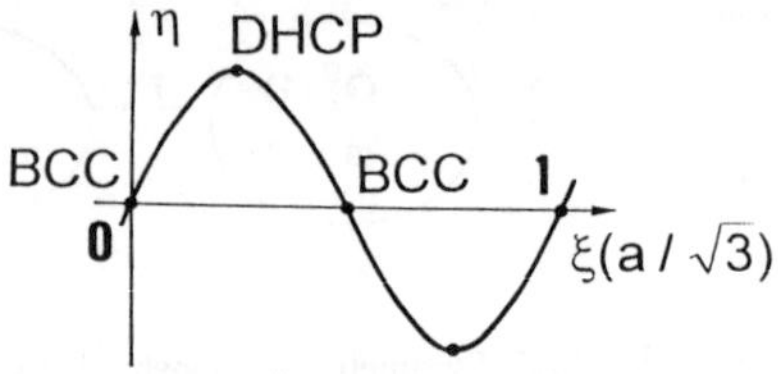

Fig. 24. (a) Shifting mechanism from the BCC unit-cell of AmV (small cubes in thick lines) to orthorhombic AmIV (short arrows) and monoclinic AmIII (long arrows). The dashed lines represent the basic translations of the monoclinic unit-cell in the direction $\vec{b}_m$. (b) Periodic dependence of the corresponding order-parameter modulus, in function of the displacements ξ along [110], scaled in $a/\sqrt{3}$ unit.

as a secondary effect, a compression of the cubic unit cell along $[001]$, and a decompression along the two other fourfold axes. For arbitrary shifts along $\pm[\bar{1}10]$ one gets the orthorhombic (AmIV) and monoclinic (AmIII) phases *for two different instabilities of the BCC lattice*. The two instabilities are respectively associated with two different points of the BCC Brillouin zone: the N point $[\vec{k}_9 = \left(\frac{\pi}{a}, \frac{\pi}{a}, 0\right)]$ leads to the orthorhombic structure of AmIV, whereas at the middle of the Σ-line $[\vec{k}_4 = \left(\frac{\pi}{2a}, \frac{\pi}{2a}, 0\right)]$, one obtains the monoclinic structure of AmIII, the unit-cell of which is quadrupled in the direction $[110]$. For the *specific atomic shifts* $a\sqrt{2}/12$ in the same directions, where a is the cubic lattice constant, *the monoclinic structure becomes the double HCP structure* of AmI, i.e. the β-monoclinic angle which is 106.1° [44] becomes 120°. Fig.24b shows the periodic dependence of the corresponding order-parameter modulus in function of the displacements along $[110]$.

Except for the BCC parent structure, the same phases than in Am are observed in Cf [45]. In Cm and Bk three of the Am structures have been found [46] (FCC, double HCP, and orthorhombic). In Es only the FCC structure is presently known [45].

5.4. The transitions mechanisms in uranium, neptunium and plutonium

Let us now focus on the complex structures which are found in intermediate actinides: uranium, neptunium, and plutonium. In their pressure -temperature phase diagrams [2, 17], the BCC structure, denoted γ-U, γ-Np, and ε-Pu, is stable below the melting line and thus appears as the natural parent structure for the three elements. Two sequences of phases:

$$(1) \quad \varepsilon\text{-Pu (BCC)} \rightarrow \delta'\text{-Pu } (D_{4h}^{17}, Z=1) \rightarrow \delta\text{-Pu (FCC)}$$

which takes place on cooling at atmospheric pressure, and:

$$(2) \quad \gamma\text{-U (BCC)} \rightarrow \alpha\text{-U } (D_{2h}^{17}, Z=2)$$

which occurs on cooling at high pressures [41], will not be discussed further, as they can respectively be interpreted in the framework of the Bain and Burgers mechanisms. All the remain structures found in U, Np and Pu require specific mechanisms and will now be described separately.

5.4.1. The formation of β-uranium

The crystallographic description of the β-phase of uranium (D_{4h}^{14}, Z=30), which is found between 941 and 1048 K at atmospheric pressure, has been the subject of thorough discussions over the past decades [41]. The connection between the γ-U and β-U structures, and our proposed displacive mechanism [36] for the transition between the two structures, is illustrated in Figs.25a-d. It can be summarized as follows: Six successive

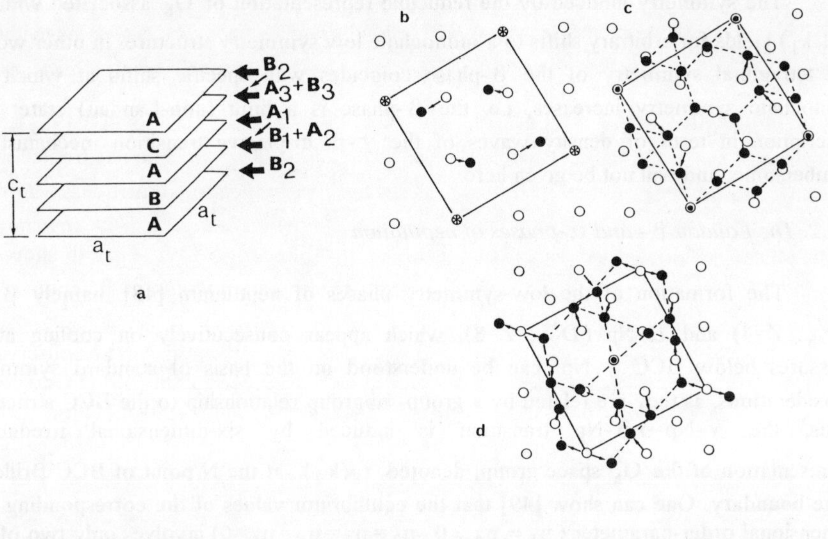

Fig. 25. Transformation mechanism from γ-U to β-U structure. (a) Schematic change from the $A_1B_1A_2B_2A_3B_3$ to the ABAC sequence of layers. (b)-(d) represent the in-plane shifts for the A, B, and C β-U layers. Open and solid circles are the γ and β positions respectively. Crossed circles in (c) indicate the atoms ejected from A_1. Centered circles in (c) and (d) underline the coincidence of the collapsed layers. The dashed lines illustrate the characteristic in-plane motif.

layers of the (100) type, of the BCC bilayer structure $A_1B_1A_2B_2A_3B_3...$, generate the four-layer ABAC β-U structure [41] in two correlated steps (Fig.25a): (i) Each couple of BCC layers ($B_1 + A_2$) and ($A_3 + B_3$) *collapse* and form respectively the B and C β-U layers; (ii) One atom of each of A_1 and B_2 BCC layers are ejected and complete the B and C layers. The remaining atoms of A_1 and B_2 constitute the two A layers of β-U, which are shifted one with respect to the other by $(\frac{a}{2},\frac{a}{2},0)$. Figs.25b-d decompose the corresponding displacements of the uranium atoms from their BCC β-U positions.

Macroscopically, the entire transformation mechanism can be understood as a drastic compression of the BCC planes along the Z axis, and a simultaneous large decompression within the collapsed planes. The numbers found for the spontaneous strain components induced by the transformation, calculated from the numerical values of the lattice constants in the two phases [47], are e_{zz}=+0.365, and $e_{xx} = e_{yy}$=-0.465. They are indicative of the huge magnitude of the displacements in the assumed mechanisms. Two different irreducible representations, six and twenty-four dimensional, associated with the two wave-vectors $\vec{k}_8 = \frac{2}{3}\left(0,0,\frac{\pi}{a}\right)$, and $\vec{k}_1 = \frac{2}{5}\left(\frac{2\pi}{a},\frac{\pi}{a},0\right)$ (in Kovalev's notations [9]), located inside the BCC Brillouin zone, express the symmetry of the two preceding mechanisms.

The symmetry induced by the reducible representation of O_h^9 associated with $\vec{k}_8$ and $\vec{k}_1$) yields for arbitrary shifts to a monoclinic low symmetry structure. In other words, the tetragonal symmetry of the β-phase coincides with specific shifts at which the monoclinic symmetry increases, i.e. the β-phase is a limit (non-Landau) state. The description in term of density waves of the γ-β displacive transition mechanism is cumbersome, and will not be given here.

5.4.2. The Landau β- and α-phases of neptunium

The formation of the low-symmetry phases of neptunium [48], namely β-Np (D_{4h}^7, Z=4) and α-Np (D_{2h}^{16}, Z=8), which appear consecutively on cooling at all pressures below BCC γ-Np, can be understood on the basis of standard symmetry considerations, as they are related by a group-subgroup relationship to the BCC structure. Thus, the γ-Np$\rightarrow\beta$-Np transition is induced by six-dimensional irreducible representation of the O_h^9 space group, denoted $\tau_8(\vec{k}_9^*)$, at the N point of BCC Brillouin zone boundary. One can show [49] that the equilibrium values of the corresponding six-dimensional order-parameter ($\eta_1 = \eta_4 \neq 0$, $\eta_2=\eta_3=\eta_5=\eta_6=0$) involves only two of the six branches of the star of the wave vector $\vec{k}_9$. The connection between the β and γ unit-cells is shown in Fig.26.

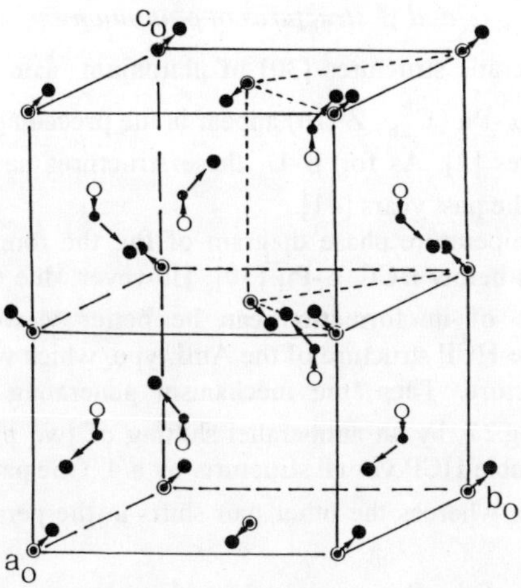

Fig. 26. Formation mechanism of β- and α-Np. Open circles, small and large solid circles, represent the atomic positions in γ–, β– and α–Np respectively. Dashed lines show half of the unit-cell.

The lowering of symmetry which leads to orthorhombic α -Np is associated to an additional six-dimensional irreducible representation, denoted $\tau_1(\vec{k}_8^*)$, where the wave-vector $\vec{k}_8 = \left(0,0,\frac{\pi}{2a}\right)$ is located on the Δ -line of the BCC Brillouin zone. The coupling of $\tau_4(\vec{k}_9)$ and $\tau_1(\vec{k}_8)$ results in the eightfold orthorhombic unit-cell represented in Fig.26. In the same figure are indicated the displacements of the Np atoms which transform successively γ -Np in β - and α -Np. One can see that the tetragonal structure corresponds to an antiparallel shift along the Z axis of two among the four Np atoms contained in each tetragonal unit-cell. The magnitude of the displacements, deduced from the lattice parameters of γ - and β -Np [48], is $\pm\Delta z = 0.42A° \cong \frac{1}{8}a$. The formation of the orthorhombic structure requires an additional antiparallel shift of all eight atoms contained in each orthorhombic cell, simultaneously along the c-tetragonal and b-orthorhombic axis. For four atoms that were immobile in the tetragonal structure, the magnitude of the shifts at the γ - α transformation is about 0.3 A°, whereas for the four remaining atoms it is 0.6 A°. The calculated relative jump in volume is $\Delta V/V \cong -2.6\%$ for the γ - β transition, and $\Delta V/V \cong -2.9\%$ at the β - α transition. With respect to the BCC structure the calculated spontaneous strains are $\Delta a/a = -0.038$, $\Delta c/c = -0.017$ for the β -phase, and $\Delta a/a = -0.05$, $\Delta b/b = -0.018$, $\Delta c/c = -0.017$ for the α -phase. All these numbers show that the β and α structures of neptunium are the result of a moderate compression of the γ structure, in agreement with the preservation of the group-subgroup relationship between the phases.

5.4.3. The idiosyncratic γ, α and β structures of plutonium

The three idiosyncratic structures [30] of plutonium, namely γ-Pu (D_{2h}^{24}, Z=2), β-Pu (C_{2h}^{3}, Z=17), and α-Pu (C_{2h}^{2}, Z=16) appear in the preceding order on cooling over a wide range of pressures [2]. As for β-U, these structures have been the subject of intensive discussion over the past years [41].

In the pressure-temperature phase diagram of Pu, the four-layered orthorhombic γ-Pu structure arises just below BCC ε-Pu [50]. However, due to its pseudohexagonal character the mechanism of its formation can be better viewed by considering an intermediate virtual double-HCP structure of the AmI type, which was shown in Sec.5.3 to form from a BCC structure. Then, the mechanism generating γ-Pu can be simply described, as shown in Fig.27, by an antiparallel shifting of two pairs of alternate planes forming the four-layer double HCP virtual structure, by a/4. One pair of planes is shifted in the $[001]$ BCC direction, whereas the other pair shifts in the perpendicular $[1\bar{1}0]$ BCC direction.

The displacements of the Pu atoms in three directions forming angles of 120° in the (110) plane, produces at the macroscopic level a shear deformation corresponding to a compression along the a and c orthorhombic directions, and a decompression along b. In other words the angles between the diagonals in the (110) BCC plane, which are 70.5° and 109.5° in ε-Pu, become respectively 60° and 120° in γ-Pu. The numbers obtained from the lattice parameters of the two phases ($\Delta a/a$=-0.15, $\Delta b/b$=+0.12, $\Delta c/c$=-0.01) show that the decompression along c is one order of magnitude smaller than along a or b.

According to the preceding description, it should be underlined: (1) that the double HCP AmI structure and γ-Pu are induced by the same twelve-dimensional irreducible representation of the O_h^9 space group at $\vec{k}_4 = \left(\frac{\pi}{2a}\frac{\pi}{2a}0\right)$ (the Σ line), the two structures differing only by the magnitude of the atomic displacements from the BCC positions;

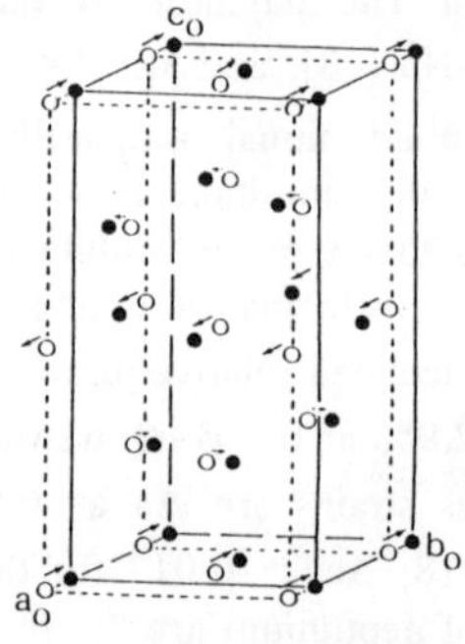

Fig. 27. Formation of γ–Pu from a double HCP virtual structure. Open and solid lines indicate the double HCP and γ unit-cell.

(2) that the γ-Pu phase is a limit state, which is not group-subgroup related to ε-Pu. Actually the orthorhombic D_{2h}^{24} (Z=2) symmetry represents an *increase* of a lower orthorhombic symmetry (C_{2v}, Z=4) corresponding to specific shifts along the $[001]$ *and* $[1\bar{1}0]$ BCC directions.

As for γ-Pu, the displacive mechanism that leads to the formation of α-Pu can be more easily represented with respect to the intermediate FCC δ-Pu structure. The shifts which transform the FCC unit-cell into the sixteenfold monoclinic α-Pu unit cell [51] are represented in Fig.28. Their nonsymmetric directions are consistent with the drastic lowering of point group symmetry between the two structures. Macroscopically, the ε-Pu$\rightarrow\alpha$-Pu transformation is induced by two, 24- and 12-dimensional irreducible representations of the O_h^9 space group, respectively associated with the wave vector $\vec{k}_1=\left(\frac{3\pi}{4a}\frac{5\pi}{4a}0\right)$ and $\vec{k}_2=\left(-\frac{\pi}{2a}\frac{\pi}{2a}0\right)$ located inside the BCC Brillouin zone. With respect to the δ-Pu FCC structure, it corresponds to a compression along the X and Y cubic axes ($\Delta a/a=-0.16$, $\Delta b/b=-0.06$) and to a decompression along Z ($\Delta c/c=+0.04$), the monoclinic angle increasing from 102° to 104°.

A realistic mechanism for the formation of the β structure of plutonium [52] remains to be found. A number of authors stresses the resemblance between β-Pu and β-U [41], but in β-Pu the structure cannot be simply described in terms of stacking of atomic planes, as its monoclinic structure (C_{2h}^3, Z=17) consists in a stacking of *seventeen*

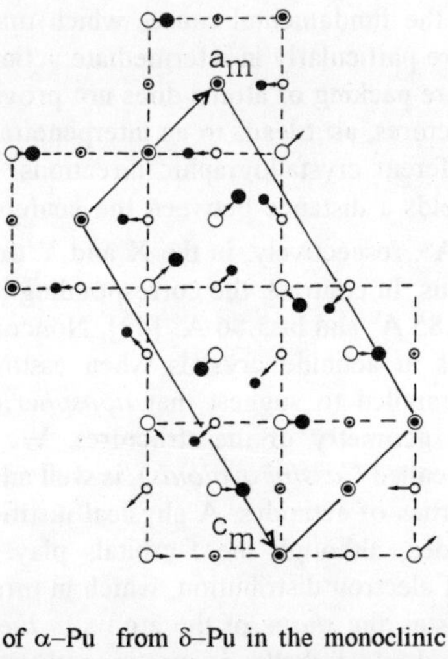

Fig. 28. Formation of α–Pu from δ–Pu in the monoclinic b plane (010). Open and solid circles are the δ and α positions. Small and large circles differ by b/2 in the monoclinic [010] direction. Dashed and solid lines are the projections of the cubic and monoclinic unit-cells.

tetragonal layers. Such a structure can be formally generated from ε-Pu using a mechanism analogous to the one proposed for the formation of β-U (i.e. a collapsing of successive sequences of three ε-Pu planes), but it should imply too large a compression of the BCC lattice to be physically sound.

5.5. Packing of "ovaloid" atoms in actinide structures

In summary, except for β-Pu, the transition mechanisms in actinide structures can be described within a unified model [35]. At the microscopic level these mechanisms are expressed in terms of atomic displacements from a common parent BCC structure, corresponding, as shown in Fig.21, to a reduced number of instabilities pertaining to the $\Sigma(\Gamma$-N)- and Δ-lines of the BCC Brillouin-zone. The phenomenological (density-wave and thermodynamic) description of the transitions has not been given in details, as it involves lengthy calculations due to the complexity of the structures, often composed by a large number of sublattices, and to the large number of degrees of freedom, which are eventually reducible, as for instance in β-U, α-Pu or α-Np. At the macroscopic level, the mechanisms correspond to large deformations (which may reach 40% as in β-U) with respect to the initial BCC structure.

The actinide group provides representative illustrations of the existence of Landau and limit states in the same system, and of the connection between the two types of structures, as in Am, U or Pu. In particular one finds in actinides the two intermediate Landau states (D_{2h}^{17} and D_{4h}^{17}) and corresponding limit states (HCP and FCC) which characterize our description of the Burgers and Bain deformation mechanisms.

It remains to clear up the fundamental causes which may be responsible for this rich polymorphism, found more particularly in intermediate actinides. In this respect let us note that a model of hard sphere packing of atoms does not provide a suitable geometrical description of the actinide structures, as it leads to an interpenetration, or to the absence of contact, of the spheres in different crystallographic directions. For example, in α-U a packing of spherical atoms yields a distance between the centers of the closest atoms of $2R=3.06$ A° and $2R\sqrt{3}=5.3$ A°, respectively, in the X and Y orthorhombic directions, R being the spherical atomic radius. In contrast, the corresponding lattice parameters in these directions are respectively a=2.85 A° and b=5.86 A° [42]. Noncoinciding numbers are also found for most of the phases in actinide crystals when assuming a close packing of *spheres*. Accordingly one is entailed to suggest that *nonspherical* atoms should give a more coherent picture of the geometry of the structures. We will now show that an *ovaloid* form of the atoms (so-called *Cassini ovaloids*), is well adapted for a description of the crystallogeometrical properties of actinides. A physical justification of this assumption lies in the fact that in actinides, although the f-orbitals play an essential part in the determination of the density of electron distribution, which in turn determines the minimal interatomic distance in the crystal, the *shape* of the atoms *in the crystal structure* results from the hybridization of s, p, d and f shells, as was recently stressed in Refs. [33, 34]. Consequently the ovaloid form, which coincides with the angular distribution

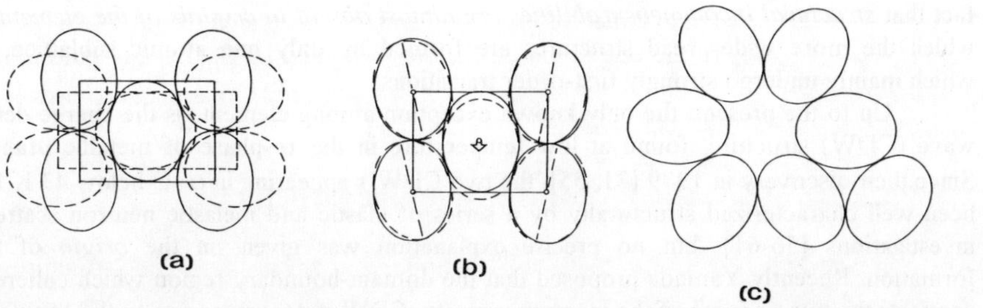

Fig. 29. Packing of ovaloid atoms in uranium and neptunium. (a) γ–α transformation in U, in projection in the BCC (110) plane. Solid and dashed lines are the projections of the γ and α atoms and unit-cells, respectively. (b) γ–β transformation in Np. Solid and dashed lines have the same meaning as in (a). The arrows symbolize the shift of the central atom of the unit-cell. (c) Characteristic motif in β–U, with a hexagonal and triangular packing of ovaloids.

corresponding to the spherical harmonics Y_2^0, provides the more suitable first approximation for a nonspherical atomic close packing.

Fig.29a shows the transition mechanism from γ-U to α-U using a close packing of hard ovaloids. It consists in a rotation by 90° of the ovaloids, which preserves the close packing structure. In Fig.29b the γ-β transition in Np is represented, in which the ovaloids are turned by about 20°. In Fig.29c details of the β-U structure, as given in Fig.25, show that this structure can be understood as a packing of pseudo six- and threefold distribution of ovaloid atoms, resulting in an intermediate state between the γ-U and α-U distributions. The γ-α (O_h^9-D_{2h}^{17}) mechanism is another example of reconstructive mechanism, in which the preservation of the group-subgroup relationship between the symmetries of the structures is due to specific geometrical constraints (the nonspherical shape of the atoms). In other words, the α-structure can be interpreted as an *unachieved limit phase*, i.e. its symmetry would be HCP, if formed by spherical atoms, and the group-subgroup relationship, with respect to the BCC phase, would be lost.

5.6. The charge density waves in α-uranium

The onset of a structural incommensurability in a crystal below a phase transition coincides generally with the requirements that: 1) there exist in each incommensurate direction a reference interatomic distance with respect to which the wave-length of the incommensurate modulation can be defined [52, 53], 2) the transition must be second-order or slightly first-order, as the inhomogeneous deformation of the lattice characterizing an incommensurate modulation is usually small, and a strongly first-order transition has the effect of "squeezing" the incommensurate structure, and to switch directly the crystal to a commensurate lock-in structure [54]. These requirements are illustrated *a contrario* by the fact that *structural incommensurabilities are almost absent in crystals of the elements*, in which the more widespread structures are formed by only one atomic sublattice, and which mainly undergo strongly first-order transitions.

Up to the present, the only known exception among elements is the charge density wave (CDW) structure, found at low temperature in the α-phase of metallic uranium. Since their discovery in 1979 [31, 55], the two CDW's appearing in α–U below 43 K have been well characterized structurally by a series of elastic and inelastic neutron scattering investigations [56-61], but no precise explanation was given on the *origin* of their formation. Recently Yamada proposed that the domain-boundary region which coherently connects the two variants of the incommensurate CDW states, gives rise to the anomalous diffraction effects [62]. Following another line, we suggest that the incommensurate structures reveals traces of the instability which contribute to the β-structure, i.e. that they can be interpreted as frustrated regimes between the α and β ovaloid distributions.

In order to justify the preceding assumption, let us consider the epitaxial relationships connecting the β and α structures. Along the fourfold axis, one has in one β-unit-cell, four stacked layers which are distant from $|\vec{c}_\beta|/4 = 1.414$ A°, where $\vec{c}_\beta$ is the tetragonal lattice vector along the Z axis. This direction coincides with the twofold X axis of the orthorhombic α structure, along which the two-layered α-unit-cell gives an interlayer distance $|\vec{a}_\alpha|/2 = 1.422$ A°. It is easy to verify that the α' structure, which is stable below 43 K at atmospheric pressure [57], *realizes in the same direction the maximal substructure compatible with both the* α *and* β *structures*. This substructure has, as shown in Fig.30, a four-layer stacking with the smallest β interlayer distance

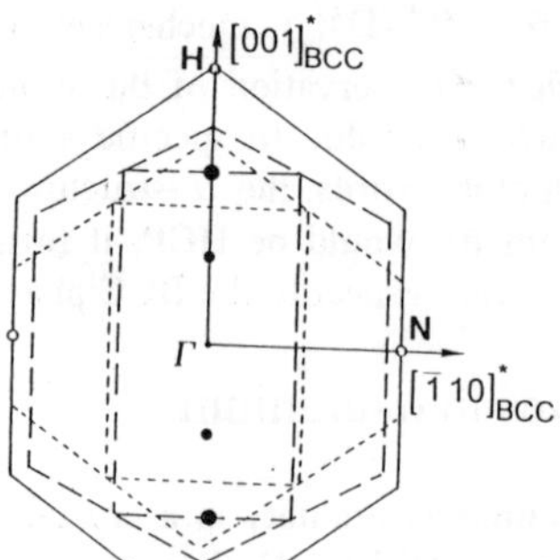

Fig. 30. (110) cubic plane in the reciprocal lattice. The pseudo-hexagonal figure in full lines is a section of the γ–structure. The pseudo-hexagons in dotted and dashed lines show the α–section before and after the macroscopic deformation. The dotted and dashed rectangles are the Brillouin zones of the corresponding Bravais unit-cells. The small solid circle is the position of the wave-vector corresponding to the γ–α transition. The large solid dot is the analogous point in the second Brillouin zone. Its position corresponds to the satellites observed by Smith et al. [57].

$|\vec{a}_\alpha| \cong 1.414$ A°. Reindexing this distance with respect to the α structure, as assumed in the experimental data, one obtains for the cell-doubling α-α' transition, an instability located on the Σ-line of the orthorhombic C Brillouin-zone at $|\vec{k}| = |\vec{c}^*_\beta| / |\vec{a}^*_\alpha| = 0.503$, which is the value found by Smith et al. [55] in the first report on superlattice reflections in uranium. As one can see in Fig.30, the position of these reflections, *also coincide*, providing a slight compression (≈ 0.008) of the β-structure along the tetragonal c-axis, *with the wave-vector* $\vec{k}_8 = \frac{2}{3}\left(0,0,\frac{\pi}{a}\right)$ of the BCC Brillouin zone, *which is the instability associated with the stacking mechanism in the* γ-β *transition*. The corresponding phonon mode was shown by Crummet et al. [58] to be a combination of longitudinal and transverse modes, respectively along the X and Y orthorhombic directions. These modes are associated to antiparallel atomic shifts, which can be depicted in terms of *combined shifting and rotations of ovaloid atoms*. The α' structure realizes the minimal distance between the centers of the ovaloids along their short axes.

An analogous description can be given for the additional modulations associated with an incommensurate wave-vector $\vec{k} = \frac{1}{2}\vec{a}^* \pm k_y \vec{b}^* \pm k_z \vec{c}^*$ (with $k_y \cong 0.176$, $k_z \cong 0.182$ at 10 K) which appear below 37 K in α-uranium. It require extensive structural considerations that will not be developed here.

6. Reconstructive phase transitions in inorganic crystals

6.1. Breaking of the group-subgroup relationship between the phase-symmetries in complex crystals

Most of the displacive reconstructive transitions discussed in the preceding sections, take place in relatively simple structures, involving often one sublattice, i.e. atoms in equivalent crystallographic positions. The simplicity of the structure is indeed an important condition for the onset of new symmetry elements at a reconstructive transition. The presence of different types of atoms and of several sublattices may have the consequence that the atomic displacements in the critical sublattice, which govern the transition, reach their limit values (at which the symmetry of the sublattice increases) but the other sublattices prevent the onset of new symmetry elements, so that in total the overall group-subgroup relationship between the initial and final structures is not broken. The preservation of the group-subgroup relationship with respect to the initial structure may also result from the splitting into unequivalent subsystems of a system of equivalent atoms, or from the formation of non trivial crystal lattices.

This is why the majority of reconstructive transitions occur in simple atomic structures, such as the structures of the elements and of binary alloys, whereas in contrast, in the complex structures of many dielectric materials , organic and inorganic compounds reconstructive transitions constitute exceptions. On the other hand, drastic formation processes which occur for a number of minerals, or in quasicrystalline alloys favor the existence of reconstructive transitions, as it will be stressed in Chapter IV and V.

There exist also a number of well established reconstructive transitions in some complex structures, due to specific crystallogeometrical conditions fulfilled by the different atoms and sublattices, as for example in the hexagonal-cubic transition of $BaTiO_3$, described in Chapter IV. For displacive reconstructive transitions, a strongly cooperative character of the critical displacements occurring in the different sublattices may also allow the onset of new symmetry elements in the system. This is the case for the reconstructive transitions in some inorganic dielectric crystals, which will be considered in this section.

6.2. Silver iodide and copper bromide

Silver iodide AgI undergoes on cooling a reconstructive transition, at 421 K, from a high-temperature α-cubic BCC phase to a β–hexagonal wurtzite-type structure (C_{6v}^4, Z=2) [63]. A similar structural change takes place in copper bromide CuBr at about 742 K [64]. The two α (B23) and β (B4) structures are shown in Figs.31a and 31b. Thus, in the high temperature phase of AgI the iode ions form a BCC lattice, while the two interstitial Ag ions pertaining to the conventional cubic cell are randomly distributed among not less than 42 different sites [65]. In the low temperature structure, the Ag ions order in octahedral positions.

The reconstructive transition mechanism proposed for AgI and CuBr follows a slightly modified Burgers mechanism scheme for the I (or Br) ions, namely an antiparallel shifting of these ions, lying in the (110) BCC planes, along $\pm[\bar{1}10]$ directions, and a correlated compression, or decompression, of the cubic unit-cell along the fourfold axes (see Sec.3.2). However, due to the simultaneous ordering of the Ag (or Cu) ions, the intermediate Landau phase has not the orthorhombic D_{2h}^{17} symmetry, but the lower C_{2v}^{12} symmetry. As a consequence the increase in symmetry which occur for the specific shifts

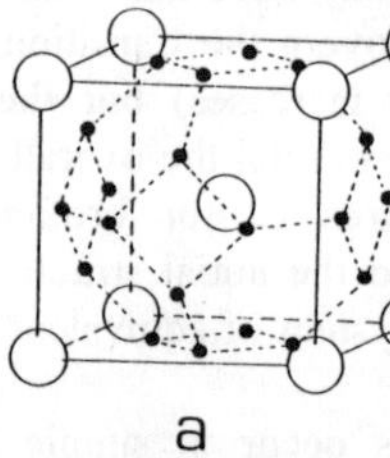

a

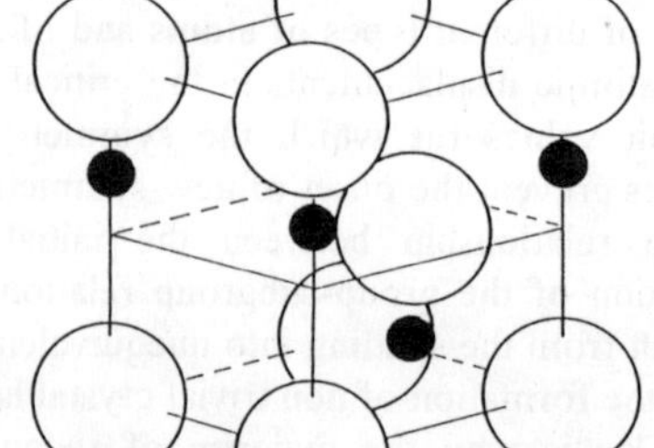

b

Fig. 31. Cubic α (a) and hexagonal β (b) unit-cells of AgI. Large open circles, and small solid circles represent the positions of the I and Ag ions.

of I (or Br) atoms of magnitude a$\sqrt{2}/12$ (where a is the cubic lattice parameter), results in the hexagonal C_{6v}^4 symmetry instead of D_{6h}^4, which correspond to a close packed structure with respect to the iode framework.

Let us emphasize that the reconstructive character of the transition is made possible by the extreme freedom of movement of the interstitial ions (Ag or Cu), which are displaced cooperatively with the (I or Br) ions. This is attested by ionic conductivity measurements [66] which show an increase by two order of magnitude from β to α in AgI. Let us also note that a confirmation of the assumed model can be found in the fact that the *polar* C_{6v}^4 phase *is not ferroelectric*. Actually, this polar symmetry does not result from a simple symmetry breaking mechanism from a non polar phase, but from a more complex process which forbids the onset of microscopic dipoles, as the close packing of the iode ions creates an electrical structure in which the positive (Ag or Cu) ions, are symmetrically surrounded by the negative (I or Br) ions.

An alternate model of the reconstructive transitions in AgI and CuBr has been proposed by Gufan et al. [67]. Although Burgers mechanism is also referred to by these authors, they claim that the initial structure cannot be the BCC structure, but is a *virtual* simple hexagonal (D_{6h}^1) structure. The corresponding theoretical mechanism, which makes use of a multidimensional order-parameter and considers general displacements in the (110) BCC plane, is however incomplete as only one atomic plane is used and the actual geometry of the structure is overlooked.

6.3. Beryllium oxide

The mechanism of the β-α transition in BeO which occur at 2107 K realizes another variant of Burgers mechanism. In the high-temperature β-structure (D_{4h}^{14}, Z=4) the oxygen ions occupy the 4(f) positions and the beryllium ions the interstitials 4(g) positions (Fig.32a) [68]. In the low-temperature α-phase (C_{6v}^4, Z=2) the oxygens are

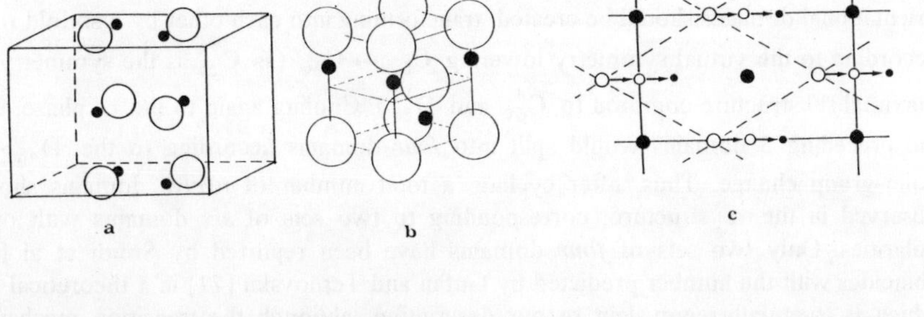

Fig. 32. Tetragonal β–structure (a), and hexagonal α–structure (b) of BeO. Large circles are oxygens. Small solid circles are Be ions. (c) β–α displacive mechanisms in the (110) tetragonal plane. Open and solid circles represent ions of two successive layers.

close packed and the Be ions occupy two-fold positions (Fig.32b) [69]. The β-α displacive mechanism, represented in Fig.32c corresponds to a shifting of the two types of ions in antiparallel directions $\pm[\bar{1}10]$ for two successive (110) tetragonal planes. In other words, we deal with the primary mechanism describing the BCC-HCP transformation (Sec.3.2), without the additional compression of the fourfold Z axis. In this respect, the tetragonal β-structure appears as a compressed BCC structure, with the ratio c/a=$\sqrt{6}/2$, or equivalently with an angle close to 60° between the diagonals in the (110) tetragonal plane.

Due to the pseudohexagonal character of the tetragonal β-phase it should appear as natural to describe the β-α transition *starting* from the hexagonal α phase. However, in the two, β and α, structures the Be ions are ordered differently. For describing this reordering process, one has to take as initial structure a virtual disordered HCP (D_{6h}^4) structure. The α and β phase can then be obtained by the following independent mechanisms:

1) The α-phase corresponds to an ordering of Be ions, following the symmetry of the one-dimensional irreducible representation $\tau_4(\vec{k}_{16})$ at the center of the hexagonal Brillouin zone.

2) The β-phase results from a combination of two mechanisms: an ordering of the Be ions corresponding to the 3-dimensional irreducible representation $\tau_3(\vec{k}_{12})$, where $\vec{k}_{12} = \left(\frac{4\pi}{3a}, \frac{2\pi}{3a}, 0\right)$ is the M point of the hexagonal Brillouin zone. It leads to an orthorhombic structure (D_{2h}^{12}, Z=4) when two among the three corresponding order-parameter components are equal to zero. Provided an additional antiparallel shifting, of magnitude $a\sqrt{3}/12$ along the $\pm[\bar{1}10]$ hexagonal directions, of the atoms located in the (001) hexagonal plane, the preceding orthorhombic symmetry increases to D_{4h}^{14} (Z=4), i.e. to the β structure.

When heating across the transition, from the β structure to the α structure 3 orientational domains should be created, transforming into each other by a sixfold rotation, according to the virtual symmetry lowering $C_{6v} \rightarrow C_{2v}$ (as C_{2v}^7 is the symmetry of the maximal substructure common to C_{6v}^4 and D_{2h}^{12}). Cooling again to the α phase, each of the preceding 3 domains would split into *four* domains according to the $D_{4h} \rightarrow C_{2v}$ point-group change. Thus, after cycling, a total number of *twelve* domains should be observed in the α structure, corresponding to two sets of *six* domains with opposite polarities. Only two sets of *four* domains have been reported by Smith et al [70]. It coincides with the number predicted by Gufan and Ternovskii [71] in a theoretical model, which is essentially equivalent to our description, although the transition mechanism is formulated in terms of rotation of BeO "quasi" molecules around the tetragonal Z axis.

Let us stress that in ref.[71], the full mechanism is not taken into account for depicting the domain structure.

6.4. The reconstructive NaCl to CsCl transition in alkali halides

Some halides of alkali metals (K, Cs, Rb) and also of NH_4, transform under hydrostatic pressure from the NaCl (O_h^5, Z=1) structure to the CsCl (O_h^1, Z=1) structure [72]. The transitions are generally strongly first-order, with a large volume jump of more than 10%. A transition mechanism between the two structures, shown in Fig.33a,b, was first proposed by Buerger [73]. It consists of a compression of the FCC unit cell along the $[111]$ cubic direction and of a dilatation perpendicular to it. However this mechanism was not confirmed by the observed orientational relationships of the crystallographic axes, and thus other improved mechanisms were proposed by Fraser and Kennedy [74] and by Watanabe et al. [75]. This latter mechanism, which is consistent with most of the experimental data on the transition, corresponds to antiparallel shifts of the atoms located in the (001) FCC planes along the $\pm[110]$ directions, and to a correlated tensile deformation, coinciding with one of the strain tensor components (3.19). It is represented in Fig.33c.

We can formulate Watanabe et al. mechanism in terms which are consistent with the general scheme developed in the preceding sections for displacive reconstructive transitions. Thus the assumed antiparallel shifting from the FCC structure is associated with a six-component order-parameter transforming as the irreducible representation $\tau_{10}(\vec{k}_{10})$, where $\vec{k}_{10} = \left(0, \frac{2\pi}{a}, 0\right)$ (X point in the FCC Brillouin zone). For a general shift involved by two non-vanishing components of the order-parameter ($\eta_1 = \eta_2 \neq 0$, $\eta_3 = ... = \eta_6 = 0$), the FCC symmetry is lowered to orthorhombic (D_{2h}^{13}, Z=2). For specific shifts of magnitudes $a\sqrt{2}/4$, $3a\sqrt{2}/4$,... in the $\pm[110]$ directions, the symmetry increases to tetragonal (D_{4h}^{17}, Z=1). Taking into account the tensile deformation ($\zeta_1 \neq 0$,

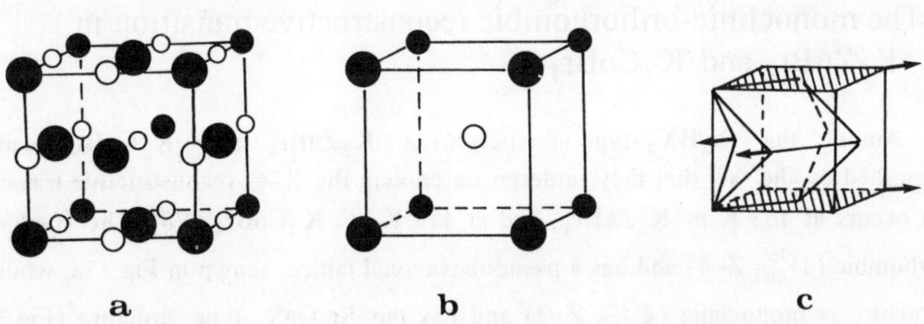

Fig. 33. NaCl-type (a) and CsCl-type (b) structures. (c) Shifting mechanism proposed by Watanabe et al. [75] for the NaCl-CsCl transition. The shifted planes are hatched.

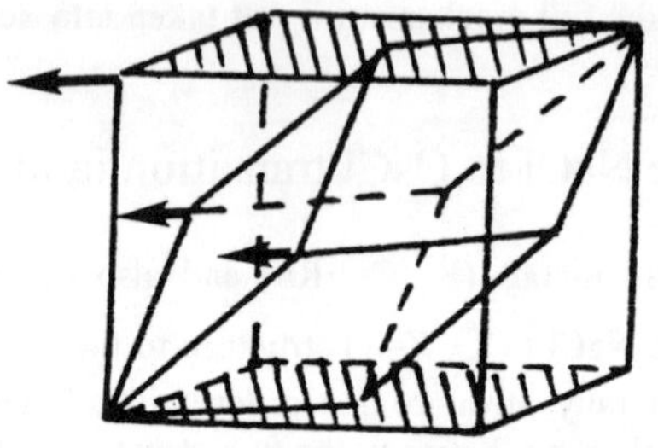

Fig. 34. Shifting mechanism corresponding to the deformation mechanism of the NaCl-CsCl transition proposed by Gufan and Ternovskii [77].

$\zeta_2 = 0$ in (3.19)) one has a further increase of the structural symmetry to the simple cubic O_h^1 (Z=1) space group. Among the confirmations of this description, let us mention the observation of phonon anomalies [75] not far from the X-point of the FCC Brillouin zone.

Another model of the NaCl-CsCl transformation was recently proposed by Gufan and Ternovskii [77]. These authors describe the transformation mechanism in terms of pure strains, as shown in Fig.34. Thus the antiparallel atomic shifting of Watanabe's model is replaced by a shear strain. When the condition $e_{xy} = e_{xz} \neq 0, e_{yz} = 0$ is realized, the FCC symmetry is lowered to monoclinic (C_{2h}^3, Z=1), then for a specific shear angle, the monoclinic symmetry increases to tetragonal (D_{4h}^{17}, Z=1) and again to primitive cubic, when the ζ_1 tensile strain is taken into account. Let us note that this model can also be expressed in terms of *parallel shifts* of the atoms lying in the (001) FCC planes, along $[110]$, so the essential difference with the model of Watanabe, is that it implies a macroscopic *change in the shape of the crystal* (from a cubic form to an oblique prism form).

6.5. The monoclinic-orthorhombic reconstructive transition in K_2ZnBr_4 and K_2CoBr_4

Among the A_2BX_4-type ferroelectrics, K_2ZnBr_4 and K_2CoBr_4 can be distinguished by the fact that they undergo on cooling the β-α reconstructive transition, which occurs at 463 K in K_2ZnBr_4, and at 473 K in K_2CoBr_4 [78]. The β-phase is orthorhombic (D_{2h}^{16}, Z=4) and has a pseudohexagonal lattice, shown in Fig.35a, while the α structure is monoclinic (C_{2h}^2, Z=2) and has the Sr_2GeS_4-type structure (Fig.35b). Takesada and Yagi [79] have observed that the orientational relationship between the α

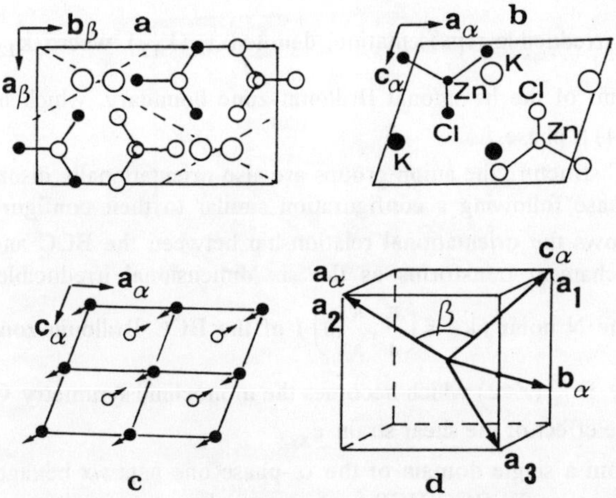

Fig. 35. (a) Orthorhombic pseudo-hexagonal β structure, and (b) monoclinic α structure of K_2ZnBr_4. (c) Direct β–α transition mechanism. (d) Orientational relationship between the cubic and monoclinic lattice vectors.

and β structures changes when cycling across the transition. This fact has been explained by the existence of different orientational domains in the two phases corresponding to the underlying existence of virtual parent structures. More precisely, Takesada and Mashiyama proposed a theoretical model [80] in which the β and α structures result respectively from a HCP and BCC structure, realizing in some respects, a low symmetry variant of Burgers mechanism.

Fig.35c represents the *direct* transition mechanism between the α and β structures. Starting from the β positions in which the complex tetrahedral anions groups ($ZnBr_4$) are close packed, with their centers in 2(c) positions, one has an antiparallel shifting of the molecular groups lying in the (001) hexagonal planes in the directions $\pm[\bar{1}10]$, and a shear deformation e_{xy} which transforms the pseudo-hexagonal 120° angle between the diagonals of the orthorhombic cell, into the 109° monoclinic angle.

The β-α transformation can be more easily understood by describing the BCC-α and HCP-β *virtual* transitions, which underlie the preceding mechanism. Thus in the HCP phase the ($ZnBr_4$) groups are orientationally disordered. The transition to the β phase involves an ordering of these groups, one of the threefold axis of the anionic tetrahedron being oriented along the six-fold direction. In addition, the two cations K which are located along the six-fold axis in positions (000, $00\frac{1}{2}$), are shifted in the antiparallel directions $\pm[010]$. This correlated ordering and shifting mechanism transforms as a single three dimensional irreducible representation, denoted $\tau_4(\vec{k}_{12})$ where $\vec{k}_{12} = \left(\frac{4\pi}{3a}, \frac{2\pi}{3a}, 0\right)$ ends at the M point of the hexagonal Brillouin zone boundary, which induces the cell-doubled D_{2h}^{16} (Z=4) β phase.

In the BCC structure the anion groups are also orientationally disordered, and they order in the α-phase following a configuration similar to their configuration in the β-phase. Fig.35d shows the orientational relationship between the BCC and α structures. The ordering mechanism transforms as the six dimensional irreducible representation $\tau_4(\vec{k}_9)$, i.e. at the N point ($\vec{k}_9 = \left(\frac{\pi}{a}, \frac{\pi}{a}, 0\right)$) of the BCC Brillouin-zone. It leads to a phase of symmetry D_{2h}^{17} (Z=2) which becomes the monoclinic symmetry C_{2h}^{2} (Z=2) of the α-phase, under the effect of the shear strain e_{xy}.

Starting from a single domain of the α-phase one gets *six* hexagonal domains in the virtual HCP phase, as the BCC-HCP transition produces *six* orientational domains (see Sec.3.5), which transform into each other by the cubic C_3^{xyz} axes. Each of these domains split into three orthorhombic domains when the HCP symmetry is lowered to the β symmetry. Accordingly a total number of *18 different domains* will be created at the α-β transition. 9 of these domains are observed by Takesada et al. [81] in a Brillouin scattering experiment in which the longitudinal and transverse (TA_2) acoustic modes show a discrete distribution of nine groups of lines corresponding to the domain distribution in direction of the six-fold axes.

7. The martensitic transformations in lithium and sodium

7.1. Experimental features of the martensitic transformations in Na and Li

A number of neutron scattering investigations of sodium and lithium by Blaschko and coworkers [82-86], and Smith and coworkers [87-92] have recently added decisive finishing touches to the complex pictures of the martensitic transformations which take place at about 35 K in Na, and 77 K in Li. One of the essential points which has been under discussion is the structure of the low symmetry phases, that were initially described from X-ray powder data. [93, 94] as faulted HCP structure coexisting with a large amount of untransformed BCC matrix. A neutron diffraction experiment on Li by McCarthy et al. [95], showed that powder data could not be well indexed by the double layered HCP stacking. Overhauser proposed [96] that a nine layered rhombohedral (9R) polytype structure could provide a better account for the preceding data. Subsequent elastic neutron scattering studies of Li single crystal [84, 90] confirmed a faulted 9R structure, but further diffuse neutron scattering experiments [85, 86] revealed that the structure could be described as a complex arrangement of close packed planes exhibiting simultaneously different orderings in the stacking sequences. This disordered polytype structure coexists, on cooling, with the 9R and BCC structures, whereas on heating above about 80 K, a FCC structure appears, and remains exclusively with the BCC phase between 120 K and about 180 K. The existence of a perfect FCC structure on warming, was independently confirmed in high resolution experiment [91], within the same temperature range, at 6.5 kbar and at atmospheric pressure.

Although Na was initially regarded as equivalent to Li [93, 94], successive studies showed that its low temperature phase was different in some respects. Thus, on the one hand, Schwartz et al. [83] found no short range ordered polytype sequence, but the coexistence of long-range close packed HCP, 9R and BCC sequences whose relative volume fractions depend on temperature and specimen characteristics. On the other hand Berliner et al. [88] interpreted the crystallography of the low temperature phase of Na, as a complex mixture of almost hexagonal, rhombohedral polytypes, with short and long periods, forming a ladder of structures connected one to another by stacking faults. In both studies [83, 88], no FCC structure was reported.

Another important point of controversy concerns the lattice instabilities revealed by inelastic neutron scattering measurements on the Σ_4 phonon branch above the transitions. Although Smith et al. [88, 90, 92] claim that no evidence of transformation precursors can be found in Na and Li, Schwartz et al [86] observe in Li a partial softening of the acoustic $[110]$ phonon branch, polarized along $[1\bar{1}0]$, extending from $\vec{k}=0.1[110]$ to the surface of the BCC Brillouin-zone (the N-point): from 200 K to 100 K the softening increases towards the zone boundary, whereas between 100 K and 80 K a small dip is observed near $\frac{1}{3}[110]$. In Na, Blaschko and Krexner [82] find the same phonon branch to soften close to $\vec{k}=0.42[110]$, a result confirmed by the recent work of Abe et al. [97].

In the two following sections we give a unified phenomenological description of the unusual static and dynamical features which characterize the martensitic transformation in sodium and lithium. More precisely, we will show that the sequence of structures reported in Na and Li below the BCC phase, can be understood by a coupling between the instabilities which are associated with the 9R, FCC and HCP phases. On the other hand a qualitative mechanism will be suggested to explain the unusually large region of the BCC Brillouin-zone (the Σ-line) in which a (weak) temperature dependence softening of the transition wave-vector is observed. More general considerations will be developed relating the phonon anomalies expected for martensitic transformations, to the topology of the metastability regions existing in the phase diagram, and to the energy barriers between the phases.

7.2. Phase diagrams of sodium and lithium

Three close packed structures appear to coexist and to dominate alternately below the BCC phase: the HCP and 9R phases in Na, and the 9R and FCC phases in Li. Let us show that the different sequences of phases observed on cooling and heating in these crystals, can be understood by simple thermodynamic considerations which involve couplings between the corresponding order-parameters. The order-parameters of the three structures, are associated with primary instabilities pertaining to the $\Gamma_5 - \Sigma_4 - N_4$ line of the Brillouin zone. They establish hexagonal ratios in (110) BCC type layers and new translational symmetry in the $[110]$ BCC directions. In addition three types of strain are involved for a realistic description of the low-symmetry structures:

$$e_6 = e_{xy}, \quad e_3 = \frac{1}{\sqrt{6}}(e_{xx} + e_{yy} - 2e_{zz}), \quad e_1 = \frac{1}{\sqrt{3}}(e_{xx} + e_{yy} + e_{zz}). \qquad (3.40)$$

e_6 provides no contact between the atoms in the $[001]$ BCC direction, but decreases the interlayer distance in the $[110]$ direction (Figs.36a and 36b). e_3 restores the close packing both inside the layers (e_{zz}) and between layers ($e_{xx} + e_{yy}$) (Fig.36c). e_1 is the non-symmetry breaking rigidity which accounts for a change in volume at the transition.

1) The BCC-HCP transition is associated with an instability located at the N-point of the BCC Brillouin zone (see Sec.3.2) and its order-parameter expansion (3.27) can be rewritten under the compact form:

$$F_1(\eta) = a_1\eta^2 + a_2\eta^4 + a_4\eta^6. \qquad (3.41)$$

The full thermodynamic potential includes, in addition, the coupling terms with the secondary strains:

$$F_1(\eta, e_1, e_3, e_6) = F_1(\eta) + \eta^2(\delta_1 e_3 + \delta_2 e_6 + \delta_3 e_1). \qquad (3.42)$$

Assuming a pure BCC-HCP transition mechanism in Na, would give for example the following numbers:

$$e_3 = -0.29015, \quad e_6 = -0.1518, \quad e_1 = 0.08625. \qquad (3.43)$$

2) The BCC-FCC transition is induced by a primary instability at the center Γ of

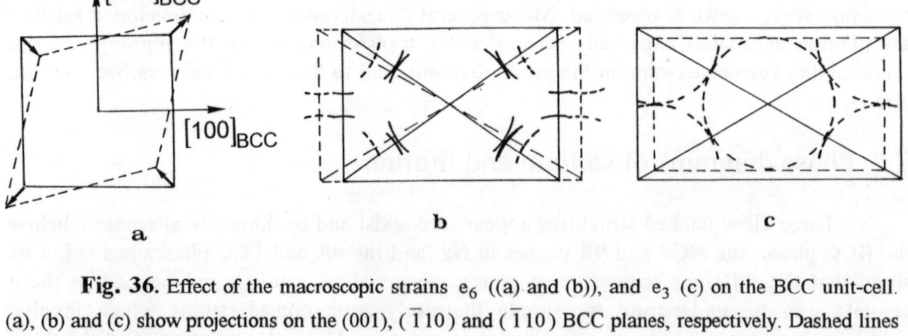

Fig. 36. Effect of the macroscopic strains e_6 ((a) and (b)), and e_3 (c) on the BCC unit-cell. (a), (b) and (c) show projections on the (001), $(\bar{1}10)$ and $(\bar{1}10)$ BCC planes, respectively. Dashed lines represent deformed states.

the BCC Brillouin zone, which is associated with the effective order-parameter expansion (3.35), in which we put $\zeta = e_3$:

$$F_2(e_3) = b_1 e_3^2 + b_2 e_3^3 + b_3 e_3^4. \qquad (3.44)$$

Taking into account the coupling of e_3 with e_6 and e_1, gives the full thermodynamic potential:

$$F_2(e_3, e_1, e_6) = F_2(e_3) + \mu_1 e_3^2 e_1 + \mu_2 e_3 e_6^2. \qquad (3.45)$$

3) The BCC-9R transition has not been considered in the preceding sections. Its current mechanism was proposed by Wilson and de Podesta [98]. It consists in a shifting in the $[110]$ BCC direction of the atoms pertaining to *nine* successive BCC layers, respectively by 0, 1, 2, 1, 2, 3, and 8, in a$\sqrt{2}$ unit (Fig.37a). However when taking into account the displacements associated with the tensile strain e_3, one obtains a displacement field with antiparallel displacements of equal magnitude $a\sqrt{2}/6$, each *third* BCC layer being unshifted. For such displacements the monoclinic symmetry C_{2h}^3 (Z=3), associated with general shifts along $[1\bar{1}0]$, *increases* to the rhombohedral 9R symmetry. Let us stress that the corresponding space group D_{3d}^5 (Z=3) *in not a subgroup* of O_h^9, as its 3-fold axis, which is parallel to the $[110]$ direction, does not coincide with the 3-fold BCC axis.

The primary symmetry breaking mechanism in the preceding description is induced by the twelve-dimensional irreducible representation $\tau_3(\vec{k}_4)$, with $\vec{k}_4 = \frac{1}{3}\left(\frac{\pi}{a}, \frac{\pi}{a}, 0\right)$. The corresponding twelve component order-parameter (ζ_i, i=1-12) takes in the 9R phase the equilibrium values $\zeta_1 = \zeta_2 = \zeta \neq 0$, and $\zeta_i = 0$ for i=3-12. Thus, the effective order-parameter expansion can be written:

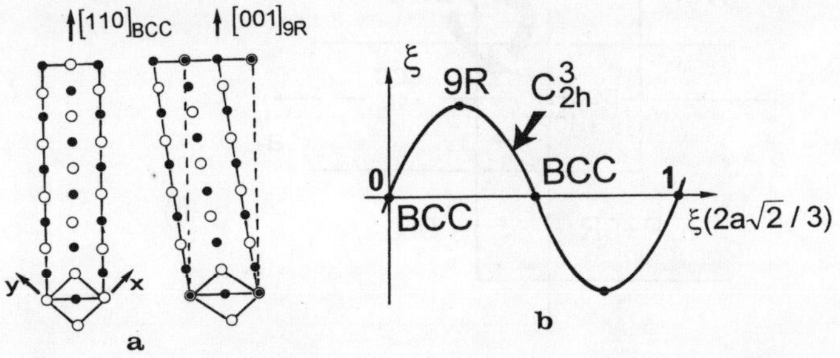

Fig. 37. (a) BCC-9R transition mechanism, following the description of Wilson and de Podesta [98]. (b) Dependence of the order-parameter in function of the shifts, at the BCC-9R transition, following Eq.(3.47).

$$F_3(\zeta) = c_1\zeta^2 + c_2\zeta^4 + c_3\zeta^6. \qquad (3.46)$$

The dependence of $\zeta(\xi)$, where ξ is magnitude of the critical displacements along $[1\bar{1}0]$, is shown in Fig.37b. It has the form:

$$\zeta(\xi) = \zeta_0 \cdot \mathrm{Sin}\frac{3\pi}{a\sqrt{2}}\xi. \qquad (3.47)$$

The coupling with the improper strains e_3, e_6 and e_1, gives the full potential:

$$F_3(\zeta, e_1, e_3, e_6) = F_3(\zeta) + v_1\zeta^2 e_1 + v_2\zeta^3 e_3 + v_3\zeta^2 e_6. \qquad (3.48)$$

The preceding description is consistent with the theoretical model of the BCC-9R transition proposed by Gooding and Krumhansl [99].

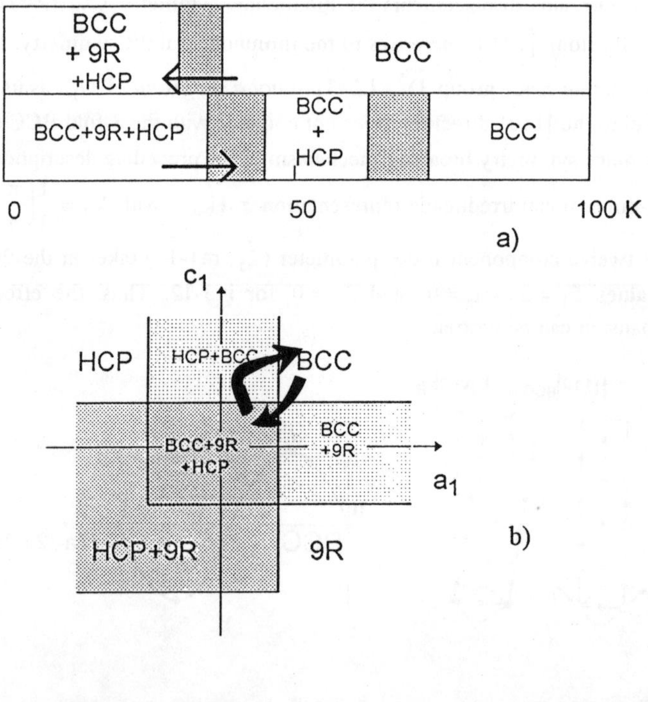

Fig. 38. (a) Sequence of phases found in sodium by Schwarz et al. [83] on cooling and heating. (b) Section of the theoretical phase diagram, corresponding to the thermodynamic potential (3.49). The arrow indicates the experimental path followed in Ref. [83].

Let us now use the preceding considerations to describe the sequence of phases reported for Na and Li. The description of these sequences for Na, following Schwartz et al. [83] is indicated in Fig.38a. In addition to the BCC parent structure, it involves the 9R and HCP phases. Accordingly, using (3.42) and (3.48), one has the full thermodynamic potential:

$$F_4(\eta,\zeta,e_i) = F_1(\eta,e_i) + F_3(\zeta,e_i) + \gamma_1\eta^2\zeta^2 \tag{3.49}$$

where the γ term expresses the coupling between the two primary order-parameters η and ζ. Fig.38b represents a section of the phase diagram corresponding to (3.49), in which one finds the limit phases (HCP and 9R) in the region close to the BCC phases. The arrows in Fig.38b show the experimental paths followed by Schwartz et al.; Thus, on cooling one goes from BCC to (BCC+9R+HCP), and on heating one gets successively (BCC+9R+HCP)$\rightarrow$(HCP+BCC) $\rightarrow$BCC.

The two preceding sequences do not reflect the more complex structures reported by Berliner et al. [88], which involve additional 27R or (15R and 45R) polytype sequence. The corresponding "short", "middle" and "long" period polytypes, which are interpreted by the preceding authors as stacking-fault affected rhombohedral polytypes of an "almost hexagonal" structure, require for their description to assume that the BCC-FCC transition is also involved in Na as a *latent mechanism*, as the FCC phase is never observed by

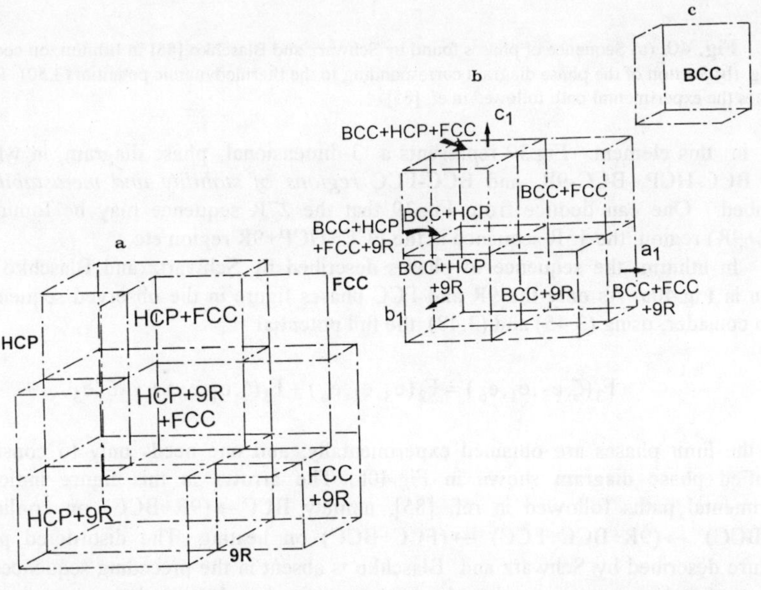

Fig. 39. Three-dimensional phase diagram, involving the BCC, FCC, HCP and 9R phases, and their coexisting regions of metastability. (b) and (c) are embedded in (a).

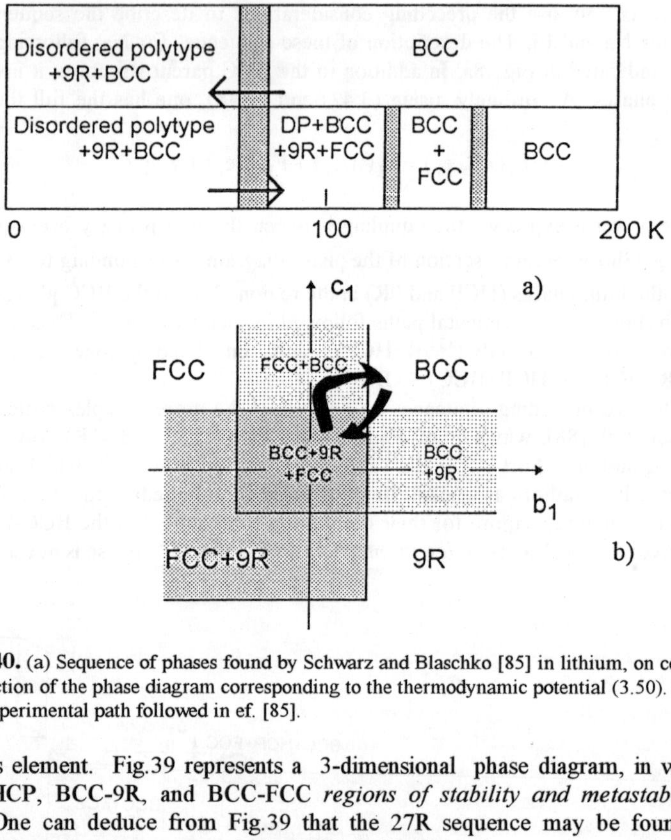

Fig. 40. (a) Sequence of phases found by Schwarz and Blaschko [85] in lithium, on cooling and heating. (b) Section of the phase diagram corresponding to the thermodynamic potential (3.50). The arrow indicates the experimental path followed in ef. [85].

itself in this element. Fig.39 represents a 3-dimensional phase diagram, in which the three BCC-HCP, BCC-9R, and BCC-FCC *regions of stability and metastability*, are described. One can deduce from Fig.39 that the 27R sequence may be found in the (FCC+9R) region, the 15R sequence in the FCC+HCP+9R region etc..

In lithium, the sequence of phases described by Schwartz and Blaschko [85] is shown in Fig.40a. As only the 9R and FCC phases figure in the observed sequence, one has to consider, using (3.45) and (3.48), the full potential:

$$F_3(\zeta, e_3, e_1, e_6) = F_2(e_3, e_1, e_6) + F_3(\zeta, e_1, e_6) + \gamma_2 \zeta^3 e_3 . \qquad (3.50)$$

Only the limit phases are obtained experimentally, and one needs only to consider the simplified phase diagram shown in Fig.40b. The arrows in this figure indicate the experimental paths followed in ref. [85], namely BCC→(9R+BCC) on cooling, and (9R+BCC) →(9R+BCC+FCC) →(FCC+BCC) on heating. The disordered polytype structure described by Schwarz and Blaschko is absent in the preceding sequences. As in the case of sodium, one may relate the existence of a disordered polytype structure to the underlying existence of a *latent HCP phase*. Along this line, Fig.41 shows that *the*

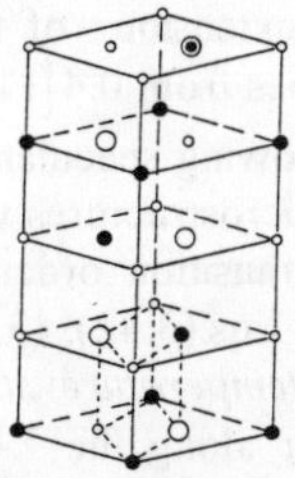

Fig. 41. Disordered polytype unit-cell, in dashed lines, corresponding to the maximal structure common to the HCP and FCC structures.

maximal substructure common to the HCP and FCC phases is a disordered polytype phase formed by a disordered stacking of hexagonal monolayer structures. This interpretation is consistent with the fact that the FCC phase is observed in Li, *on heating and not on cooling*, as we show in Chapter IV that the FCC structure may result from an ordering process from a disordered polytype *parent* phase. Thus the onset of disordered polytypes at low temperatures in Li, originating in the disintegration of latent HCP and FCC structures, creates the condition of the formation of a FCC phase on heating.

It remains to understand the transformation mechanism which results, in Li and Na, into a low temperature phase, reflecting two or three simultaneous instabilities located in different regions of the Σ–line in the BCC Brillouin-zone. The simplest explanation refers to the actual topology of the pressure-temperature phase diagrams of these elements, which still display large unexplored high-pressure regions [17]. It assumes that in both Li and Na the low temperature region in which the martensitic transformation takes place is located *not far from a triple point*, within the region of metastability of three, or even four, stable or potentially stable phases, as confirmed by pseudopotential and band calculations [100].

The pressure-temperature phase diagram of Li [17], shown in Fig.42a, reveals that the regions of stability of the BCC, FCC and martensite phase may merge at about 3 kbars. Besides the calculations of Vaks et al. [100] predict a narrow region of stability for a HCP structure, close to the 9R region. In Na, the region of stability of the martensite phase surrounds the origin of the phase diagram (Fig.42b), and the eventual stabilization of a FCC structure seems improbable for positive pressures and temperatures, although ab initio calculations [100] show its possible stability close to the 9R and HCP regions.

Two correlated series of facts suggest a more precise complementary interpretation of the transition mechanisms in Li and Na. Above the transitions, in the premartensitic region, experimental observations [101] indicate a complex nucleation process involving the existence of large microdomains, possessing different microstructures, i.e. 9R, FCC, and HCP. This indetermination in the structure of the martensitic "embryos" is consistent with the unusually large extension of the softening region observed by Schwarz and Blaschko in Li [85] for the acoustic $[110]$ phonon branch, which extends from $0.1[110]$ to $0.5[110]$ above 100 K, around $\frac{1}{3}[110]$ below 100 K, and even close to the Γ point in a deteriorated crystals of Li. In Na, the extension of the softening region for the same phonon branch [82] is smaller, but still goes from $0.4[110]$ to the N point.

These observations allow the following speculative interpretation for the transition mechanism: the appearance of distinct microstructures which involve different strain fields [102] acting inhomogeneously on the transition order-parameters (due to the coupling between the two quantities expressed by Eqs.(3.42), (3.45) and (3.48)) give *a potentially incommensurate character to the low-temperature structure*, which is attested by the temperature dependence of the k-vector along the Σ-line, and by the incommensurate values of k in the softening regions [82, 85, 89]. However, the large shifts of the atoms which take place at the transition lead the system to *a lock-in transition*, at which *the k-vector locks simultaneously* at the lock-in values corresponding to the fractional numbers 0 (FCC), 1/3 (9R) and 1/2 (HCP), reflecting the various tendencies competing above the transition.

The preceding mechanism can be formalized using the phenomenological approaches to nucleation [103] and to incommensurate "strain-induced" phases [104, 105], which are beyond the scope of this Chapter. By contrast, the unusual behavior reported for the precursor instabilities, in the BCC phase of Li and Na, can be better understood, in the framework of a more general description of the phonon spectrum of martensitic transformations.

7.3. Phonons and martensitic transformations

In the description of the transitions mechanisms discussed in this chapter, it has been stressed that the macroscopic deformations (the secondary strains) do not play a leading role. Except for the BCC-FCC transition, in which the strain e_3 constitute the primary order-parameter, the primary role in displacive reconstructive transitions is played by instabilities, which generally consist in antiparallel shiftings of the atoms. This picture holds for martensitic transformations as well, in contrast to the traditional view [18] which depicts these transitions as strain induced. However although the macroscopic strains are secondary order-parameters, they nevertheless play an important part in the mechanism of the transitions. Thus, as already noted in Sec.3.5, they are symmetry breaking quantities, since they induce the loss of group-subgroup relationship (the "increase in symmetry") between the initial and limit structures. Besides, as usually for "improper" elastic transitions, the strains must be taken into account for a realistic (numerical) description of the transition in a given system.

Let us analyze the influence of the secondary strains on the height of the energy barrier, which separates the initial and limit structures, using as an illustrative example the BCC-HCP transition. Fig.42a, which reproduces the phase diagram of Fig.12, shows the form of the energy barriers in each relevant regions of the phase diagram. Let us focus on the energy barrier corresponding to the reconstructive BCC-HCP transition, which is detailed in Fig.42b. Minimizing the order-parameter expansion associated with the transition, under its form (3.41), one gets easily the coordinates of the minima of $F_1(\eta)$, on the first-order BCC-HCP transition line:

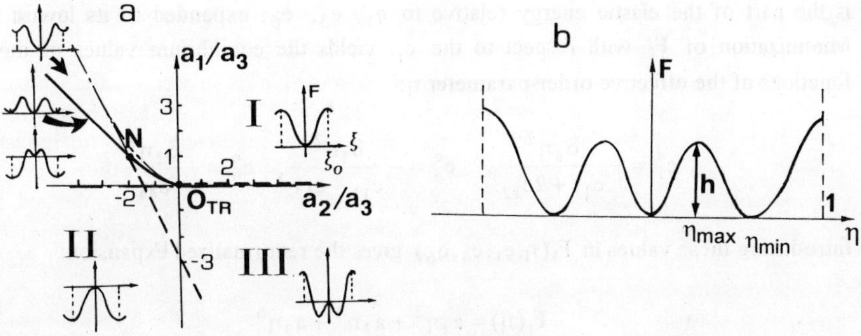

Fig. 42. (a) Energy barriers, in different regions of the BCC-HCP phase diagram. (b) Energy barrier at the BCC-HCP transition. h is the height of the barrier.

$$\eta_{min} = \pm\left(-\frac{a_2}{2a_3}\right)^{1/2} \tag{3.51}$$

and the coordinates of the maxima of the energy barrier (Fig.42b):

$$\eta_{max} = \pm\left(-\frac{a_2}{6a_3}\right)^{1/2} \tag{3.52}$$

From (3.51) and (3.52) one can deduce the height of the barrier:

$$h = F(\eta_{max}) - F(\eta_{min}) = -\frac{a_2^3}{54a_3^2} \tag{3.53}$$

which is a positive quantity, as the BCC-HCP transition takes place for negative values of a_2, in Fig.42a.

For evaluating the influence of the secondary strains (3.40) on h, let us use the full thermodynamic potential:

$$F_1'(\eta, e_1, e_3, e_6) = F_1(\eta, e_1, e_3, e_6) + F_e(e_1, e_3, e_6) \tag{3.54}$$

where $F_1(\eta, e_1, e_3, e_6)$, given by (3.42), includes the coupling with the strains, and:

$$F_e(e_1, e_3, e_6) = \frac{1}{2}(c_{11} + 2c_{12})\cdot e_1^2 + \frac{1}{2}(c_{11} - c_{12})\cdot e_3^2 + \frac{1}{2}c_{44}\cdot e_6^2 \tag{3.55}$$

is the part of the elastic energy relative to e_1, e_3, e_6, expanded to its lowest power. Minimization of F_1' with respect to the e_i, yields the equilibrium values of the e_i as functions of the effective order-parameter η:

$$e_1^e = -\frac{\delta_3\eta^2}{c_{11}+2c_{12}}, \quad e_3^e = -\frac{\delta_1\eta^2}{c_{11}-c_{12}}, \quad e_6^e = -\frac{\delta_2\eta^2}{c_{44}}. \tag{3.56}$$

Introducing these values in $F_1(\eta, e_1, e_3, e_6)$ gives the renormalized expansion:

$$\widetilde{F}_1(\eta) = a_1\eta^2 + \widetilde{a}_2\eta^4 + a_3\eta^6 \tag{3.57}$$

where $\widetilde{a}_2 = a_2 - \frac{\delta_1^2}{c_{11}+2c_{12}} - \frac{\delta_2^2}{c_{44}} - \frac{\delta_3^2}{c_{11}-c_{12}}$ is *smaller* than a_2. Accordingly, one gets the renormalized height of the energy barrier (3.53):

$$\widetilde{h}(e_i) = -\frac{\widetilde{a}_2^3}{54a_3^2} \tag{3.58}$$

which is *lower* than h. In other words, the strains which couple to the order-parameter favor the transition from the BCC to the HCP phase. Let us emphasize, that this result, which is consistent with the conclusion reached by Gooding et al. [106] for BCC-9R transition, is strictly valid only if a_3 is assumed to remain constant. Taking into account higher degree terms in the elastic energy F_e, or higher degree coupling terms between η and e_i, would lead to a less simple result, and may correspond, in specific situations, to an *increase* of the height of the energy barrier.

In contrast to another presupposed view, which assumes that strongly first-order transitions correspond to a strong anharmonicity of the lattice, let us now show that displacive reconstructive transitions actually realize a *harmonic* situation. Using the experimental data available, for instance for the BCC-HCP transition in Ba [2], namely the numerical value found for the jump in entropy: $\Delta S_{Ba} = -\frac{\partial a_1}{\partial T}\cdot\eta^2(T_c)$, and the equation of the BCC-HCP transition line, which is $a_1 = -(a_2 + a_3)$, one gets the following numbers for the phenomenological coefficients:

$$a_1 = -1.084 \ \text{cal}\cdot\text{mole}^{-1}\cdot\text{K}^{-1}, \quad a_2 = -0.084 \ \text{cal}\cdot\text{mole}^{-1}\cdot\text{K}^{-1} \tag{3.59}$$

where the normalizing value $|a_3| = 1$ is assumed. It yields to the value of the potential barrier height $h \cong 5$ meV/atom. On the other hand, a measure of the *depth* d of the potential *well* within the HCP phase (Fig.43) is obtained using the value of the order-

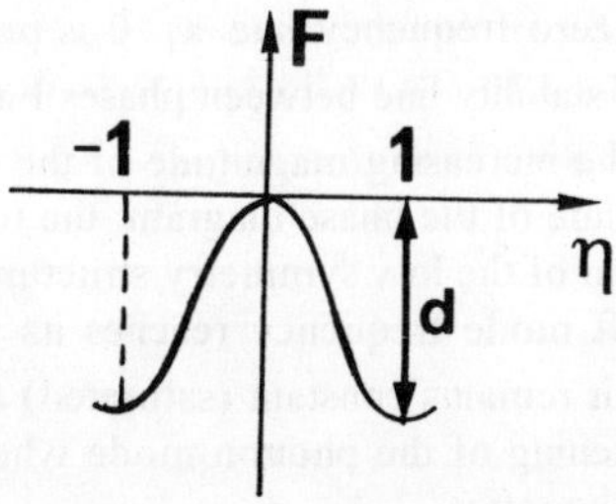

Fig. 43. Depth d of the potential well within the HCP phase.

parameter (η=1) which minimizes $F_1(\eta)$ in this phase. So $F_1(1) = a_1 + a_2 + a_3$. The *minimal* value of d will be reached on the transition line II-III, which has the equation: $a_1 = -(2a_2 + 3a_3)$, so $d_{min} = -a_2 - 2a_3$. It gives: $d_{min} \cong$ -25 meV/atom. Comparing h and d_{min} shows that the potential depth is about five times larger that the potential barrier revealing a tendency to a destabilization of the HCP phase. On the other hand the experimental value found for the frequency of the $T_1(N)$ phonon mode is 3.1 meV [107], i.e. the atoms are in reasonable harmonic situation, and not far from the three phase point. Let us underline, that the numbers found using our approach are very close from those found by Chen et al. [107] by first principle total energy calculations, which are: $h \cong 4$ meV/atom, and $d \cong 27.2$ meV/atom, demonstrating the consistency of our estimations.

The harmonicity of the lattice at reconstructive transitions can be verified by analogous calculations, for the other types of transitions. For example, at the α−ω transition in Zr, one has [2] ΔS=-1.37 $\text{cal} \cdot \text{mole}^{-1} \cdot K^{-1}$, from which one can deduce $h \cong 14.5$ meV/atom and $d_{min} \cong$ -69 meV/atom (which is close to the value $d \cong$ -45 meV/atom found from pseudopotential calculations [108]). Here again, the depth of the well is about five times the barrier height, reflecting a strong harmonic regime, as compared to the frequency of the soft mode which is 4.17 meV.

The more convincing data can be found for the β−α transition in titanium [15]. Here, our calculated numbers, deduced from the jump in entropy at the transition, which is ΔS=0.866 $\text{cal} \cdot \text{mole}^{-1} \cdot K^{-1}$ [2], correspond to a barrier height of 53.5 meV/atom while the frequency of the soft mode is 5 meV. It shows again that Ti is in full harmonic regime, even at the transition.

Following the preceding considerations, one can conclude that the various transitions taking place in the phase diagrams of Fig.42a, including the BCC-HCP reconstructive transition, correspond to a jump between two harmonic wells. Let us use this result to discuss the soft mode behavior for the various thermodynamic paths, parallel to the a_1 (temperature) axis, indicated in Fig.42a.

On the right hand side of the phase diagram, one has the I-III second-order transition for $a_1 \propto \omega^2 = 0$, where ω is the soft mode frequency: phase III loses its stability at zero frequency $\omega = 0$. The zero frequency line $a_1 = 0$ is prolongated beyond the origin ($a_1 = 0$, $a_2 = 0$) into the limit of stability line between phases I and III. The *hardening* of the preceding soft mode reflects the increasing magnitude of the structural deformation when moving towards the left hand side of the phase diagram: the transition line II-III represents the line of maximal deformation of the low symmetry structure with respect to phase I. On the I-III transition line the soft mode frequency reaches its maximal value ω_{max} at the three phase point, after which it remains constant (saturated) across the I-II transition line. One can thus evaluate the softening of the phonon mode when approaching the transition lines from the BCC phase, by the ratio:

$$\frac{\delta}{\delta_{max}} = \left(\frac{\Delta\omega}{\omega}\right)^2 \tag{3.60}$$

where δ is the length of the thermodynamic path between the limit of stability lines I-III and II-III, along the a_1 axis (Fig.42a). δ_{max} is the maximal value reached by δ at $a_2 = 0$. One finds, using the difference of coordinates between the preceding lines:

$$\delta = \frac{a_2^2 + 8a_2a_3 + 12a_3^2}{4a_3}, \quad \delta_{max} = \frac{(a_2 + 3a_3)^2}{3a_3} \quad \text{and} \quad \frac{\delta}{\delta_{max}} = \frac{3}{4} + \frac{3a_3(2a_2 + 3a_3)}{4(a_2 + 3a_3)^2}.$$

Let us note that this ratio varies from 0 (i.e. no softening at the three-phase point) to 1 (maximal softening at $a_1 = a_2 = 0$).

Such a description expresses quantitatively the reciprocal result, already quoted in a qualitative formulation, that when an incomplete softening of the phonon mode is observed, the state of the system is not far from the three-phase point whereas far from this latter point, on the left hand side of the phase diagram, no softening is expected. The contradictory results reported for Li and Na: weak softening [82, 86], or absence of precursor effects [91, 109], may be understood in this framework, i.e. as corresponding to different experimental paths on either parts of the softening limit $\delta/\delta_{max} = 0$, in agreement with the thermodynamic paths proposed hereabove for the two elements.

In summary, a number of suggestions have been pointed out in this section to explain the features of the martensitic transformations in Li and Na: coexistence of structures related to different order-parameters in metastability regions of the phase diagram, critical influence of secondary strains, harmonicity of the lattice vibrations at strongly first-order transitions, inherent incommensurate character of the martensite phase. These suggestions find more striking confirmations in the experimental results reported for the martensitic transformations in a number of alloys, in which the Σ_4 instability line is involved, the parent structure being either BCC (β-phase alloys) or of the CsCl type. Thus an incommensurate modulation has been found between the CsCl high-temperature structure and the low-temperature 7M martensite phase of NiAl alloys [110-113]. The transition corresponds to another point of the Σ-line, around $k = \frac{1}{7}\left[\frac{2\pi}{a}, \frac{2\pi}{a}, 0\right]$, at which a softening of the transverse acoustic branch is clearly observed [110]. Incommensurate modulations have been also reported in TiNi(Fe) [114-116] and VR_4 alloys [117]. The theoretical interpretations of these data [118-121] stress the importance of the coupling between the primary instability and the macroscopic strains, but overlook the limit state character of the martensitic phase.

8. References

1. S. K. Sikka, Y. K. Vohra and R. Chidambaram, *Progress in Materials Science* **27** (1982) 245.
2. E. Tonkov, *Phase Diagrams of the Elements Under Pressure* (Nauka, Moscow, 1979).
3. D. J. Cometto, G.L. Houze Jr. and R. F. Hehemann, *Trans. Metall. Soc. AIME* **233** (1965) 30.
4. J. Silcock, *Acta Met.* **6** (1958) 481.
5. Y. Bagariatskii, T. Tagunova and G. Nosova, *Dokl. Acad. Nauk SSSR* **105** (1955) 1225; *Acta Cryst.* **14** (1961) 1087.
6. B. Hatt and J. Roberts, *Acta Met.* **8** (1960) 575.
7. S. Sass, *J. Less Common Metals* **28** (1972) 157; *Acta Met.* **17** (1969) 813.
8. D. De Fontaine, N. Paton and J. Williams, *Acta Met.* **19** (1971) 1153.
9. O. V. Kovalev, *Irreducible Representations of the Space-Groups* (Gordon and Beach, New York, 1965).
10. V. P. Dmitriev, S. B. Rochal, Yu. M. Gufan and P. Toledano, *Phys. Rev. Lett.* **60** (1988) 1958.
11. D. De Fontaine and O. Buck, *Phil. Mag.* **27** (1973) 967.
12. H. Cook, *Acta Met.* **21** (1973) 1445; *Acta Met.* **22** (1974) 239; H. Cook and W. Pardee, *Acta Met.* **25** (1977) 1043.
13. S. C. Moss, D. T. Keating and J. D. Axe, *Solid State Commun.* **13** (1973) 1465; *Phys. Rev. Lett.* **35** (1975) 530.
14. C. Stassis, J. Zaretsky and N. Wakabayashi, *Phys. Rev. Lett.* **41** (1978) 1726.
15. W. Petry et al., *Phys. Rev.* **B43** (1991) 10933; A. Heiming et al., *Phys. Rev.* **B43** (1991) 10948; J. Trampenau et al., *Phys. Rev.* **B43** (1991) 10963.
16. S. L. Sass and B. Borie, *J. Appl. Crystallogr.* **5** (1972) 236.
17. D. A. Young, *Phase Diagrams of the Elements* (Univ. California Press, Berkeley, 1991).
18. Z. Nishiyama, *Martensitic Transformations* (Academic Press, New York, 1978).
19. W. G. Burgers, *Physica* **1** (1934) 561.
20. V. P. Dmitriev, Yu. M. Gufan and P. Toledano, *Phys. Rev.* **B44** (1991) 7248.
21. J. C. Toledano and P. Toledano, *The Landau Theory of Phase Transitions* (World Scientific, Singapore, 1987), Chap. 3.
22. Landolt-Bornstein Tables: *Metals: Phonon States, Electron States and Fermi Surfaces* (Springer-Verlag, Berlin, 1983), Vols. 13a, b, c.
23. E. C. Bain, *Trans. AIME* **70** (1924) 25.

24. H. Warlimont and L. Delaey, *Martensitic Transformations in Copper-Silver and Gold Based Alloys* (Pergamon, Oxford, 1974), Chap. 3.
25. H. M. Strong, R. E. Tuft and R. E. Hanneman, *Metal. Trans.* **4** (1973) 2657.
26. E. G. Ponyatovskii, *Kristallografiya* **4** (1959) 257.
27. A. Jayaraman, W. Klement, R. C. Newton and G. C. Kennedy, *J. Phys. Chem. Solids* **24** (1963) 7.
28. J. Donohue, *The Structure of the Elements* (Wiley, New York, 1974).
29. *The Chemistry of the Actinide Elements*,ed. J.J. Katz, G. T. Seaborg and L. R. Morss (Chapman and Hall, London, 1986), vol.1.
30. W. B. Pearson, *A Handbook of Lattice Spacings and Structures of Metals and Alloys* (Pergamon, Oxford, 1967), vols. 1 and 2.
31. W. P. Crummet, H. G. Smith, R. M. Nicklow and N. Wakabayashi, *Phys. Rev.* **B19** (1979) 6028.
32. H. G. Smith, N. Wakabayashi, W. P. Crummet, R. M. Nicklow, G. H. Lander and E. S. Fisher, *Phys. Rev. Lett.* **44** (1980) 1612.
33. 0. Ericksson, P. Söderlind and J. M. Wills, *Phys. Rev.* **B45** (1992) 12588.
34. J. M. Wills and O. Ericksson, *Phys. Rev.* **B45** (1992) 13879.
35. B. Mettout, V. P. Dmitriev, M. Ben Jaber and P. Toledano, *Phys. Rev.* **B48**, (1993) 6908.
36. L. G. Chachkiani, V. P. Dmitriev and P. Toledano, *Sov. Phys. Solid State* **32** (1989) 1334.
37. W. H. Zachariasen, *Acta Crystallogr.* **5** (1952) 19; *ibid* **12** (1959) 698.
38. J. A. G. Marples, *Acta Crystallogr.* **18** (1965) 815.
39. L. B. Asprey, R. D. Fowler, J. D. G. Lindsay and R.W.White, *Inorg. Nucl. Chem. Lett.* **7** (1971) 977.
40. P. Chiotti and K. J. Gill, *Trans. Metall. Soc. AIME* **221** (1961) 573.
41. U. Benedict, in *Handbook on the Physics and Chemistry of the Actinides*, ed. A. J. Freeman and G. H. Lander (Elsevier Science Publishers, Amsterdam, 1987), p.
42. F. L. Oelting, M. H. Rand and R. J. Ackerman, *The Chemical Thermodynamics of Actinide Elements and Compounds* (IAEA, Vienna,1976), part 1.
43. R.B. Roof, *J. Appl. Crystallogr.* **14** (1981) 447.
44. R. B. Roof, *Z. Kristallogr.* **158** (1982) 307.
45. R. G. Haire and R. D. Baybarz, *J. Phys. (France)* **I40** (1979) C4, 101.
46. R. G. Haire and J.R. Peterson, *J. Less-Common Met.* **102** (1984) 119.
47. C. W. Tucker, H. Senio, J. Thewlis and H. Steeple, *Acta Crystallogr.* **9** (1956) 472.
48. W. H. Zachariasen, *Acta Crystallogr.* **5** (1952) 660; *ibid* **5** (1952) 664.
49. J. C. Toledano and P. Toledano, *Phys. Rev.* **B21** (1980) 1172.
50. W. H. Zachariasen and F. H. Ellinger, *Acta Crystallogr.* **16** (1963) 777.
51. W. H. Zachariasen and F. H. Ellinger, *Acta Crystallogr.* **16** (1963) 369.
52. I. E. Dzialoshinskii, *Sov. Phys. JETP* **19** (1964).
53. A. P. Levanyuk and D. G. Sannikov, *Sov. Phys. Solid State* **18** (1976) 245.

54. W. L. Mc Millan, *Phys. Rev.* **B14** (1976) 1496.
55. J. C. Marmeggi and A. Delapalme, *Physica B+C* **120B** (1980) 309.
56. J. C. Marmeggi, A. Delapalme, G. H. Lander, C. Vettier and N. Lehner, *Solid State Commun.* **43** (1982) 577.
57. H.G. Smith and G. H. Lander, *Phys. Rev.* **B30** (1984) 5407.
58. G. H. Lander, *J. Magn. and Magn. Mater.* **29** (1982) 271.
59. S. Van Smaalen and T. F. George, *Phys. Rev.* **B35** (1987) 7939.
60. G. Grübel, J. D. Axe, D. Gibbs, G. H. Lander, J. C. Marmeggi and T. Brückel, *Phys. Rev.* **B43** (1990) 8803.
61. J. C. Marmeggi, G. Lander, S. Van Smaalen, T. Brückel and C. M. E. Zeyen, *Phys. Rev.* **B42** (1990) 9365.
62. Y. Yamada, *Phys. Rev.* **B47** (1993) 5614.
63. S. Hoshino, *J. Phys. Soc. Japan* **12** (1957) 315.
64. J. Nölting and D. Rein, *Z. Phys. Chem. N. F.* **66** (1966) 150.
65. L. W. Strock, *Z. Phys. Chem.* **B25** (1934) 441; *ibid* **B31** (1935) 132.
66. C. Tubant and E. Lorentz, *Z. Phys. Chem.* **87** (1914) 513 and 543.
67. Yu. M. Gufan, I. N. Moshchenko and V. I. Snezhkov, *Phys. Solid State* **35** (1993) 1037.
68. G. Aminoff, *Z. Kristallogr.* **62** (1925) 113.
69. *Physicochemical Properties of Oxides*, ed. G. V. Samsonov (Metallurgiya, Moscow, 1978).
70. D. K. Smith, C. F. Cline and S. B. Austerman, *Acta Crystallogr.* **18** (1965) 393.
71. Yu. M. Gufan and I. V. Ternovskii, *Phys. Solid State* **35** (1993) 1165.
72. P. W. Bridgman, *Z. Kristallogr. Kristallgeom. Kristallphys. Kristallchem.* **67** (1928) 363.
73. M. J. Buerger, in *Phase Transformation in Solids*, ed. R. Smoluchowski, J. E. Mayer and W. A. Weyl (Wiley, New York, 1951).
74. W. L. Fraser and S. W. Kennedy, *Acta Crystallogr.* **A30** (1951) 13.
75. M. Watanabe and S. W. Kennedy, *Acta Crystallogr.* **A33** (1977) 294.
76. O. Blaschko, G. Ernst, G. Quittner and G. Pepy, *Phys. Rev.* **B23** (1981) 3017.
77. Yu. M. Gufan and I. V. Ternovskii, *Phys. Solid State* **35** (1993) 639.
78. H. Mashiyama, H. Kasano and T. Yamaguchi, *J. Phys. Soc. Japan* **60** (1991) 45.
79. M. Takesada and T. Yagi, *J. Phys. Soc. Japan* **63** (1994) 1241.
80. M. Takesada and H. Mashiyama, *J. Phys. Soc. Japan* **63** (1994) 2618.
81. M. Takesada, M. Yamaguchi and T. Yagi, *Ferroelectrics* **152** (1994) 391.
82. O. Blaschko and G. Krexner, *Phys. Rev.* **B30** (1984) 1667.
83. W. Schwarz, O. Blaschko and I. Gorgas, *Phys. Rev.* **B46** (1992) 14448.
84. G. Ernst, C. Artner, O. Blaschko and G. Krexner, *Phys. Rev.* **B33** (1986) 6465.
85. W. Schwarz and O. Blaschko, *Phys. Rev. Lett.* **65** (1990) 3144.
86. W. Schwarz, O. Blaschko and I. Gorgas, *Phys. Rev.* **B44** (1991) 6785.

87. H. G. Smith, R. Berliner, J. D. Jorgensen and J. Trivisonno, *Phys. Rev.* **B43** (1991) 4524.
88. R. Berliner, H. G. Smith, J. R. D. Copley and J. Trivisonno, *Phys. Rev.* **B46** (1992) 14436.
89. H. G. Smith, *Phys. Rev. Lett.* **58** (1987) 1228.
90. H. G. Smith, R. Berliner, J. D. Jorgensen, M. Nielsen and J. Trivisonno, *Phys. Rev.* **B41** (1990) 1231.
91. H. G. Smith, R. Berliner and J. Trivisonno, *Phys. Rev.* **B49** (1994) 8547.
92. R. Berliner, O. Fajen, H. G. Smith and R. L. Hitterman, *Phys. Rev.* **B40** (1989) 12086.
93. C. S. Barrett and O. R. Trauz, *Trans. Am. Inst. Min. Metall. Pet. Eng.* **175** (1948) 579.
94. C. S. Barrett, *Acta Crystallogr.* **9** (1956) 671.
95. C. M. Mc Carthy, C. W. Tomson and S. A. Werner, *Phys. Rev.* **B22** (1980) 574.
96. A. W. Overhauser, *Phys. Rev. Lett.* **53** (1984) 64.
97. H. Abe, K. Ohshima, T. Suzuki, S. Hoshino and K. Kakurai, *Phys. Rev.* **B49** (1994) 3739.
98. J. A. Wilson and M. de Podesta, *J. Phys. F: Met. Phys.* **16** (1986) L121.
99. R. Gooding and J. A. Krumhansl, *Phys. Rev.* **B38** (1988) 1697.
100. V. Vaks, M.I. Katsnelson, V. G. Koreshkov, A. I. Likhtenstein, O. E. Parfenov, V. F. Skok, V. A. Sukhoparov, A. V. Trefilov and A. A. Chernyshov, *J. Phys.:Condens. Matter* **1** (1989) 5319.
101. H. Abe and K. Ohshima, private communication.
102. W. Pichl, *Scripta Metall.* **31** (1994) 1593.
103. A. A. Bulbich, V. P. Dmitriev, O. A. Zhelnova and P. E. Pumpyan, *Sov. Phys. JETP* **71** (1991) 619.
104. T. A. Aslanyan and A.P. Levanyuk, *JETP Lett.* **28** (1978) 70.
105. T.A. Aslanyan, A. P. Levanyuk, M. Vallade and J. Lajzerowicz, *J. Phys. C: Solid State Phys.* **16** (1983) 6705.
106. R. G. Gooding, Y. Y. Ye, C. T. Chan, K. M. Ho and B. N. Harmon, *Phys. Rev.* **B43** (1991) 13626.
107. Y. Chen, K. M. Ho and B. N. Harmon, *Phys. Rev.* **B37** (1988) 283.
108. K. M. Ho, C. L. Fu and B. N. Harmon, *Phys. Rev.* **B29** (1984) 1575.
109. R. Berliner, H. G. Smith, J. R. D. Copley and J. Trivisonno, *Phys. Rev.* **B51** (1995) 16467.
110. S. M. Shapiro, J. Z. Larese, Y. Noda, S. S. Moss and L. E. Tanner, *Phys. Rev. Lett.* **57** (1986) 3199.
111. S. M. Shapiro, B. X. Xang, G. Shirane, Y. Noda and L. E. Tanner, *Phys. Rev. Lett.* **62** (1989) 1298.
112. S. M. Shapiro, B. X. Yang, Y. Noda, L. E. Tanner and D. Schayvers, *Phys. Rev.* **B44** (1991) 9301.
113. Y. Noda, S. M. Shapiro, G. Shirane, Y. Yamada and L. E. Tanner, *Phys. Rev.* **B42** (1990) 10397.

114. S. M. Shapiro, Y. Noda, Y. Fujii and Y. Yamada, *Phys. Rev.* **B30** (1984) 4314.
115. S. K. Satija, S. M. Shapiro, M. B. Salamon and C. M. Wayman, *Phys. Rev.* **B29** (1984) 6031.
116. P. Moine, J; Allain and B. Reuker, *J. Phys. F: Met. Phys.* **14** (1984) 2517.
117. Y. Yamada, *J. Phys. Soc. Japan* **59** (1990) 2517.
118. I. Folkins and B. M. Walker, *Phys. Rev.* **B40** (1989) 255.
119. R. J. Gooding and J. A. Krumhansl, *Phys. Rev.* **B39** (1989) 1535.
120. Y. Yamada, Y. Noda, M. Takimoto and K. Furukawa, *J. Phys. Soc. Japan* **54** (1985) 2940.
121. Y. Yamada, Y. Noda and K. Fuchizaki, *Phys. Rev.* **B42** (1990) 10405.

CHAPTER IV

ORDERING TYPE RECONSTRUCTIVE PHASE TRANSITIONS

1. Introduction

In the introductory example of order-disorder transition considered in Chapter II (Sec.2.3), it has been shown that the loss of group-subgroup relationship between the disordered and ordered phases, could only be realized when assuming a fractional concentration of the initial structure. It resulted in the creation of regularly distributed vacancies in the ordered phase, and subsequently to an increase of symmetry of the actual distribution of atoms with respect to the symmetry of the distribution function of the sites. It is needless to say that the preceding situation is not the more realistic one, as condensed matter has a strong tendency to form fully occupied close packed atomic structures. Hence, structures with regularly distributed vacancies constitute exceptions, and order-disorder transitions do not have, in general, a reconstructive character: they are in most cases second-order or slightly first-order, and fulfill the group-subgroup relationship criterion [1, 2].

There exist however another type of ordering transitions, which is compatible with close packing atomic configurations, and is, in some respect, *inherent* to such configurations, namely *transitions between close packed structures*. Indeed, in the simplest hexagonal close packing of identical atoms the stacking of identical layers which form polytypes (defined as structures formed by different sequences of identical layers) implies the possibility of changing the sequences in more ordered or more disordered structures. This fact favors naturally the formation of *stacking faults* in ordered structures, and conversely, the formation of *disordered polytype structures*, which are random sequence (statistically disordered) of hexagonal close packed layers. The existence of disordered polytypes is well established in materials such as ZnS or SiC [3]. Their ordering yields ordered polytypic structures which are formed by short sequences of stacked planes, as the HCP (2H, 4H, 6H) and FCC (3R) structures, or by middle (7R, 9R,...) or longer sequences (27R, 45R,...) of hexagonal planes.

Along the preceding line, there exist a number of reconstructive transitions, the mechanisms of which can be described by an ordering process from a disordered polytype parent phase. This is for example the case for the FCC-HCP transition which takes place in Co, He, Pr or Nd, i.e. when the FCC and HCP phases are not both adjacent to a parent BCC phase in the pressure-temperature phase diagram [4]. The HCP-FCC transitions in the preceding elements displays a weakly first-order character [5-7] excluding the possibility of large displacements that would be required in a displacive reconstructive mechanism. In Co and C a reordering mechanism between the phases is suggested by the observation of a strong tendency to form disordered polytype structures above the regions of stability of the FCC (or diamond) and HCP phases [8, 9]. However, in most of the examples discussed in this Chapter, a disordered polytype structure is *assumed* to be the parent structure, i.e. it is a *latent* phase, as it provides the best crystallogeometrical correlation between the observed phases surrounding a reconstructive transition. In this case the choice of an ordering mechanism, instead of a displacive one, is based also on strong experimental indices, such as the existence of regular distributions of stacking faults or clearcut indication of atomic ordering as in some phases of ice [10].

From the point of view of our approach, there exist precise additional criteria which help to recognize an ordering-type reconstructive transition, as the absence of intermediate Landau-type phases, which appear in displacive mechanisms, but are forbidden in ordering mechanisms. Let us note in this respect, that the diffusion process usually assumed for ordering transitions, has a partially correlated character for reconstructive transitions, due to the in-layer close packing of atoms. Another indicative property, is the existence of different types of stacking faults which result, in our interpretation, from the distribution of domain walls between antiphase domains.

In the *purely* ordering type of reconstructive transitions discussed in this Chapter, the density-wave formalism does not play an important role, as we deal only with a variation of the waves *amplitudes*. By contrast when both ordering and displacive mechanisms are involved, such as in the graphite-diamond transition, the description of the transition in terms of density waves is essential for the understanding of the transition process.

As in Chapter III, we will start from the simplest situation found among ordering-type reconstructive transitions: the FCC-HCP transition mechanism (and its variants), which possesses among these transitions the same universal character as Burgers mechanism among displacive-type reconstructive transitions. The more complex cases, such as the mechanisms underlying the phase diagrams of Ice or SiO_2, will be treated later in the chapter.

2. The FCC-HCP reconstructive transition

2.1. Crystallographic description and transition order-parameter

Among the elements, the FCC-HCP transition is found in ^3He, ^4He, Co, Fe, Tl, Am and in most lanthanides [4]. The FCC and HCP structures represent the simplest close packing configurations of hard sphere atoms. Their unit-cells are formed by *three* and *two* stacked hexagonal layers, respectively, which are shown in Figs. 1a and 1b. Each layer is shifted with respect to another by $a_h/\sqrt{3}$ in the [120] HCP direction, where a_h is the basic vector of the (rhombohedral) FCC unit-cell, in hexagonal coordinates.

For a better insight of the FCC and HCP structures, let us underline that, although a close packing of atoms represented by hard spheres occupying a minimal volume may be realized in several ways [3, 11], in real crystals close packing always corresponds to a

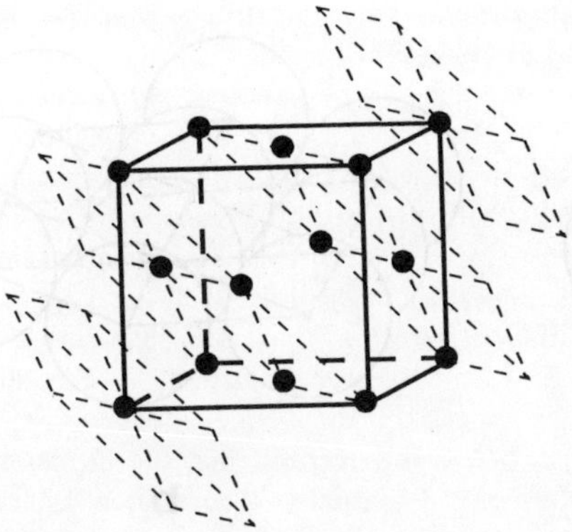

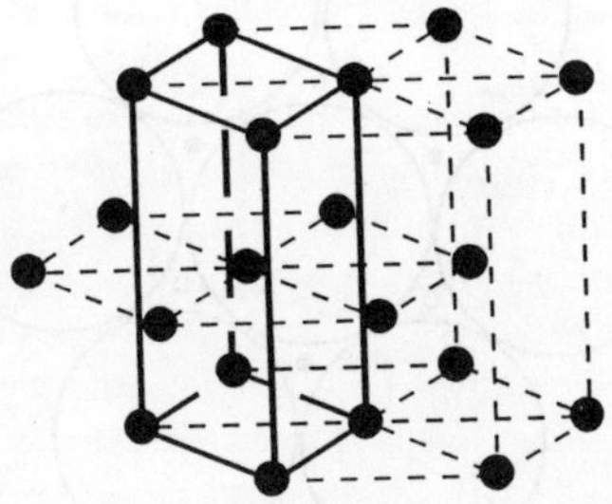

a b

Fig. 1. (a) Three-layered FCC structure. (b) Two-layered HCP structure.

layered configuration, which gives the possibility of isolating planes of atoms packed in the closest manner. These planes are stacked according to some rules of filling up space, and they represent hexagonal packing of spheres, each of which is in contact with six nearest neighbors (position A in Fig.2a). In the centers of the triangles formed by neighboring atoms exist geometrically equivalent sites B and C (Fig.2a). A spatial close packing is realized when each of the successive layers occupy the free spacings left by the preceding layer, related to positions of the B or C type. The stacking order of the layers determines the close packed structure.

Fig.2b shows a layer of equiradii (R) close packed spheres. Within the layer, the close packing condition is $a_h = b_h = 2R$, where a_h and b_h are the hexagonal lattice parameters. For stacked layers one has an additional condition reflecting the close packing in the third dimension, namely: $c_h = a_h \sqrt{8/3}$ for a two layer stacking (HCP) structure, $c_h = a_h \sqrt{6}$ for a three layer FCC structure, etc...

The basic vectors of the HCP unit-cell shown in Fig.1b, can be expressed as functions of the basic vectors of the FCC rhombohedral (primitive) unit-cell as:

$$\vec{a}_h = \vec{b}_c - \vec{a}_c, \quad \vec{b}_h = \vec{a}_c - \vec{c}_c, \quad \vec{c}_h = \frac{2}{3}\left(\vec{a}_c + \vec{b}_c + \vec{c}_c\right). \tag{4.1}$$

From Eq.(4.1), or directly from the structures shown in Fig.1, one can easily deduce that *the maximal substructure*, common to the HCP and FCC structures, is composed by a *monolayer hexagonal structure*. It is also easy to verify that this substructure, which will be denoted L, has the symmetry D^1_{6h}, and a unit-cell volume $V_L = V_h/6 = V_c/3$, where V_h and V_c are the volumes of the HCP and FCC unit-cells, respectively. The L unit-cell, which is represented within the FCC and HCP structures in Figs.3a and 3b, is *filled by 1/3*

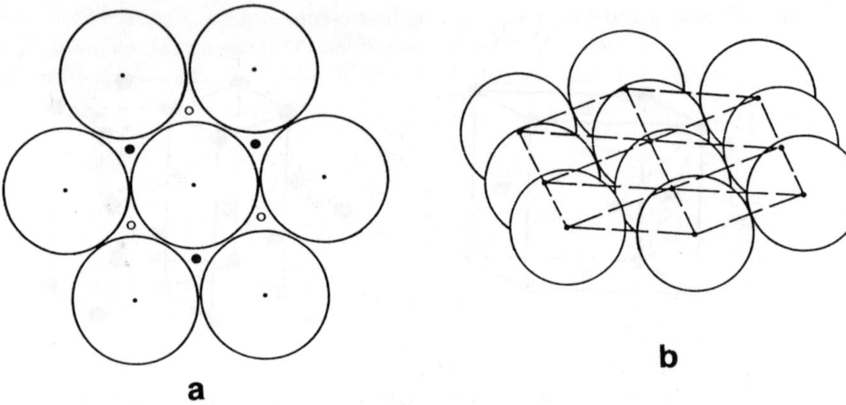

Fig. 2. Close packing of hard spheres. (a) Projection of the layer on the (001) plane. Centers of the circles are in A position, small solid dots represent the B position and small open circles denote the C positions. (b) Layer of equi-radii close packed spheres.

atom, i.e. it corresponds to an occupancy (a concentration) x=1/3 for the L-structure. This fractional number must be understood as follows: in a given L monolayer the atoms occupy the positions 1(a):(000). Only one among the three A, B, and C positions in Fig.2a, is occupied in the monolayer. In the next layer, the atoms cannot be again in position A, but only in position B *or* C, let say B. In the following layers they will thus occupy the position A or C, and so forth. In other words the L-layer stacking corresponds to a statistically disordered polytype structure in which the A, B, and C sites are equivalent: the 1(a) positions are occupied with equal probabilities by atoms and vacancies.

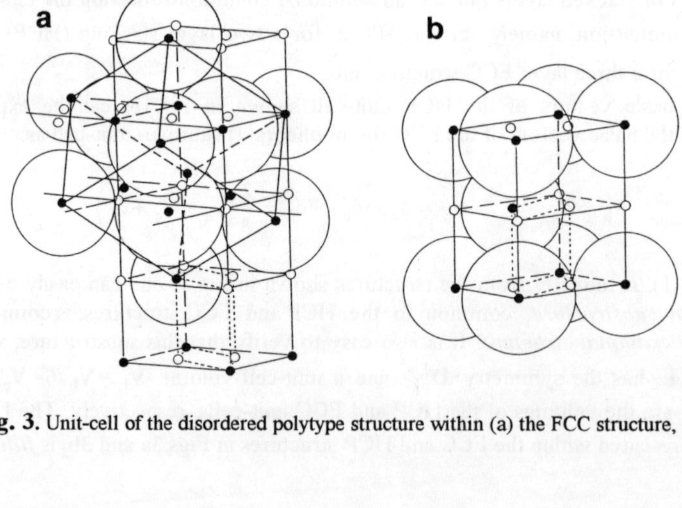

Fig. 3. Unit-cell of the disordered polytype structure within (a) the FCC structure, (b) the HCP structure.

We will now assume that the polytype L-structure is the *initial structure* for our description of the FCC-HCP reconstructive transition. With that goal, let us write the relationships between the basic vectors of the HCP, and FCC unit-cell, and the basic vectors ($\vec{a}_L, \vec{b}_L, \vec{c}_L$) of the L structure:

$$\vec{a}_h = 2\vec{a}_L = \vec{b}_L, \quad \vec{b}_h = \vec{a}_L + 2\vec{b}_L, \quad \vec{c}_h = 2\vec{c}_L; \tag{4.2}$$

$$\vec{a}_c = \vec{a}_L + \vec{b}_L + \vec{c}_L, \quad \vec{b}_c = -\vec{a}_L + \vec{c}_L, \quad \vec{c}_c = -\vec{b}_L + \vec{c}_L. \tag{4.3}$$

From (4.2) and (4.3) one can deduce the wave-vectors expressing the breaking of translational symmetry at the *virtual* L-HCP and L-FCC transitions. One finds, respectively: $\vec{k}_{15} = \frac{1}{3}\left(\vec{a}_L^* + \vec{b}_L^*\right) + \frac{1}{2}\vec{c}_L^*$, and $\vec{k}_{10} = \frac{1}{3}\left(\vec{a}_L^* + \vec{b}_L^* + \vec{c}_L^*\right)$, where $\vec{a}_L^*$, $\vec{b}_L^*$, and $\vec{c}_L^*$ are the reciprocal lattice vectors of the L-hexagonal Brillouin-zone. $\vec{k}_{15}$ coincides with the H point of the preceding zone and its corresponding invariance group is $\hat{G}(\vec{k}_{15}) = D_{3h}$. Hence, the star $\vec{k}_{15}^*$ has two branches: $\vec{k}_{15}^1 = -\vec{k}_{15}^2 = \left(\frac{4\pi}{3a}, 0, \frac{\pi}{c}\right)$. Since, on the other hand the small representation $\hat{\tau}_1$ of D_{3h}, which describes the permutation of atoms in the L-structure is the identity irreducible representation, one can construct the 2×2 matrices generating the irreducible representation $\tau_1(\vec{k}_{15}^*)$ of the D_{6h}^1 space-group:

$$\begin{array}{cccc} \{C_6|000\}\ \{I|000\} & \{C_1|3\vec{c}_L\} & \{C_1|\vec{a}_L\} & \{C_1|4\vec{a}_L\} \\ \begin{bmatrix} 0 & 1 \\ 1 & 0 \end{bmatrix} & \begin{bmatrix} -1 & 0 \\ 0 & -1 \end{bmatrix} & \begin{bmatrix} \varepsilon & 0 \\ 0 & \varepsilon^* \end{bmatrix} & \begin{bmatrix} 1 & 0 \\ 0 & 1 \end{bmatrix} \end{array} \tag{4.4}$$

with $\varepsilon = e^{2\pi i/3}$. Using the transformation properties of the matrices (4.4), one gets the possible symmetries of the phases corresponding to the two-component order-parameter (η_1, η_2) associated with $\tau_1(\vec{k}_{15}^*)$. Three symmetries are found: I) $D_{6h}^1 (V\times 6)$ for $\eta_1 = \eta_2 \neq 0$; II) $D_{6h}^4 (V\times 6)$ for $\eta_1 = -\eta_2 \neq 0$ and III) $D_{3h}^1 (V\times 6)$ for $\eta_1 \neq \eta_2 \neq 0$. The second solution corresponds to the six-fold unit-cell and space-group symmetry of the HCP structure. Thus, the virtual L→HCP transition is associated with the direction of the two-component order-parameter in the ε_2 space, given by $\eta_1 = -\eta_2 = \eta \neq 0$. Let us emphasize that this transition is of the standard Landau-type, the D_{6h}^4 (Z=2) symmetry being group-subgroup related to the latent L symmetry (D_{6h}^1 (Z=1/3)).

From a structural point of view the L-HCP transition involves a splitting of the initial one fold site 1(a) into three two-fold positions shown in Fig.3a: 2(b): $(000, 00\frac{1}{2})$, 2(c): $\pm(\frac{1}{3}\frac{2}{3}\frac{1}{4})$, and 2(d): $\pm(\frac{1}{3}\frac{2}{3}\frac{3}{4})$. As the concentration in the initial structure is x=1/3, after the transition one of the preceding positions has to be occupied by two $(1/3\times 6)$

atoms in the ordered HCP structure. As a result of close packing, the hard sphere system of atoms can occupy either positions 2(c) or 2(d), which are crystallographically equivalent (Fig.3b). Therefore the ordered crystal has a two-layered hexagonal close packed structure with the standard ratio for the unit-cell parameters $|\vec{c}_h|/|\vec{a}_h| = \sqrt{8/3} \cong 1.63$.

The wave vector $\vec{k}_{10}$, which is associated with the L-FCC translational relationship (4.3), is located on *the edge* of the hexagonal Brillouin zone of the L structure, along the K-H line. As its invariance group is $\hat{G}(\vec{k}_{10}) = C_{3v}$, the star $\vec{k}_{10}^*$ has *four* branches:

$$\vec{k}_{10}^1 = \left(\frac{4\pi}{3a}, 0, \frac{2\pi}{3c}\right) = -\vec{k}_{10}^2, \quad \vec{k}_{10}^3 = \left(\frac{4\pi}{3a}, 0, -\frac{2\pi}{3c}\right) = -\vec{k}_{10}^4.$$

From the identity irreducible representation of C_{3v} (which describes the permutation of atoms in the L-structure), one can construct the generating 4×4 matrices associated with $\tau_1(\vec{k}_{10}^*)$:

$$\begin{matrix} \{C_6|000\} & \{I|000\} & \{C_1|3\vec{c}_h\} & \{C_1|\vec{a}_h\} & \{C_1|4\vec{a}_h\} \\ \begin{bmatrix} 0 & 0 & 0 & 1 \\ 0 & 0 & 1 & 0 \\ 0 & 1 & 0 & 0 \\ 1 & 0 & 0 & 0 \end{bmatrix} & \begin{bmatrix} 0 & 1 & 0 & 0 \\ 1 & 0 & 0 & 0 \\ 0 & 0 & 0 & 1 \\ 0 & 0 & 1 & 0 \end{bmatrix} & \begin{bmatrix} 1 & 0 & 0 & 0 \\ 0 & 1 & 0 & 0 \\ 0 & 0 & 1 & 0 \\ 0 & 0 & 0 & 1 \end{bmatrix} & \begin{bmatrix} \varepsilon & 0 & 0 & 0 \\ 0 & \varepsilon^* & 0 & 0 \\ 0 & 0 & \varepsilon & 0 \\ 0 & 0 & 0 & \varepsilon^* \end{bmatrix} & \begin{bmatrix} \varepsilon & 0 & 0 & 0 \\ 0 & \varepsilon^* & 0 & 0 \\ 0 & 0 & \varepsilon^* & 0 \\ 0 & 0 & 0 & \varepsilon \end{bmatrix} \end{matrix} \qquad (4.5)$$

from which we can deduce the possible symmetries induced by $\tau_1(\vec{k}_{10}^*)$ and the equilibrium values of the corresponding four-component order-parameter, denoted $(\varsigma_1, \varsigma_2, \varsigma_3, \varsigma_4)$, which are summarized in Table 1. One can see in this Table that two solutions involve a three-fold multiplication of the L-unit-cell, in agreement with Eq.(4.3). They possess the space group C_{3v}^5 and D_{3d}^5. This latter solution actually coincides with the FCC symmetry, due to the specific crystallogeometrical conditions fulfilled by the L-unit-cell, which are: 1) a concentration x=1/3 of the L structure, 2) a ratio $|\vec{c}_L|/|\vec{a}_L| = \sqrt{2}$, corresponding to a close-packed structure.

In order to understand why the fulfillment of the preceding conditions lead to a coincidence of the D_{3d}^5 and O_h^5 symmetries let us clarify some crystallographic aspects of the $D_{6h}^1 \rightarrow D_{3d}^5$ mechanism. In this mechanism, the initial 1(a) position splits into the 1(a) and 2(c): $\pm(\frac{1}{3}\frac{1}{3}\frac{1}{3})$ positions. Since atoms occupy only one of the 1(a) positions in the rhombohedral structure (Fig.3b), they are localized at lattice nodes but not in sites located inside the unit-cell. This appears in Figs.4a and 4b, where the position denoted X, for example is empty. As a consequence, additional four-fold axes are created in the structure,

Table 1. Stable states associated with the irreducible representation $\tau_1(\vec{k}_{10})$. Columns (b), (c) and (d) give, for the seven states numbered in column (a), the equilibrium values of the four-component order-parameter, the corresponding space-group, and the multiplication of the L-unit-cell. $\varepsilon=\exp\{2i\pi/3\}$.

(a)	(b)				(c)	(d)
1	ς	ς	ς	ς	D^1_{6h}	9
2	ς_1	ς_2	ς_3	ς_4	C^1_{3v}	9
3	ς_1	ς_1	ς_2	$\varepsilon^*\varsigma_2$	D^3_{3d}	9
4	ς	ς	0	0	D^5_{3d}	3
5	0	0	ς_1	ς_2	C^5_{3v}	3
6	ς_1	ς_2	ς_1	ς_2	D^1_{3h}	9
7	ς_1	ς_2	ς_2	$\varepsilon^*\varsigma_2$	C^1_{6v}	9

which are connected with the six-fold axis of the latent structure by: $\left\{C^z_6|00c_L\right\}_L$ acting as $\left\{C^z_4|000\right\}_c$. This correspondence is illustrated in Fig.4a, which shows that a fourfold axis exists only if atoms of the X-type are absent. If, on the contrary, such atoms are present the four fold rotations would bring the X atom in position Y, which is an empty position.

The coincidence $D^5_{3d} \leftrightarrow O^5_h$ is also determined by the close packing condition $|\vec{c}_L|/|\vec{a}_L|=\sqrt{2}$ which implies a 60° angle between the vectors $\vec{a}_c$, $\vec{b}_c$ and $\vec{c}_c$. It is notorious that a rhombohedral lattice possessing equal angles α between its basic vectors, has the FCC cubic symmetry for α=60°. Let us stress that the preceding close packing

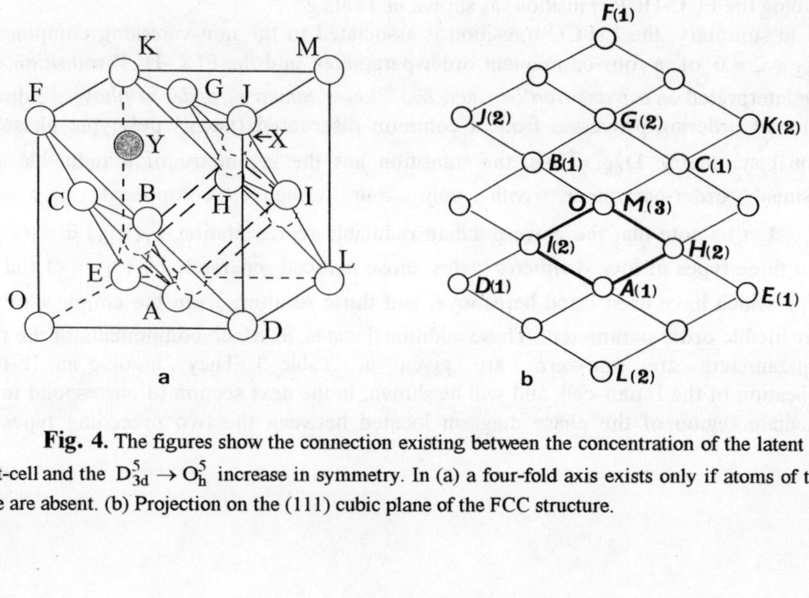

Fig. 4. The figures show the connection existing between the concentration of the latent phase unit-cell and the $D^5_{3d} \rightarrow O^5_h$ increase in symmetry. In (a) a four-fold axis exists only if atoms of the X-type are absent. (b) Projection on the (111) cubic plane of the FCC structure.

Table 2. Fulfillment of the close packing condition for elemental crystals undergoing the FCC-HCP transition.

	a_h (A°)		c_h (A°)		
	Hexagonal (experimental)	Cubic (recalculated from experimental)	Hexagonal (experimental)	Cubic (recalculated from experimental)	$\frac{c_h}{a_h}$
Tl	3.457	3.392	5.525	5.518	1.52
Pb	3.265	3.514	5.387	5.717	1.65
Fe	2.461	2.589	3.952	4.412	1.61
La	3.770	3.770	6.080	6.133	1.61
Ce	3.673	3.444	5.901	5.602	1.61
Pr	3.672	3.465	5.918	5.636	1.61
Nd	3.658	3.408	5.899	5.544	1.61
Yb	3.880	3.895	6.386	6.336	1.64
He	3.450	3.010	5.519	4.897	1.62
Co	2.505	2.516	4.089	4.093	1.63
Sm	3.618	3.621	5.830	6.880	1.61
Am	3.468	3.475	5.620	5.652	1.62

condition (which corresponds to $|\vec{c}_h|/|\vec{a}_h| \cong 1.63$) is well satisfied for the elements undergoing the FCC-HCP transition, as shown in Table 2.

In summary, the L-FCC transition is associated to the non-vanishing components $\zeta_1 = \zeta_2 = \zeta \neq 0$ of a four-component order-parameter, and the FCC-HCP transition can thus be interpreted as *a transition between two "low-symmetry" ordered phases*, induced by different ordering processes from a common disordered (latent) polytypic phase of hexagonal symmetry D^1_{6h}. Thus, the transition has the symmetry of a reducible six-dimensional order-parameter, with only four effective components $\eta_1 = -\eta_2$, $\zeta_1 = \zeta_2$. Let us note that the corresponding reducible representation $\tau_1(\vec{k}_{10}) \oplus \tau_1(\vec{k}_{15})$ leads to three types of low symmetry states: those induced separately by (η_1, η_2) and by the (ζ_i), which have been listed hereabove, and those resulting from the coupling of the two irreducible order-parameters. These additional states, in which components of the two order-parameters are non-zero, are given in Table 3. They involve an 18-fold multiplication of the L unit-cell, and will be shown, in the next section to correspond to an intermediate region of the phase diagram located between the two preceding types of states.

Table 3. Stable states resulting from the coupling of the irreducible representations $\tau_1(\vec{k}_{10})$ and $\tau_1(\vec{k}_{15})$. Column (b) gives the equilibrium values of the order-parameter components, whereas columns (c) and (d) provide the corresponding space-groups and multiplication of the L-unit-cell. ε=exp{2iπ/3}, ω=exp{iπ/3}.

(a)	(b)						(c)	(d)
1	ς_1	ς_2	ς_2	ς_1	η_1	η_1	C_{6v}^1	18
2	ς_1	ς_2	ς_1	ς_2	η_1	η_2	D_{3h}^1	18
3	ς_1	ς_2	ς_2	ς_1	η_1	$-\eta_1$	C_{6v}^4	18
4	ς_1	ς_1	$\varepsilon^*\varsigma_1$	$\varepsilon\varsigma_2$	η_1	$\omega^*\eta_1$	D_{6h}^4	18
5	ς_1	ς_1	$\varepsilon^*\varsigma_1$	$\varepsilon\varsigma_1$	η_1	$\varepsilon\eta_1$	D_{6h}^1	18
6	ς_1	ς_1	ς_2	$\varepsilon^*\varsigma_2$	η_1	$\omega^*\eta_1$	D_{3d}^3	18
7	ς_1	ς_1	ς_2	$\eta^*\varsigma_2$	η_1	$\varepsilon\eta_2$	D_{3d}^3	18
8	ς_1	ς_2	ς_3	ς_4	η_1	η_2	C_{3v}^1	18

2.2. Thermodynamic description: The phase diagrams of helium and praseodymium

As a verification of the thermodynamic implications of the FCC-HCP transition mechanism described hereabove, let us construct the theoretical diagram predicted for the latent L, FCC and HCP phases, and compare it to the experimental pressure-temperature phase diagram obtained for ^{4}He [12, 13], Praseodymium [14], Neodymium [15] and Lanthanum [16].

The integrity basis corresponding to the six-dimensional reducible representation $\tau_1(\vec{k}_{10}) \oplus \tau_1(\vec{k}_{15})$ is given in Table 4. It contains sixteen invariants. However, when taking into account the equilibrium relations corresponding to the two-component (η_1, η_2) and four-component $(\zeta_1, \zeta_2, \zeta_3, \zeta_4)$ order-parameters associated with the L-HCP and L-FCC transitions, respectively, which are $\eta_1 = -\eta_2 = \eta \neq 0$, and $\zeta_1 = \zeta_2 = \zeta \neq 0$, $\zeta_3 = \zeta_4 = 0$, one gets the effective potential:

$$F(\eta,\zeta) = a_1\zeta^2 + a_2\zeta^3 + a_3\zeta^4 + b_1\eta^2 + b_2\eta^4 + c\zeta^2\eta^2 \qquad (4.6)$$

which has been truncated to the fourth degree. The corresponding equations of state:

$$\zeta \cdot \left(2a_1 + 3a_2\zeta + 4a_3\zeta^2 + 2c\eta^2\right) = 0, \qquad (4.7)$$

$$\eta \cdot \left(2b_1 + 4b_2\eta^2 + 2c\zeta^2\right) = 0 \qquad (4.8)$$

can be solved following the examples discussed in Chapter I (Sec.4.4). It leads to different phase diagram topologies depending on the signs of c, a_2 and $\Delta = 4a_3b_2 - c^2$. Fig.5

Table 4. Integrity basis corresponding to the six-dimensional reducible representation $\tau_1(\vec{k}_{10}) \oplus \tau_1(\vec{k}_{15})$.

Degree	Invariants
2	$I_1 = \zeta_1\zeta_2 + \zeta_3\zeta_4;\ I_2 = \eta_1\eta_2$
3	$I_3 = \zeta_1^3 + \zeta_2^3 + \zeta_3^3 + \zeta_4^3$
4	$I_4 = \zeta_1\zeta_3\eta_2^2 + \zeta_2\zeta_4\eta_1^2;\ I_5 = \zeta_1\zeta_2\zeta_3\zeta_4$
5	$I_6 = \zeta_1\zeta_2(\zeta_3^3 + \zeta_4^3) + \zeta_3\zeta_4(\zeta_1^3 + \zeta_2^3);\ I_7 = \eta_1^2(\zeta_1^2\zeta_4 + \zeta_2^2\zeta_3)$
6	$I_8 = \zeta_1^3\zeta_3^3 + \zeta_2^3\zeta_4^3;\ I_9 = \zeta_1\zeta_3\eta_1^4 + \zeta_2\zeta_4\eta_2^4;\ I_{10} = \zeta_1^2\zeta_3^2\eta_1^2 + \zeta_2^2\zeta_4^2\eta_2^2;$ $I_{11} = \eta_1^6 + \eta_2^6;\ I_{12} = \zeta_1^3\zeta_4^3 + \zeta_2^3\zeta_3^3$
7	$I_{13} = \zeta_2\zeta_4\eta_1^2(\zeta_2^3 + \zeta_4^3) + \zeta_1\zeta_3\eta_2^2(\zeta_1^3 + \zeta_3^3);$ $I_{14} = \eta_1^4(\zeta_1\zeta_4^2 + \zeta_2\zeta_3^2) + \eta_2^4(\zeta_1^2\zeta_4 + \zeta_2^2\zeta_3)$
8	$I_{15} = \eta_1^2(\zeta_1^2\zeta_4^4 + \zeta_2^4\zeta_3^2) + \eta_2^2(\zeta_1^4\zeta_4^2 + \zeta_2^2\zeta_3^4)$
9	$I_{16} = \eta_1^2(\zeta_1^2\zeta_4^4 + \zeta_2^4\zeta_3^2) + \eta_2^2(\zeta_1^4\zeta_4^2 + \zeta_2^2\zeta_3^4)$

shows the phase diagram corresponding to $c>0$ and $\Delta<0$. It contains four phases denoted 0, I, II, III: 0 is the L-phase obtained for $\eta=0$, $\zeta=0$; I is the HCP phase, which verifies $\eta \neq 0$, $\zeta=0$; II is the FCC phase ($\eta=0$, $\zeta \neq 0$), and III is a six-layered structure (triple HCP) stabilized for $\eta \neq 0$, $\zeta \neq 0$.

In Fig. 5, the boundaries of the L phase with respect to the HCP and FCC phases

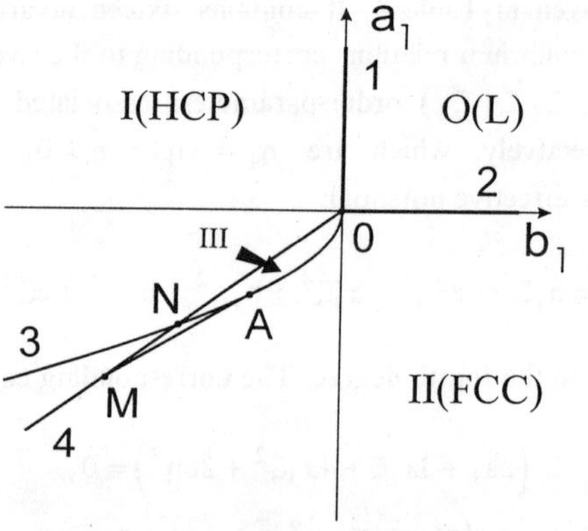

Fig. 5. Phase diagram corresponding to the potential (4.6) when $\Delta = 4a_3b_2 - c^2 < 0$ and $c>0$. Only the limit of stability lines are represented. The diagrams correspond to a positive value of ζ.

correspond to $b_1=0$ (line 0-1) and $a_1=0$ (line 0-2) respectively. The limit of stability of the HCP phase with respect to the FCC phase is also a straight line $a_1 = \frac{cb_1}{2b_2}$ (line 0-4), and the value of the HCP order-parameter within phase I is given by $\eta^2 = -\frac{b_1}{2b_2}$. The boundaries of the FCC phase II, are determined by the equations:

$$\zeta \cdot (3a_2 + 6a_3\zeta) > 0, \text{ and } c \cdot \zeta + b_1 > 0. \tag{4.9}$$

Therefore, for positive values of ζ, one has the situation assumed in Fig.5, in which the boundaries of phase II correspond to the parabolic branch 3-A-0, and to the $a_1=0$ line. In this case the stability region comprised between the lines 3-A-0 and 0-4 contains the domain of stability of phase III, and the line AM represents the limit of stability of phase III with respect to phase II.

The topology of the phase diagram is more complicated if the value of the FCC order-parameter ζ changes sign. Figs.6a and 6b (corresponding to $c>0$ and $c<0$ respectively) show that, in this case, there are two disjuncted regions of stability for the FCC phase, denoted II and II', which are separated by a forbidden (unstable) interval for the ζ values: $-\frac{3a_2}{4a_3} < \zeta < 0$, where a_2 and a_3 are assumed to be positive (i.e. taking $a_2 < 0$ is equivalent to changing the sign of ζ in our considerations). Fig.7 shows the phase diagram corresponding to the same conditions $c > 0$, $\Delta < 0$, but with two regions of stability

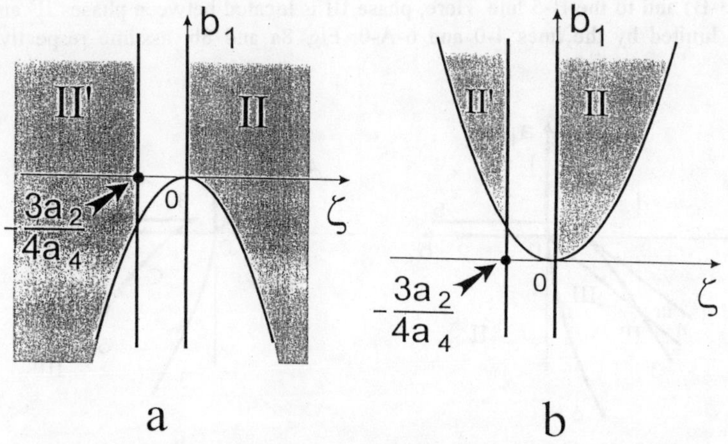

Fig. 6. Regions of stability of the FCC phase as a function of the ζ order-parameter for: (a) c>0 and (b) c<0.

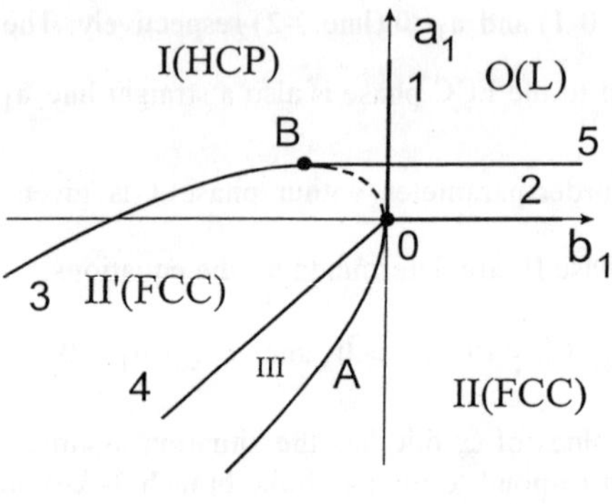

Fig. 7. Phase diagram corresponding to the potential (4.6), for $\Delta<0$ and $\Delta>0$, taking into account two regions of stability for ζ.

for the FCC phase, denoted II and II', which are separated by a forbidden (unstable) interval for the ζ values: $-\frac{3a_2}{4a_3}<\zeta<0$, where a_2 and a_3 are assumed to be positive (i.e. taking $a_2<0$ is equivalent to changing the sign of ζ in our considerations). Fig.7 shows the phase diagram corresponding to the same conditions $c>0$, $\Delta<0$, but with two regions of stability of the FCC phase: II ($\zeta>0$) and II' ($\zeta<-\frac{3a_2}{4a_3}$), and limited by the parabola $b_1=\pm c\zeta^2$. One can see that the boundaries of phase II' extend to another portion of the parabola (3-B) and to the B-5 line. Here, phase III is located between phases II' and II, in the region limited by the lines 4-0 and 6-A-0. Figs.8a and 8b, assume respectively the

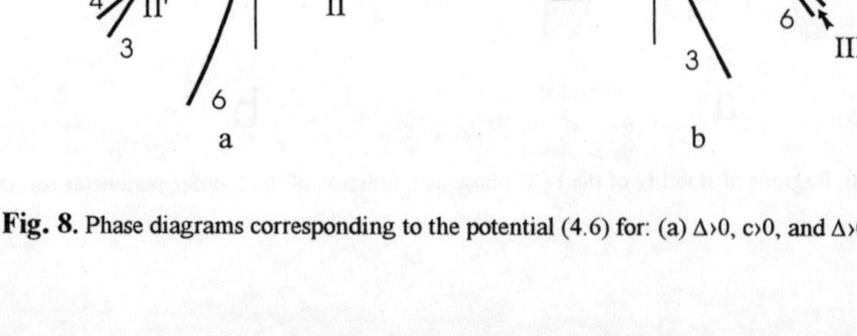

Fig. 8. Phase diagrams corresponding to the potential (4.6) for: (a) $\Delta>0$, $c>0$, and $\Delta>0$, $c<0$.

conditions ($\Delta>0$, $c>0$) and ($\Delta>0$, $c<0$). One can verify, that for such conditions phase III also splits into separated stability regions determined by different signs of ζ.

We can verify the reality of the preceding theoretical phase diagrams by comparison with the experimental phase diagrams obtained in a number of elements. The experimental points in Fig.9a have been obtained by Franck [12], and Franck and Daniels [13] for the FCC-HCP transition in ^4He, on cooling and heating up to 9 kbar. They can be compared to the metastability region located near the N point in Fig.5. Fig.9b reproduces the surrounding of this region, together with the pressure and temperature axes, used for the linear transformation of the three coefficients a_1, b_1 and a_2:

$$a_1 = \gamma_1 P + \delta_1 T + \varepsilon_1, \quad b_1 = \gamma_2 P + \delta_2 T + \varepsilon_2, \quad a_2 = \alpha a_1 + \beta b_1 + B. \tag{4.10}$$

It yields the analytical equations of the 3-N line:

$$a_1 = \left(3\beta c\tau^3 - 4a_3\tau^2 - 3B\tau\right)\cdot\left(2+\alpha\tau\right)^{-1}, \quad b_1 = c\tau^2 \tag{4.11}$$

where τ is a parameter. The 4-N line has the linear equation:

$$a_1 = \frac{cb_1}{2b_2} \tag{4.12}$$

Fig.9a shows the best fit obtained by the least squares method for Eqs.(4.11) and (4.12),

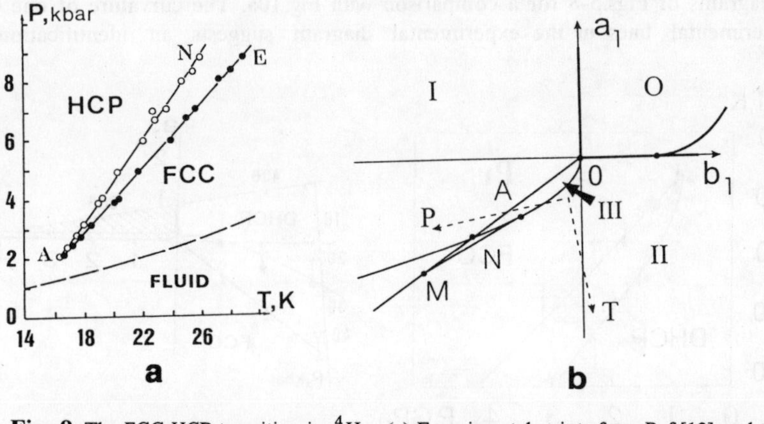

Fig. 9. The FCC-HCP transition in ^4He. (a) Experimental points from Ref.[12] and the best fit obtained from our calculations. The points are represented by small open circles on cooling and small solid circles on heating. (b) Details of the phase diagram of Fig.5 in the region of measurements. Dashed lines are the pressure-temperature coordinates used for the fit.

with the experimental points. The numerical values taken by the adjusting constants, assuming a dimensionless thermodynamic potential (4.6) and the normalizing condition $b_2=1$, are:

$$\alpha=-0.5,\ \beta=0.556,\ a_3=3.286,\ B=-0.648,\ c=3.76,\ \varepsilon_1=4.89,\ \varepsilon_2=-2.59,$$

$$\delta_1=-0.471\ K^{-1},\ \delta_2=0.159\ K^{-1},\ \gamma_1=-0.327\ \text{kbar}^{-1},\ \gamma_2=-0.942\ \text{kbar}^{-1}.$$

These values allow to evaluate the distance of the N point from the melting line, which is about 0.25 K, and the largest width of phase III, parallel to the T-axis, which is less than 0.15 K. The disordered polytype region in Fig.9b, lies outside from the experimental region of stability of the ^{4}He crystal.

The decrease of the hysteresis region between the FCC and HCP phases, with decreasing temperature and pressure, as shown in Fig.9a, illustrates the fact that reconstructive transitions of the reordering type can be close to second-order, i.e. they can take place with a small change in the probability of occupancy of the atoms between the ordered structures. This decrease cannot be attributed to specific microscopic properties of ^{4}He, and to the Nernst principle, as our model is general and free from any microscopic assumptions on the transition mechanism. This is confirmed by the phase diagrams of Pr, Nd and La, in which the reduction of the hysteresis between the FCC and HCP phases is observed when increasing T and P.

Fig.10a shows the hysteretic region located between the FCC and double-HCP (DHCP) phases found by Nikolaev and Khvostantsev in Praseodymium, below 5 GPa and 650 K [14]. In Sec.7 of this Chapter we show that the DHCP-FCC transition is described by the same expansion (4.6) than the HCP-FCC transition. Accordingly one can use the phase diagrams of Figs.5-8 for a comparison with Fig.10a. The curvature of one of the two experimental lines in the experimental diagram suggests an identification with

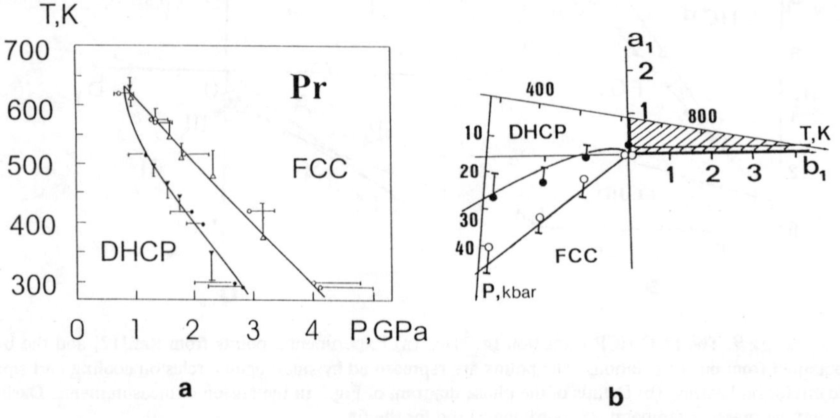

Fig. 10. (a) Phase diagram of praseodymium from Ref.[14]. (b) Best fit with the experimental points following our model. The hatched region corresponds to the disordered polytype structure.

the metastability region 3-B-0-4 of Fig.7. One needs here to express only the two coefficients a_1 and b_1 as linear functions of P and T:

$$a_1 = \gamma_1 P + \delta_1 T + \varepsilon_1, \quad b_1 = \gamma_2 P + \delta_2 T + \varepsilon_2$$

since we possess, in addition, equations (4.12) for the 0-4 line, and equation for the parabolic branch 0-B-3, which can be written:

$$a_1 = \frac{2a_3}{c} b_1 - \frac{3a_2}{2}\left(-\frac{b_1}{c}\right)^{1/2}. \tag{4.13}$$

The best fit between the calculated and experimental curves is shown in Fig.10b. It corresponds to the values:

$$a_3 = 1,\ b_2 = 2,\ a_2 = -0.8,\ c = 3,\ \varepsilon_1 = 1.85,\ \varepsilon_2 = -5.84,$$
$$\delta_1 = -0.00155\ \mathrm{K}^{-1},\ \delta_2 = 0.00948\ \mathrm{K}^{-1},\ \gamma_1 = -0.884\ \mathrm{kbar}^{-1},\ \gamma_2 = -0.093\ \mathrm{kbar}^{-1}.$$

It is interesting to note that the preceding fit predicts a disordered polytype structure (the L-phase) in an experimentally accessible region near above 650 K, and at atmospheric pressure, which has still not be explored by diffraction methods.

Figs.11a and 11b represent the experimental phase diagrams of Nd [15] and La [16], which display similar hysteretic regions between their DHCP and FCC structures. The description of the more strongly curved transition lines, obtained on heating and cooling between the preceding phases, requires to take into account six degree terms in η, in expansion (4.6). Let us finally note that the first-order character of the reconstructive reordering transitions in Pr, Nd and La, decreases with decreasing pressure, and becomes,

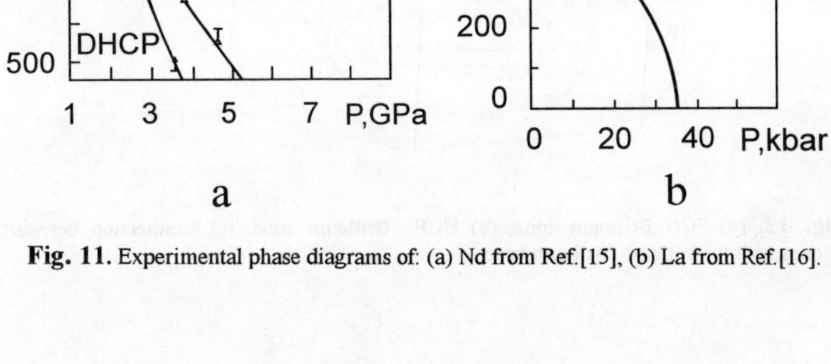

Fig. 11. Experimental phase diagrams of: (a) Nd from Ref.[15], (b) La from Ref.[16].

near the merging of the lines, close to second-order.

2.3. The phonon spectrum of cobalt

Let us show that another indirect confirmation of our theoretical model of the FCC-HCP transition, and of the underlying existence of a latent polytypic phase, can be found in the phonon spectra reported for cubic [17, 18] and hexagonal [19] cobalt. In order to disclose the structural relationship between the FCC and HCP structures following our model, let us represent the connection between the reciprocal lattices of the three FCC, HCP and L-phase structures. Figs. 12a-12c show the FCC (Fig. 12a) and HCP (Fig. 12b) Brillouin zones, and their embedding within the L hexagonal Brillouin zone (Fig. 12c). This latter figure reveals that the coinciding symmetry directions for the three zones are $\Gamma A_L = \Gamma L_c = 2\Gamma A_h$, and $\Gamma M_L \| \Gamma K_h \| \Gamma K_c$, where the indices L, c and h refer to the L, FCC and HCP phases, respectively.

Figs. 13a and 13b show the phonon spectra obtained for the HCP and FCC phases of Co in the preceding directions. Fig. 13c represents the superposition of the two

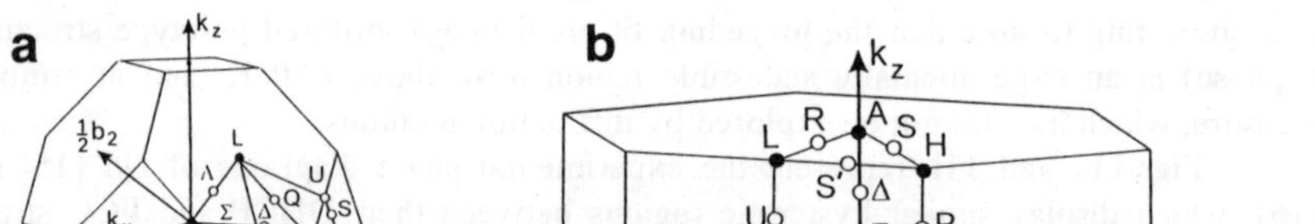

Fig. 12. (a) FCC Brillouin zone, (b) HCP Brillouin zone, (c) Connection between the preceding zones with the L-phase Brillouin zone.

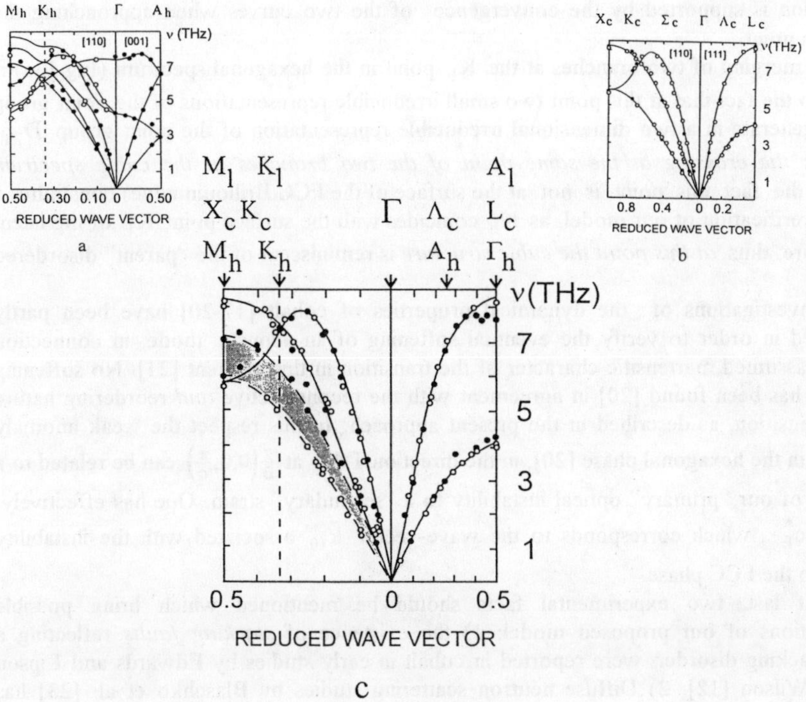

Fig. 13. Phonon spectra of: (a) hexagonal cobalt from Ref.[19], (b) cubic cobalt from Refs.[17,18]. (c) Superposition of the two preceding spectra. In (c), black and white dots correspond to hexagonal and cubic cobalt, respectively.

spectra, the black dots corresponding to hexagonal cobalt [19] and white dots to cubic cobalt [17, 18]. We can verify the following:

(1) In the direction ΓA_L there is a perfect coincidence for the black and white dots which lie on the same phonon branches, in agreement with the cell-doubling of the cubic unit-cell in this direction.

(2) Along the ΓKM reciprocal space direction there is a good agreement for two phonon branches, whereas for the third branch, the two structures give different curves. This discrepancy can be explained by the fact that experiments for FCC and HCP cobalt were performed at temperatures differing by about 700 K [17-19], and by the observation by Frey et al. [20] that the slope of this branch increases when temperature decreases. This

explanation is supported by the convergence of the two curves when approaching the transition point.

(3) The merging of two branches at the K_h point in the hexagonal spectrum (Fig.13a) is related to the fact that at this point two small irreducible representations of the point group C_{2v} degenerate in a two dimensional irreducible representation of the point group D_{3h}. However *the crossing at the same point of the two branches in the cubic spectrum* (despite the fact this point is not at the surface of the FCC Brillouin zone) constitutes a striking verification of our model, as K_h coincides with the surface point K_L of the latent L-structure: thus, *at this point the cubic structure* is reminiscent of the "parent" disordered phase.

Investigations of the dynamical properties of cobalt [17-20] have been partly performed in order to verify the eventual softening of an acoustic mode, in connection with the assumed martensitic character of the transition in this element [21]. No softening behavior has been found [20] in agreement with the reconstructive *and* reordering nature of the transition, as described in the present approach. In this respect the weak anomaly reported in the hexagonal phase [20], in the direction ΓA_L at $\frac{2}{3}\left(0,0,\frac{\pi}{c}\right)$ can be related to a coupling of our "primary" optical instability to a "secondary" strain. One has effectively: $\frac{2}{3}c_h^* = \frac{1}{3}c_L^*$, which corresponds to the wave-vector $\vec{k}_{10}$ associated with the instability leading to the FCC phase.

At last, two experimental facts should be mentioned which bring possible confirmations of our proposed model: 1) the existence of *stacking faults* reflecting a strong stacking disorder, were reported in cobalt in early studies by Edwards and Lipson [8] and Wilson [12]. 2) Diffuse neutron scattering studies by Blaschko et al. [23] has revealed the presence of satellites around the (111) FCC reflections, within a temperature range where the FCC and HCP phases coexist. This result was interpreted [23] as the occurrence of an *intermediate six-layered structure* where nuclei of the HCP phase are coherently inserted into the FCC matrix. One may suggest other alternate interpretations in the framework of a (non-martensitic) reordering mechanism. Thus, the detected six-layered structure, may coincide either with the six-layered phase III predicted between the HCP and FCC phases (Fig.5 and Fig.7), or with the complex mixture of FCC and HCP phases shown in Fig.8a. The preceding data confirm the importance of stacking faults and structural disorder in the FCC-HCP mechanism. The different types of stacking faults, which are predicted in the framework of the reordering FCC-HCP mechanism, are described in the following section.

2.4. Stacking faults, antiphase domains, and incomplete ordering

Let us analyze the domain structure resulting from our description of the FCC-HCP transition as a transition between two, differently ordered, structures generated from a disordered polytype structure. At the L-HCP (D_{6h}^1 - D_{6h}^4) transition, the point-group symmetry is not modified, but one has a six-fold decrease of the translational symmetry, expressed by Eq.(4.2). Therefore *antiphase domains* are created, which transform one

Table 5. Antiphase domains corresponding to the L-HCP transition. The meaning of the columns is given in the text. $\omega=\exp\{i\pi/3\}$, $\varepsilon=\exp\{2i\pi/3\}$.

(a)	(b)	(c)	(d)
1	$\eta\ -\eta$	CBCB...	$\{C_1\|000\}$
2	$-\eta\ \eta$	BCBC...	$\{C_1\|\vec{c}_L\}$
3	$\varepsilon\eta\ \omega\eta$	ABAB...	$\{C_1\|\vec{a}_L\}$
4	$\omega^*\eta\ \varepsilon^*\eta$	BABA...	$\{C_1\|\vec{a}_L+\vec{c}_L\}$
5	$\varepsilon^*\eta\ \omega^*\eta$	ACAC...	$\{C_1\|-\vec{a}_L\}$
6	$\omega\eta\ \varepsilon\eta$	CACA...	$\{C_1\|-\vec{a}_L+\vec{c}_L\}$

into another by the translations lost at the transition. Table 5 lists the equilibrium values of the two component order-parameter (η_1,η_2) for each of the six types of antiphase domains (column (b)), as well as the possible sequences of layers corresponding to each type of domain (column (c)). For example, the first domain ($\eta, -\eta$) implies a layer sequence of the type CBCB... whereas the domain ($-\eta$, η) yields the sequence BCBC.... The symmetry operators which transform the antiphase domains denoted 1 into the other given by (4.4), are listed in column (d). Let us note that the atoms belonging to the domains forming the pairs (1, 2), (3, 4), and (5, 6) in Table 5, correspond to the crystallographic positions 2(c) and 2(d).

The domain *texture* resulting from the existence of antiphase domains, must be compatible with the atomic close packing. This requirement excludes certain contacts between domains , that would bring, for example, sequences of the type AA, BB, or CC, as such type of contacts would increase the energy of the domain walls. Consequently, there exist a limited number of allowed contacts between the domains. Using the standard notation [25] which labels h (hexagonal) an hexagonal layer having an identical surrounding (e.g. B in ABA) and c (cubic) an hexagonal layer having a non-symmetric surrounding (e.g. B, in ABC), one finds only *two* sorts of contacts: ...hhhchhh... and ...hhhcchhh... which both correspond to *deformation stacking faults* in the current classification [24].

At the L-FCC (D_{6h}^1 - O_h^5) transition, the underlying change in point-group symmetry (D_{6h} - D_{3d}) and the corresponding threefold multiplication of the initial unit-cell, expressed by Eq.(4.3), lead to two different types of domains: (1) *Two* orientational domains which transform into one another by the lost six-fold rotation. (2) *Three* antiphase domains. Table 6 lists the *six* different types of domains, with respect to the equilibrium values of the four-component order-parameter (ς_i). In column (c), one can find the layer sequences associated with each domain, e.g. ACBACB... for (ζ, ζ, 0, 0) and BACBAC... for ($\varepsilon\zeta$, $\varepsilon^*\zeta$, 0, 0). Column (d) indicates the symmetry operations transforming the first domain into the others.

As for the L-HCP transition, the contacts between the domains in the FCC structure must preserve the atomic close packing. At variance with the HCP structure, one

Table 6. Antiphase and orientational domains corresponding to the L-FCC transition. The meaning of the columns is given in the text. $\varepsilon=\exp\{2i\pi/3\}$.

(a)	(b)	(c)	(d)
1	$\zeta\ \zeta\ 0\ 0$	ACBACB...	$\{C_1 \vert 000\}$
2	$\varepsilon\zeta\ \varepsilon^*\zeta\ 0\ 0$	BACBAC...	$\{C_1 \vert \vec{b}_L\}$
3	$\varepsilon^*\zeta\ \varepsilon\zeta\ 0\ 0$	CBACBA...	$\{C_1 \vert -\vec{a}_L\}$
4	$0\ 0\ \zeta\ \zeta$	ABCABC..	$\{\sigma_z \vert 000\}$
5	$0\ 0\ \varepsilon\zeta\ \varepsilon^*\zeta$	CABCBA...	$\{\sigma_z \vert \vec{a}_L\}$
6	$0\ 0\ \varepsilon^*\zeta\ \varepsilon\zeta$	BCABCA...	$\{\sigma_z \vert 2\vec{a}_L\}$

finds here *two types of stacking faults*: (1) deformation types stacking faults which result for example from the contact between domains denoted 1 and 3 in Table 6. It gives the layer stacking ACBABACB... which corresponds to ...ccchhccc... . (2) *Twining (growth) stacking faults* corresponding to the contacts between domains 1 and 4, for example. The layer sequence, in this case is ACBABCABC... which gives ...ccchccc... .

In summary, in the framework of the approach proposed for the FCC-HCP transition [25], and more generally for transitions between close-packed structures, the domain structure and the structure of the domain walls show that *these structures are intrinsically faulted.* The corresponding stacking faults are symmetry induced and temperature independent, i.e. the stacking faults cannot be annihilated by external fields (ageing, annealing,...), as we deal with a probabilistic mechanism with no conjugated field, that may lead to a single domain structure.

Another type of, *temperature dependent*, defects can be found in close packed structures, which originates from the *non-maximal* character of the order-parameter, which is inherent to ordering-type reconstructive transitions. In the disordered polytype structure each close packed layer corresponds to one stacking fault. In the ordering mechanism to close-packed phases one can characterize the transition process by the number:

$$\Delta = 1 - \frac{N_d}{N} \tag{4.14}$$

where N is the total number of close packed layers, and N_d the number of stacking faults. N_d/N represents the concentration of defects (stacking faults). Δ=0 in the disordered polytype phase and Δ=1 in the ideal fully ordered close-packed phases. Intermediate states, will correspond to $0<\Delta<1$. The value of Δ at all temperatures and pressures is determined by, and gives an estimation of , the number of defects. Far from the transition, within the close-packed phases, the asymptotic value of Δ will reflect the symmetry induced type of stacking faults, whereas close to the transition Δ will account also for temperature dependent defects. In this respect it is interesting to note that using similar considerations, Edwards and Lipson [8] measure, near the melting line of cobalt, a concentration of stacking faults of about 0.6, a number which reflects a strong tendency to the formation of a disordered polytype structure.

3. Polytypic structures in zinc sulfide and barium titanate

3.1. The wurtzite-sphalerite transition in ZnS

The wurtzite-sphalerite type of reconstructive transition occurs in a number of sulfides (MnS, ZnS, CdS, Ga_2S_3,...) and selenides (MnSe, CdSe,...). Among these compounds ZnS is known for the variety of its polytypic structures [3]. When cooling ZnS from the melt, its high-temperature wurtzite type structure (C_{6v}^4, Z=2) undergoes a transition at about 1024 °C to a cubic sphalerite type structure (T_d^2, Z=2), which is never fully accomplished, the final product being a polytype "microtwinned" structure, containing about 10% of stacking faults [26].

Let us show that the disordered polytype latent phase model formulated for the FCC-HCP transition, applies with only slight modification to the transition in ZnS. Figs.14a and 14b represent the hexagonal and cubic structures of ZnS. In both structures the large anions S are close packed and occupy positions 2(b):($\frac{1}{3}$ $\frac{2}{3}$ 0) in the wurtzite structure, whereas in the sphalerite structure they are in positions 1(a). The relatively small atoms Zn are respectively in positions 2(b):($\frac{1}{3}$ $\frac{2}{3}$ $\frac{3}{8}$) and ($\frac{1}{12}$ $\frac{1}{12}$ $\frac{1}{12}$) in the two preceding structural modifications. In the unit-cells shown in Figs.14a and 14b, the maximal substructure common to the hexagonal and cubic structure is represented. It is formed by a monolayer structure of symmetry C_{6v}^4, in which the anions are in position 1(a), and the cations in positions (0 0 $\frac{3}{4}$). As the concentration of this substructure is x=1/3, i.e. the two ions Zn and S are randomly distributed, occupying one among three possible positions in

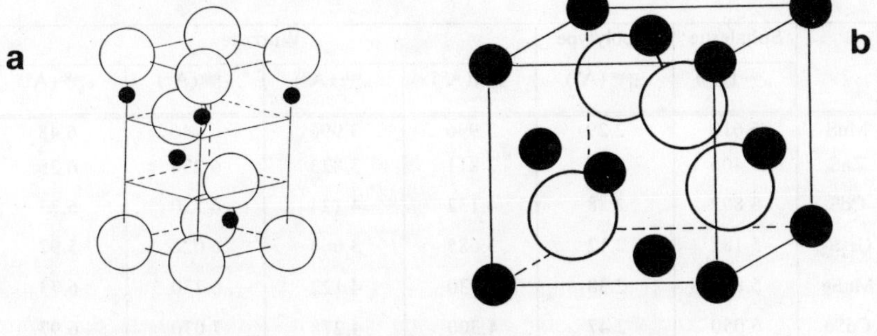

Fig. 14. The wurtzite (a) and sphalerite (b) type of structures of ZnS.

each layer, we deal with a disordered polytypic structure, which plays the same role as the L-phase in the FCC-HCP transition.

The same relationship (4.1) exist between the wurtzite and sphalerite basic unit-cell vectors, and the relations (4.2) and (4.3) also hold between, respectively, the cubic or hexagonal unit-cells, and the "latent" polytypic unit-cell. These relations, which derive from the close packed character of the phases, imply the following ratios between the lengths of the lattice parameters:

$$a_c = a_L\sqrt{6}, \quad a_h = a_L\sqrt{3}, \text{ and } c_h = 2c_L = 2\sqrt{2}a_L \tag{4.15}$$

and *between* the hexagonal and cubic lattice parameters:

$$a_h = \frac{a_c}{\sqrt{2}}, \text{ and } c_h = \frac{2a_c}{\sqrt{3}}. \tag{4.16}$$

Table 7 shows that for most of the compounds undergoing the wurtzite-sphalerite transition, that the experimental values reported for their lattice parameters fulfill the preceding ratios. It means that the FCC-HCP model applies, without adding any important secondary macroscopic deformation.

These preliminary considerations show that the wurtzite-sphalerite transition can be described as a transition between two low-symmetry differently ordered structures, both derived from a common parent disordered polytypic structure. The group theoretical formalism remains the same as for the FCC-HCP transition, but *the symmetries of the structures are lowered.* Thus, the two- and four-dimensional irreducible representations (denoted $\tau_1(\vec{k}_{15})$ and $\tau_1(\vec{k}_{10})$) of the parent space group C_{6v}^4, induce respectively the $C_{6v}^4(Z=1/3) \rightarrow C_{6v}^4(Z=2)$ six-fold wurtzite structure, and the $C_{6v}^4(Z=1/3) \rightarrow T_d^2(Z=1)$ threefold sphalerite structure. The effective equilibrium values of the corresponding two-

Table 7. Experimental values for the lattice parameters of the compounds listed in the first column. The calculated values use the ideal ratios (4.15) and (4.16).

	Sphalerite	L-polytype	Wurtzite			
	a_c^{exp} (A°)	a_L^{calc} (A°)	a_h^{exp} (A°)	a_h^{calc} (A°)	c_h^{exp} (A°)	c_h^{calc} (A°)
MnS	5.610	2.29	3.996	3.996	6.446	6.48
ZnS	5.406	2.21	3.811	3.823	6.234	6.25
CdS	5.892	2.38	4.132	4.124	6.730	6.73
Ga_2S_3	5.182	2.12	3.685	3.664	6.028	5.97
MnSe	5.830	2.38	4.130	4.122	6.730	6.73
CdSe	6.050	2.47	4.300	4.278	7.070	6.97

and four-component order-parameter are $(\eta,-\eta)$ and $(\zeta, \zeta, 0, 0)$ respectively, the form of the thermodynamic potential (4.6), and of the phase diagrams of Figs.5-8, remaining unchanged.

Note that the virtual transition (non-polar→polar) between the two hexagonal polytype forms: D_{6h}^1 (Z=1/3) $\rightarrow C_{6v}^4$ (Z=1/3) corresponds to a shifting along the six-fold axis of magnitude $c_L/4$, which brings the Zn cations in the center of tetrahedra formed by S anions. The same shifting mechanism lowers the FCC and HCP symmetries to T_d^2 (Z=1) and C_{6v}^4 (Z=2). Note also that ZnS provides a clear-cut verification of the description of stacking faults given in Sec.2.4 as one observes the two sorts of stacking faults (deformation and twinning types), under the form of various polytypic sequences on the one hand, and of microdomains, on the other hand [27]. These symmetry-induced stacking faults never disappear completely, and tend asymptotically to 10% of the sphalerite low-temperature structure [27]. By contrast, the temperature dependent type of defects, related in our approach to the ordering parameter (4.14), are increased by cycling across the trabsition, and on the contrary, reduced by annealing far from the transition temperature [27].

3.2. The cubic-hexagonal transition in $BaTiO_3$

In addition to the classical sequence of Landau phases which take place at lower temperatures, barium titanate undergoes at about 1460°C a reconstructive transition between the cubic perovskite type structure (O_h^1, Z=1) and an hexagonal close packed modification (D_{6h}^4, Z=6). We will show that this transition can be described, provided some additions, in a similar way than the FCC-HCP transition [28].

The cubic $BaTiO_3$ structure, which is represented in Fig.15a is composed by a packing of oxygens and barium ions, of almost equal sizes, and in some octahedral coordinated interstitials, by small titanium ions. The radii of the atoms are: R_i=1.36 A° for O^{2-}, R_i=1.38 A° for Ba^{2+}, R_i=0.6 A° for Ti^{4+}, which fulfill the standard relation $R_{int}=0.41 R_i$, characterizing close packed structures. In hexagonal $BaTiO_3$, which is shown in Fig.15b, the large ions are also close packed, and form a six-layered structure, somehow analogous to a triple HCP structure, although the existence of two types of large ions produces a specific in-layer structure.

According to the preceding description, cubic and hexagonal $BaTiO_3$ are close packed structures, respectively formed by *three and six hexagonal close packed layers.* The maximal subcell common to the cubic and hexagonal unit-cells, which is shown in Figs.15a and 15b, is a monolayer unit-cell of volume:

$$V_L = \frac{V_c}{3} = \frac{V_h}{18} \tag{4.17}$$

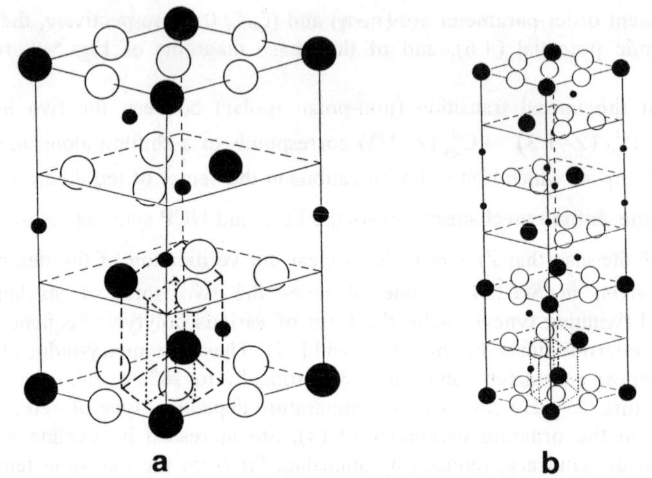

Fig. 15. Three-layered cubic (a) and six-layered hexagonal (b) structures of $BaTiO_3$.

corresponding to the D^1_{6h} space group, and a number of formula unit 1/12. As atoms occupy randomly 3 equivalent positions in each layer, this maximal substructure pertains to a disordered polytype structure (L). However at variance with the preceding cases considered in this Chapter, we will not use this L-structure as a parent structure for the cubic and hexagonal phases of $BaTiO_3$ but use another polytype structure (L'), having the same symmetry D^1_{6h}, but 1/3 formula unit in the unit-cell. It is obtained from the L-structure by in-layer ordering (the stacking of layers remains disordered) with respect to the oxygen and barium ions. More precisely, the Ba and O ions that where in identical positions 1(a) in the L-polytype, will occupy the positions 1(a) and 3(f), respectively, in the L' polytype. The L-L' partial ordering mechanism, corresponds to a 3-dimensional irreducible representation of the D^1_{6h} space group, denoted $\tau_1(\vec{k}_{12})$ where $\vec{k}_{12}=(\frac{4\pi}{3a},\frac{2\pi}{3a},0)$ is the M point of the hexagonal Brillouin zone. The matrice generators of this representation which can be constructed from Kovalev's tables [29], are:

$$\{C_1|\vec{a}_L\} \quad \{C_1|\vec{b}_L\} \quad \{C_6|000\} \quad \{U_y|000\}$$

$$\begin{bmatrix}-1&0&0\\0&1&0\\0&0&-1\end{bmatrix}\quad\begin{bmatrix}-1&0&0\\0&-1&0\\0&0&1\end{bmatrix}\quad\begin{bmatrix}0&0&1\\1&0&0\\0&1&0\end{bmatrix}\quad\begin{bmatrix}1&0&0\\0&0&1\\0&1&0\end{bmatrix}. \tag{4.18}$$

For equal equilibrium values of the corresponding three-component order-parameter: $\xi_1 = \xi_2 = \xi_3 = \xi \neq 0$, one has a quadrupling of the L unit-cell, with the new L' basic translations:

$$\vec{a}_{L'} = 2\vec{a}_L, \quad \vec{b}_{L'} = 2\vec{b}_L, \quad \vec{c}_{L'} = \vec{c}_L \tag{4.19}$$

and a space-group symmetry D^1_{6h}, with 1/3 formula per unit-cell.

Starting from the *parent* L'-polytypic structure, the phenomenological description of the cubic-hexagonal transition in $BaTiO_3$ can be formulated in terms of two *interdependent* ordering processes. In order to make this point clear, let us first consider the connections between the lattice vectors of the L' and hexagonal $BaTiO_3$ unit-cells. One has:

$$\vec{a}_h = 2\vec{a}_{L'} + \vec{b}_{L'}, \quad \vec{b}_h = \vec{a}_{L'} + \vec{b}_{L'} \quad \vec{c}_h = 6\vec{c}_{L'}. \tag{4.20}$$

It corresponds to an *eighteen-fold* multiplication of the L'-unit-cell, associated with the wave-vector:

$$\vec{k}'_{10} = \tfrac{1}{3}(\vec{a}^*_{L'} + \vec{b}^*_{L'}) + \tfrac{1}{6}\vec{c}^*_{L'} \tag{4.21}$$

located on the same P-line of the hexagonal Brillouin-zone than the wave-vector $\vec{k}_{10}$ considered for the FCC-HCP mechanism. Accordingly the 4×4 generating matrices, associated with the irreducible representation $\tau_1(\vec{k}'_{10})$ differ from the matrices (4.5) only by the matrix corresponding to the translation along the six-fold axis, which is here:

$$\{C_1|\vec{c}_{L'}\} \equiv \begin{bmatrix} \omega & 0 & 0 & 0 \\ 0 & \omega^* & 0 & 0 \\ 0 & 0 & \omega^* & 0 \\ 0 & 0 & 0 & \omega \end{bmatrix} \quad \text{with } \omega = e^{2\pi i/6}. \tag{4.22}$$

On the other hand, one has the same relationship (4.3) between the primitive cubic $BaTiO_3$ lattice vectors and the L' primitive translations, which correspond to the same wave-vector: $\vec{k}_{10} = \frac{1}{3}(\vec{a}^*_{L'} + \vec{b}^*_{L'} + \vec{c}^*_{L'})$ than for the L-FCC transition. This similarity is due to the fact that the cubic primitive structure of $BaTiO_3$ can be understood as an *ordered FCC* structure, i.e. when O and Ba are not distinguished the O^1_h symmetry of $BaTiO_3$ transforms into O^5_h. Accordingly the same irreducible representation $\tau_1(\vec{k}_{10})$, given by the generating matrices (4.5), induces the $L' \rightarrow O^1_h$ transition.

The preceding considerations show that the cubic and hexagonal symmetries in $BaTiO_3$ can be obtained from the 8-dimensional *reducible* representation

Table 8. Low-symmetry states associated with the reducible representation $\tau_1(\vec{k}_{10}) \oplus \tau_1(\vec{k}'_{10})$. Column (b): relationship between the order-parameter components. Column (c): corresponding space groups. Column (d): multiplication of the D^1_{6h} unit-cell.

(a)	(b)								(c)	(d)
1	η	η	0	0	0	0	0	0	D^5_{3d}	3
2	η	η	η	η	0	0	0	0	D^1_{6h}	9
3	0	0	η_1	η_2	0	0	0	0	C^5_{3v}	9
4	η	η	η	η	ζ	ζ	ζ	ζ	D^1_{6h}	18
5	η	$\varepsilon^*\eta$	$\varepsilon^*\eta$	η	ζ	$\omega^*\zeta$	$\omega^*\zeta$	ζ	D^1_{6h}	18
6	η_1	η_2	η_2	η_1	0	0	0	0	C^1_{6v}	9
7	η_1	η_2	η_1	η_2	0	0	0	0	D^1_{3h}	9
8	η	η	η	η	ζ	$-\zeta$	ζ	$-\zeta$	D^4_{6h}	18
9	η	η	0	0	0	0	ζ	ζ	D^5_{3d}	6
10	η_1	η_1	η_2	η_2	0	0	0	0	D^3_{3d}	9
11	η	η	η	η	ζ	ζ	$-\zeta$	$-\zeta$	D^4_{6h}	18
12	η	$\varepsilon^*\eta$	0	0	0	0	ζ	$\omega\zeta$	D^5_{3d}	6
13	η_1	η_2	η_2	η_1	ζ_1	ζ_2	ζ_2	ζ_1	C^1_{6v}	18
14	η_1	η_2	η_1	η_2	ζ_1	ζ_2	ζ_1	ζ_2	D^1_{3h}	18
15	η_1	η_1	η_2	η_2	ζ_1	ζ_1	ζ_2	ζ_2	D^3_{3d}	18
16	η_1	η_2	$\varepsilon^*\eta_1$	$\varepsilon\eta_2$	ζ_1	ζ_2	$\omega^*\zeta_1$	$\omega\zeta_2$	D^1_{3d}	18
17	η_1	η_1	$\varepsilon^*\eta_2$	$\varepsilon\eta_2$	ζ_1	$\omega^*\zeta_1$	ζ_2	$\omega\zeta_2$	D^3_{3d}	18
18	η_1	η_2	0	0	0	0	ζ_1	ζ_2	C^5_{3v}	6
19	η_1	η_2	0	0	0	0	0	0	C^1_{3v}	9
20	η_1	η_2	η_2	η_1	ζ_1	ζ_2	$-\zeta_2$	$-\zeta_1$	C^4_{6v}	18
21	η_1	η_2	η_3	η_4	ζ_1	ζ_2	ζ_3	ζ_4	C^1_{3v}	18

$\tau_1(\vec{k}_{10}) \oplus \tau_1(\vec{k}'_{10})$, for specific equilibrium values of the corresponding 8-component order-parameter ($\eta_1,\eta_2,\eta_3,\eta_4,\zeta_1,\zeta_2,\zeta_3,\zeta_4$). Table 8 gives the full set of low symmetry groups induced by the preceding representation. One can verify that the hexagonal structure of $BaTiO_3$, which coincides with solution 8 of Table 8, is associated with non-zero values of the components pertaining to the sets η_i and ζ_j, namely:

$$\eta_i = \eta \neq 0 \ \ (i{=}1\text{-}4), \ \ \zeta_1 = -\zeta_2 = \zeta_3 = -\zeta_4 = \zeta \neq 0 . \qquad (4.23)$$

Note that the solution numbered 11 in Table 8, also leads to an eighteen-fold D^4_{6h} structure, but one can show that it corresponds to an ordering of atoms, different than the one found in hexagonal $BaTiO_3$.

The only three-fold multiplication of the L' unit-cell in Table 8, corresponds to solution 1, and to the change in symmetry: D^1_{6h} (Z=1/3) $\rightarrow$ D^5_{3d} (Z=1) associated with the

equilibrium values of the order-parameter components:

$$\eta_1 = \eta_2 = \eta \neq 0, \quad \eta_3 = \eta_4 = 0, \quad \zeta_i = 0 \quad (i=1\text{-}4). \tag{4.24}$$

Using similar arguments than for the L-FCC transition (Sec.2.1), one can identify the rhombohedral D^5_{3d}(Z=1) symmetry to the primitive cubic O^1_h symmetry of $BaTiO_3$, due to the following crystallogeometrical constraints fulfilled by the L' cubic structure: a ratio $a_{L'} / c_{L'} = 1/\sqrt{2}$, which corresponds to a close packed structure, or equivalently an angle of 90° between the cubic lattice vectors $\vec{a}_c$, $\vec{b}_c$ and $\vec{c}_c$; 2) a 1/3 formula per unit-cell in the L' structure.

Let us work out the theoretical phase diagram containing the L', cubic and hexagonal phases. From the matrices (4.5) and (4.22) one can construct the order-parameter invariants, as well as the coupling invariants between the η_i and ζ_j. When restricting to the maximal degrees 4 and 6, respectively for the η_i and ζ_j components, and to the lowest degree coupling, one gets the following invariants (in complex order-parameter coordinates):

$$I_1 = \eta_1\eta_2 + \eta_3\eta_4, \quad I_2 = \zeta_1\zeta_2 + \zeta_3\zeta_4, \quad I_3 = \eta_1^3 + \eta_2^3 + \eta_3^3 + \eta_4^3,$$
$$I_4 = \eta_1\eta_2\eta_3\eta_4, \quad I_5 = \zeta_1\zeta_2\zeta_3\zeta_4, \quad I_6 = I_2 \cdot I_5, \quad I_7 = I_2^3,$$
$$I_8 = (\eta_1 + \eta_2 + \eta_3 + \eta_4)(\zeta_1 - \zeta_2 + \zeta_3 - \zeta_4)^2. \tag{4.25}$$

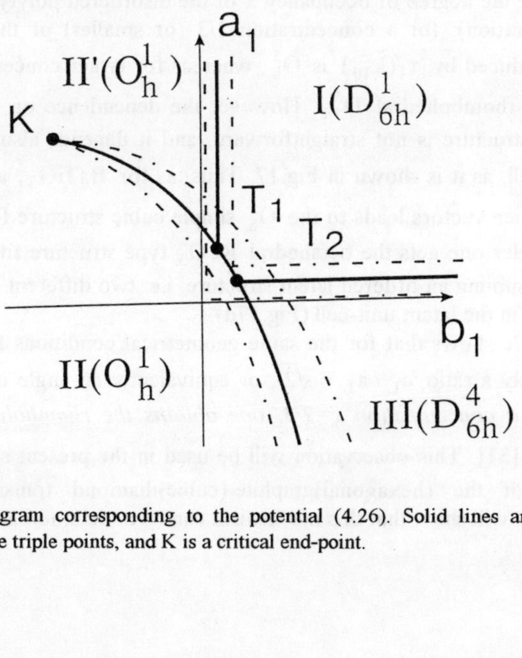

Fig. 16. Phase diagram corresponding to the potential (4.26). Solid lines are first-order transition lines. T_1 and T_2 are triple points, and K is a critical end-point.

When assuming *the same magnitudes* η *and* ζ for the η_i and for the ζ_j (i,j=1-4), respectively, a condition which is realized in both equilibrium conditions (4.23) and (4.24), one gets the effective thermodynamic potential:

$$F(\eta,\zeta) = a_1\eta^2 + a_2\eta^3 + a_3\eta^4 + b_1\zeta^2 + b_2\zeta^4 + b_3\zeta^6 + c\eta\zeta^2 \qquad (4.26)$$

which has been studied in Chapter I (Sec.4.4.3). The corresponding phase diagram worked out by Gufan et al. [30], is shown in Fig.16. It contains three stable phases: phase I for $\eta=\zeta=0$ (the L' phase), phase II for $\eta \neq 0$, $\zeta=0$ (the cubic phase of $BaTiO_3$), and phase III, for $\eta \neq \zeta \neq 0$ (the hexagonal phase of $BaTiO_3$). All the phases are separated by first-order transition lines which merge at two triple points T_1 and T_2. Besides, there exist two distinct regions of stability of the cubic phases (II and II') which are separated by a line of isostructural transition, ending at a critical point K.

4. The graphite-diamond reconstructive transition

4.1. Degree of occupancy of the disordered polytype cell, and symmetry of ordered cubic structures

In the phenomenological approach to the FCC-HCP transition, and in the derivative mechanisms proposed for the transitions in ZnS and $BaTiO_3$, an essential parameter appears to be the degree of occupancy x of the disordered polytype unit-cell by the atoms (the concentration): for a concentration 1/3 (or smaller) of the polytype structure the symmetry induced by $\tau_1(\vec{k}_{10})$ is O_h^5, whereas for larger concentrations, let say 1/2, the symmetry is rhombohedral D_{3d}^5. However the dependence on x of the preceding "low symmetry" structure is not straightforward, and it depends also on the *geometry* of the latent unit-cell, as it is shown in Fig.17. Thus, as for $BaTiO_3$, an angle of 90° between the cubic lattice vectors leads to the O_h^1 simple cubic structure for x=1/12 (Fig.17b). For the same angles one gets the octahedral ReO_3 type structure for x=1/4 (Fig.17c), or for x=1/3 but assuming an ordered latent structure, i.e. two different crystallographic positions are occupied in the latent unit-cell (Fig.17d).

Fig.17e shows that for the same geometrical conditions fulfilled in the FCC-HCP model, namely a ratio $c_L / a_L = \sqrt{2}$, or equivalently an angle of 60° between $\vec{a}_c$, $\vec{b}_c$, $\vec{c}_c$, *but for a concentration x=2/3, one obtains the rhombohedral graphite structure* (D_{3d}^5, Z=2) [31]. This observation will be used in the present section, as a basis for the description of the (hexagonal)graphite-(cubic)diamond transition mechanism. More precisely, we will show that this mechanism can be understood as a transition between two

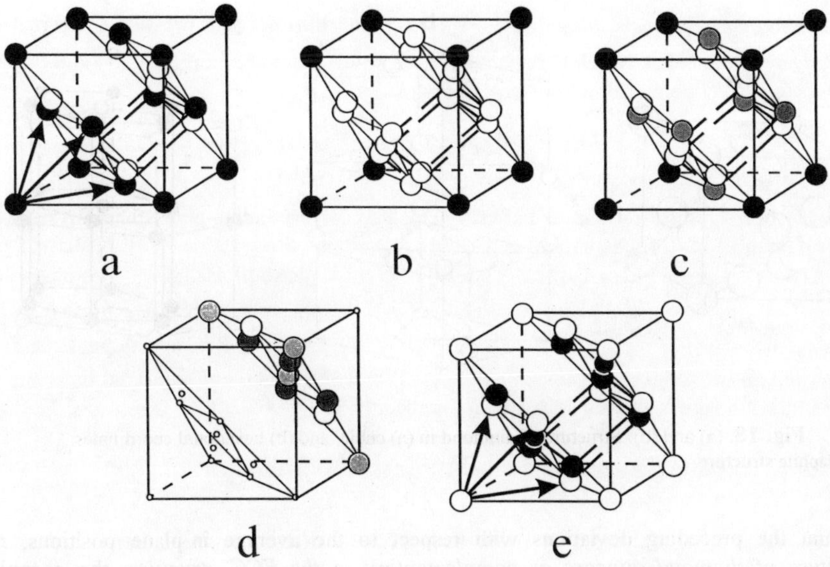

Fig. 17. Relation between the symmetry of the cubic structure and the crystallogeometrical parameters (angle between the cubic lattice vectors, concentration of the L-unit-cell). Comments on figures (a)-(e) are in the text.

ordered phases, which are deduced from a parent polytypic disordered phase, by combined ordering and shifting mechanisms for the diamond structure, and by ordering plus decompression for the graphite structure [25].

4.2. Crystallogeometrical properties of hexagonal graphite and cubic diamond

The diamond structure corresponds to the space group O_h^7 with two atoms in the primitive (rhombohedral) unit-cell, and eight atoms in the conventional cubic cell (Fig.18a). The cubic parameter is a_c=3.567 A°, and the minimal distance between atoms is 1.445 A° [32]. Although the packing of carbon atoms is compact, one cannot speak of close packing, as the structure is covalently bonded. Fig.18b shows the diamond structure in hexagonal configuration. The lattice parameters of the hexagonal cell (which is three times larger than the rhombohedral cell) are a_d=2.52 A°, and c_d=6.3 A°, with six atoms per unit cell. The structure is three-layered, as the FCC structure, but *the layers are not flat*: they are composed by atoms half of which are shifted up, and half shifted down, with respect to the average atomic planes, by about 0.26 A°. The c_d / a_d ratio is 2.50 which differs by less than 2% from the ideal FCC ratio of $\sqrt{6} \cong 2.45$. So, if we do not take into

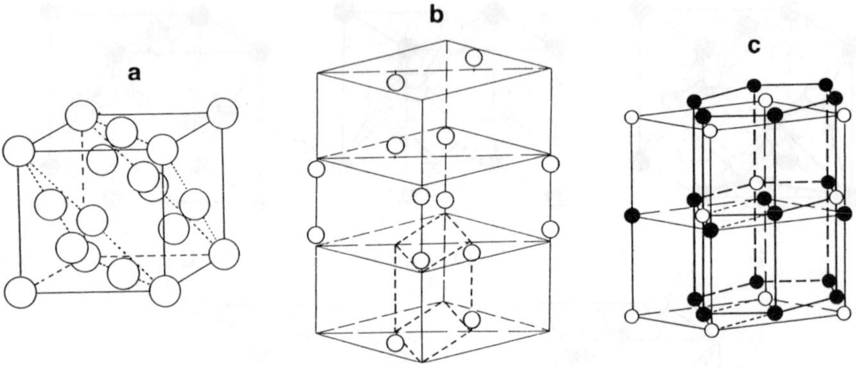

Fig. 18. (a) and (b): Structure of diamond in (a) cubic, and (b) hexagonal coordinates. (c) Graphite structure.

account the preceding deviations with respect to the average in-plane positions, *the structure of diamond appears as complementary to the FCC structure*: the occupied positions in diamond correspond to empty position in a FCC structure, and vice versa.

The structure of graphite, shown in Fig.18c, has the space group D_{6h}^4 with four atoms per unit-cell. The hexagonal parameters are a_g =2.46 A° and c_g =6.7076 A° [33]. Hence, one has less than 2% difference between the a_g and a_d parameters: the (001) hexagonal planes of the two structures (or the (111) cubic plane) are nearly identical. Along the same line, the two-layered graphite structure can also be interpreted as complementary to the HCP structure: the occupied positions in graphite are empty in HCP and vice versa.

Aside the mentioned similarities between the graphite and diamond structures, there exists a considerable difference in the interlayer distance for the two structures: d_d =2.1 A° in diamond, and d_g =3.39 A° in graphite. This property plays an essential role in the description of the graphite-diamond transition mechanism, which is proposed in the next section [25].

4.3. The graphite-diamond transition mechanism

Figs.19a and 19b represent respectively one unit-cell of hexagonal graphite and a tripled unit-cell of rhombohedral graphite. The maximal substructure common to these two structures is represented in the two figures: it is formed by a monolayer structure of

Table 9. Mechanisms involved in the transition from the latent structure (D_{6h}^1 , Z=2/3) to the hexagonal graphite (b), diamond (c), and rhombohedral graphite (d) structures.

(a) Structure / Mechanism	(b) Hexagonal Graphite (D_{6h}^4 , Z=4)	(c) Diamond (O_h^7 , Z=2)	(d) Rhombohedral Graphite (D_{3d}^5 , Z=4)
Ordering $\tau_1(\bar{k}_{15})$	●		
Ordering $\tau_1(k_{10})$		●	●
Shifting $\tau_1(k_{10})$		●	
Decompression $\tau_1(k_{16})$	●		●

symmetry D_{6h}^1 , with a 2/3 occupancy of the unit-cell, i.e. it is complementary to the maximal substructure common to the HCP and FCC structures, and also represents a statistically disordered polytype structure. As in the preceding sections, we will use this latent structure as a parent phase (denoted L) for the graphite and diamond phases.

Columns (b) and (c) of Table 9 summarize the irreducible representations and mechanisms, which are respectively involved in the $L \rightarrow$ graphite and $L \rightarrow$ diamond

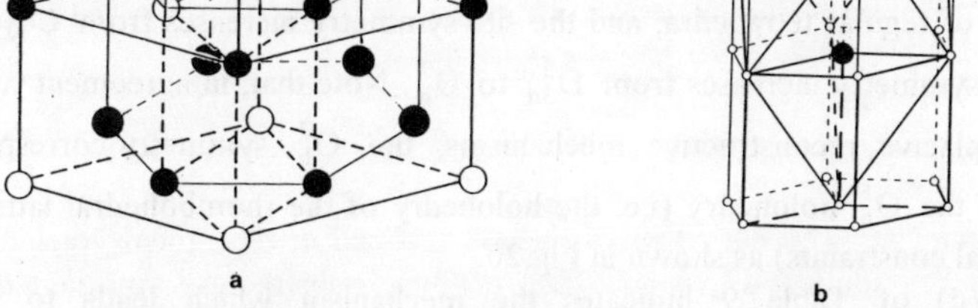

Fig. 19. (a) One unit-cell of hexagonal graphite. (b) Tripled unit-cell of rhombohedral graphite. The L-maximal substructure to the two preceding structures, is represented in dashed line.

transitions. One can see that the same wave-vector $\vec{k}_{15} = \frac{1}{3}(\vec{a}_L^* + \vec{b}_L^*) + \frac{1}{2}\vec{c}_L^*$ (the H point) and $\vec{k}_{10} = \frac{1}{3}(\vec{a}_L^* + \vec{b}_L^* + \vec{c}_L^*)$, of the hexagonal L-Brillouin zone, which have been used for the FCC-HCP transition, take place for the graphite and diamond structures. From Table 9, one can give the following description:

1) The hexagonal graphite structure results from an ordering mechanism, L(D_{6h}^1,Z=2/3) → D_{6h}^4(Z=4), associated with a two-component order-parameter (η_1, η_2) transforming as $\tau_1(\vec{k}_{15})$. For the equilibrium values $\eta_1 = -\eta_2 = \eta$, an hexagonal six-fold structure of symmetry D_{6h}^4, with four atoms per unit-cell, is formed, which represents a compressed structure with respect to hexagonal graphite, as the ratio $c_h / a_h \cong 1.63$ (since $c_L / a_L = \sqrt{2}$) differs from $c_g / a_g = 2.73$. Accordingly, an additional decompression along Z is necessary, corresponding to an "improper" strain e_{zz}, which transforms as the totally symmetric irreducible representation ($\tau_1(\vec{k}_{16})$) of the D_{6h}^1 space group, at the center (Γ-point) of the hexagonal Brillouin zone. Let us stress that while e_{zz} is not a symmetry breaking order-parameter (i.e. it couples to η via the "improper" coupling $\eta^2 e_{zz}$), it can not be considered as a "secondary" order-parameter as it produces an enlargement of the interlayer distance by a factor 1.67.

2) The diamond structure is obtained by correlated ordering *and* shifting mechanisms from the L-structure. The two mechanisms transform as the same four-component order-parameter ζ_i (i=1-4), which has the symmetry of $\tau_1(\vec{k}_{10})$, for the same equilibrium values $\zeta_1 = \zeta_2 = \zeta \neq 0$, $\zeta_3 = \zeta_4 = 0$, i.e. they are bilinearly coupled in a "pseudoproper" coupling. The ordering process leads to a three-layered rhombohedral structure of symmetry D_{3d}^5(Z=2) which has a tripled unit-cell with respect to the L structure. However, this phase plays the role of an intermediate Landau state in the framework of our approach. To realize the diamond structure, one needs additional antiparallel shifts of the carbon atoms along the [111] rhombohedral directions (the [001] hexagonal directions), which transform the flat layer stacking into the framework-type diamond structure. For the specific displacements of magnitude c_L/8≅0.26 A°, the atoms reach the centers of regular tetrahedra, and the site symmetry increases from C_{3v} to T_d, while the overall symmetry increases from D_{3d}^5 to O_h^7. Note that, in agreement with our description of displacive reconstructive mechanisms, the O_h^7 symmetry corresponds to a special point of the O_h^5 holohedry (i.e. the holohedry of the rhombohedral lattice with special geometrical constraints) as shown in Fig.20.

Column (d) of Table 9 indicates the mechanism which leads to another polymorphic form of carbon: rhombohedral graphite, which has been found by Lipson and Stokes [31]. It is obtained by the same ordering mechanism ($\tau_1(\vec{k}_{10})$) than diamond, but requires an additional decompression e_{zz} ($\tau_1(\vec{k}_{16})$) as hexagonal graphite. The

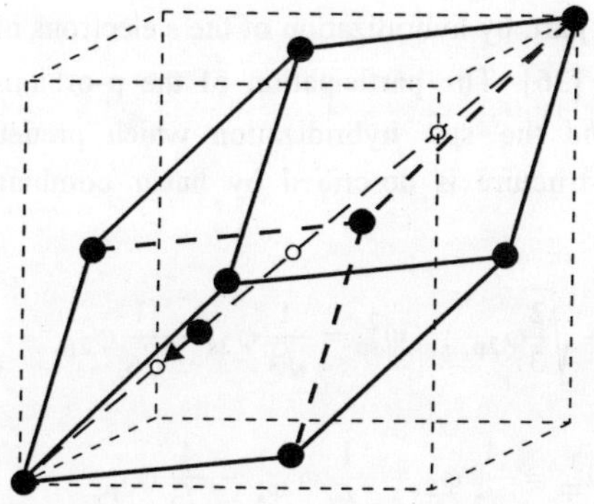

Fig. 20. Holohedry of $D_{3d}^5(O_h^5)$ and special point corresponding to the increase in symmetry to O_h^7.

disordered wurtzite-type structure of hexagonal diamond, synthesized by Bundy and Kasper [34], involves different combined mechanisms of ordering and shifts, and is not present in Table 9. Let us emphasize, that in our presentation of the mechanisms, we have chosen arbitrarily to start from a "compressed" latent structure for the L-polytype. Starting from a "decompressed" polytype structure would lead to describe the transitions to the hexagonal and rhombohedral graphite structures, by simple ordering mechanisms. By contrast the L$\rightarrow$diamond transition mechanism would require an additional compression e_{zz}.

To conclude, let us stress that an intensive effort has been made, during the last decades, to understand the nature of the graphite-diamond transition mechanism, and also to reproduce this transition to a metastable state, in laboratory. Our description, which provides a crystallographic description of the mechanism, and the symmetry of the corresponding order-parameter, suggests that the transition should take place more easily through the energetically advantageous path, passing by the rhombohedral graphite structure.

4.4. Changes in the chemical bonds at the graphite-diamond transition: Physical nature of the order-parameter

Following Buerger's classification [35], the graphite-diamond transition is a typical representative example of "bond-type" transformation, i.e. a reconstructive transition involving a disruption of the bonds between nearest neighbor atoms. In order to illustrate this diagnostic, and to connect it to our proposed mechanism, let us investigate the modifications of chemical bonding which occur at the graphite-diamond transition. With that goal, we can start from the well known facts that the carbon atoms have the electronic ground state $1s^2 2s^2 2p_x 2p_y$ and that in the chemical bonding, the excited states

$1s^2 2s2p_x 2p_y 2p_z$ also take part, by hybridization of the s electrons of one atom with the p electrons of its neighbors [36]. The participation of the p-orbitals give their covalent character to the bonds, and the sp^2 hybridization which predetermines the layered character of the graphite structure is described by linear combinations of the atomic functions [36]:

$$\psi^1_{sp} = \frac{1}{\sqrt{3}}\varphi_{2s} + \sqrt{\frac{2}{3}}\varphi_{2p_x}; \quad \psi^2_{sp} = \frac{1}{\sqrt{3}}\varphi_{2s} - \frac{1}{\sqrt{6}}\varphi_{2p_x} + \frac{1}{\sqrt{2}}\varphi_{2p_y};$$

$$\psi^3_{sp} = \frac{1}{\sqrt{3}}\varphi_{2s} - \frac{1}{\sqrt{3}}\varphi_{2p_x} - \frac{1}{\sqrt{2}}\varphi_{2p_y}. \tag{4.27}$$

The p_z-orbitals are symmetry independent from the p_x and p_y orbitals, and their effect on the hybridization of orbitals determines the interlayer bonding, and thus the interlayer distance. In contrast, the effect of the p_x and p_y orbitals fix the interatomic distance *within* the layers. In diamond, one has three equivalent pairs of atoms producing an hybridization of the sp^3 type, which corresponds to a linear combination of the basic functions [36]:

$$\psi^1_{sp} = \frac{1}{2}(\varphi_{2s} + \varphi_{2p_x} + \varphi_{2p_y} + \varphi_{2p_z}); \quad \psi^2_{sp} = \frac{1}{2}(\varphi_{2s} + \varphi_{2p_x} - \varphi_{2p_y} - \varphi_{2p_z});$$

$$\psi^3_{sp} = \frac{1}{2}(\varphi_{2s} - \varphi_{2p_x} + \varphi_{2p_y} - \varphi_{2p_z}); \quad \psi^4_{sp} = \frac{1}{2}(\varphi_{2s} - \varphi_{2p_x} - \varphi_{2p_y} + \varphi_{2p_z}). \tag{4.28}$$

From (4.28) one can deduce that the electron density distribution has maxima in the directions of the apexes of tetrahedra, a fact which determines a framework type of structure. Besides as the s-basis functions have a spherical distribution, the basis functions (4.28) have transformation properties coinciding with that of the p-orbitals, which transform as *bivectors* [36]. We can therefore image the effect of the electron density distribution as a set of bivectors localized in the positions of the carbon atoms.

According to the preceding considerations, one can represent the chemical bonds in the parent polytype structure, as shown in Fig.21a. Due to the orientational disorder and to the six-fold symmetry of the L-structure, the bivectors form a six-fold planar star, each branch of the star being oriented in the direction of the nearest neighbor in-layer position. In addition, bivectors will be directed along the six-fold axis. Thus there will be two sets of orthogonal bivectors forming the basis of the irreducible representation $\tau_1(\vec{k}_{15})$.

Fig.21b shows the equilibrium configuration of ordered bivectors in hexagonal graphite. One can verify that no mixture of p_z and (p_x, p_y) orbitals is needed. The corresponding sp^2 hybridization determines the layered structure of hexagonal graphite. By contrast the ordered diamond structure, shown in Fig.21c corresponds to a sp^3 type of hybridization, in which the mixed p_x, p_y and p_z orbitals form the basis of the irreducible representation $\tau_1(\vec{k}_{10})$ and result in a non-layered structure, of the framework type. Let us stress, in this respect, that the transition which transforms the disordered polytype into the diamond structure can be equivalently interpreted by three distinct physical mechanisms: an ordering or a shifting mechanism, or a change in the electronic distribution. All three mechanisms possess the symmetry of the irreducible representation $\tau_1(\vec{k}_{10})$, and only a complete set of measurements may allow to disclose which of them plays the leading role.

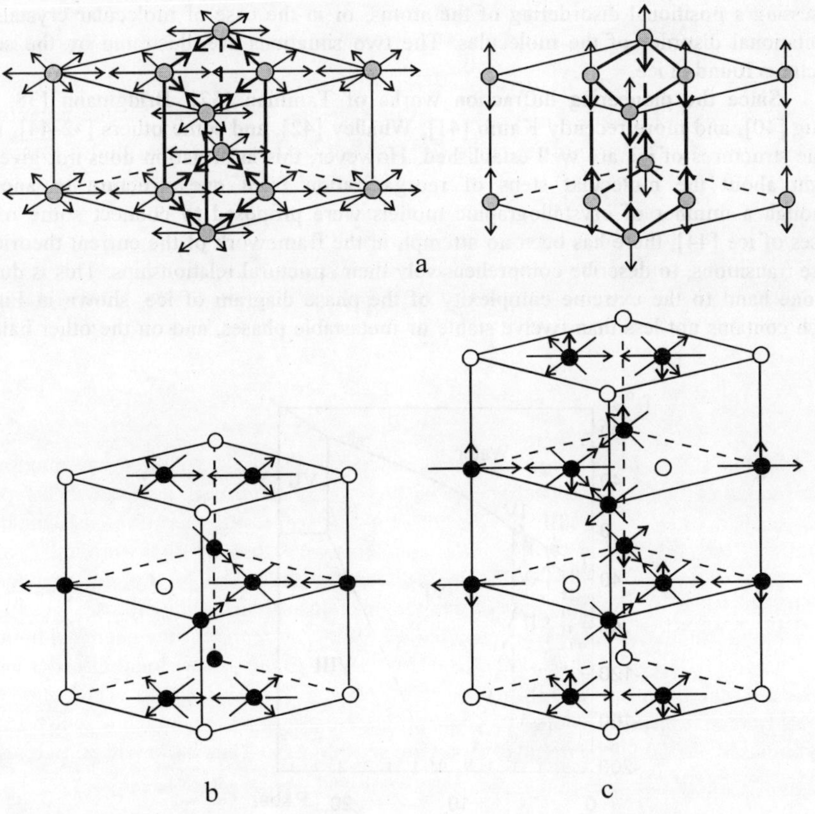

Fig. 21. Distribution of p-orbitals represented by bivectors (a) in the L-structure, (b) in hexagonal graphite, (c) in diamond.

5. Theory of ice structures

5.1. The phase diagram of ice: The parent I_0 structure

In the preceding examples, we have considered reconstructive transitions between close packed, or almost close packed, structures, and we have shown that a disordered polytype phase could be assumed as a latent (virtual) parent structure for understanding the connections between the observed structures. For reconstructive transitions between non-close packed structures, a similar approach can be used when the parent phase is not present in the phase diagram, i.e. one can take as a parent structure, the maximal substructure common to the structures surrounding the reconstructive transition line. However in this case the latent phase will not be a disordered polytypic phase, but a phase possessing a positional disordering of the atoms, or in the case of molecular crystals, an orientational disorder of the molecules. The two situations are illustrated by the set of structures found in ice.

Since the pioneering diffraction works of Tamman [37], Bridgmann [38, 39], König [40], and more recently Kamb [41], Whalley [42], and some others [42-44], most of the structures of ice are well established. However, this information does not give any insight about the molecular steps of reorganization from one structure to another. Although a number of crystallographic models were proposed to connect some of the phases of ice [44], there has been no attempt, in the framework of the current theories of phase transitions, to describe comprehensively their structural relationships. This is due on the one hand to the extreme complexity of the phase diagram of ice, shown in Fig.22, which contains not less than twelve stable or metastable phases, and on the other hand to

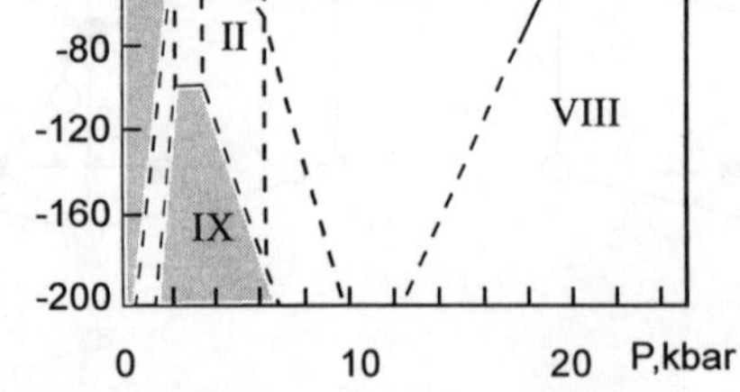

Fig. 22. Pressure-temperature phase diagram of ice.

the presupposed opinion that, as the transitions between the phases are of reconstructive type, their mechanisms cannot be conveniently accounted by a standard phenomenological or microscopic scheme.

In this section, we will show, on the contrary, that at least nine of the ice structures can be understood within the same unified approach. More precisely, we will describe the phases of ice as the result of ordering and displacive mechanisms from *a single parent disordered BCC structure possessing different fractional concentration of water molecules*.

In order to introduce the assumed parent structure in this approach let us consider the normal hexagonal I_h (ice I) phase, which is found in the natural environment. The oxygen framework of its tridymite-type structure (D_{6h}^4, Z=4) is represented in Fig.23. It consists of puckered hexagonal layers of water molecules forming hexagonal rings [45]. The four oxygen atoms contained in one unit-cell are in crystallographic positions 4(f), the tetrahedral coordination of each molecule shifting the oxygen atoms, with respect to the average positions of the layers along Z, in two opposite directions by $\vec{c}_h/16$, where $\vec{c}_h$ is the hexagonal lattice vector. In Fig.23 a substructure of I_h corresponding to a BCC structure (O_h^9) is indicated by dashed lines. Its unit-cell is filled by 2/3 water molecule, i.e. the oxygen atoms are randomly distributed among the sites. We will use this disordered structure, denoted hereafter I_0, to describe the formation of ice I_h as the result of a virtual I_0 to I_h phase transition. Furthermore, I_0 will be shown to be *the maximal substructure common to all ice structures*, provided different concentrations of water molecules are used, and accordingly will be taken as the common parent phase for describing the formation of all the considered ice phases.

5.2. Ordering of the oxygen atoms: The transition mechanisms to the I_h, I_c, ice III and ice VI phases

The mechanism transforming I_0 into I_h can be described in two correlated steps, which are illustrated in Fig.24: an ordering of the oxygen atoms which fill the 4(f)

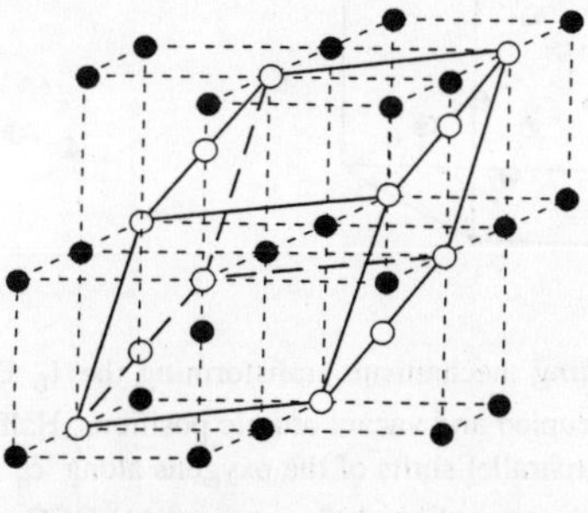

Fig. 23. Oxygens in the unit-cell of I_h (thick lines) and I_0 (small cubes).

positions, with vacancies in 2(a) positions, and an antiparallel displacement of two pairs of atoms by $5\vec{c}_h/48$ and $11\vec{c}_h/48$, which is shown in Fig.24b, within the $(0\bar{1}1)$ BCC plane. Figs.24c and 24d give the corresponding distribution of hydrogen atoms within the I_0 and I_h unit-cells, in which they occupy respectively the positions 16(f) and 12(k)+4(f), with a probability of occupancy 1/6 and 1/3, respectively. The preceding coupled mechanism is accompanied by a large decompression of about 27% along the six-fold hexagonal axis.

In the group-theoretical formalism, the ordering and shifting of the oxygens, which constitute the primary transformation mechanism is associated with a single eight-dimensional irreducible representation of the parent O_h^9 space group, denoted $\tau_1(\vec{k}_7)$ in Kovalev's notations [29]. In terms of the reciprocal BCC lattice vectors, $\vec{k}_7 = \frac{1}{6}(\vec{a}_1^* + \vec{a}_2^* + \vec{a}_3^*)$. It corresponds in direct space to the following relationships between the I_h and I_0 unit-cell basic vectors:

$$\vec{a}_h = \vec{a}_1 - \vec{a}_2, \quad \vec{b}_h = \vec{a}_2 - \vec{a}_3, \quad \vec{c}_h = 2(\vec{a}_1 + \vec{a}_2 + \vec{a}_3). \tag{4.28}$$

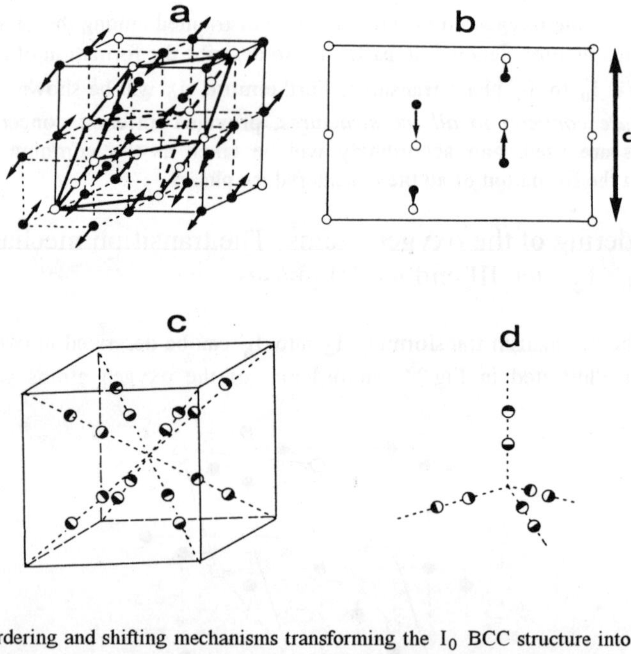

Fig. 24. Ordering and shifting mechanisms transforming the I_0 BCC structure into I_h. Full and open circles are, respectively, occupied and vacant atomic positions. Half filled circles correspond to an occupancy 1/2 of the sites. (a) Antiparallel shifts of the oxygens along $\vec{c}_h$. Long and short arrows are two types of shifts. (b) Projection of the preceding shifts on the (011) BCC plane. (c) and (d) Hydrogen distributions around oxygen sites within the I_0 and I_h structures, respectively.

Let us now show that the formation of the I_c structure of ice, which is stable at very low pressure and low temperatures, can be understood as the result of an analogous ordering process from the I_0 structure. Fig.25a indicates in dashed line the I_0 unit-cell within the cubic cristobalite-type structure of I_c (space-group O_h^7, Z=2) [46]. The three axes spanning the BCC primitive unit-cell coincide with the axes of the conventional FCC unit-cell of I_c, with a doubling of the three corresponding lattice parameters. The relationship between the primitive cubic lattice parameters of the I_c and I_0 unit-cells is:

$$\vec{a}_c = 2\vec{a}_1 + \vec{a}_2 + \vec{a}_3, \quad \vec{b}_c = \vec{a}_1 + 2\vec{a}_2 + \vec{a}_3, \quad \vec{c}_c = \vec{a}_1 + 2\vec{a}_3. \tag{4.29}$$

Thus, the number of water molecules per unit-cell in I_0 is 1/2. The molecules occupy in I_c the 2(a) crystallographic positions, the vacancies filling the positions 2(b). The hydrogens occupy the positions 8(e), and their distribution is given in Fig.25b. Since the symmetries of I_0 and I_c are related by a group-subgroup relationship, the $I_0 \to I_c$ ordering transition is described by a standard Landau-type approach: It is induced by the two-dimensional irreducible representation, denoted $\tau_1(\vec{k}_{10})$ [29], of the O_h^9 space group, at the P point ($\vec{k}_{10} = \frac{1}{4}(\vec{a}_1^* + \vec{a}_2^* + \vec{a}_3^*)$) of the BCC Brillouin-zone surface.

Ice III, which is made from ice I_h by raising up the pressure to above 2070 bars below -22°C, has a tetragonal structure [47] (D_4^4, Z=12) usually described from ice I_h. It can also be deduced from the I_c structure by a simultaneous rotation of the eight oxygen atoms belonging to the spirals which surround the four-fold cubic axes, and by a shifting of the four remaining oxygens in the plane perpendicular to these axes. Fig.26a shows the connection between I_0 and Ice III. The relationship between the unit-cell basic vectors is:

$$\vec{a}_{III} = 2(\vec{a}_2 + \vec{a}_3), \quad \vec{b}_{III} = 2(\vec{a}_1 + \vec{a}_3), \quad \vec{c}_{III} = 2(\vec{a}_1 + \vec{a}_2) \tag{4.30}$$

and the I_0 unit-cell contains 3/4 of disordered oxygen atoms, which order in the sixteen-

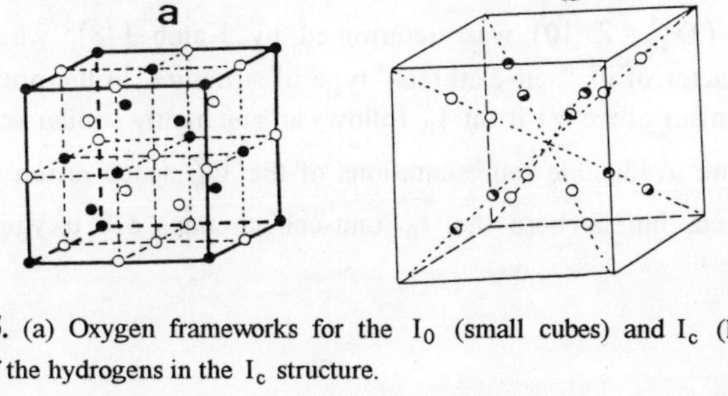

Fig. 25. (a) Oxygen frameworks for the I_0 (small cubes) and I_c (large cube) structures. (b) Occupancy of the hydrogens in the I_c structure.

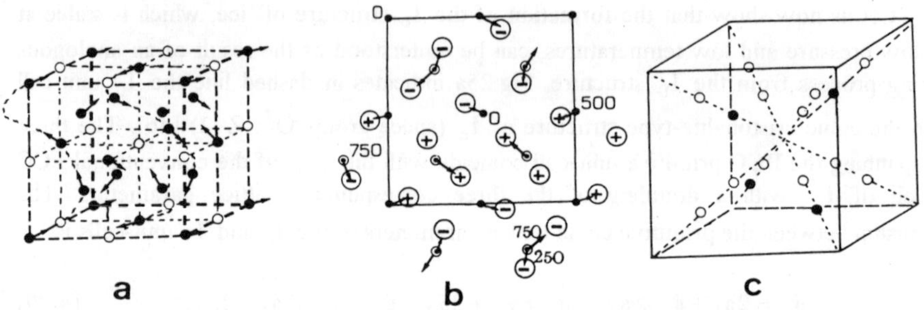

Fig. 26. Shifting of the oxygens from I_0 to ice III (a) and their projection in the (001) BCC plane (b). The signs + and - correspond to the directions of the shifts along Z. The numbers in (b) indicate the level of the atoms with respect to the basic plane. (c) Hydrogen distribution in ice III unit-cell.

fold tetragonal cell, with eight oxygen filling the tetragonal positions 8(b), the four remaining oxygens occupying the 4(a) positions. In addition to this ordering mechanism, the formation of ice III from I_0, requires two combined sets of displacements, which are represented in Fig.26a. They consist in a tilting around the fourfold axes of each group of eight oxygen atoms surrounding the middle of the faces of the tetragonal unit-cell, and a shifting of the two pairs of remaining oxygens in the antiparallel tetragonal directions $\pm[110]$ and $\pm[\bar{1}10]$. A projection of the shifts in the (001) BCC I_0 plane is represented in Fig.26b. Fig.26c shows the hydrogen atoms distribution in the unit-cell of Ice III (positions 8(b)).

The group-theoretical description of the combined ordering and displacive mechanism which transforms I_0 in ice III requires two different, four- and six-dimensional irreducible representations of the O_h^9 group denoted $\tau_4(\vec{k}_{10})$ and $\tau_4(\vec{k}_9)$, at the P and N points respectively of the BCC Brillouin zone [29], with $\vec{k}_9 = \frac{1}{2}(\vec{a}_1^* + \vec{a}_2^*)$.

Ice VI occupies a large area in the phase diagram of ice, as it is stable from about 6 to 20 kbars and from the melt to very low temperatures. The oxygen framework of its tetragonal structure (D_{4h}^{15}, Z=10) was determined by Kamb [48], who stressed the interpenetrating character of its "self-clathrate" type of structure. In the present approach, the formation mechanism of ice VI from I_0 follows an apparently similar scheme than for ice III. Thus the same irreducible representations of the O_h^9 space group ($\tau_4(\vec{k}_{10})$ and $\tau_4(\vec{k}_9)$) are involved, but as here the I_0 unit-cell contains 5/8 oxygen atoms, the

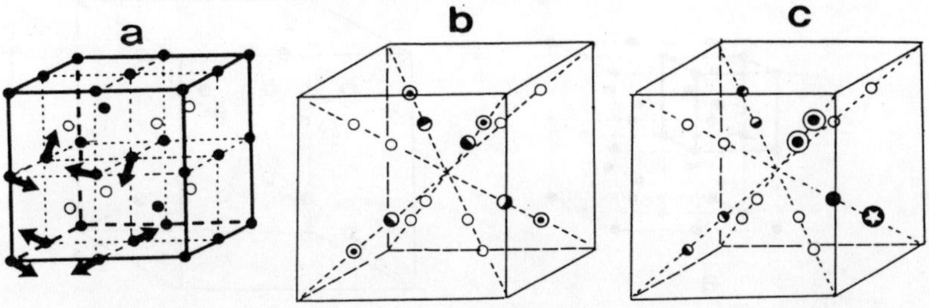

Fig. 27. Ordering and shifting mechanisms transforming I_0 into ice VI. (a) I_0-ice VI mechanism. (b) and (c): Two types of hydrogen distributions around oxygen sites in ice VI. The three types of hydrogen sites with an occupancy 1/2 are represented by half-filled, centered and star-shaped circles.

ordering and displacive mechanisms transforming I_0 in ice VI occur differently. This isillustrated in Fig.27a, where one can see that in the sixteen-fold tetragonal unit-cell of ice VI the oxygens occupy all the vertices of the initially disordered BCC cubes, as well as two among eight cube centers. Besides, the four pairs of oxygen atoms located on the faces of the BCC cubes carry out antiparallel displacements along general low-symmetric directions. Figs.27b and 27c describe the two types of hydrogen distributions in the ice VI unit-cell: Three hydrogens in positions 8(g) and one in position 16(h) are associated with the oxygens in positions 2(a) and 8(g).

5.3. Orientational and positional ordering of the hydrogen atoms: The structures of ice VII, ice VIII, ice IX and ice XI

In the structures of ice discussed in the preceding section, the oxygen framework were mainly considered, since the distribution of hydrogen atoms was assumed to follow the symmetry created by the transformation mechanisms. By contrast in ice VII (O_h^4, Z=2), which is the high pressure densest form of ice, the structure of which was analyzed by Kamb and Davis [49] in terms of two interpenetrating frameworks of ice I_c, the transformation mechanism from I_0 involves exclusively the hydrogen bonds. Fig.28 shows that in the I_0 unit-cell, within the ice VII unit-cell, all the positions are filled, i.e. the concentration of water molecules is equal to 1 and I_0 is a fully ordered structure. The transformation to ice VII will consist in a reorientation of the hydrogen bonds: The water molecules located at the vertices and at the center of the I_0 unit-cell become respectively orientationally inequivalent. The hydrogens which are fully disordered in the I_0 structure become tetrahedrally ordered, the effective symmetry of water molecules being lowered

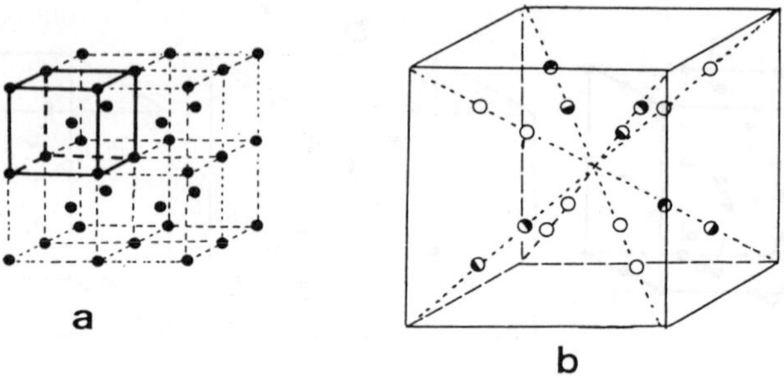

Fig. 28. (a) Oxygens in the I_0 (small cube) and ice VII (large cube) unit-cells. (b) Occupancy of the hydrogens in ice VII.

from O_h to T_d. Fig.28b shows the distribution of the hydrogens which occupy the positions 8(e).

The $I_0(O_h^9) \rightarrow$ Ice VII(O_h^4) cell-doubling transition can be described within a standard Landau-type approach: The one-dimensional irreducible representation of the O_h^9 space-group $\tau_4(\vec{k}_{12})$ induces the corresponding breaking of translational symmetry, where $\vec{k}_{12} = \frac{1}{2}(\vec{a}_1^* + \vec{a}_2^* + \vec{a}_3^*)$ is the H point of the BCC Brillouin-zone surface. The basic unit-cell vectors are related by the equations:

$$\vec{a}_{VII} = \vec{a}_2 + \vec{a}_3, \quad \vec{b}_{VII} = \vec{a}_1 + \vec{a}_3, \quad \vec{c}_{VII} = \vec{a}_1 + \vec{a}_2. \tag{4.31}$$

The formation of three other phases of ice, namely ice VIII, ice IX, and ice XI, can also be understood on the basis of a positional or orientational ordering of the sole hydrogen atoms, since these phases have similar oxygen frameworks than ice VII, ice III, and ice I_h, respectively. Thus, ice VIII, which is the low temperature form of ice VII, results from a tetrahedral to tetragonal ordering of the hydrogen bonds [50] within the oxygen framework of ice VII. The hydrogens occupy the positions 16(h) as illustrated in Fig.29a. The corresponding cell-doubling transition ice VII$\rightarrow$ice VIII is induced by the two-dimensional irreducible representation $\tau_8(\vec{k}_{12})$ of the O_h^4 space-group, at the R-point of the primitive cubic Brillouin zone [29].

Along the same line, orthorhombic ice XI (C_{2v}^{12}, Z=4) is the ordered hydrogen form of ice I_h, from which it is obtained by cooling below -200°C, provided ice I_h is doped with specific impurities [51]. The ordering of the hydrogen atoms is here a *positional* one, as shown in Fig.29b: In I_h the hydrogens occupy equivalent positions on each part of the hydrogen bonds, while in ice XI only one position 8(b) is occupied. The ferroelectric-ferroelastic ice $I_h \rightarrow$ice XI transition is associated with the two-dimensional irreducible representation $\tau_{12}(\vec{k}_{16})$ of the D_{6h}^4 space group at the center (Γ-point) of the

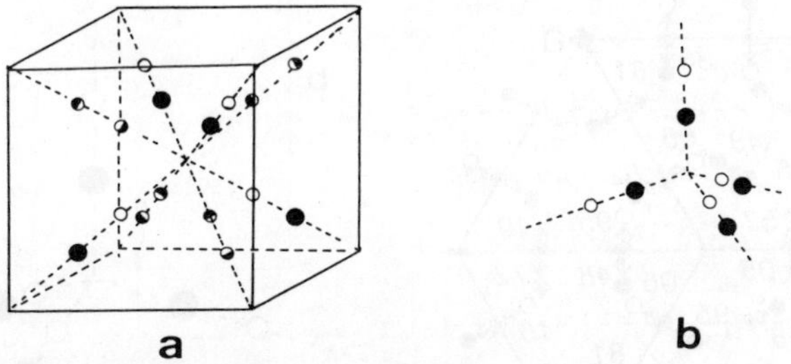

Fig. 29. Hydrogen sites in (a) ice VIII, and (b) ice XI.

hexagonal Brillouin zone. It is still unclear if ice IX, which has the same symmetry as ice III (D_4^4, Z=12) [52], constitutes a distinct low temperature form, more ordered with respect to the hydrogen bonds than phase III, as suggested by some authors [52]. To our knowledge, no clear-cut experimental data have proved the existence of an actual isostructural transition between the two phases.

5.4. Coupled ordering and displacive mechanisms for the formation of ice II

The assumed mechanisms leading to the phases of ice considered herein were analyzed in terms of one symmetry breaking order-parameter for purely ordering processes and with two coupled order-parameters for combined ordering and displacive mechanisms. A description of the formation of ice II from I_0 is more complex. High pressure ice II (D_{3d}^6, Z=12) is obtained on cooling from ice I_h below -50°C and is stable between about 3 and 6 kbars [53]. Although it possesses some resemblance with ice I_h, with similar puckered rings of water molecules, its structure is more compact, as was stressed by Kamb [19], and is ordered with respect to the hydrogens.

The transition from I_0 to the ice II structure can be described in three stages. The first stage is an ordering of the oxygens which occurs as for the formation of ice I_h : The initial disordered I_0 unit-cell contains 2/3 oxygens which order in the 4(f) crystallographic positions, resulting in a bilayered rhombohedral structure with four water molecules per unit-cell and a space group D_{3d}^6. The second stage, which is shown in Fig.30a, is a combined mechanism involving a tilting in two opposite directions of the hexagonal rings forming the two layers, and an antiparallel shifting of the rings along the [111] rhombohedral direction (the Z hexagonal axis) by an amount of $\pm\vec{c}_{rh}/3$, where $\vec{c}_{rh}$ is the lattice vector of the rhombohedral unit-cell, along [111]. It produces a decompression of about 10% of the distance between the rhombohedral layers, and a resulting structure which has the same symmetry as ice II (D_{3d}^6, Z=12), but with flat in-plane structures. In order to obtain puckered layers, a third mechanism is needed, also shown in Fig.30a,

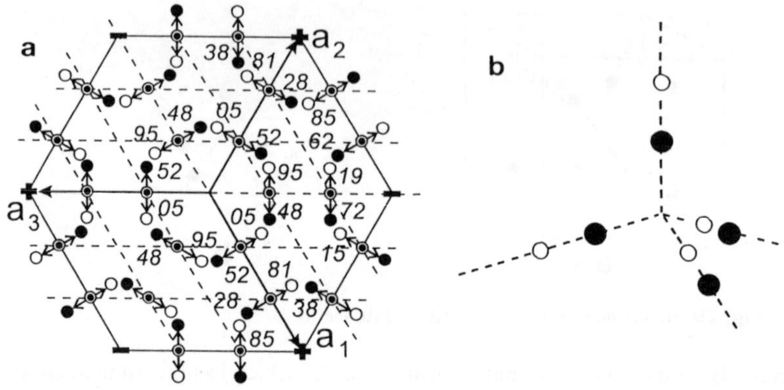

Fig. 30. (a) Mechanism leading to the formation of ice II. Comments are in the text. (b) Hydrogen sites in ice II.

which consists in antiparallel displacements of the oxygens along [111], the magnitude of which is about 10% of $|\vec{c}_{rh}|$.

Fig.30b gives the hydrogen positions in the rhombohedral unit-cell of ice II with two hydrogens in 4(a) positions and two others in positions 8(b). Despite the complexity of the proposed mechanism in crystallographic terms, the corresponding group theoretical description is relatively simple. Only two distinct irreducible representations of the O_h^9 space group are involved: $\tau_1(\vec{k}_7)$ for the ordering process, and $\tau_1(\vec{k}_2)$ for the displacive *and* tilting mechanisms, where $\vec{k}_2 = \frac{1}{9}(2\vec{a}_1^* - 2\vec{a}_2^* - \vec{a}_3^*)$ lies inside the BCC Brillouin zone.

5.5. The phase diagram of ice: Connection between pressure, densities, and the fractional concentrations of I_0

Fig.31 summarizes the connections assumed in our approach [55], between the nine structures of ice for which a formation mechanism was proposed, and the I_0 parent structure. Three phases of ice, namely, ice IV, ice V, and ice X, are not connected to I_0. The structure of ice X, which has been reported above 40 GPa [54] is still unknown, and more information is also needed for the metastable rhombohedral ice IV which takes place inside the stability range of ice V. We were unable to find a realistic mechanism for the formation of this latter monoclinic structure (C_{2h}^6, Z=28) from I_0.

Two summarizing remarks can be made on the phase diagram of Fig.31.

1) Except I_h and Ice II, which are typical non-Landau states with respect to I_0, i.e. they are not group-subgroup related to I_0, all the other phases of ice are Landau states,

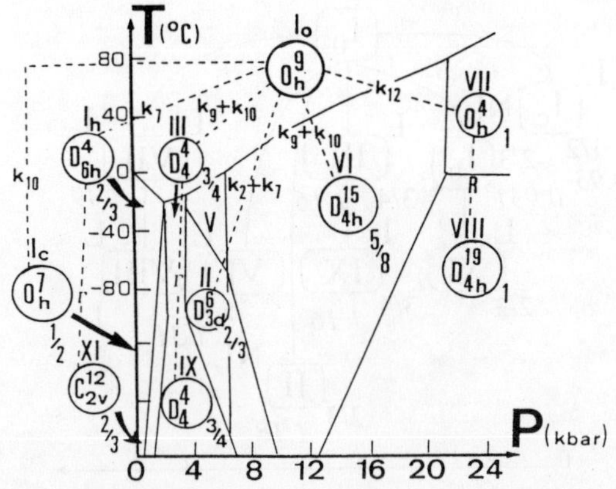

Fig. 31. Relationship between I_0 and the various structures of ice, for which are indicated the respective concentrations of I_0, and the corresponding wave-vectors of the BCC Brillouin zone.

corresponding to subgroups of O_h^9. They can be obtained either by one (I_c, Ice VII) or by two (Ice III, Ice VI) irreducible representations of I_0, or indirectly across intermediate structures (Ice VIII, Ice IX, Ice XI). This relatively simple connection to I_0, and the fact that there exist no group-subgroup relationship *between* the phases which border the melting line, justify our assumption of a parent *latent* phase for the ice structures. The BCC symmetry chosen for this latent structure, stems from the fact that it is the more compact packing that can be proposed for water molecules when preserving their identity. The formation of close packed FCC or HCP structures should require the absence of covalent bonds and fully ionic interactions between oxygen and hydrogen atoms, as was noted by Kamb [56]. Thus unlinked BCC clusters constitute the more easily formed structures in water, when approaching solidification.

2) Fig.32 indicates the densities ρ of the ice structures with increasing pressure P. One can see that ρ increases by more that 1.5 from the low pressure phases (I_c, I_h) to the highest ones (Ice VII, Ice VIII). The density reflects, though not directly, the concentration x of the parent I_0 unit-cell. However one can see in Fig.32 that the connection between P and x and the correlated influence of these parameters on the structural changes in ice is not straight. Thus, ice I_c, which is the lowest pressure structure, coincides with the lowest concentration x=1/2. Ice VII corresponds to the highest pressure and maximal concentration x=1. Between the preceding extremes, the relation between P and x is not monotonous: x=2/3 for ice I_h, and ice II, x=3/4 for ice III, but for the higher pressure structure of ice VI, the concentration decreases to x=5/8. This can be explained by the fact that for the formation of ice VI, one needs not only the introduction of additional molecules, but also an exchange of the atoms between filled and

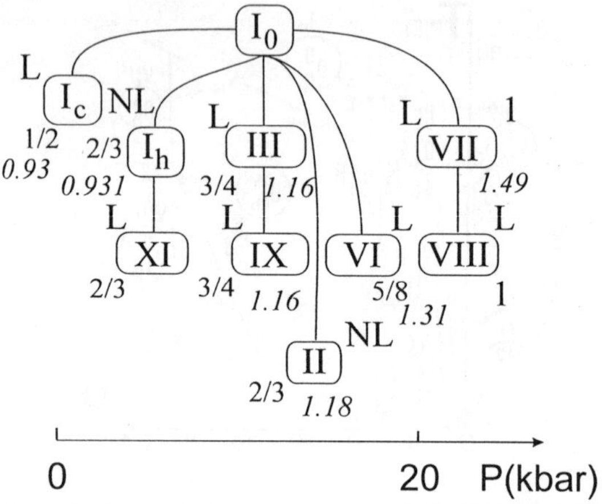

Fig. 32. Connection between pressure, the phase densities and the degree of occupancy of the I_0 structure.

vacant positions, which requires additional energy and thus larger pressures.

As a final remark, let us note that the transformation of ice with pressure has a *morphotropic character* as concentration also changes. Thus one should have for the high pressure phases of ice, inclusion of interstitials water phases surrounding microcrystallites.

6. Reconstructive transitions in SiO_2

6.1. The phase diagram of SiO_2

Minerals constitute, *despite* the complexity of their structures (see Sec.6.1 of Chapter III) but also *due to* this complexity, a preeminent field of application for the approach developed in this book. Indeed, a large number of the polymorphic transitions found in minerals *are* reconstructive transitions [57, 58]. However, the corresponding mechanisms are, in majority unknown partly because the density of the polymorphic structures notably increases with increasing pressure, and thus the transitions appear as having an *implicit morphotropic character*, with a discontinuity in the topology of the structures. In this respect, the approach used for ice structures in the preceding section, which assumes a parent structure with variable (fractional) concentrations, seems to be well adapted for the description of the different phases in a given mineral. We will illustrate this proposition, by a brief discussion of one of the most classical example of reconstructive transitions in minerals: the transitions between the SiO_2 structures.

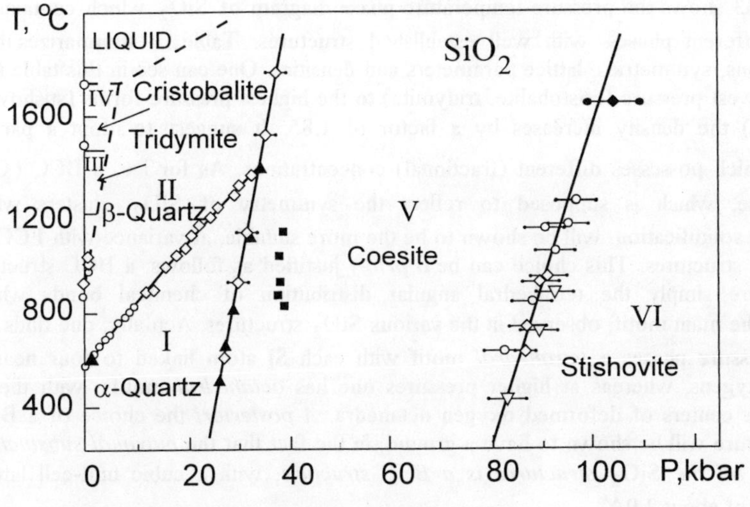

Fig. 33. Pressure-temperature phase diagram of SiO_2 .

Table 10. The stable and metastable phases of SiO_2.

Number	Denomination	Space-group	Lattice parameters (A°)	ρ (g/cm³)
I	α-quartz	D_3^4 (Z=3)	a=4.4, c=5.29	2.655
II	β-quartz	D_6^4 (Z=3)	a=4.44, c=5.45	2.655
III	β-tridymite	D_{6h}^4 (Z=4)	a=5.03, c=8.22	2.37
IV	β-cristobalite	O_h^7 (Z=2)	a=7.05	2.33
IV'	α-cristobalite	D_4^4 (Z=4)	a=4.97, c=6.92	2.33
V	Coesite	C_{2h}^6 (Z=16)	a=b=7.17, c=12.38, γ=120°	3.01
VI	Stishovite	D_{4h}^{14} (Z=2)	a=4.179, c=2.665	4.29
VII	Fe_2N - or ($\alpha-PbO_2$)-type	D_{6h}^3 (Z=3) or D_{2h}^{14} (Z=4)	a=2.561, c=4.112 or a=4.181, b=4.617, c=3.824	4.26
(VIII)	Keatite	D_4^4(Z=12)	a=7.456, c=8.604	2.50
(IX)	$CaCl_2$ -type	D_{2h}^{12} (Z=2)	a=4.30, b=4.77, c=4.50	4.35

Fig.33 shows the pressure-temperature phase diagram of SiO_2 which contains at least *ten* different phases, with well established structures. Table 10 summarizes their denominations, symmetries, lattice parameters and densities. One can see in this table that from the lowest pressure (cristobalite, tridymite) to the highest pressure forms (stishovite, $CaCl_2$-type) the density increases by a factor of 1.85. It suggests to adopt a parent structure which possesses different (fractional) concentrations. As for ice, a BCC (O_h^9) parent phase, which is supposed to reflect the symmetry of SiO_2 clusters when approaching solidification, will be shown to be the more suitable, at variance with FCC or HCP parent structures. This choice can be *a priori* justified as follows: a BCC structure realizes more simply the tetrahedral angular distribution of chemical bonds which constitutes the main motif, observed in the various SiO_2 structures. Actually, one finds for the low pressure phases a *tetrahedral* motif with each Si atom linked to four nearest neighbor oxygens, whereas at higher pressures one has *octahedral* motifs, with the Si atoms at the centers of deformed oxygen octahedra. *A posteriori* the choice of a BCC parent structure will be shown to have a ground, in the fact that the *maximal substucture common to all the* SiO_2 *structures is a BCC structure*, with a cubic unit-cell lattice parameter a of about 2.9A°.

Fig.34 shows the connections assumed in our model between the assumed BCC parent structure and the actual SiO_2 structures. Thus, in terms of the concentration x (number of oxygen atoms in the parent BCC unit-cell) one finds x=1/2 for cristobalite and tridymite, x=1 for stishovite and x=2/3 for coesite (and also possibly for quartz). Fig.34 indicates the points of the BCC Brillouin-zone corresponding to the irreducible representations of O_h^9, associated with the transition mechanism to the SiO_2 structures, which are now briefly depicted.

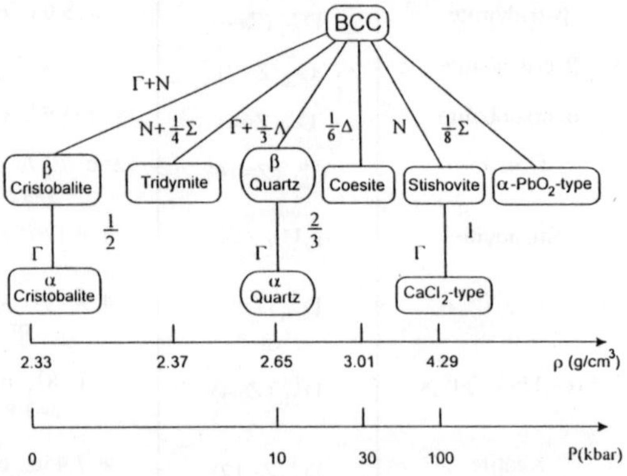

Fig. 34. Connection between the parent BCC structure and the SiO_2 phases, their densities, and the wave-vectors associated with the proposed transition mechanisms from the BCC phase.

6.2. The BCC-stishovite transformation

The stishovite structure, which is of the rutile-type [59], is obtained by ordering and shifting processes, from a disordered BCC structure represented in Fig.35. In this latter structure the oxygens are ordered (x=1) in positions 1(a), and the Si atoms occupy randomly (1 atom for three sites) the 3(b) positions. The ordering mechanism brings the Si atoms in positions 2(a), and correlatively the oxygens are slightly shifted by about 0.08 a_s (where a_s=4.179 A°) in the cubic directions [100] and [010], reaching the positions 4(f).

In group-theoretical terms the ordering and shifting of the atoms transform as the same six-dimensional irreducible representation ($\tau_4(\vec{k}_9)$ in Kovalev's notations [29]) at the N-point of the BCC Brillouin-zone. It induces a fourfold multiplication of the BCC unit-cell, corresponding to two non-vanishing components of the six-dimensional order-parameter. The relationship between the stishovite and BCC lattice parameters is:

$$\vec{a}_s = \vec{a}_1 + \vec{a}_2 + 2\vec{a}_3, \quad \vec{b}_s = \vec{a}_1 - \vec{a}_2, \quad \vec{c}_s = \vec{a}_1 + \vec{a}_2 \tag{4.32}$$

where $\vec{a}_1$, $\vec{a}_2$ and $\vec{a}_3$ are the primitive (rhombohedral) BCC lattice vectors. Starting from a cubic parameter a=2.9 A°, the preceding mechanism gives the following numbers for the tetragonal stishovite unit-cell: a_s=4.1 A°, c_s=2.9 A°, which, as compared to the

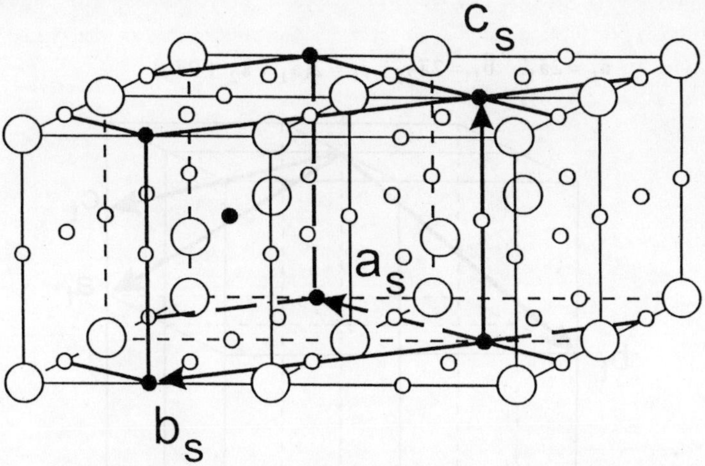

Fig. 35. The rutile-type stishovite unit-cell (in solid lines), embedded in four conventional cubic cells.

experimental values (a_s=4.179 A°, c_s=2.665 A°) imply an additional (secondary) compression (e_{zz}) along the $\vec{c}$ axis, of about 8%.

6.3. The BCC-tridymite transformation

Fig.36 shows the connection between the BCC parent unit-cell, which is filled by 1/2 oxygen atom, and the hexagonal structure (D_{6h}^4, Z=4) of β-tridymite [60]. In the transition from one structure to the other both the oxygen and the silicon atoms *order* and undergo *displacements* from their virtual BCC positions. The ordering mechanism brings the oxygens, initially distributed in positions 1(a), with 1 oxygen for two sites, to the positions 2(c) and 6(g), whereas the Si atoms initially in 3(b) positions (one atom per 6 sites) order in positions 4(f).

The displacive mechanism corresponds to a variant of Burgers mechanism which leads to a *double* HCP structure, i.e. it corresponds to antiparallel shifts of the atoms located in the (110) planes, along the $[1\bar{1}0]$ cubic directions. The tridymite structure is realized for shifts of magnitude $a\sqrt{2}/12$.

As for the BCC-stishovite transition, the preceding ordering and displacive mechanisms are bilinearly coupled ("pseudo-proper" coupling), and transform as the same twelve-dimensional irreducible representation associated with the wave-vector $(\frac{\pi}{2a}, \frac{\pi}{2a}, 0)$ located *inside* the BCC Brillouin zone, at the *middle* of the Σ-line. The corresponding relationship between the lattice-parameters of the tridymite and BCC unit-cells is:

$$\vec{a}_t = 2\vec{a}_1, \quad \vec{b}_t = 2\vec{a}_2, \quad \vec{c}_t = 2(\vec{a}_1 + \vec{a}_2 + 2\vec{a}_3) \tag{4.33}$$

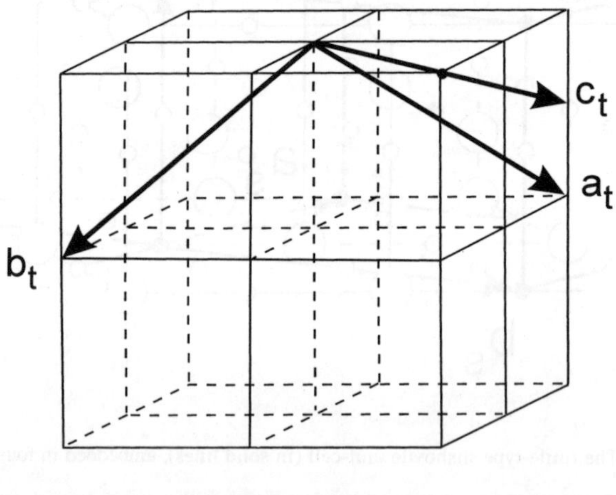

Fig. 36. Primitive translations of the β-trydimite unit-cell within eight cubic cells.

which yields a *sixteen-fold* multiplication of the BCC unit-cell. Using the initial cubic parameter a=2.9 A°, one gets: a_t=5.02 A° and c_t=8.20 A°, which are very close from the experimental values of 5.03 A° and 8.22 A° respectively, i.e. no additional deformation is needed in the mechanism.

6.4. The BCC-β-cristobalite, and β-α cristobalite transformations

The cubic structure of β-cristobalite is shown in Fig.37 [61], as well as the corresponding disordered BCC parent structure, which is filled by 1/2 oxygen, as for the tridymite structure. The BCC-β-cristobalite transformation can be decomposed in two successive steps: a Bain-type deformation, which transforms the BCC structure into FCC (see Chapter III, Sec.4), and an ordering mechanism which brings the oxygen and silicon atoms in positions 4(c) and 2(a), respectively. Accordingly, two distinct order-parameters are needed: two combinations of strain-tensor components which are involved in the macroscopic deformation, and a six-dimensional irreducible representation, corresponding again to the N-Brillouin zone point which relates to the ordering process. The relationship between the β-cristobalite lattice parameters and the BCC translations of the conventional cubic unit-cell ($\vec{a}_c, \vec{b}_c, \vec{c}_c$) is:

$$\vec{a}_{cr} = 2(\vec{a}_c + \vec{b}_c), \quad \vec{b}_{cr} = 2(\vec{a}_c - \vec{b}_c), \quad \vec{c}_{cr} = 2\vec{c}_c \tag{4.34}$$

corresponding to an *eight-fold* multiplication of the initial BCC unit-cell.

The transition from β- to α-cristobalite [62] is a standard Landau-type transition, as the α symmetry (D_4^4, Z=4) corresponds to a subgroup of the β space-group (O_h^7, Z=2). The β–α mechanism can be described in terms of tilts of the SiO_4 tetrahedra which bring the oxygens from the positions 4(f) to 8(b), and the silicons from 2(a) to 4(a). This transition between ordered phases, transforms as the three-dimensional irreducible representation $\tau_4(\vec{k}_{10})$ [29] at the X point $(0,0,\frac{\pi}{a})$ of the FCC Brillouin-zone, which

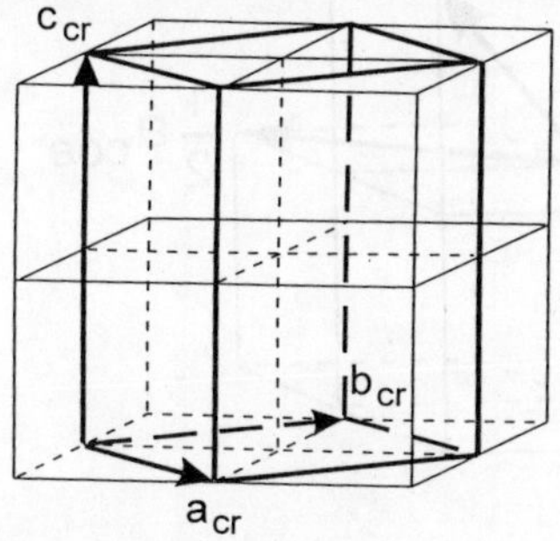

Fig. 37. β-Cristobalite unit-cell inscribed in eight cubic cells.

involves a doubling of the β-cristobalite unit-cell, and a sixteen-fold multiplication, of the initial BCC unit-cell. With respect to the conventional BCC lattice parameters one has:

$$\vec{a}_\alpha = \vec{a}_c, \quad \vec{b}_\alpha = \vec{b}_c, \quad \vec{c}_\alpha = 2\vec{c}_c. \tag{4.35}$$

6.5. The BCC-coesite transformation

The BCC cube embedded in the monoclinic coesite structure (C_{2h}^6, Z=16) [63] is shown in Fig.38. It contains 2/3 oxygens. The transformation from BCC to coesite requires two, independent, ordering and shifting mechanisms. Ordering of both O and Si atoms yield a six-fold rhombohedral (D_{3d}^3) unit-cell with the lattice parameters:

$$\vec{a}_r = \vec{a}_1 - \vec{a}_2, \quad \vec{b}_r = \vec{a}_2 - \vec{a}_3, \quad \vec{c}_r = 2(\vec{a}_1 + \vec{a}_2 + \vec{a}_3) \tag{4.36}$$

associated with a point of the BCC Brillouin-zone located at 1/3 of the Λ-line. Additional shifts are necessary to realize the monoclinic structure of coesite, through a doubling of the rhombohedral parameters:

$$\vec{a}_{coe} = 2\vec{a}_r, \quad \vec{b}_{coe} = 2\vec{b}_r, \quad \vec{c}_{coe} = 2\vec{c}_r \tag{4.37}$$

which give the calculated parameters: $a_{coe} = b_{coe}$=8.48 A°, c_{coe}=10.4 A°. These numbers show, by comparison with the experimental values $a_{coe} = b_{coe}$=7.17A° and c_{coe}=12.38A°, that additional compression, within the (111) BCC plane and decompression, along the [111] BCC direction, have to be taken into account.

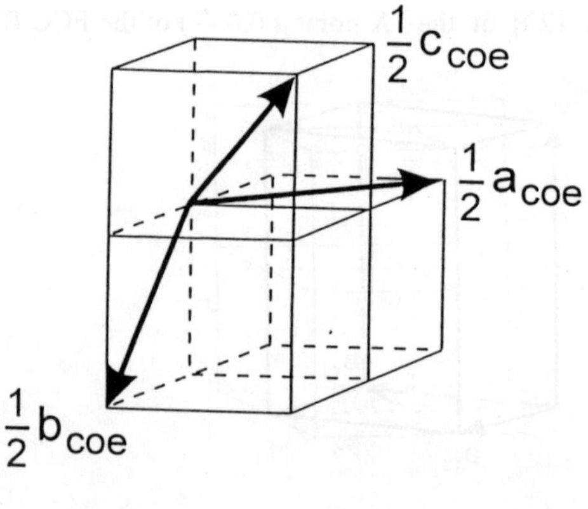

Fig. 38. Half of the coesite primitive translations and three cubic cells.

7. Ordering-type and displacive-type mechanisms in lanthanides

7.1. Different approaches to lanthanide structures

The difficulty in deciding, in some cases, which type of mechanism, displacive or ordering-type, is responsible of a reconstructive transition, has already been stressed. This difficulty also stems from the arbitrariness which may arise in the choice of the parent structure. For instance, the FCC and HCP phases of iron have been deduced (Chapter III, Sec.4.5) from δ- or α-BCC iron, through displacive mechanisms. At variance the FCC and HCP phases of Co or He have been derived from a disordered polytype phase, by ordering mechanisms. The choice of the suitable mechanism was not arbitrary, but based on the structural, and physical properties of these elements. Besides, we have seen that for the realization of the diamond structure, or for a number of ice structures correlated ordering and displacive processes had to be assumed.

A good illustration of the interplay existing between ordering and displacive mechanisms, in the interpretation of reconstructive transitions, is given by the members of the lanthanide family [64]. Fig.39 shows the distribution of structures among the fifteen elements which form this family. The existence of a BCC structure in the phase diagram of eleven lanthanides suggests at the first glance that a similar scheme than the one proposed for actinides (Chap. III, Sec.5) applies to lanthanides, i.e. displacive mechanisms from the

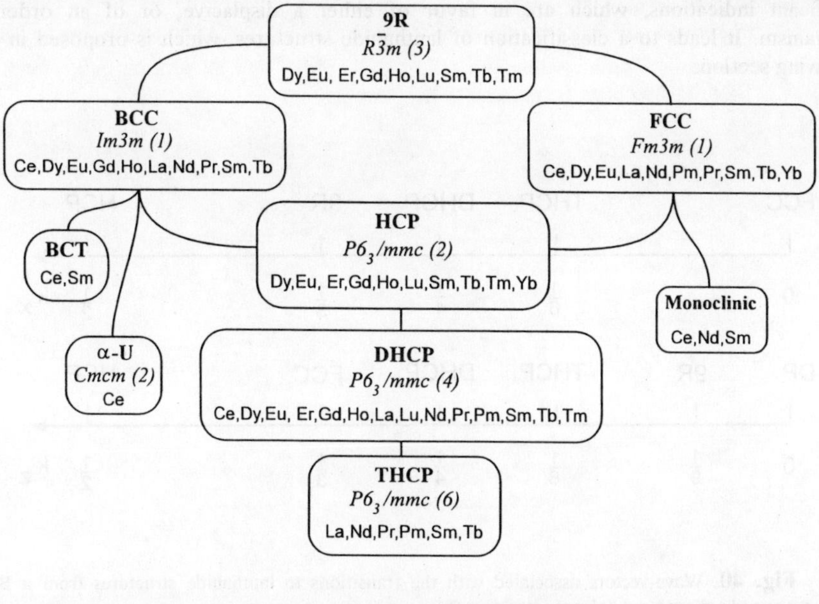

Fig. 39. Distribution of structures among lanthanide elements.

BCC structure. However, the existence of the double and triple HCP, or of a 9R structure in most of the lanthanide phase diagrams denote a strong tendency to polytypism, which is associated to ordering processes.

Fig.40 gives the set of wave-vectors and corresponding symmetries, which are associated to lanthanide structures, assuming either a parent BCC structure (Fig.40a) or a parent disordered polytypic phase (Fig.40b). Thus, starting from a BCC structure the corresponding wave-vectors all pertain to the Σ-line of the BCC Brillouin zone, namely $\vec{k} = (k_x, k_x, 0)$. For k_x=0 (Γ-point) one gets the FCC phase (Bain deformation) while for k_x=1/6, 1/4 and 1/3 one obtains respectively the triple HCP, double HCP and 9R structures. For k_x=1/2 (the N point) the HCP is stabilized (Burgers mechanism). Furthermore the body centered tetragonal structure found in Ce and Sm [65], as well as the orthorhombic, or monoclinic, structures [66] of the α-U type reported in Pr, Sm, Ce or Nd, can be simply interpreted as Landau-type phases, which occur as intermediate structures in our description of the Bain deformation and Burgers mechanisms.

However, one can see from Fig.40b, that an alternative unified scheme can be assumed for lanthanide structures in terms of ordering mechanisms. Fig.40b shows that starting from a disordered polytype structure, the lanthanide symmetries are associated with wave-vectors of the hexagonal Brillouin zone which all pertain to the same P-line , corresponding to $\vec{k} = (\frac{4\pi}{3}, \frac{2\pi}{3}, k_z)$. For k_z=1/9, 1/6, 1/4, 1/3 and 1/2 one gets respectively the 9R, triple HCP, double HCP, FCC and HCP structures.

In order to decide which type of mechanism is involved, one has to consider other significant indications, which are in favor of either a displacive, or of an ordering mechanism. It leads to a classification of lanthanide structures, which is proposed in the following section.

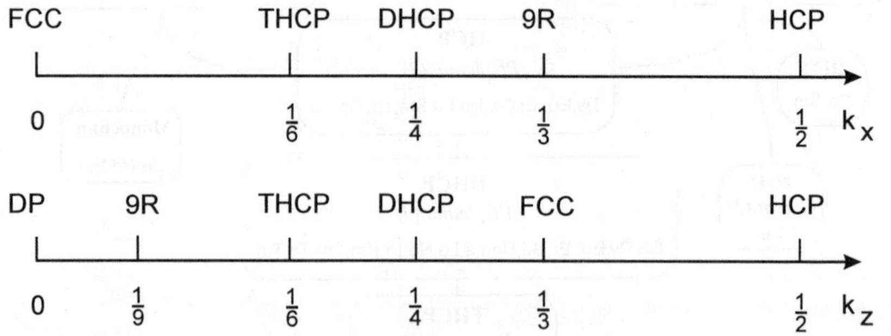

Fig. 40. Wave-vectors associated with the transitions to lanthanide structures from a BCC structure (a) and a disordered polytype structure (b).

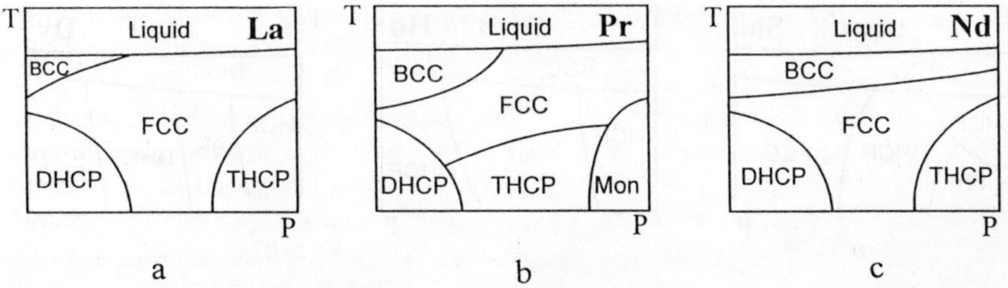

Fig. 41. Schematic phase diagrams of La, Pr, and Nd.

7.2. Classification of lanthanide structures

On the basis of the presently available experimental data one can, tentatively, distinguish three groups of lanthanide elements, with respect to their transition mechanisms.

In La, Nd, Pr and Pm, two facts are in favor of assuming ordering mechanisms: 1) The possibility, already mentioned in Sec.2, of undergoing a slightly first-order FCC-DHCP transition, which is incompatible with the large displacements, that would be necessary to obtain one structure from the other. 2) The existence of a six-layered triple HCP structure, which is a distinctive feature of these four elements [4]. It represents a clear tendency to polytypism, and can be interpreted as a disintegration of the HCP structure (which is absent in the four elements) in contrast with the three-layered FCC structure. One can see in Figs.41a,b,c that the FCC phase surrounds the hexagonal phases and separates them from the BCC phase which occupies a small region below the melt. In Pm, the pressure-temperature phase diagram is incomplete, and no BCC structure has been reported. The FCC structure is found [4] between the double and triple HCP phases.

In the heavy lanthanides Sm, Eu, Dy, Ho, Er, Tm, Yb and Lu, the BCC phase occupy a large fraction of the phase diagram, and is directly in contact with the hexagonal phases, as shown in Figs.42a-d. A direct displacive mechanism from the parent BCC structure to the other structures is suggested by the strongly first-order character of the transitions found in these elements [16, 65]. The 9R structure is, with the exception of Yb, a common feature of their phase diagrams and has been described by a displacive reconstructive mechanism in Chapter III (Sec.7). The different structures of Ce (Fig.42e) can also be described by displacive mechanisms from the BCC phase, which also lead to the intermediate tetragonal and orthorhombic Landau states, found in this element.

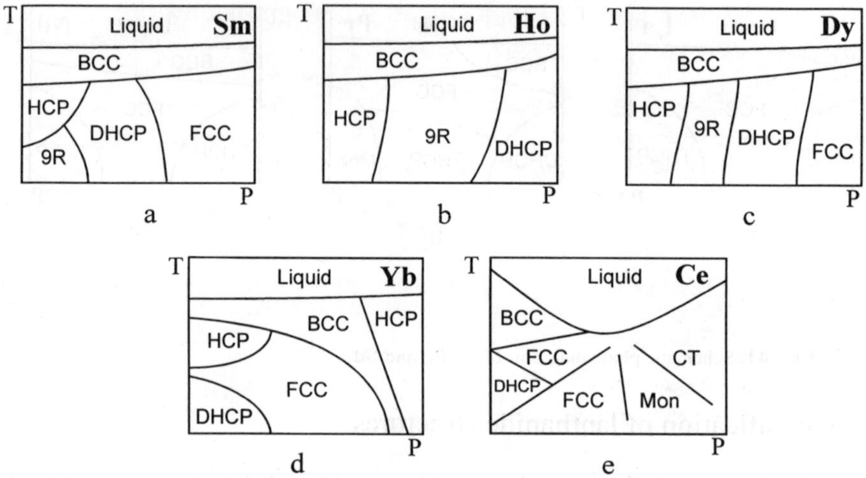

Fig. 42. Schematic phase diagrams of Sm, Ho, Dy, Yb, and Ce.

The phase diagram of Tb and Gd, shown schematically in Fig.43 possess features which belong to the two preceding groups of elements. Thus on cooling at low pressures from the BCC phase, one obtains the 9R and HCP phases, but increasing at higher pressure from the HCP phase, on gets the polytypic double and triple HCP structures. Accordingly, one is entailed to suggest that in these two elements, one has either displacive or ordering processes, depending on which of the preceding thermodynamic paths is followed.

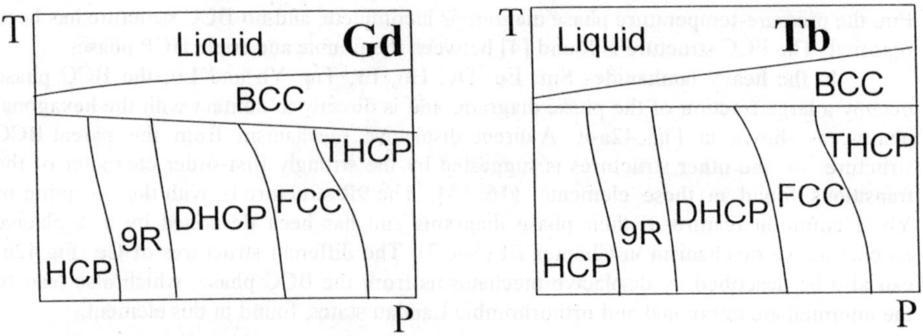

Fig. 43. Schematic phase diagrams of Gd, and Tb.

8. Reconstructive phase transitions in a cryogenic crystal: Nitrogen

In the case of the lanthanides Tb and Gd, we have suggested that the set of phases found in their phase diagrams could follow two different theoretical schemes, with two different parent structures (a BCC structure and a disordered polytype structure). This situation can be encountered in a more obvious manner in some crystals, in which different regions of the phase diagram display different types of interactions. This is for example the case of nitrogen where the high ($P \geq 4$ GPa) and low pressure regions correspond respectively to non-quadrupolar and quadrupolar types of ordering [66], the set of structures being related in the two regions, in an essentially different way [67].

Let us first consider the high pressure part of the phase diagram of N_2, which is represented in Fig.44. It contains three phases, denoted δ-N_2 (O_h^3, Z=8), ε-N_2 (D_{3d}^6, Z=8) and ζ-N_2 (C_{3v}^6, Z=8) [68-70]. Since the ε and ζ structures correspond to subgroups of the δ-structure, the δ–ε–ζ sequence of phases can be straightforwardly interpreted in the framework of the standard Landau theory. The ordering mechanism which transforms δ-N_2 in ε-N_2 is shown in Figs.45a and 45b. It consists in an orientational ordering of the N_2 molecules, which in the δ-cubic structure realizes two types of disordered configurations [68]: spherical (X) and disk-like (Y) (Fig.45a). In the ε structure the molecules have fixed orientations with the X-type molecules along the 3-fold axis of the δ-cube, and the Y-type molecules in planes perpendicular to the axes of the Y-molecular rows.

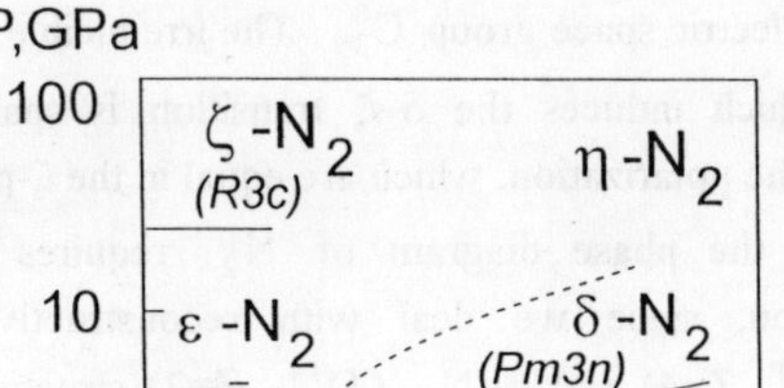

Fig. 44. The pressure-temperature phase diagram of nitrogen.

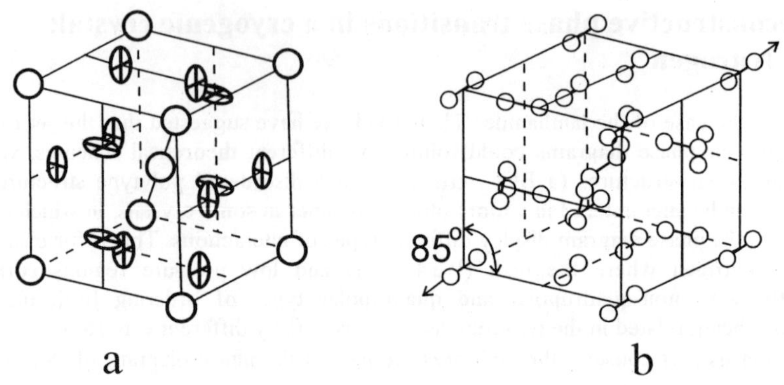

Fig. 45. Ordering mechanism transforming δ-N_2 (a) in ε-N_2 (b). Comments are in the text.

The δ-ε transition occurs without multiplication of the δ unit-cell, and it coincides with a three-dimensional irreducible representation (τ_4) of the O_h^3 space-group, at the center of the primitive-cubic Brillouin zone. The corresponding order-parameter is the set of shear strains (e_{yz}, e_{xz}, e_{xy}), which become equal in the ε-phase. Let us note that τ_4 does not satisfy the Landau symmetric-cube criterion (see Chapter I, Sec.4.2), in agreement with the first-order character reported for the δ-ε transition, and with the observation of a coexistence region between the δ and ε phases [71].

The ζ-N_2 phase represents a slight distortion of ε-N_2, which lowers the ε-N_2 symmetry to the polar ferroelectric space group C_{3v}^6. The irreducible representation (τ_{10}) of the O_h^3 space-group which induces the δ–ζ transition is spanned by the three-components (P_x, P_y, P_z) of the polarization, which are equal in the ζ-phase.

The lowest part of the phase diagram of N_2, requires a more elaborated phenomenological description, since we deal with reconstructive transitions: the β-N_2 (D_{6h}^4, Z=2), α-N_2 (T_h^6, Z=4) and γ-N_2 (D_{4h}^{14}, Z=2) structures [72-75] *are not* group-subgroup related. The β-phase is an orientationally (spherically) disordered HCP structure, whereas the α and γ phases are ordered. Fig.46 represents the orientations of the N_2-molecules in the α-phase, along the three-fold cubic axis, while Fig.46b shows the orientations of the molecules along the X or Y axes of the cube, in the γ-phase.

It is clear from the preceding figures that the centers of the molecules in the α and γ structures form a FCC lattice. We will thus describe the reconstructive transitions to these structures *from* the β structure via a virtual intermediate FCC structure. Thus the HCP-FCC transition is a reordering type transition corresponding to the mechanism depicted in Sec.2, i.e. using a parent disordered polytypic structure. The FCC$\rightarrow$ α-N_2, and FCC$\rightarrow$ γ-N_2 transitions are of the Landau-type: they are respectively associated with a six-

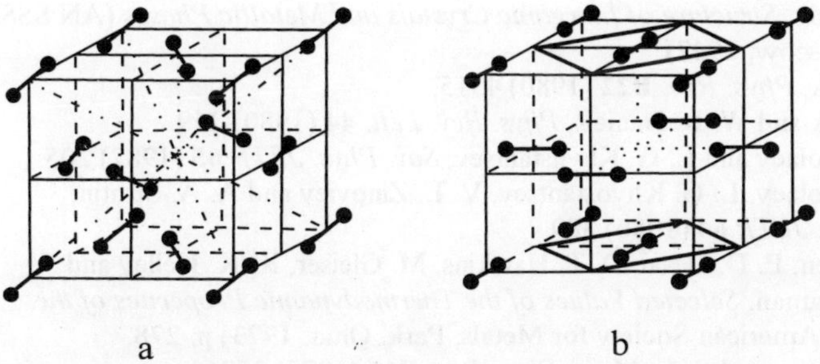

Fig. 46. Orientations of the N_2 molecules: (a) in α-N_2, (b) in γ-N_2.

dimensional ($\tau_9(\vec{k}_{10})$) and three-dimensional ($\tau_3(\vec{k}_{10})$) irreducible representation of the O_h^5 space group at the X-point of the FCC Brillouin-zone. The corresponding cell-quadrupling and cell-doubling transitions involve three non-vanishing components ($\eta_1 = -\eta_3 = -\eta_5 \neq 0$, $\eta_2 = \eta_4 = \eta_6 = 0$) and one non-vanishing component ($\zeta_1 \neq 0$, $\zeta_2 = \zeta_3 = 0$) for the respective six- and three-dimensional order-parameters.

9. References

1. N. G. Parsonage and L. A. K. Staveley, *Disorder in Crystals* (Clarendon Press, Oxford, 1978).
2. *Order-Disorder Transformations in Alloys*, ed. H. Warlimont (Springer Verlag, Berlin, 1974).
3. A. R. Verma and P. Krishna, *Polymorphism and Polytypism in Crystals* (John Wiley Inc., New York, 1966).
4. D. A. Young, *Phase Diagrams of the Elements* (University of California Press, Barkcley, 1991).
5. J. C. Dugdale and J. P. Frank, *Phil. Trans. Roy. Soc. (London)* **A257** (1964) 1.
6. A. Jayaraman, *Phys. Rev.* **A139** (1965) 690.
7. G. J. Piermarini and C. E. Weir, *Science* **144** (1964) 69.
8. O. S. Edwards and H. Lipson, *Proc. Roy. Soc. (London)* **A180** (1942) 268.
9. K. Lonsdale, *Nature* **155** (1945) 572; C. V. Raman, *Nature* **156** (1945) 22.
10. P. V. Hobbs, *Ice Physics* (Clarendon Press, Oxford, 1974).

11. N. V. Belov, *Structure of Inorganic Crystals and Metallic Phases* (AN SSSR Publ., Moscow, 1947).
12. J. P. Frank, *Phys. Rev.* **B22** (1980) 4315.
13. J. P. Frank and W. B. Daniels, *Phys. Rev. Lett.* **44** (1980) 259.
14. N. A. Nikolaev and L. G. Khvostantsev, *Sov. Phys. JETP* **65** (1987) 205.
15. N. A. Nikolaev, L. G. Khvostantsev, V. E. Zinoviev and A. A. Sirotin, *Sov. Phys. JETP* **64** (1986) 389.
16. R. Hultgren, P. D. Desai, D. T. Hawkins, M. Gleiser, K. K. Kelley and D. D. Wagman, *Selected Values of the Thermodynamic Properties of the Elements* (American Society for Metals, Park, Ohio, 1973) p. 278.
17. S. M. Shapiro and S. C. Moss, *Phys. Rev.* **B15** (1977) 2726.
18. E. C. Svensson, B. M. Powell, A. D. B. Woods and W. D. Teuchert, *Can. J. Phys.* **57** (1979) 253.
19. N. Wakabayashi, R. H. Scherm and H. G. Smith, *Phys. Rev.* **B25** (1982) 5122.
20. F. Frey, W. Prandl, J. Schneider, C. Zeyen and K. Ziebeck, *J. Phys. F: Metal Phys.* **9** (1979) 603.
21. G. Shirane, *Rev. Mod. Phys.* **46** (1974) 437.
22. A. J. C. Wilson, *Proc. Roy. Soc. London* **A180** (1942) 277.
23. O. Blaschko, G. Krexner, J. Pleschiutschnig, G. Ernst, C. Hitzenberger, H. P. Karnthaler and A. Korner, *Phys. Rev. Lett.* **60** (1988) 2800.
24. H. Jagodzinski, *Acta Cryst.* **7** (1954) 300.
25. V. P. Dmitriev, S. B. Rochal, Yu. M. Gufan and P. Toledano, *Phys. Rev. Lett.* **62** (1989) 2495.
26. M. P. Kulakov and S. Z. Shmurak, *Phys. Stat. Sol.* **A59** (1980) 147.
27. O. Brafman and I. T. Steinberger, *Phys. Rev.* **143** (1966) 501.; B. A. Abdikamalov, M. P. Kulakov, V. Sh. Shekhtman and S. Z. Shmurak, *Fiz. Tverd. Tela* **17** (1975) 3463.
28. V. P. Dmitriev, S. B. Rochal, Yu. M. Gufan and P. Toledano, *Ferroelectrics* **79** (1988) 305.
29. O. V. Kovalev, *Irreducible Representations of the Space Groups* (Gordon and Breach, New York, 1965).
30. Yu. M. Gufan, E. I. Kut'in, V. L. Lorman and E. N. Sidorenko, *Sov. Phys. Solid State* **29** (1987) 432.
31. H. Lipson and A. R. Stokes, *Nature* **149** (1942) 328.
32. W. H. Bragg, *Proc. Phys. Soc. (London)* **33** (1921) 304; W. Ehrenberg, *Z. Krist.* **63** (1926) 320.
33. J. D. Bernal, *Proc. Roy. Soc. (London)* **106A** (1924) 749; O. Hassel and H. Mark, *Z. Physik* **25** (1924) 317.
34. F. P. Bundy and J. S. Kasper, *J. Chem. Phys.* **46** (1967) 3437.
35. M. J. Buerger, in *Phase Transphormations in Solids*, eds. E. Smoluchowski, J. Mayer and W. A. Weil (Wiley, New York, 1951), p. 183.
36. L. Pauling, *The Nature of the Chemical Bond* (Cornell University Press, Ithaca, 1960), Chapter 4.
37. G. Tamman, *Kristallisieren und Schmelzen* (Barth, Leipzig, 1903), p. 315.

38. P. W. Bridgman, *Proc. Am. Acad. Arts Sci.* **47** (1912) 44.
39. P. W. Bridgman, *J. Chem. Phys.* **5** (1937) 964.
40. H. Z. König, *Krist.* **105** (1943) 279.
41. B. Kamb, in *Physics and Chemistry of Ice*, eds. E. Whalley, S. J. Jones and L. W. Gold (Royal Society of Canada, Ottawa, 1973), p. 28.
42. E. Whalley, *Physics of Ice*, eds. N. Riehl, B. Bullemer and E. Englehardt (Plenum, New York, 1969), p. 19.
43. N. H. Fletcher, *The Chemical Physics of Ice* (Cambridge Univ. Press, New York, 1970).
44. A. Von Hipper and E. F. Farrel, *Mater. Res. Bull.* **8** (1973) 127.
45. P. G. Owston, *Adv. Phys.* **7** (1958) 171.
46. K. Lonsdale, *Proc. Roy. Soc. (London)* **A247** (1958) 424.
47. B. Kamb and A. Prakash, *Acta Crystallogr.* **B24** (1968) 1317.
48. B. Kamb, *Science* **150** (1965) 206.
49. B. Kamb and B. L. Davis, *Proc. Natl. Acad. Sci. USA* **52** (1964) 1433.
50. W. F. Kuhs, *J. Chem. Phys.* **81** (1984) 3612.
51. R. Howe and R. W. Whitworth, *J. Chem. Phys.* **90** (1989) 4450.
52. J. E. Bertie and E. J. Whalley, *J. Chem. Phys.* **40** (1964) 1646.
53. B. Kamb, *Acta Crystallogr.* **17** (1964) 137.
54. A. Polian and M Grimsditch, *Phys. Rev. Lett.* **52** (1984) 1312.
55. V. P. Dmitriev, S. B. Rochal and P. Toledano, *Phys. Rev. Lett.* **71** (1993) 553.
56. B. Kamb, *J. Chem. Phys.* **43** (1965) 3917.
57. L. Bragg and G. F. Claringbull, *Crystal Structures of Minerals* (G. Bell and Sons, London, 1965).
58. Lin-gun Liu and W. A. Basset, *Elements, Oxides, Silicates* (Oxford University Press, New York, 1986).
59. E. C. T. Chao, J. J. Fahey, J. Littler and D. J. Milton, *J. Geophys. Research* **67** (1962) 419.
60. J. E. Fleming and H. Lynton, *Physics and Chemistry of Glasses*, **1** (1960) 148.
61. R. W. G. Wyckoff, *Z. Krist.* **62** (1925) 189.
62. N. Neuwenkamp, *Z. Krist.* **92** (1935) 82.
63. T. Zoltai and M. J. Buerger, *Z.Krist.* **111** (1959) 129.
64. *Systematics and the Properties of the Lanthanides*, ed. S. P. Sinha (D. Reidel Dordrecht, 1983).
65. R. A. Stager and H. G. Duckamer, *Phys. Rev.* **133** (1964) 830:
E. A. Perez-Albuerne, R. L. Clendenen, R. W. Lynch and H. G. Drickamer, *Phys. Rev.* **142** (1966) 392.
66. C. A. English and J. A. Venables, *Proc. Roy. Soc. (London)* **A340** (1974) 5.
67. R. L. Mills, B. Olinger and D. T. Cromer, *J. Chem. Phys.* **84** (1986) 283.
68. B. Olinger, *J. Chem. Phys.* **80** (1984) 1309.
69. D. Schiferl, S. Buschbaum and R. L. Mills, *J. Phys. Chem.* **89** (1985) 2324.
70. D. T. Cromer, R. L. Mills, D. Schiferl and L. A. Schwalbe, *Acta Crystallogr.* **37** (1981) 8.

71. R. L. Mills, D. Schiferl, A. I. Katz and B. W. Olinger, *J. Phys. (Paris)* **45** (1982) C8-187.
72. J. A. Venables and C. A. English, *Acta Crystallogr.* **B30** (1974) 929.
73. B. M. Powell, G. Dolling and H. F. Nieman, *J. Chem. Phys.* **79** (1983) 982.
74. R. L. Mills and A. F. Schuch, *Phys. Rev. Lett.* **23** (1969) 1154.
75. A. F. Schuch and R. L. Mills, *J. Chem. Phys.* **52** (1970) 6000.

CHAPTER V

CRYSTAL-QUASICRYSTAL RECONSTRUCTIVE PHASE TRANSITIONS

1. Introduction

Crystal-quasicrystal transitions provide one of the best illustrations of the usefulness and applicability of the density-wave approach to reconstructive phase transitions. In contrast to reconstructive transitions between crystal phases, considered in Chapters III and IV, in which the loss of group-subgroup relationship concerned the point-group symmetry of the phases *but not their translational group* (at least when neglecting the "secondary" macroscopic strains), at the crystal-quasicrystal transition *the translational symmetry is completely lost from one structure to the other:* The long-range order of the quasicrystal structure is no longer crystalline.

Although the point-group symmetry of the quasicrystal contains in some cases the point-group symmetry of the crystal, obviously the (three-dimensional) space-group concept cannot be used for a complete description of the transition mechanism between the two states. In this respect let us emphasize that higher dimensional crystallography [1], which provides an efficient support for a geometrical description of the quasicrystal structure [2] and for its diffractional analysis [3], gives no insight into the mechanism of reorganization of the crystal structure, which results in the structure of the quasicrystal. On the contrary, we will show that the description of the preceding structure in terms of deformation of a family of density waves associated with the crystal's atomic planes [4-8] is well adapted to understand the way a quasicrystal can form, and clarifies in the same time the connections with higher dimensional crystallography.

The phenomenological models which have been proposed initially to explain the stability of icosahedral quasicrystal phases [9-12] are different extensions of the Landau theory of crystallization [13-14]. The choice of the isotropic liquid as the parent phase for the icosahedral structures stemmed from the fact that the first quasicrystals phases found experimentally were obtained by rapid cooling and direct crystallization below the range of stability of the melt. However, further experiments revealed that one could get a quasicrystalline structure under as equilibrium conditions [15], and that a quasicrystal phase could coexist with, or could be obtained from, a crystal phase [16, 17]. A large number of experimental evidence is presently known which shows the close connection existing between the crystalline and quasicrystalline solid states: These include the similarity of their local order [18] and of their electron and X-ray diffraction spectra [19, 20], the proximity of the two states in the phase diagram of quasicrystalline alloys, in which they often coexist in similar stoichiometries [21], and the reversibility found for some crystal-quasicrystal transitions [22, 23]. These results show indeed that quasicrystal phases should be inserted in a more complex phase diagram than the current liquid-to-icosahedral model initially proposed.

Following the preceding considerations, the approach presented in this chapter starts from a solid crystalline phase for describing the reconstructive transition to a quasicrystal structure. The transition mechanism is shown to be essentially displacive, though there exists necessarily a symmetry induced disorder in the quasicrystal state. In the framework of the density wave theory presented in Chapter II, the choice of the crystal structure as the initial structure is, in the present case, arbitrary, and is not the "natural" parent structure from the point of view of the Landau theory. However, since the crystalline order can be described by well established mathematical tools, and can be unequivocally verified by standard diffractional methods, it is clearly more practical to describe the unknown quasicrystal structure, in reference to the (known) crystal structure.

Starting from the atomic distribution found in the crystal structure expanded in Fourier series, the density wave approach to the crystal-quasicrystal transition, consists in finding a finite number of harmonics forming a basic system of density waves which express the inhomogeneous deformation of the crystal structure leading to the quasicrystal structure. As for reconstructive transitions in crystals the approach shows that the stability of the quasicrystal phase depends on specific displacements characterizing icosahedral (or decagonal, octagonal, dodecagonal, etc.) density waves, and not only on general symmetry arguments. The possible formation of a quasicrystal structure is predetermined by the existence of *almost* icosahedral (or decagonal, octagonal, etc.) *lattice complexes* in the parent crystal structure, a fact which constitute a *necessary* (but not sufficient) condition for a quasicrystal structure to take place. An important consequence of this approach is that *the atomic positions in the quasicrystal structure can be entirely specified with respect to their positions in the crystal structure, in the three-dimensional space*.

The density wave description of crystal-quasicrystal transition will be presented in three successive steps. At first we will introduce the general features of the transition mechanism using the example of the icosahedral structures in AlMn(Fe)Si and AlLiCu alloys, in which the mechanism can be simply decomposed. A second step will be to show the full applicability of the density-wave approach to the more complex situations found in the icosahedral $FeTi_2$ and AlCuFe alloys. At last , a *local* description of the transition mechanism, relating the density wave formalism to the formalism used in high-dimensional crystallography, will be exposed through the illustrative examples of the octagonal MnSi alloys, the icosahedral AlCuLi alloys, and the decagonal structure found in MnAl alloys.

2. The crystal-quasicrystal transition mechanism

2.1. Symmetrization process and local icosahedral order

Let us first examine the connections existing between the crystal and icosahedral structures in AlMnSi, starting from the cubic α-AlMnSi structure described by Cooper [24]. The 138 atoms contained in one (primitive) unit-cell of the α-phase (space-group T_h^1) decompose into 11 orbits which, according to the labeling of Ref. [24], can be shared in three groups (taking into account the fact that the diffraction spectra make no difference

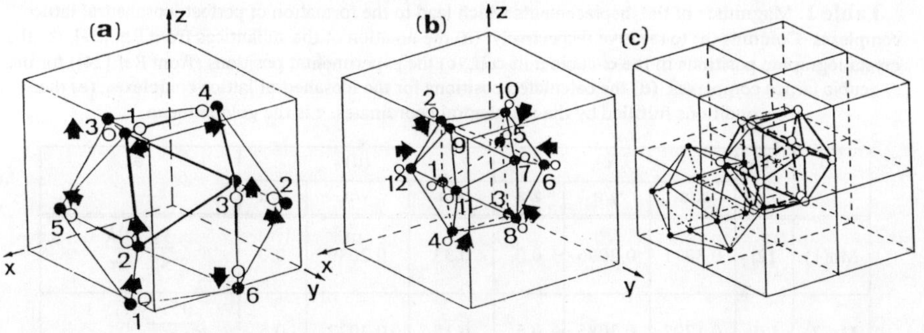

Fig. 1. Icosahedra formed by the atoms belonging to the Al(7) (a) and Mn(2) (b) sublattices. Open and full circles represent the cubic and icosahedral positions respectively. Fig.(c) shows two icosahedral domains of the Al(7) lattice complex corresponding to different displacements indicated in Fig.2. The arrows indicate the antiparallel displacements.

between the Al and Si atoms): Al(1), Al(2) and Al(7) are characterized by a 6-fold position; Al(3), Al(4), Al(8), Al(9), Mn(1) and Mn(2) possess a 12-fold degeneracy, whereas Al(5) and Al(6) have a 24-fold degeneracy.

Let us focus on the atoms of the Al(7) orbit. From Fig.1a one can see that *antiparallel displacements* of each pair of atoms, denoted (1, 4), (2, 5) and (3, 6) in Fig.1a, along the respective cubic directions [100], [010] and [001], bring the twelve atoms forming the Al(7) orbit, i.e. the three preceding pairs plus the translationally equivalent atoms, *to form an icosahedron* for the specific displacements of magnitude 0.244 A°=0.0194a, where a=12.56 A° is the lattice parameter of the cubic unit-cell [24]. The same *displacement field* applied to the atoms of the orbits Mn(1), Mn(2), Al(3) and Al(4) also gives icosahedral *lattice complexes*. Thus, Al(3), Mn(1) and Al(7) form three embedded icosahedra of increasing volume, centered at the origin of the unit-cell, whereas Al(4) and Mn(2) form two embedded icosahedra centered at ($\frac{1}{2}\frac{1}{2}\frac{1}{2}$) (Fig.1b).

Table 1 summarizes the magnitude of the displacements corresponding to each of the preceding orbits. Besides, one can show that under the same displacement field, the atoms belonging to the Al(1) and Al(2) orbits give *octahedral* lattice complexes (respectively centered at ($\frac{1}{2}\frac{1}{2}\frac{1}{2}$) and (000)). Al(9) and Al(11) become *cuboctahedra* (centered at ($\frac{1}{2}\frac{1}{2}\frac{1}{2}$)) whereas Al(5) and Al(6) remain regular polyhedra with 24 vertices.

The *symmetrization process* which transforms a crystal displaying a local order with slightly deformed icosahedral lattice complexes into a crystal with *perfect local icosahedral lattice complexes*, can be described in the framework of the Landau theory, as an *isostructural transition* (see Sec.4.5 of Chapter I) since only the local arrangement of atoms are changed but not the macroscopic symmetry. The basis functions spanning the

Table 1. Magnitude of the displacements which lead to the formation of perfect icosahedral lattice complexes. Columns (a) to (e) give respectively: (a) the notation of the sublattices from Ref.[24], (b) the crystallographic positions in the α-cubic unit-cell, (c) the experimental positions (from Ref.[24]) for the cubic lattice complexes, (d) the calculated positions for the icosahedral lattice complexes, (e) the conditions fulfilled by the icosahedral coordinates. τ is the golden mean.

(a)	(b)	(c)			(d)			(e)
		x/a	y/a	z/a	x/a	y/a	z/a	
Mn(1)	12j	0.3271	0.2006	0.0	0.33	0.2039	0.0	$\frac{y_{ic}}{x_{ic}} = \frac{1}{\tau}$
Mn(2)	12k	0.1797	0.3085	0.5	0.18	0.3022	0.5	$\frac{05 - y_{ic}}{05 - x_{ic}} = \frac{1}{\tau}$
Al(3)	12j	0.1636	0.0997	0.0	0.165	0.102	0.0	$\frac{y_{ic}}{x_{ic}} = \frac{1}{\tau}$
Al(4)	12k	0.3342	0.3990	0.5	0.335	0.398	0.5	$\frac{05 - y_{ic}}{05 - x_{ic}} = \frac{1}{\tau}$
Al(7)	6f	0.2897	0.0	0.5	0.3091	0.0	0.5	$x_{ic} = 0.3091$ $z_{ic} = 0.5$

antiparallel displacement field of the Al(7) atoms can be expressed in terms of symmetric coordinates, using the notation of Fig. 1a, as

$$\varphi_7 = x_1 - x_4 + y_2 - y_5 + z_3 - z_6 \tag{5.1}$$

where φ_7 transforms as the identity representation, denoted Γ_1, of the T_h^1 space-group, at the center of the primitive cubic Brillouin-zone. Furthermore, it is easy to verify that although the atomic displacements associated with the other orbits are different (i.e. they are spanned by different functions $\varphi_{i \neq 7}$) they transform as the same representation Γ_1. Accordingly, the whole displacement field is associated with *a one-component order-parameter* η, having the Γ_1-symmetry. Expressing the distributions of atoms for each sublattice before and after the displacement field in terms of lattice functions ψ(x, y, z) (see Chapter II), the corresponding increment δψ provides the dependence of the order-parameter as a function of the *distance* $\Delta x = x_i - x_j$ between antiparallel atoms (e.g. atoms 1 and 4 in Fig. 1a) *which belong to the same cluster*. One finds:

$$\eta = \eta_0 \left\{ \sin\left[\frac{2\pi}{a}\left(x_i - x_j\right) - \frac{\pi}{2} \right] + \cos\frac{2\pi}{\tau} \right\} \tag{5.2}$$

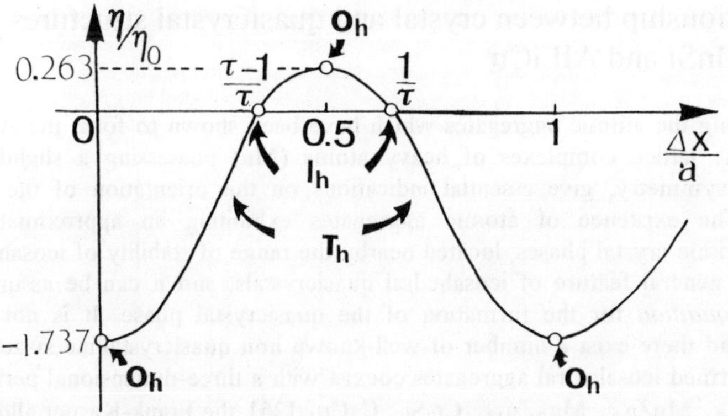

Fig. 2. Periodic dependence of the order-parameter associated with the isostructural T_h^1 (cubic lattice complexes) - T_h^1 (icosahedral lattice complexes) transition for the atoms belonging to the Al(7) sublattice.

where $\tau = (1+\sqrt{5})/2$ is the golden mean. The form of the function (5.2) which holds for any displacement of the atoms, is represented in Fig.2. The points on the curve correspond however to the specific displacements associated with the Al(7) atoms. Thus, for the specific displacements $\Delta x = a\tau^{-1}$ and $a(\tau-1)\tau^{-1}$ (modulo a) the preceding atoms form two sets of icosahedral lattice complexes which can be deduced one from another by a shift of [a/2 a/2 a/2] *and* a rotation of 90° around the [001] cubic axis. For the specific displacements corresponding to such icosahedral *domains* the point-group symmetry of the Al(7) lattice complex increases from T_h to I_h. Fig.1c shows two differently oriented domains of the Al(7) complex. Fig.2 shows that the extrema of the $\eta(\Delta x)$ curve the point-group symmetry of the lattice complex increases from T_h to O_h for the antiparallel displacements $\Delta x=(2n+1)a/2$ and $\Delta x=na$, the lattice complex acquiring a fourfold axis.

At this point it should be stressed that the assumed displacive field mechanism takes place *within the initial cubic unit-cell* and has to be complemented by an additional mechanism in order to lose the cubic periodicity and stabilize a three-dimensional quasicrystalline order. In reality the two mechanisms occur *correlatively*, and the formation of ideal lattice complexes within a cubic structure does not correspond to a realistic situation, and cannot be separate from the full transition mechanism. Nevertheless the symmetrization process described herebefore will be used in the density-wave description of the crystal-quasicrystal transition, as it will be shown to reduce the number of critical wave-vectors involved in the transition (see Sec.3).

Before dealing with the second part of the crystal-quasicrystal transition mechanism, some preliminary considerations are needed on the relationship existing between the crystal and quasicrystal structures.

2.2. Relationship between crystal and quasicrystal structures in AlMnSi and AlLiCu

Among the atomic aggregates which have been shown to form the AlMnSi cubic structure, the lattice complexes of heavy atoms (Mn) possessing a slightly deformed icosahedral symmetry, give essential indications on the orientation of the quasicrystal structure. The existence of atomic aggregates exhibiting an approximate five fold symmetry within crystal phases, located nearby the range of stability of icosahedral phases constitute a general feature of icosahedral quasicrystals, and it can be assumed to be a *necessary condition* for the formation of the quasicrystal phase. It is not a sufficient condition, and there exist a number of well-known non quasicrystalline systems in which slightly deformed icosahedral aggregates coexist with a three-dimensional periodicity such as β-Mn, B_{12}, $MgZn_2$, Mg_2Zn_{11}, Cr_3Si, $CaCu_5$ [25], the Frank-Kasper alloys [26] etc..

In AlMnSi the deformed Mn icosahedra define a BCC lattice with the following relationships:

$$[100]_c \| A_2 \,,\ [111]_c \| A_3 \,,\ \left[\tau^2 01\right]_c \| A_3 \,,\ [10\tau]_c \| A_5 \tag{5.3}$$

where A_2, A_3 and A_5 denote two, three and five fold axes of the icosahedral AlMnSi structure [27, 28]. As shown in Fig.3a, (5.3) expresses the simple situation of an icosahedron inscribed into a cube (in the reciprocal space) due to the fact that the point group of α-AlMnSi (T_h) is a subgroup of icosahedral AlMnSi (Y_h). These relations reflect

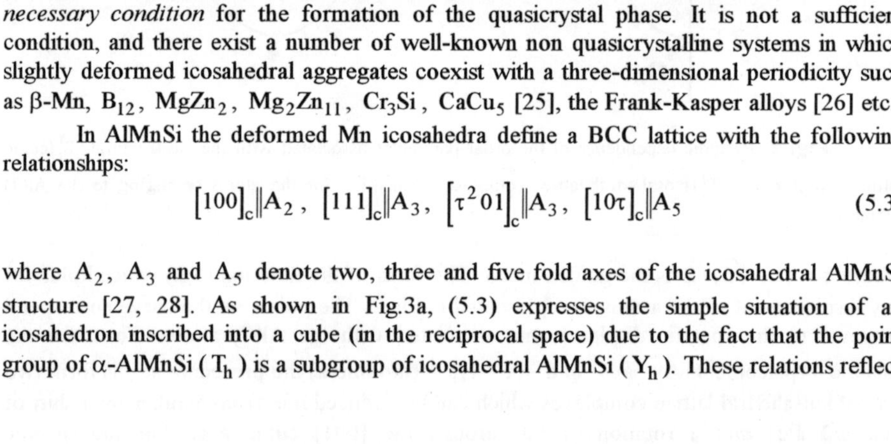

Fig. 3. (a) Orientational relationship between the cubic and icosahedral axes in AlMnSi. (b) Projection along [001] of the sublattice of Mn atoms. d represents half the distance between the pentagonal planes.

Table 2. More intense reflections from the diffraction spectrum of AlMnSi, corresponding to axes of order 5, 3, and 2 of the basic icosahedron of heavy atoms (from Ref.[29]).

Index	Planes	Intensity (arb. units)	d (A°)
(1 1 0 0 0 1)	*h* 2	22	3.8499
$(1\ 1\ 1\ 0\ \bar{1}\ 1)$	*c* 2, 3, 5	8	3.34924
$(3\ 1\ \bar{1}\ \bar{1}\ \bar{1}\ 1)$	2	1.5	
(1 0 0 0 0 0)	*a* 2	100	2.1696
(1 1 0 0 0 0)	*d* 2, 3, 5	78	2.0648
(2 2 0 0 0 2)	*j* 2	1	1.9392
(1 1 1 0 0 0)	2	11	
(1 0 1 0 0 0)	*f* 2, 3, 5	20	1.2749
(2 0 0 0 0 0)	*r* 2	7	1.0852
(2 2 1 0 0 2)	2	3	
(2 2 0 0 0 0)	*r* 2, 3, 5	1	1.0368

equivalently the existence in the crystal structure of atomic planes into which the atoms form quasi-pentagonal motifs. In AlMnSi these planes belong to the family $(035)_c$ which slightly differs from the icosahedral planes $(0\frac{5}{\tau}5)$=(0 3.095 5) that are parallel to the icosahedra formed by Mn atoms. As a consequence, the main interplanar distances of the quasicrystal, namely those associated with the more intense reflections of the diffraction spectra in the icosahedral structure [29], correspond to planes which are perpendicular to the symmetry axes of order 5, 3 and 2 of the *basic icosahedron of heavy atoms*. This is illustrated in the results shown in Table 2, which is extracted from Ref.[29], showing that the more intense reflections in the diffraction spectrum of AlMnSi are indeed the ones perpendicular to the preceding planes. It is also illustrated in Fig.3b where one can see that the main reflections of AlMnSi (corresponding to the interatomic distance d=2.169 A°) coincide with half the distance between the pentagonal plane of the basic icosahedron of Mn atoms.

The connection between the crystal and icosahedral structures in AlLiCu provides another illustration of the preceding considerations. The orientational relationship between the two structures is given in Fig.4a, and Fig.4b shows the icosahedral lattice complexes which are found in the cubic structure of BCC symmetry T_h^5, as described in Ref.[36]. Thus in the smallest icosahedra, denoted 1 in Fig.4b 89% of the vertices are occupied by Al atoms and only 11% by Cu atoms, whereas the embedding (twice larger) triacontahedra, denoted 3 in Fig.4b, have their vertices occupied in nearly equal proportions (54%Al and 46%Cu [30]). Thus it is this latter structure, the richest in heavy Cu atoms, which will constitute the basic structure determining the orientation of the

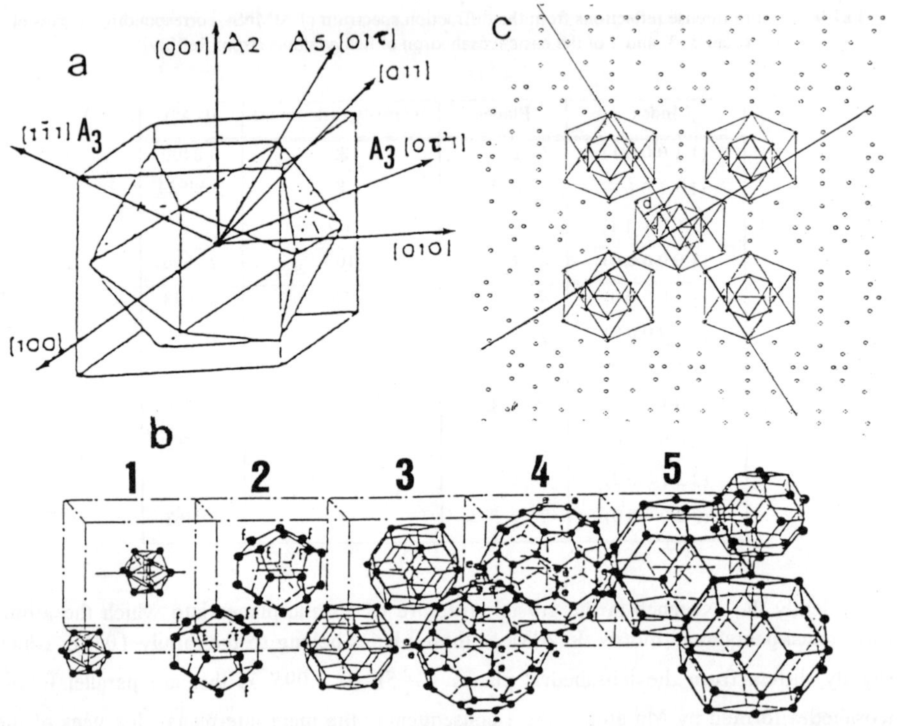

Fig. 4. Structure of AlLiCu from Ref.[29]. (a) Orientational relationship between the crystal and quasicrystal axes. (b) Lattice complexes in the BCC structure: (1) Small icosahedra Al-Cu centered at (000) and (1/2 1/2 1/2), (2) Li dodecahedron, (3) Al-Cu triacontahedra, (4) Truncated icosahedra formed by Al and Al-Cu mixture (Samson polyhedra), (5) Large triacontahedra connected by their faces along [100]. (c) Projection along [001] of the Cu sublattices. d is half the distance between the pentagonal planes of the Al icosahedra.

quasilattice. As in AlMnSi, the distance between the pentagonal planes of the preceding Al-Cu triacontahedra corresponds to the double of the distance between the quasiperiodic planes (Fig.4c) which itself coincide with the interplanar distance associated with the more intense reflections of the X-ray spectrum of the icosahedral structure [31, 32].

The preceding remarks have stressed the influence of the local structure of the crystal on the structure of the quasicrystal. This is verified by a number of observations showing that the quasicrystal structure is largely reminiscent of the crystal structure: deformed fragments of the initial structure are observed in the quasicrystal structure [33, 34], which also reflects the symmetry operations of the crystal which have been *lost* in the quasicrystal, i.e. one can find antiphase domains and antiphase domain walls corresponding to the loss of translational symmetry [35, 36]. The actual traces of the crystal structure in the quasicrystal are investigated on a theoretical ground in Sec.4.

2.3. Breaking of the translational crystal symmetry: Long-range quasicrystalline order in AlFeSi

In the present section we will develop our theoretical model of the crystal-quasicrystal transition through the example of AlFeSi which has a BCC symmetry (T_h^5), instead of the primitive cubic symmetry of AlMnSi (T_h^1), since the lattice complexes located at the corner and center of the primitive cell (Fig.1) are equivalent, due to different positions occupied by Al atoms [24].

2.3.1. The incommensurate wave-vector associated with the crystal-quasicrystal transition

The orientational relationships (5.3) between the BCC and icosahedral AlFeSi structures correspond to *almost but not fully coinciding directions*. One of the basic ideas of our model is that the transition mechanism between the two preceding structures consists in an *inhomogeneous deformation* of the initial structure which *brings about the cubic directions along the icosahedral ones*. The deformation is inhomogeneous as it involves a *rotation* of the cubic directions and simultaneously a change in the *lengths* of the basic vectors of the cubic structure. At the atomistic level the mechanism brings into coincidence the cubic planes of the system $\{035\}_c$ and the icosahedral system of planes $\{0\frac{5}{\tau}5\}$ which, in the six-dimensional notation, correspond to the planes [100000] associated with the maximal intensity in the diffraction pattern.

The system of planes {035} are characterized in the reciprocal (FCC) space of the BCC structure by the six wave vectors:

$$\vec{q}_1^c = 4\vec{b}_1 + \vec{b}_2 - \vec{b}_3, \quad \vec{q}_2^c = -\vec{b}_1 + 4\vec{b}_2 + \vec{b}_3, \quad \vec{q}_3^c = \vec{b}_1 - \vec{b}_2 + 4\vec{b}_3,$$
$$\vec{q}_4^c = \vec{b}_1 + 4\vec{b}_2 - 4\vec{b}_3, \quad \vec{q}_5^c = 4\vec{b}_1 - \vec{b}_2 - 4\vec{b}_3, \quad \vec{q}_6^c = 4\vec{b}_1 - 4\vec{b}_2 - \vec{b}_3 \qquad (5.4)$$

where $\vec{b}_1$, $\vec{b}_2$ and $\vec{b}_3$ are the basic vectors of the reciprocal space:

$$\vec{b}_1 = \left(0, \frac{2\pi}{a}, \frac{2\pi}{a}\right), \quad \vec{b}_2 = \left(\frac{2\pi}{a}, 0, \frac{2\pi}{a}\right), \quad \vec{b}_3 = \left(\frac{2\pi}{a}, \frac{2\pi}{a}, 0\right).$$

Let us note that the $\vec{q}_i^c$ (i=1-6) are the branches, transforming by the symmetry operations of T_h, of the star $\vec{q}_1^*$, the modulus $\left|\vec{q}_1^c\right|^{-1}$ corresponding to d/2π where d is the distance between two cubic planes of the {035} system. On the other hand, as shown in Fig.5a, the average rotations of the planes (035) which brings these plane along the icosahedral planes

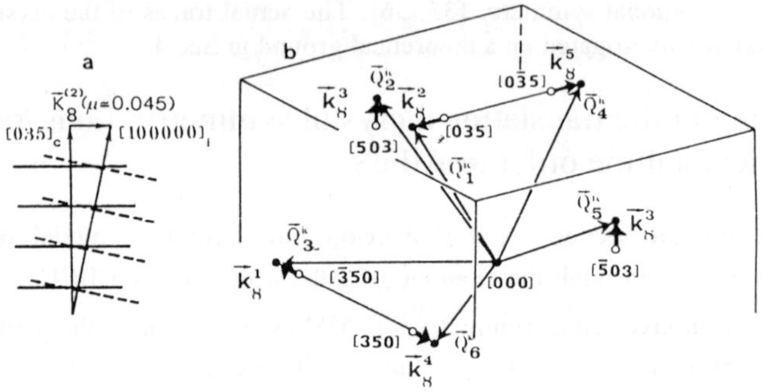

Fig. 5. (a) Reorientation of the cubic $[035]_c$ direction into the icosahedral $[100000]_i$ direction, via the deformation wave-vector $k_8^2 = 0.045(0,2\pi/a,0)$. Full and dashed lines symbolize layers of atoms perpendicular to the preceding directions. (b) Orientation of the $\vec{q}_i^{ic}$ icosahedral basis, and of the arms of $\vec{k}_8^*$ in reciprocal space.

$(0\frac{5}{\tau}5)$ is accounted for by a wave-vector oriented along the cubic $[010]_c$ direction, i.e. having the coordinates:

$$\vec{k} = \mu\left(0,\frac{2\pi}{a},0\right) \tag{5.5}$$

the value of μ being given by half the number obtained by substracting the coordinates corresponding to $(0\frac{5}{\tau}5)$ and (035), namely $\mu=(5/\tau-3)/2\cong0.045$. This value, which shows that $\vec{k}$ is close to the center of the BCC Brillouin-zone, corresponds to a deformation wave of about 11.1 cubic-cell wave length: $\vec{k}$ *is the incommensurate wave-vector which breaks the crystal translational symmetry and transforms the cubic structure into a long-range icosahedral structure.*

The star $\vec{k}^*$ (which is denoted $\vec{k}_8$ in Kovalev's notations [37]) possesses six arms which are:

$$\vec{k}_8^1 = \mu(-\vec{b}_1+\vec{b}_2+\vec{b}_3),\ \ \vec{k}_8^2 = \mu(\vec{b}_1-\vec{b}_2+\vec{b}_3),\ \ \vec{k}_8^3 = \mu(\vec{b}_1+\vec{b}_2-\vec{b}_3),$$
$$\vec{k}_8^4 = -\vec{k}_8^1,\ \ \vec{k}_8^5 = -\vec{k}_8^2,\ \ \vec{k}_8^6 = -\vec{k}_8^3. \tag{5.6}$$

Using (5.4) and (5.6) one can thus define the six-dimensional basis which will be used to construct the *set of basic density waves necessary for determining the icosahedral structure*:

$$\vec{q}_1^{ic} = \vec{q}_1^{c}+\vec{k}_8^2,\ \ \vec{q}_2^{ic} = \vec{q}_2^{c}+\vec{k}_8^3,\ \ \vec{q}_3^{ic} = \vec{q}_3^{c}+\vec{k}_8^1,$$
$$\vec{q}_4^{ic} = \vec{q}_4^{c}+\vec{k}_8^5,\ \ \vec{q}_5^{ic} = \vec{q}_5^{c}+\vec{k}_8^6,\ \ \vec{q}_6^{ic} = \vec{q}_6^{c}+\vec{k}_8^4. \tag{5.7}$$

Fig.5b shows the orientation of the $\vec{q}_i^{ic}$ wave-vectors within the cubic reciprocal space. One can note that these vectors lie *outside* the first Brillouin-zone. This is not surprising as they express the "propagation" of icosahedral lattice complexes which have a smaller volume than the BCC unit-cell, and thus correspond to a larger volume in reciprocal space. Let us also underline that the $\vec{q}_i^{ic}$ as well as the $\vec{k}_i$ vectors, fulfill the sufficient conditions selecting wave-vectors that may create icosahedral density waves, namely: 1) to be formed by at least six independent arms, and 2) to allow construction of twelve vectors pointing towards the vertices of a regular icosahedron in reciprocal space, by combination with the reciprocal lattice vectors. The preceding conditions are necessary for wave-vectors corresponding to *irreducible* stars, but not to *reducible* stars, as it will be illustrated in Sec.3.

2.3.2. Deformation of the system of stationary density-waves. Atomic positions in the icosahedral structure

Using the density-wave formalism introduced in Chapter II (Sec.3.2), one can write the lattice function in the crystal structure as:

$$\psi_c^e(\vec{r}) = \frac{1}{NV}\sum_{i=1}^{N}\sum_{j=1}^{6} e^{i\vec{q}_j^c(\vec{r}_i^c - \vec{r})} + \text{c.c.} \tag{5.8}$$

where N is the total number of atoms contained in a sample of volume V. The $\vec{r}_i^c$ are the equilibrium vector positions of the atoms in the cubic structure and the $\vec{q}_j^c$ are the six wave-vectors of the BCC reciprocal lattice given by (5.4). The system of six basic density waves contained in (5.8) is deformed into the system of icosahedral density waves, expressed by the lattice function:

$$\psi_{ic}^e(\vec{r}) = \frac{1}{NV}\sum_{i=1}^{N}\sum_{j=1}^{6} e^{i\vec{q}_j^{ic}(\vec{r}_i^{ic} - \vec{r})} + \text{c.c.} \tag{5.9}$$

where the $\vec{r}_i^{ic}$ are the equilibrium vector positions of the atoms in the icosahedral structure.

In order to obtain the critical displacements associated with the cubic-to-icosahedral transition, and therefore to deduce the icosahedral position vectors $\vec{r}_i^{ic}$ from the cubic positions $\vec{r}_i^c$, one has to maximize with respect to the $\vec{r}_i$, the amplitude $A_{ic}(\vec{r}_i)$ of the non-equilibrium lattice function:

$$\psi_{ic}(\vec{r}, \vec{r}_i) = \frac{1}{NV}\sum_{i=1}^{N}\sum_{j=1}^{6} e^{i\vec{q}_j^{ic}(\vec{r}_i - \vec{r})} + \text{c.c.} \tag{5.10}$$

which is obtained by pairing in (5.10) the basis functions $\varphi_j^{ic}(\vec{r}) = \exp\{-i\vec{q}_j^{ic}\vec{r}\}$ with their complex conjugate. One gets:

$$A_{ic}(\vec{r}_i) = \frac{1}{NV}\sum_{i=1}^{N}\sum_{j=1}^{6}\cos\left(\vec{q}_j^{ic}\vec{r}_i\right). \quad (5.11)$$

Let us stress that the maximization of $A_{ic}(\vec{r}_i)$ has to be performed *in the vicinity of the cubic positions*, as the function $A(\vec{r}_i)$ possesses several maxima and only one coincides with the critical displacements corresponding to the BCC-icosahedral transition. In other words one assumes that only the maxima involving the *smallest* displacements from the cubic positions (compatible with the symmetry of the collective mechanism of deformation) are taken into account.

The experimental values found for the atomic positions in the BCC phase of AlFeSi [24] are reported in Table 3 column (c). They determine a maximum of the amplitude

$$A_c(\vec{r}_i) = \frac{1}{NV}\sum_{i=1}^{N}\sum_{j=1}^{6}\cos\left(\vec{q}_j^{c}\vec{r}_i\right) \quad (5.12)$$

associated with the cubic non-equilibrium lattice function (which has the form (5.8) in which the equilibrium vector positions $\vec{r}_i^{\,c}$ are replaced by the non-equilibrium vector-positions $\vec{r}_i$). Columns (d) and (e) of Table 3 list respectively the calculated icosahedral positions and the corresponding magnitudes of the average critical displacements from the initial (BCC) to the final (icosahedral) structure. One can note from column (e) that the largest displacements, which correspond to the Al(9) and Al(2) atomic sublattices, are of the order of 4% of the cubic lattice parameter a number which is commonly found among first-order reconstructive transitions in crystals. Along the same line one finds for the variation of volume at the transition $\Delta V/V \cong 1.6\%$.

The maximization procedure corresponding to Eqs.(5.8)-(5.12) implies the *preliminary knowledge of the wave-vectors* $\vec{q}^{\,c}$ *and* $\vec{k}$ which have been deduced from the analysis of the diffraction spectra of AlFeSi. It consists in the selection of the length and orientation of the relevant deformation (incommensurate) wave-vector $\vec{k}$, and results in the construction of the icosahedral wave-vectors $\vec{q}_j^{ic}$. Formally the critical value and orientation found for $\vec{k}$ should maximize a variational lattice function $\psi(\vec{k},\vec{r},\vec{r}_k)$ the expression of which was used in the cubic ($\psi_c(\vec{k}^c = 0,\vec{r},\vec{r}_k^{\,c})$) and icosahedral ($\psi_{ic}(\vec{k}^{ic},\vec{r},\vec{r}_i^{\,ic})$) phases. This preliminary deduction of essential informations on the transition mechanism from the experimental data is typical of the phenomenological approach to reconstructive transitions. It is equivalent, for example, to the knowledge of the direction of the critical displacements assumed in the descriptions of the displacive-type reconstructive transitions (see Chapter III).

Table 3. Magnitude of the average critical displacements (column (e)) which lead to the formation of the icosahedral structure in AlFeSi. (a) Notation of the sublattices. (b) Crystallographic positions. (c) Experimental positions in the BCC structure from Cooper [24]. (d) Calculated positions in the icosahedral structure following the maximization of Eq.(5.10).

(a)	(b)	(c)			(d)			(e)
		x/a	y/a	z/a	x/a	y/a	z/a	Δ(A°)
Fe(1)	24g	0.3243	0.1981	0.0	0.3095	0.1916	0.0	0.203
Fe(2)		0.1757	0.3019	0.5	0.1905	0.3084	0.5	
Al(1)	12d	0.3777	0.0	0.0	0.3838	0.0	0.0	0.077
Al(2)		0.1223	0.5	0.5	0.1162	0.5	0.5	
Al(3)	24g	0.1651	0,1006	0.0	0.1550	0.0960	0.0	0.139
Al(4)		0.3349	0.3994	0.5	0.3450	0.4040	0.5	
Al(5)	48h	0.1146	0.1872	0/3003	0.1167	0.1915	0.3114	0.152
Al(6)		0.3854	0.3128	0.1997	0.3833	0.3085	0.1886	
Al(7)	12e	0.2990	0.0	0.5	0.3078	0.0	0.5	0.111
Al(8)		0.2010	0.5	0.0	0.1922	0.5	0.0	
Al(9)	24(g)	0.3266	0.4030	0.0	0.3674	0.3916	0.0	0.532
Al(12)		0.1734	0.0970	0.5	0.1326	0.1084	0.5	
Al(10)	24g	0.3725	0.3830	0.0	0.3773	0.3881	0.0	0.088
Al(11)		0.1275	0.1170	0.5	0.1227	0.1119	0.5	

Figs.6-8 show computer simulations which have been performed using a cluster of $N=10^4$ Fe atoms. It corresponds to homogeneous crystallites of about 10^3 A° in consistence with the correlation lengths of some hundred of A° reported by Bancel et al. [29] in AlFe(Mn)Si alloys. Furthermore the number N is large enough to obtain relative intensities of electron scattering within an accuracy not worst than 5%.

Figs.6a and 6b show the calculated electron diffraction patterns for the Fe atoms for the initial cubic structure (6a) after the symmetrization process, and for the final icosahedral structure (6b). From one figure to the other one can verify the disappearance of some weak reflections characterizing the cubic structure, as well as a modification of the positions and intensities of the reflections resulting in the onset of a typical five-fold icosahedral pattern.

Fig.7 illustrates the coincidence of the calculated pattern with the high-resolution electron microscopy (HREM) pattern obtained by Audier and Guyot [28]. A good agreement is found between the calculated positions (open circles) and the schematic representation of the experimental image (full circles), although there exist a number of additional calculated positions (defects).

Fig.6b corresponds to the “critical” length of the wave-vector k_8 (μ=0.045) and to the critical displacements obtained by maximization of Eq.(5.11). For non-critical values of μ or for non-maximal amplitudes $A_{ic}(\vec{r}_i)$ one obtains non-icosahedral superstructure reflections which are represented in Figs.8a-c. Thus Figs.8a and 8b correspond to the non

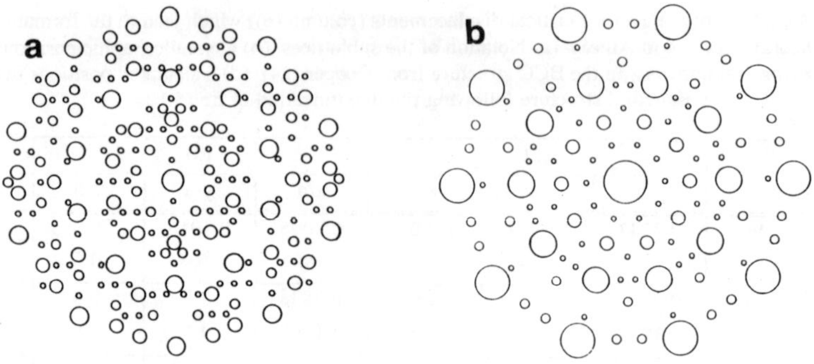

Fig. 6. Calculated electron diffraction patterns for the Fe atoms: (a) cubic structure after the symmetrization process, along the $[035]_c$ direction. (b) Final icosahedral structure along the [100000] direction. A doubled scale of intensities is used in (a) with respect to (b).

critical wave vectors with μ=0.022 and μ=0.011, whereas Fig.8c is obtained for a critical wave vector but for non-maximal amplitude. One can see that when getting away from the central spot, the five fold symmetry is progressively destroys, i.e. the typical icosahedral angles (36° and 72°) which are found in Fig.6b are lost and the intensities become disymmetrical in the deformed five and tenfold patterns.

Since the value of μ determines both the angles between the arms of the star $\vec{k}^*$ and their magnitude, and that the maximum value of $A(\vec{r}_i)$ specifies the orientation of $\vec{q}$, one can say that the icosahedral structure represents a *limit situation to an incommensurate regime*, both in the magnitude and orientation of the density-wave vector. Only for a *determined orientation and magnitude* of $\vec{q}_j^{\,ic}$ the point-group symmetry of the structure *increases* to I_h, while for all other orientations and (or) magnitudes the point-group of the corresponding incommensurate structures (i.e. the symmetry of the average structure) is T_h. By analogy with the situations described in the preceding chapters, the

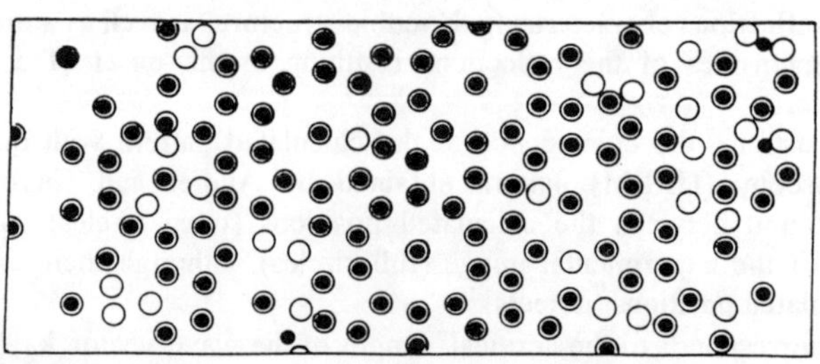

Fig. 7. Coincidence of the calculated electron diffraction pattern of AlFeSi (open circles) along [100000], with the HREM pattern obtained by Audier and Guyot [28].

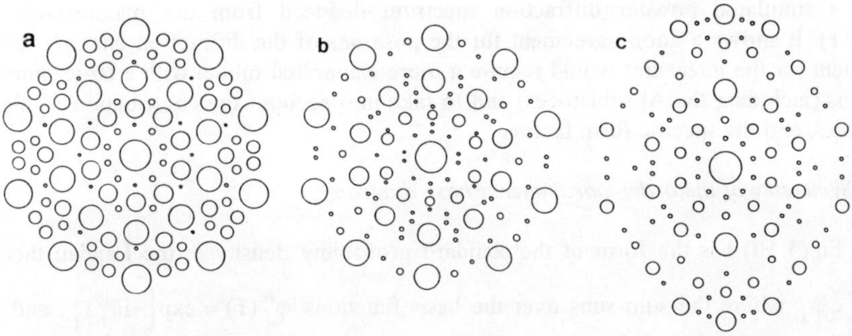

Fig. 8. Calculated electron diffraction patterns for AlFeSi, along [100000] for three non-icosahedral structures corresponding to (a) μ=0.022 and $A(\vec{\tau}_1)$ maximal, (b) μ=0.111 and $A(\vec{\tau}_1)$ maximal, (c) μ=0.045 and $A(\vec{\tau}_1)$ non-maximal.

icosahedral state thus represents a non-Landau state, in which the reflections and intensities should remain constant as a function of temperature and pressure, within the range of stability of the icosahedral phase. In this respect it has to be stressed that incommensurate (non-icosahedral) structures found as non-limit situations, differ in an essential manner from incommensurate structures in insulators [38] or long period modulations in alloys [39] by the property that the associated propagation vectors correspond to *irrational directions* with respect to the cubic axes, and not only to irrational lengths in terms of reciprocal basis vector units. Besides, two classes of incommensurate modulations can be distinguished depending if $\vec{k}$ is non critical or critical. In the first case the modulation should vary with external variables whereas it should be fixed in the second case. Fig.9

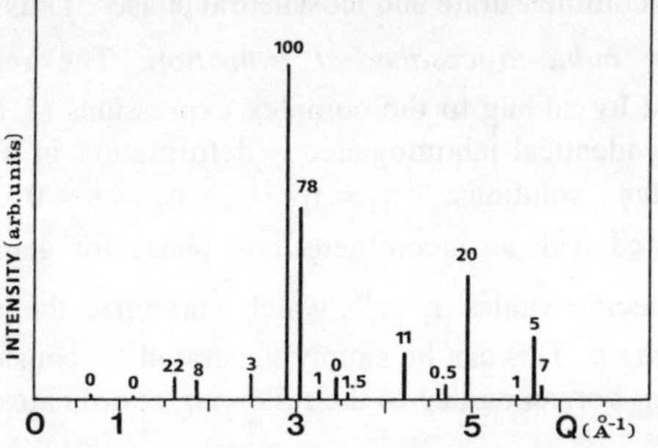

Fig. 9. Simulated powder diffraction spectrum, taking into account only the Fe sublattices. The numbers on the top of the diffraction lines are the experimental values found by Bancel et al. [29].

shows a simulated powder diffraction spectrum deduced from the maximization of (Eq.5.11). It shows a good agreement for the positions of the diffraction lines. A similar agreement for the intensities would require a more elaborated model with a larger number of atoms (including the Al sublattices) and to take into account the contribution of higher harmonics, and the specific form factors.

2.3.3. Symmetry of the order-parameter. Phase diagram

Eq.(5.10) has the form of the standard probability density in the Landau theory: $\rho = \sum_{j=1}^{6} c_j \varphi_j$ where the sum runs over the basis functions $\varphi_j^{ic}(\vec{r}) = \exp\left\{-i\vec{q}_j^{ic}\vec{r}\right\}$, and the coefficient $c_j(r_i) = \frac{1}{NV}\sum_{i=1}^{N}\exp\left\{iq_j^{ic}r_i\right\}$ form the basis of the six-dimensional irreducible representation $\tau_1(\vec{k}_8^*)$ of the T_h^5 space group. One can define the 6 functions:

$$\eta_1(\vec{r}_i) = \frac{1}{NV}\sum_{i=1}^{N}\left[e^{i\vec{q}_3^{ic}(\vec{r}_i-\vec{r}_i^{c})} + e^{i\vec{q}_6^{ic}(\vec{r}_i-\vec{r}_i^{c})}\right], \quad \eta_4(\vec{r}_i) = \eta_1^*(\vec{r}_i),$$

$$\eta_2(\vec{r}_i) = \frac{1}{NV}\sum_{i=1}^{N}\left[e^{i\vec{q}_1^{ic}(\vec{r}_i-\vec{r}_i^{c})} + e^{i\vec{q}_4^{ic}(\vec{r}_i-\vec{r}_i^{c})}\right], \quad \eta_5(\vec{r}_i) = \eta_2^*(\vec{r}_i),$$

$$\eta_3(\vec{r}_i) = \frac{1}{NV}\sum_{i=1}^{N}\left[e^{i\vec{q}_2^{ic}(\vec{r}_i-\vec{r}_i^{c})} + e^{i\vec{q}_5^{ic}(\vec{r}_i-\vec{r}_i^{c})}\right], \quad \eta_6(\vec{r}_i) = \eta_3^*(\vec{r}_i). \qquad (5.13)$$

The η_i also form the basis of $\tau_1(\vec{k}_8^*)$. Besides one has $\eta_i=0$ for all i=1-6 in the cubic phase and $\eta_i \neq 0$ in the incommensurate and icosahedral phases. Thus the $\eta_j(\vec{r}_i)$ *define the order-parameter of the cubic-to-icosahedral transition.* The real form of the η_i components are obtained by adding to the complex expressions (5.13) the corresponding complex conjugates. An identical inhomogeneous deformation in all directions of space will correspond to the solutions: $\eta_1 = \eta_2 = ... = \eta_6 = \eta \neq 0$. These equilibrium relationships are associated with an incommensurate phase for general values of the r_i whereas only for the specific values $r_i = r_i^{ic}$ which maximize the amplitude (5.11) one obtains the icosahedral phase. This can be simply illustrated by considering the orbit of 24 Fe atoms in position 24(g) corresponding to the following coordinates:

$$r_1 = -r_{13} = (x,y,z),\ r_2 = -r_{14} = (x,\bar{y},\bar{z}),\ r_3 = -r_{15} = (\bar{x},y,\bar{z}),\ r_4 = -r_{16} = (\bar{x},\bar{y},z),$$
$$r_5 = -r_{17} = (y,z,x),\ r_6 = -r_{18} = (y,\bar{z},\bar{x}),\ r_7 = -r_{19} = (\bar{y},z,\bar{x}),\ r_8 = -r_{20} = (\bar{y},\bar{z},x),$$
$$r_9 = -r_{21} = (z,x,y),\ r_{10} = -r_{22} = (z,\bar{x},\bar{y}),\ r_{11} = -r_{23} = (\bar{z},x,\bar{y}),\ r_{12} = -r_{24} = (\bar{z},\bar{x},y).$$

Introducing these coordinates in (5.11) one gets:

$$A \cong \frac{1}{n}\sum_{i=1}^{n}\left[\cos(10\pi y)\cos(10\pi x/\tau)+\cos(10\pi z)\cos(10\pi y/\tau)+\cos(10\pi x)\cos(10\pi z/\tau)\right] \quad (5.14)$$

where n is the total number of different orbits of Fe atoms. Minimizing with respect to x, y, and z, one gets the equations of state:

$$\sin(10\pi x/\tau)\cos(10\pi x)+\tau\sin(10\pi x)\cos(10\pi z/\tau)=0 \quad (5.15)$$

plus the two analogous equations obtained by circular permutation. The solutions of these equations, which maximize the amplitude (5.11), *provide the values* (x^{ic}, y^{ic}, z^{ic}) *corresponding to the positions of the Fe atoms in the icosahedral phase*. Equations similar to (5.14) can be found for the other atomic orbits.

The *effective form* of the Landau potential deduced from the symmetry $\tau_1(\vec{k}_8^*)$ and of the equality of the order-parameter components (which insure the identity between the various icosahedral planes) $\eta_i = \eta$ (j=1-6) is:

$$F(\eta(\vec{r}_i)) = a_1[\eta(\vec{r}_i)]^2 + a_2[\eta(\vec{r}_i)]^4 + a_3[\eta(\vec{r}_i)]^6. \quad (5.16)$$

The corresponding phase diagram which has been discussed in Chapter III (Sec.2.4 and Fig.3) is represented in Fig.10. One can see that the icosahedral phase can be reached from the cubic phase across a line of first-order transitions, while the incommensurate phase is obtained through a second-order or a first-order transition. The two, icosahedral and incommensurate phases are separated by a second-order transition line. However these results possess only a general indicative value as they correspond to a drastic schematization of the experimental and theoretical situation. In order to develop our model, which aims to describe the icosahedral phase, we have assumed a hypothetic

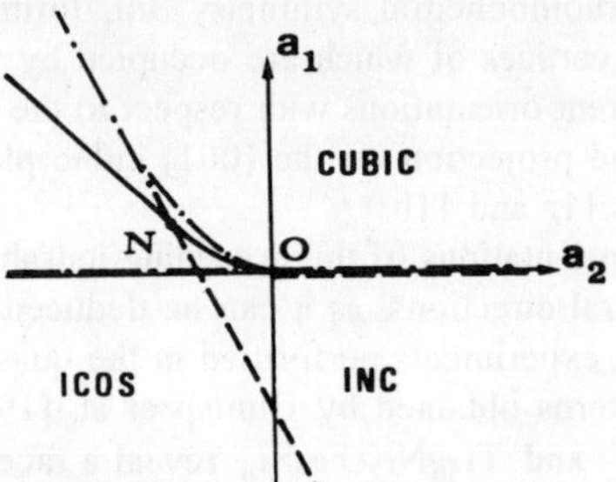

Fig. 10. Theoretical phase diagram of AlMnSi corresponding to the potential (5.16) in the (a_1, a_2) plane. Full, dashed and dashed-dotted lines are respectively first-order transition, second-order transition and limit of stability lines. N is a three-phase point. O is a tricritical point.

transition between the two phases ignoring the surrounding regions of the actual phase diagram of AlMnSi [40, 41]. Accordingly, a realistic theoretical description would have to take into account the concrete environment of the two phases. It would also require to consider the connections with the other equilibrium solutions of the six-dimensional order-parameter, and not only its effective form.

3. Density-wave theory of the FCC-icosahedral transitions in $FeTi_2(NiTi_2)$ alloys

We will now illustrate the applicability of the approach presented in the preceding section by considering the more complex situation found in $FeTi_2$ alloys [6, 7]. In this quasicrystal system the orientational relationship does not correspond, in the reciprocal space, to the simplest configuration of an icosahedron inscribed into a cube, as in AlMn(Fe)Si or AlLiCu alloys, or to a group-subgroup relationship between the point-group symmetries of the crystal and icosahedral structures.

3.1. Orientational relationship between the FCC and icosahedral structures in $FeTi_2$

In the FCC of $FeTi_2$ alloys [42, 43] which have the space group Fd3m (O_h^7) with a=11.3 A°, the 64 titanium atoms are in positions 48f and 16c and the 32 iron atoms are in position 32e. The structure was commonly described [42, 43] as a stacking of 8 octahedral and 8 tetrahedral groups, with 16 atoms in equivalent positions 16c located between two neighboring tetrahedral faces, which are turned from one to another by 60°. The coordination polyhedra of these atoms are two deformed pentagonal prisms with additional vertices. However, as it is shown in Figs.11a and 11b, the preceding description of the structure is not a unique one, as it can also be described by a single type of adjacent icosahedral complexes, of local rhombohedral symmetry 3m, formed by the 12 nearest neighbors of an Fe atom, the 12 vertices of which are occupied by 9 Ti and 3 Fe atoms. The complexes possess four different orientations with respect to the cubic axes, forming a spiral along the fourfold axis. The projection on the (001) cubic plane of a sequence of four clusters is represented in Figs.11a and 11b.

Let us compare the local orientations of the preceding icosahedral complexes with respect to the cubic and icosahedral directions, as it can be deduced from X-ray [19, 44] and electron microscopy [45, 46] experiments performed in the quasicrystal state of these alloys. The X-ray diffraction patterns obtained by Dunlap et al. [19] and Chatterjee and O'Handley [44] on $FeTi_2$, $NiTi_2$ and $Ti_{56}Ni_{23}Fe_5Si_6$ reveal a face-centered icosahedral quasilattice [47] in the reciprocal space with the more intense Bragg peak along the icosahedral (110000) direction. This type of quasilattice can be described by 15 basis vectors, denoted $\vec{q}_i$, pointing towards *the centers of the edges* of a regular icosahedron in the reciprocal space [47] and corresponding to the two-fold axes of the icosahedron. The length of these vectors can be deduced from the interlayer distance d in the direction

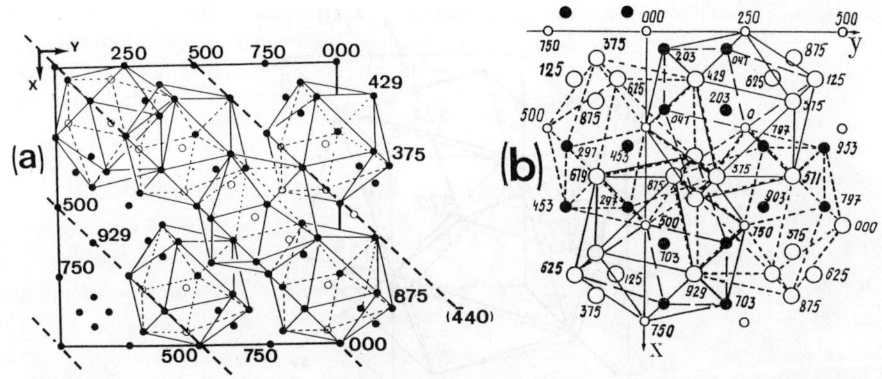

Fig. 11. Projection on the (001) cubic plane of a sequence of four icosahedral clusters pertaining to the conventional FCC unit-cell (a), and to one-quarter of the FCC unit-cell (b). The numbers refer to the z cubic coordinate. In (a) the dashed lines denote the system of $(\bar{4}40)$ planes. In (b) large and small open circles represent Ti atoms in positions 48f and 16c respectively.

[110000]. One finds for $FeTi_2$ $q_i = 2\pi/d = 2.88\ (A°)^{-1}$ [45].

The relative orientations of the crystal and quasicrystal structures was discussed by Dong et al. [45] using a TEM micrography for a sample in which the two phases coexist. The almost coinciding directions worked out by these authors from the sets of cubic and icosahedral spots are:

$$A_5 \| [110]_c,\ A_2(1) \| [1\bar{1}0]_c,\ A_2(2) \| [1\bar{1}1]_c,\ A_2(3) \| [1\bar{1}4]_c\ /[1\bar{1}5]_c \qquad (5.17)$$

and are shown in Fig. 12 in the reciprocal space.

3.2. Basic system of density waves associated with the inhomogeneous deformation mechanism

As already pointed out for AlMn(Fe)Si, the set of inhomogeneous deformations which bring about the cubic directions along the icosahedral axes coincide with correlated rotations of atomic planes. In $FeTi_2$ the rotations involve the four systems of cubic planes (224), $(4\bar{4}0)$, $(1\bar{1}5)$, and $(3\bar{3}3)$, into which are distributed the 96 atoms contained in the FCC unit-cell. In order to express and formalize quantitatively the preceding mechanism, let us write the lattice function expressing the distribution of atoms in the cubic phase, as a

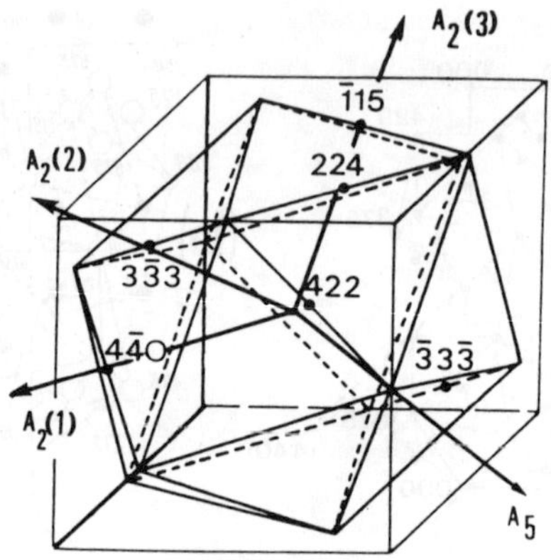

Fig. 12. Relative orientations of the relevant cubic and icosahedral directions in reciprocal space. The almost coinciding directions are given in (5.17).

set of *four systems of density waves* associated with the preceding four systems of cubic planes:

$$\psi_c^e(\vec{r},\vec{r}_i^{\,c}) = \frac{1}{NV}\sum_{k=1}^{N}\left[f_1\sum_{j=1}^{3} e^{i\vec{q}_j^{\,c}(\vec{r}_k^{\,c}-\vec{r})} + f_2\sum_{j=4}^{6} e^{i\vec{q}_j^{\,c}(\vec{r}_k^{\,c}-\vec{r})} + f_3\sum_{j=7}^{12} e^{i\vec{q}_j^{\,c}(\vec{r}_k^{\,c}-\vec{r})} + f_4\sum_{j=13}^{15} e^{i\vec{q}_j^{\,c}(\vec{r}_k^{\,c}-\vec{r})} \right] + \text{c.c.} \tag{5.18}$$

where N is the total number of atoms contained in the volume V of the sample, $\vec{r}$ and $\vec{r}_k^{\,c}$ are, respectively, the space variables and the equilibrium cubic positions of the atoms in direct space. The $\vec{q}_j^{\,c}$ (j=1-15) are the 15 vectors of the FCC reciprocal space, which share into four orbits characterizing the four systems of cubic planes, namely:

$$\begin{aligned}
&j=1\text{-}3 &&: (2\,2\,4),\ (2\,4\,2),\ (4\,2\,2);\\
&j=4\text{-}6 &&: (4\,\bar{4}\,0),\ (4\,0\,\bar{4}),\ (0\,4\,\bar{4});\\
&j=7\text{-}12 &&: (1\,5\,\bar{1}),\ (5\,1\,\bar{1}),\ (\bar{1}\,1\,5),\ (5\,\bar{1}\,1),\ (\bar{1}\,5\,1),\ (1\,\bar{1}\,5);\\
&j=13\text{-}15 &&: (3\,3\,\bar{3}),\ (3\,\bar{3}\,3),\ (\bar{3}\,3\,3).
\end{aligned} \tag{5.19}$$

f_i (i=1-4) are the respective weights of each orbit.

3.3. The critical transition wave-vector. Symmetrization process

In the icosahedral state the four *inequivalent* set of cubic density waves will become equivalent, i.e. will form a single orbit of icosahedral density waves. The transformation must obey the following requirements to have icosahedral symmetry in the final state, namely:

1) The wave-vectors $\vec{q}_j^{\,ic}$ in the icosahedral phase must represent the 15 basis vectors of a face-centered icosahedral (FCI) quasilattice.
2) The weights f_i must all become equal.

Using the observations of Dong et al. [45], one can show using Fig. 12 that the four cubic orbits become in the icosahedral state:

$$
\begin{aligned}
&j=1\text{-}3 &&: (\alpha\,\alpha\,\beta),\ (\alpha\,\beta\,\alpha),\ (\beta\,\alpha\,\alpha);\\
&j=4\text{-}6 &&: (\gamma\,\bar{\gamma}\,0),\ (\gamma\,0\,\bar{\gamma}),\ (0\,\gamma\,\bar{\gamma});\\
&j=7\text{-}12 &&: (\delta\,\varepsilon\,\phi),\ (\varepsilon\,\delta\,\phi),\ (\phi\,\delta\,\varepsilon),\ (\varepsilon\,\phi\,\delta),\ (\phi\,\varepsilon\,\delta),\ (\delta\,\phi\,\varepsilon);\\
&j=13\text{-}15 &&: (\psi\,\psi\,\upsilon),\ (\psi\,\upsilon\,\psi),\ (\upsilon\,\psi\,\psi).
\end{aligned}
\tag{5.20}
$$

with

$$
\alpha = n_1(\tau - 1/\tau\sqrt{2}) \cong 2.0456,\quad \beta = n_1(\tau + \sqrt{2}/\tau) \cong 4.316,\quad \gamma = 3.674,
$$
$$
\delta = n_2\left[(\tau-2)/\sqrt{2}+1\right] \cong 1.264,\quad \varepsilon = n_2\left[(\tau+1)/\sqrt{2}+1\right] \cong 4.938,
$$
$$
\phi = n_2\left[(1-2\tau)/\sqrt{2}+1\right] \cong -1.006,\quad \psi = n_3(1/\tau + \tau/\sqrt{2}) \cong 3.052,
$$
$$
\psi = n_3(1/\tau - \tau/\sqrt{2}) \cong -2.893
$$

where τ is the golden mean, and n_1, n_2, n_3 are three normalization coefficients which are determined from the interplanar distances in the cubic structure.

Substracting the corresponding density wave vectors of the two structures (i.e. $\vec{q}_j^{\,ic} - \vec{q}_j^{\,c}$ for each j) one gets the wave vectors $\vec{k}_i$ which govern the inhomogeneous deformation mechanism transforming one structure into the other:

$$
\begin{aligned}
\vec{k}_1 &\cong 0.316(\vec{b}_1 + \vec{b}_2) - 0.215\vec{b}_3,\\
\vec{k}_2 &\cong 0.348(\vec{b}_2 - \vec{b}_1)\\
\vec{k}_3 &\cong -0.332\vec{b}_1 + 0.320\vec{b}_2 + 0.208\vec{b}_3\\
\vec{k}_4 &\cong 0.117(\vec{b}_1 + \vec{b}_2) - 0.013\vec{b}_3
\end{aligned}
\tag{5.21}
$$

where $\vec{b}_1$, $\vec{b}_2$ and $\vec{b}_3$ are the reciprocal basis vectors of the FCC Brillouin-zone:

$$
\vec{b}_1 = \left(\frac{2\pi}{a}, \frac{2\pi}{a}, -\frac{2\pi}{a}\right),\quad \vec{b}_2 = \left(\frac{2\pi}{a}, -\frac{2\pi}{a}, \frac{2\pi}{a}\right),\quad \vec{b}_3 = \left(-\frac{2\pi}{a}, \frac{2\pi}{a}, \frac{2\pi}{a}\right).
$$

The *incommensurate* wave vectors $\vec{k}_1$-$\vec{k}_4$ express the respective *rotations* of the four systems of cubic planes when they transform into the icosahedral structure. At this point the preceding four mechanisms appear as independent, i.e. the cubic-to-icosahedral transformation in $FeTi_2$ appears to be associated with *four distinct order-parameters*, which transform respectively as irreducible representations corresponding to the linearly independent stars $\vec{k}_1^*$-$\vec{k}_4^*$. We will now show a contrario that the transformation is actually governed by a *single primary* (symmetry breaking) *order-parameter*. On that goal, let us note that when deforming the FCC structure into the icosahedral structure, the FCC rhombohedron which possess an angle 60° between its twofold axes, becomes a rhombohedron in which the twofold axes form an angle 63.43° (i.e. $\cos\alpha = 1/\sqrt{5}$). This homogeneous deformation of the cubic rhombohedron can be expressed by the equation:

$$\vec{r}_{ic} = \vec{r}_c + m\vec{r}_c \tag{5.22}$$

where $\vec{r}_c$ and $\vec{r}_{ic}$ represent the position vector within, respectively, the cubic and icosahedral rhombohedra. m is a numerical coefficient which accounts for the magnitude of the compression along the diagonal of the cubic rhombohedron, i.e. the [111] cubic direction. When choosing the origin of the space coordinates in the center of the icosahedron (solid dot in position 0.203 on Fig.11b) one finds for $FeTi_2$:

$$m = \frac{\sqrt{2\tau^2} - 4}{12} \cong -0.0248. \tag{5.23}$$

When taking into account the preceding deformation of the FCC rhombohedron, the four $\vec{k}_i$ vectors can be rewritten in the cubic Brillouin zone:

$$\vec{k}_1' = \left[\frac{2\tau - 3}{\tau}\right]\vec{b}_1 \cong 0.15\vec{b}_1,$$
$$\vec{k}_2'\left[\frac{2\tau - 3}{\tau}\right](\vec{b}_1 - \vec{b}_2) \cong 0.15(\vec{b}_1 - \vec{b}_2),$$
$$\vec{k}_3' = \vec{k}_1', \quad \vec{k}_4' = 0. \tag{5.24}$$

It must be noted that the basis vectors $\vec{b}_1$, $\vec{b}_2$ and $\vec{b}_3$ used here have the same length than the one used for the definition of $\vec{k}_1$-$\vec{k}_4$ but due to the assumed deformation the angle between them changes from 60° (cubic) to 63.43° (icosahedral).

As $\vec{k}_2'$ is a linear combination of two branches of the $\vec{k}_1' \equiv \vec{k}_3'$ star, it is clear that *only one independent non-zero wave vector governs the cubic-to-icosahedral transformation*. Let us emphasize that the *homogeneous deformation* expressed by Eqs.(5.22) and (5.23) corresponds *also* to a *displacement field*, i.e. the reduction of the number of critical wave-vectors from four to one results from a symmetrization process equivalent to the one performed in AlMnSi in Sec.2.1.

3.4. Atomic positions in the icosahedral $FeTi_2$ structure

Because all four systems of planes (5.20) become part of the same symmetrically related system of planes in the icosahedral structure, we must have $f_1 = f_2 = f_3 = f_4 = 1$. The lattice function (5.18) giving the distribution of atoms in the icosahedral phase will be written as:

$$\psi_{ic}(\vec{r}, \vec{r}_k) = \frac{1}{NV} \sum_{k=1}^{N} \sum_{j=1}^{15} e^{iq_j^{ic}(\vec{r}_k - \vec{r})} + \text{c.c.} \tag{5.25}$$

In order to obtain *equal* and *maximal* intensities for the Bragg reflections in the icosahedral diffraction patterns, one has to maximize the amplitude of ψ_{ic} with respect to the current position $\vec{r}_k$, namely to maximize the function:

$$A_{ic}(\vec{r}_k) = \frac{1}{NV} \sum_{k=1}^{N} \sum_{j=1}^{15} \cos\left(2\pi q_j^{ic} r_k\right) \tag{5.26}$$

with respect to the N vector variables $\vec{r}_k$.

The experimental values found for the atomic positions in the FCC phase [43] are reported in Table 4, column (e). In order to find the atomic positions in the icosahedral phase, one has (as for AlFeSi) to maximize (5.26) in the vicinity of these cubic positions. Column (e) and (g) of Table 4 list respectively the calculated icosahedral atomic positions and the corresponding magnitudes of the average displacements from the initial FCC to the final icosahedral structure. One can note, as for AlFeSi, that the largest displacements (column (g)) are of the order ($8.5 \cdot 10^{-2}$ a), which is a usual number for first-order displacive transitions of the reconstructive type in alloys.

Figs. 13a-d show the simulated diffraction and HREM patterns for the FCC and icosahedral structures, in the plane $(110)_c$ perpendicular to the five fold axis, using a sample of about 5×10^4 atoms. One can verify that the deformed decagons of the initial FCC phase become perfect in the icosahedral phase. Besides, the icosahedral patterns also reveal the existence of a *regular array of point defects*, which can be well understood by considering the connection between the icosahedral and FCC structures as assumed in our description: In the icosahedral structure, the *first coordination shell*, formed by the atoms denoted 2-4 in Table 4, constitute a regular icosahedron with the 20 vertices filled by atoms. However, the *second coordination shell* formed by atoms 5-7 is a dodecahedron in which only 16 vertices among the 20 are filled. The *third* and *fourth* coordination shells corresponding to atoms 8 and 9, and 10-12 are again defected icosahedron and dodecahedron, which possess respectively *three* and *five* vacant vertices.

Figs. 13e and 13f show, as for AlFeSi, that for non critical values of the wave-vector $\vec{k}_1'$, one obtains incommensurate non-icosahedral structures. The two figures illustrate the two types of precedingly mentioned incommensurabilities, namely

Table 4. The columns indicate (a) the notation of the $FeTi_2$ atoms as used in the text and (b) their crystallographic positions. (c) Atomic positions in the FCC structure from Ref.[43] and (d) distances of the atoms from the atom number 1 located in (0.203 0.203 0.203) cubic position. (e) Calculated icosahedral atomic positions, and (f) distance of the atoms from the center of one icosahedron. (g) Magnitude of the average displacements from the FCC to the icosahedral positions.

(a)	(b)	(c)			(d)	(e)			(f)	(g)
1	32e	0.203	0.203	0.203	0.0	0.0	0.0	0.0	0.0	0.0
2	48f	0.429	0.203	0.203	0.251	0.225	-0.072	-0.072	0.247	0.0085
3	16c	0.0	0.25	0.25	0.214	-0.225	-0.072	-0.072	0.247	0.1697
4	32e	0.047	0.203	0.047	0.221	-0.175	0.009	-0.175	0.247	0.0283
5	32e	0.453	0.297	-0.047	0.366	0.310	0,115	-0.202	0.388	0.0796
6	48f	0.071	-0.125	0.375	0.393	-0.115	-0.31	0.202	0.388	0.0389
7	16c	0.0	0.0	0.0	0.352	-0.224	-0.224	-0.224	0.388	0.0364
8	32e	0.297	-0.203	0.297	0.427	0.126	-0.393	0.126	0.431	0.0471
9	48f	0.179	-0.125	-0.125	0.464	0.016	-0.305	-0.305	0.431	0.0516
10	16c	0.5	0.0	0.0	0.466	0.27	0.27	-0.26	0.46	0.0686
11	48f	0.375	0.375	0.375	0.441	0.382	0.181	0.181	0.46	0.0189
12	32e	-0.047	-0.047	-0.203	0.531	-0.181	-0.181	-0.382	0.46	0.1005

corresponding to non-critical $\vec{k}_1'$ (Fig.13e) and critical $\vec{k}_1'$ (but non-critical displacements). In these incommensurate states the point-group symmetry is rhombohedral D_{3d}, and it increases to Y_h only at the limit icosahedral state, the incommensurability becoming quasiperiodicity.

3.5. The phase diagram

The star of $\vec{k}_1'$ has *eight* branches [37] and one can show that only six of the branches are active in the transformation. In other words, the symmetry-breaking *primary* order-parameter associated with the cubic-to-icosahedral transition has the symmetry of an eight-dimensional irreducible representation (denoted $\tau_1(\vec{k}_2)$ in Kovalev's tables [37]) and the equilibrium values of the corresponding order-parameter components are:

$$\eta_1 = \eta_2 = ... = \eta_6 = \eta \neq 0, \quad \eta_7 = \eta_8 = 0. \tag{5.27}$$

From these equilibrium relations, one can work-out the effective thermodynamic potential, which has the non-symmetric form:

$$F = a_1\eta^2 + a_2\eta^3 + a_3\eta^4. \tag{5.28}$$

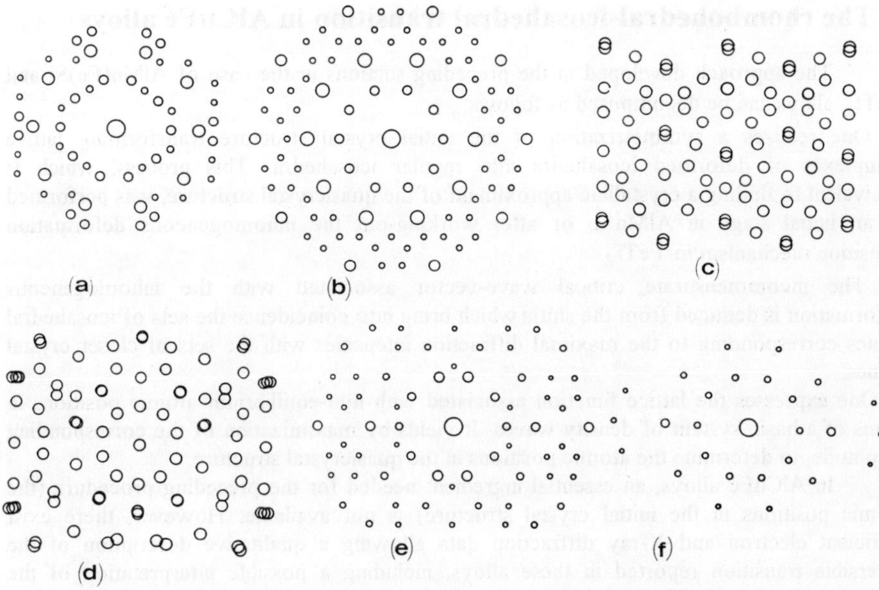

Fig. 13. (a) and (b) are simulated electron diffractograms, respectively for the (110) cubic and (100000) icosahedral planes. (c) and (d) are simulated HREM images for the (110) and (100000) planes. (e) and (f) show diffractograms with incommensurate structures corresponding to the non critical wave-vector $0.5\vec{k}_1'$ (e) and to $\vec{k}_1'$ but for non critical displacements (f).

Keeping in mind the remarks of Sec.2.3.3 (on the applicability to the real experimental diagram) the theoretical phase diagram corresponding to (5.28) is shown in Fig.13'. One can see that the cubic-to-icosahedral and cubic-to-incommensurate transitions are always first-order, whereas the incommensurate-to-icosahedral transition is of the second-order. The phase diagram also reveals the existence of icosahedral domains that were already noted in the description of the AlMn(Fe)Si system.

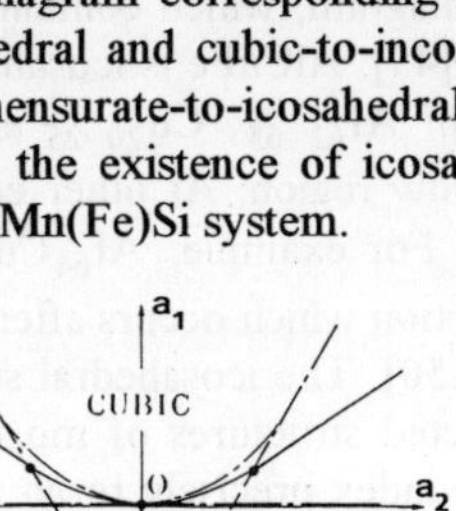

Fig. 13'. Schematic phase diagram corresponding to the potential (5.28). Solid, dashed and dot-dashed lines correspond to first-order transitions, second-order transitions, and limit of stability lines.

4. The rhombohedral-icosahedral transition in AlCuFe alloys

The approach developed in the preceding sections in the case of AlMn(Fe)Si and $FeTi_2$ alloys can be decomposed as follows:

1) One realizes a symmetrization of the initial crystal structure transforming lattice complexes of deformed icosahedra into regular icosahedra. This process, which is equivalent to finding a crystalline approximant of the quasicrystal structure, was performed as an initial stage in AlMnSi, or after working-out the inhomogeneous deformation transition mechanism in $FeTi_2$.

2) The incommensurate critical wave-vector associated with the inhomogeneous deformation is deduced from the shifts which bring into coincidence the sets of icosahedral planes corresponding to the maximal diffraction intensities with the sets of closer crystal planes.

3) One expresses the lattice function associated with non-equilibrium atomic positions in terms of a basic system of density waves. It yields by maximization of the corresponding amplitude, to determine the atomic positions in the quasicrystal structure.

In AlCuFe alloys, an essential ingredient needed for the preceding procedure (the atomic positions in the initial crystal structure) is not available. However, there exist sufficient electron and X-ray diffraction data allowing a qualitative description of the reversible transition reported in these alloys, including a possible interpretation of the various stages found in the transition process.

4.1. Reversible transition between microcrystalline-rhombohedral and ideal icosahedral phases

The phase diagram of AlCuFe system, first studied by Bradley and Goldshmidt [48] shows eleven distinct phases. The icosahedral phase lies in an intricate corner of the subsolidus projection of the phase diagram, which contains a number of additional phases related to the icosahedral structure [49]. AlCuFe is icosahedral over a composition range of several atomic percent, namely for Al_{62-65}, Cu_{20-25} and Fe_{12-15} [49], and is stable at all temperature only in a very narrow region. At other composition varying temperature induces a succession of transitions. For example, $Al_{64}Cu_{24}Fe_{12}$ becomes icosahedral at 838°C at a strongly first-order transition which occurs after two other first-order transitions taking place below the liquid state [50]. The icosahedral structure is considered as *perfect* [51] (compared to the highly defected structures of most quasicrystals) since diffraction peaks of simple grains of the phase index precisely to an icosahedral face centered lattice [52], and the width of the peaks is sharper than the instrumental resolution of high resolution single crystal diffractometer, providing a typical example (as AlPdMn or AlCuRu) of a quasicrystal ordering free of topological ("phason") disorder [53]. The coherence length is about 8000 A°.

At about 700°C a *reversible* transition to a microcrystalline state takes place [54] in which the structure of the micrograins, of several hundred Angströms, corresponds to a rhombohedral $R\bar{3}m$ symmetry with a lattice parameter a=32.16 A° and α=36°. Such a unit-cell is related to the prolate (Amman) rhombohedron [54] which constitutes the unit-cell of the approximant of the icosahedral phase corresponding to a'=37.81 A° and α=63.43°. The icosahedral-microcrystalline transition occurs as a weakly first-order transition [50] although the two phases are found to coexist at about 620°C within a two phase region [54]. Assembling 20 microcrystals built of these large rhombohedral unit-cells in a polyhedron is able to preserve a pseudo icosahedral orientational order, over large distances, indicating a coherent decomposition of the icosahedral structure into the microcrystallites. It thus yields electron diffraction patterns differing slightly from those of the icosahedral phase.

Between the two preceding phases, an intermediate stage is found [56] which manifests as a modulation of the electron scattering intensity connecting the diffraction spots, regularly spaced in the rhombohedral crystal pattern, and aperiodically placed in the perfect icosahedral phase. An in-situ X-ray study of single grains of AlCuFe by Bancel [53] has also identified an additional intermediate stage in which resolution limited Bragg peaks decrease in intensity as the temperature is lowered from 700°C to 600°C. Diffuse scattering appears as the peaks broaden and diminish, with "Lorentzian" line shapes.

The preceding description shows that the transition from a crystal to a quasicrystal, although appearing as a complex process, occurs indeed in successive reversible steps between two solid state phases. Furthermore it will be shown herebelow that these successive steps find a coherent explanation in the framework of a density-wave description of the transition.

4.2. Orientational relationship between the rhombohedral and icosahedral structures

High resolution electron microscopy [54, 55] and X-ray measurements [51] reveal the following orientational relationships between the icosahedral and rhombohedral axes in AlCuFe:

$$A_2 \| [110]_R, \quad A_3 \| [111]_R, \quad A_5 \| [100]_R \tag{5.29}$$

corresponding to *almost* coinciding directions. Figs. 14a and 14b show the connection between the primitive translations of the rhombohedral unit-cell and the icosahedral directions in direct (a) and reciprocal (b) spaces. Thus in direct space the vectors $\vec{a}_1$-$\vec{a}_3$ which form an angle α=36° point towards the center of the edges of the icosahedron with:

$$\vec{a}_1 = \frac{a}{2}\left(\tau, 1, \tau^{-1}\right), \quad \vec{a}_2 = a(1,0,0), \quad \vec{a}_3 = \frac{a}{2}\left(\tau, 1, -\tau^{-1}\right) \tag{5.30}$$

and a=32.21 A°. In reciprocal space the basic vectors $\vec{b}_1$-$\vec{b}_3$ of the rhombohedron are along the directions of the vertices of the icosahedron (Fig. 14b), with:

$$\vec{b}_1 = \frac{2\pi\tau}{a}\left(0, \tau^{-1}, 1\right), \quad \vec{b}_2 = \frac{2\pi\tau}{a}\left(\tau^{-1}, -1, 0\right), \quad \vec{b}_3 = \frac{2\pi\tau}{a}\left(0, \tau^{-1}, -1\right) \tag{5.31}$$

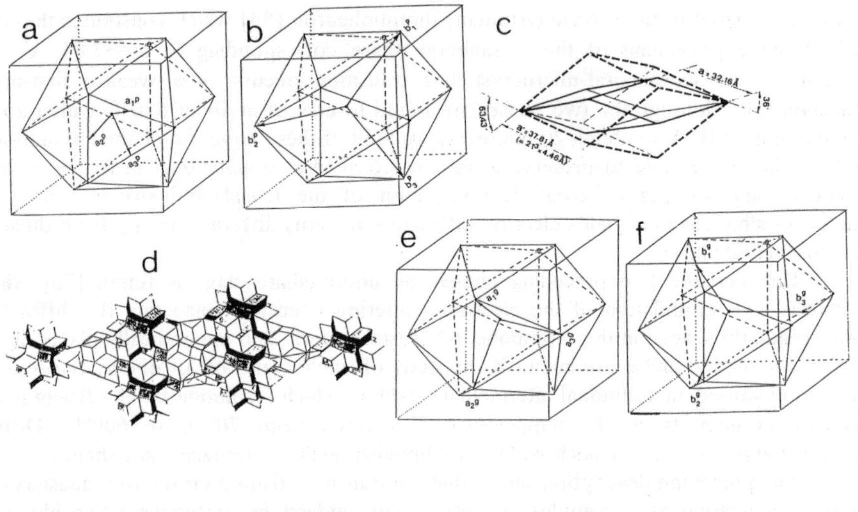

Fig. 14. (a) and (b): Orientations of the basic vectors of the rhombohedral unit-cell of the microcrystallites with respect to the icosahedral directions, in direct (a) and reciprocal (b) spaces. (c) Relationship between the preceding rhombohedral unit-cell and the largest (approximant) rhombohedron. (d) Decoration scheme proposed for the rhombohedral unit-cell by Audier et al. [54]. (e) and (f): Orientations of the basic vectors of the approximant rhombohedron with respect to the icosahedral directions in direct (a) and reciprocal (b) spaces.

and form an angle $\alpha^*=116.57°$.

On the other hand the relationship between the preceding (experimentally observed) rhombohedron (a, α) and the larger rhombohedron (a', α') of the face-centered unit-cell, which has the same form than the prolate Amman approximant rhombohedron, is shown in Fig.14c, together with the decoration scheme proposed by Audier et al. [54] and represented in Fig.14d. The connection between the small and large (approximant) rhombohedra are in direct space (Fig.14.e):

$$\vec{a}_1' = \vec{a}_1 + \vec{a}_2 - \vec{a}_3 = a'\left(1,0,\tau^{-1}\right), \quad \vec{a}_2' = -\vec{a}_1 + \vec{a}_2 + \vec{a}_3 = a'\left(1,0,-\tau^{-1}\right),$$
$$\vec{a}_3' = \vec{a}_1 - \vec{a}_2 + \vec{a}_3 = a'\left(\tau^{-1},1,0\right) \qquad (5.32)$$

with $\alpha'=63.43°$ and a'=37.86 A°. In reciprocal space (Fig.14f) one has:

$$\vec{b}_1' = \frac{1}{2}\left(\vec{b}_1 + \vec{b}_2\right) = \frac{\pi}{a'}\left(1,-\tau^{-1},\tau\right), \quad \vec{b}_2' = \frac{1}{2}\left(\vec{b}_2 + \vec{b}_3\right) = \frac{\pi}{a'}\left(1,-\tau^{-1},-\tau\right),$$

$$\vec{b}'_3 = \frac{1}{2}\left(\vec{b}_1 + \vec{b}_3\right) = \frac{2\pi}{a'}(0,1,0) \tag{5.33}$$

with an angle between the $\vec{b}'_i$ of $(\alpha^*)'=108°$. Eqs.(5.30)-(5.33) are compatible with the experimentally observed orientational relationship given by (5.29) *which also hold for the approximant rhombohedron*. These equations will be used to determine: 1) the coordinates of the system of rhombohedral planes which are inhomogeneiusly deformed into the icosahedral planes, 2) the critical wave-vectors associated with the transition mechanism.

4.3. Density-wave description of the inhomogeneous deformation mechanism: Primary and secondary order-parameters

As the rhombohedral structure is unknown we will *deduce* the coordinates of the families of rhombohedral planes, which are closest to the icosahedral planes, from the coordinates of these latter planes. On that goal we can use the X-ray diffraction data of Calvayrac et al. [51] which give the positions of the rhombohedral approximant. From these data, one can see the following: 1) The more intense peaks correspond to the basic vectors of the quasilattice and the icosahedral orbit of planes decompose into four types of rhombohedral orbits corresponding to the icosahedral directions (in rhombohedral coordinates and up to a permutation):

$$q_j^{ic} \rightarrow j=1\text{-}3: (1,0,0);\ j=4\text{-}6: (1/\tau,1,1);\ j=7\text{-}9: (1/\tau,-1/\tau,0);\ j=10\text{-}15: (1/\tau,0,1). \tag{5.34}$$

2) The peak having a maximal intensity is along the two-fold icosahedral axis. It corresponds to the icosahedral wave-vector of length:

$$\left|\vec{q}_j^{\,ic}\right| = 0.4983\ (A^\circ)^{-1} \tag{5.35}$$

and to four rhombohedral wave-vectors, corresponding to the four orbits (5.34) respectively:

$$q_1^c = 0.49680\ (A^\circ)^{-1},\quad q_2^c = 0.49679\ (A^\circ)^{-1},\quad q_3^c = 0.50239\ (A^\circ)^{-1},$$
$$q_4^c = 0.49787\ (A^\circ)^{-1}. \tag{5.36}$$

In order to find the coordinates of the families of rhombohedral planes one can consider an icosahedral P-time larger unit-cell, where P is a Fibonacci number, and express the coordinates of the 15 icosahedral planes corresponding to (5.34) as:

$$j=1\text{-}3: (P,0,0), (0,P,0), (0,0,P);\ j=4\text{-}6: (P/\tau,P,P), (P, P/\tau,P), (P,P, P/\tau);$$
$$j=7\text{-}9: (P/\tau, -P/\tau,0), (0, P/\tau,- P/\tau), (P/\tau,0,- P/\tau);$$

j=10-15: (P/τ,0,P), (0, P/τ,P), (P/τ,P,0), (P,0, P/τ), (0,P, P/τ), (P, P/τ,0). (5.37)

Then, replacing τ by P/Q where P and Q are two consecutive numbers in the Fibonacci set one gets the 15 planes of an approximant of the icosahedral structure, which can be written up to a permutation:

j=1-3: (P,0,0); j=4-6: (Q,P,P); j=7-9: (Q,-Q,0); j=10-15: (Q,0,P). (5.38)

From the experimental data [51] one gets the best approximation:

$$P=16, \quad Q=10. \tag{5.39}$$

(5.37) and (5.38) also represent the wave-vectors of the icosahedral and rhombohedral approximant structures, respectively, associated with the equivalent systems of planes. Subtracting (5.38) from (5.37) will therefore yield the set of wave-vectors k_i which govern the inhomogeneous deformation mechanism transforming one structure into the other. One finds:

$$k_1=0,\ k_2=-\left(\frac{Q\tau-P}{\tau},0,0\right),\ k_3=-\left(\frac{Q\tau-P}{\tau},\frac{-Q\tau-P}{\tau},0\right),\ k_4=-\left(\frac{Q\tau-P}{\tau},0,0\right). \tag{5.40}$$

Accordingly, the rhombohedral directions pertaining to the first orbit of planes (j=1-3) coincide with the analogous icosahedral directions, whereas orbit 2 (j=4-6) and 4 (j=10-15) are deformed in icosahedral directions by the same deformation wave-vector $k_2 = k_4$. The critical wave-vector corresponding to the deformation of the plane of the third orbit (j-7-9) is a linear combination of two branches of the star of k_2. Consequently one can conclude that the transition mechanism corresponds to *a single irreducible critical wave-vector* k_2, although the full mechanism requires to take into account an homogeneous deformation (corresponding to the Brillouin-zone-center wave vector k_1=0) and a secondary inhomogeneous deformation corresponding to k_3. In other words one has a *primary* order-parameter (associated with k_2) and two secondary order-parameters, associated with k_1 and k_3. Using (5.39) one finds in Kovalev's notations [37]:

$$\vec{k}_2 = 0.1115\vec{b}_1 \quad \text{and} \quad \vec{k}_3 = 0.1115(\vec{b}_1 - \vec{b}_2). \tag{5.41}$$

The stars $\vec{k}_2^*$ and $\vec{k}_3^*$ have both six branches and correspond to six-dimensional (primary and secondary) order-parameters.

The set of basic vectors of the density waves associated with the icosahedral and rhombohedral structures are given respectively by (5.38) and (5.34). One may thus deduce the atomic positions in the icosahedral structure by maximizing the amplitude of the lattice function:

$$\psi(\vec{r}_i) = \frac{1}{NV}\left[\sum_{i=1}^{N}\sum_{j=1}^{15} e^{i\vec{q}_j^{io}(\vec{r}_i-\vec{r})} + c.c.\right] \tag{5.42}$$

in the vicinity of the rhombohedral positions. As these positions are not known, we will restrict ourselves to discuss the transition process described in Sec.4.1, at the light of the obtained results.

4.4. Incommensurate modulation of the rhombohedral aggregates and lock-in icosahedral phase

Let us first note that the five-fold axis of the icosahedral structure, as well as the pseudo icosahedral aggregates formed by rhombohedral grains, coincides with the direction $\vec{b}_1 \| \vec{k}_2$ of the side of the approximant rhombohedron, which is also the 2-fold axis of the rhombohedral crystallites. For part of the grains however, the coincidence may be along the direction $(\vec{b}_1 - \vec{b}_2)$ corresponding to the "secondary" critical wave vector k_3. Therefore the multiple superposition of the reflections of the rhombohedral grains appear as a modulation of the electron scattering intensity, corresponding to $\vec{k}_2 = \mu\vec{b}_1$ and $\vec{k}_3 = \mu(\vec{b}_1 - \vec{b}_2)$, along the five-fold axis. The icosahedral phase arises for the *locking* of the preceding modulation at μ=0.1115.

Accordingly, one can give the following qualitative picture of the four different stages reported for the transition process in AlCuFe. Starting from the microcrystalline phase at low temperature, one gets an incommensurate modulation with the temperature-dependent wave-vectors $\vec{k}_2 = \mu\vec{b}_1$ and $\vec{k}_3 = \mu(\vec{b}_1 - \vec{b}_2)$. This dependence is reflected by the observed shifts of the X-ray peaks [51]. When the wave-vectors *lock* to their limit values with μ=0.1115, the satellites disappear but leave place to diffuse icosahedral spots, indicating that the lock-in transition is not achieved: the diffuse character of the peaks express the fact that the atomic displacements are *not maximal*. When these displacements reach values which maximize the amplitude of $\psi(\vec{r}_i)$ as given by (5.42), the form of the peaks become narrow gaussians associated with the positions of the perfect face-centered icosahedral lattice.

5. Local structural mechanism for crystal-quasicrystal transitions: The T-lattice

In this section we will deal with the description of the crystal-quasicrystal transition mechanism at the level of the atomic sites. We will show that this transition preserves a lattice of points where the inhomogeneous deformation *is minimal* (zero), and in the vicinity of which the crystal and quasicrystal structures are *similar*, with an *homogeneous dilatation*. This will be first demonstrated in a one-dimensional model of crystal-quasicrystal transition, then in the concrete case of the tetragonal-octagonal transition in MnSi alloys [56].

5.1. Connection between the density-wave and cut-and-projection descriptions. One-dimensional example of T-lattice

Let us first give the terminology which establishes the connection between the density-wave description of a quasicrystal and the cut-and-projection method [57, 58]. The lattice function used for depicting the atomic density distribution can be written for a quasicrystal under the compact form:

$$\psi(\vec{r}) = \sum_{k} f_k e^{ik\vec{r}} \tag{5.43}$$

where $\vec{r}$ is the radius vector and the f_k are complex structural amplitudes. The sum (5.43) runs over the wave-vectors $\vec{k}$ which are linear integer combinations of basic vectors $\vec{e}_i^{\,q}$ the number N of which is defined by the quasicrystal symmetry, namely N=6 for icosahedral, N=4 for planar octagonal, N=5 for decagonal etc.. The variables:

$$n_i = \vec{e}_i^{\,q} \cdot \vec{r} \tag{5.44}$$

are the *phases* of the basic density waves describing the quasicrystal structure. They span an N-dimensional *phase-space* E. For any $\vec{k}$ the scalar product $\vec{k}\cdot\vec{r}$ can be expanded in function of the n_i, so that the quasicrystal structure is defined either by the *aperiodic* function $\psi(\vec{r})$ or by the *periodic* function $\psi(n_i)$. There exist a linear dependence between the n_i which allows us to construct mutually orthogonal combinations of the n_i, decomposing the phase space into *subspaces* $E_\parallel$ and $E_\perp$.

Let us consider the simple illustrative example of a one-dimensional model of the crystal-quasicrystal transformation. In this case, any wave-vector $\vec{k}$ is a linear combination of *at least* two basic vectors of the crystal structure. For example, $e_1^c = \langle 2\rangle$ and $e_2^c = \langle 3\rangle$, which corresponds to the density-wave phases: n_1=2x and n_2=3x, where x is the space variable. n_1 and n_2 fulfill the identity:

$$3n_1 - 2n_2 = 0. \tag{5.45}$$

This defines the $E_\parallel^c$ subspace of the crystal, which passes by the origin of the phase space (n_1, n_2), as shown in Fig.15.

In the quasicrystal structure, the corresponding basic vectors have an irrational ratio, which can be assumed to be close to 3/2, i.e. $e_2^q / e_1^q = \tau$, where τ is the golden mean. The equation defining the subspace $E_\parallel^q$ of the quasicrystal is thus:

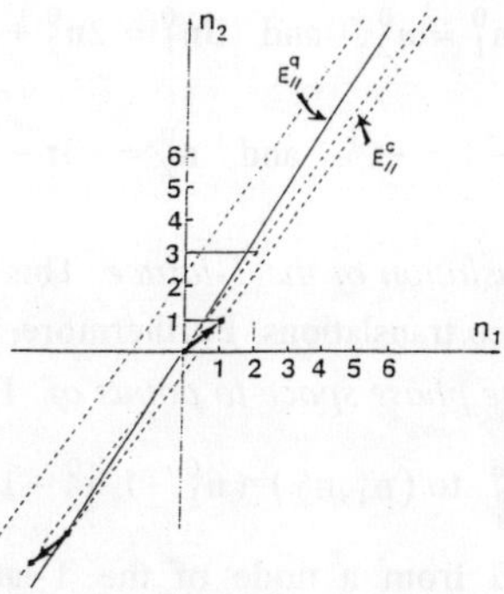

Fig. 15. Intersection of $E_{\|}^{q}$ with the system of planes $E_{\|}^{c}$ (2,3) in the phase space $E(n_1,n_2)$. Straight arrows indicates the translation (1,1).

$$\tau n_1 - 2n_2 = 0. \tag{5.46}$$

Since $E_{\|}^{q}$ intersects $E_{\|}^{c}$ at the origin of the phase space, one has at this point:

$$\psi^{c}(0,0) = \psi^{q}(0,0). \tag{5.47}$$

Indeed, while the lattice functions of the crystal and quasicrystal, as functions of the radius vector $\vec{r}$ *are different*, they can be equal as functions of the phase space variables n_1, n_2. Strictly speaking, this equality is rigorous for the quasicrystal and its *approximant* but the symmetrization process (see Sec.2.1) modifies only slightly $\psi^{c}(n_i)$.

We will now show that in addition to the origin, *there exist an infinite set of pairs of points belonging respectively to the phase spaces* $E_{\|}^{c}$ *and* $E_{\|}^{q}$ *for which the probability densities* $\varphi^{c}(n_i')$ *and* $\varphi^{q}(n_i)$ *coincide*. In order to find all such points, let us consider a system of equidistant planes, parallel to $E_{\|}^{c}$ and passing by integer nodes of the phase space lattice. These planes, which are deduced one from another (see Fig.15) by a translation Z(1,1) of the phase space (where Z is an integer) are defined by the equations:

$$3n_1 - 2n_2 = Z. \tag{5.48}$$

Their intersection with $E_{\|}^{q}$ determines a lattice of points of $E_{\|}^{q}$ which we term *the T-lattice*.

The coordinates of the node of the T-lattice, which is the closest to the origin are given by the equations:

$$\tau n_1^0 = n_2^0, \quad \text{and} \quad 3n_1^0 = 2n_2^0 + 1 \tag{5.49}$$

namely for:

$$n_1^0 = -2\tau - 1 = -\tau^3, \quad \text{and} \quad n_2^0 = -3\tau - 1 = -\tau^4. \tag{5.50}$$

(n_1^0, n_2^0) *defines the basic translation of the T-lattice*. This translation is *incommensurate* with respect to the crystal lattice translations. Furthermore, *the nodes of the T-lattice are translationally equivalent in the phase space to points of* $E_{\parallel}^c$. This follows from the fact that (n_1^0, n_2^0) corresponds in $E_{\parallel}^c$ to (n_1', n_2')=$(n_1^0 - 1, n_2^0 - 1)$, and more generally that any node of $E_{\parallel}^c$ can be deduced from a node of the T-lattice by a translation Z(1,1). Accordingly, at the nodes of the T-lattice the probability densities $\rho^c(n_1', n_2')$ and $\rho^q(n_1, n_2)$ are equal, and consequently: *In the vicinity of these nodes the crystal and quasicrystal have similar structures*, i.e. one can be deduced from the other by a *local homogeneous deformation.*

Let us emphasize that, due to the translational symmetry of the crystal, an infinite set of points in the crystal may correspond to one node of the T-lattice. On the other hand, a phase transition is a correspondence (mapping) between the initial (crystal) and final (quasicrystal) positions. Thus, this correspondence is not here necessarily one-to-one (bijective). However, in a real crystal-quasicrystal transition, the transition mechanism will establish a correspondence between one node of the T-lattice and one point of the crystal structure, which will be determined not from symmetry but from *energetical considerations*. In this respect, one can assume in the spirit of the Lifshitz argument [59], that the coinciding points of the two structures will correspond to the minimal variation of the transition wave-vector, i.e. to the minimal deformation of the crystal structure. Such a deformation is obtained for the translation (1,1) of the phase space. In this case, one gets for the first coinciding point of the T-lattice in the crystal structure: $x=\tau-1$, and for the lengths of the basic vectors associated with the crystal-quasicrystal transition

$$e_1^q = 2 - \frac{1}{\tau^2}, \quad e_2^q = 3 - \frac{1}{\tau^2} \tag{5.51}$$

which shows that the transition wave-vector is equal to $1/\tau^2$.

In summary, the crystal-quasicrystal transition preserves the infinite set of nodes of the T-lattice, in the vicinity of which the crystal and quasicrystal structures are similar. Reciprocally, point of the quasicrystal structure which do not belong to the T-lattice cannot be deduced from $E_{\parallel}^c$ by a translation of the phase space. Consequently, between two nodes of the T-lattice there exists a region where the similarity of the two structures *is lost*. In order to analyze more concretely the local structure of the quasicrystal in the vicinity of the nodes of the T-lattice, and between these nodes, we will consider the real example of the tetragonal-octagonal transition which has been disclosed in MnSi alloys [56].

5.2. Two-dimensional example of T-lattice: The tetragonal-to-octagonal transition in MnSi alloys

For an octagonal quasicrystal one has four basis vectors $\vec{e}_i$, which can be written:

$$\vec{e}_1=(0, P),\ \vec{e}_2=(Q, Q),\ \vec{e}_3=(P, 0),\ \vec{e}_4=(-Q, Q) \tag{5.52}$$

with $P/Q=\sqrt{2}$. The phase space E is spanned by the four variables $n_i = \vec{e}_i \cdot \vec{r}$ (i=1-4), and its subspaces $E_\parallel$ and $E_\perp$ transform as two vector representations of the C_{8v} point-group. The components of $\vec{r}$ can be expressed as:

$$x_\parallel = \frac{Pn_1 + Q(n_2 - n_4)}{P^2 + 2Q^2},\ y_\parallel = \frac{Pn_3 + Q(n_2 + n_4)}{P^2 + 2Q^2} \tag{5.53}$$

and one has the identities:

$$x_\perp = \frac{-2Qn_1 + P(n_2 - n_4)}{PQ} = 0,\ y_\perp = \frac{-2Qn_3 + P(n_2 + n_4)}{2PQ^2} = 0. \tag{5.54}$$

Let us analyze the body centered tetragonal (BCT) structure of $Mn_{12}Si_5$ (a=8.9 A°, c=5.0 A°) and its transformation into an octagonal structure, reported by Cao et al. [56]. The projection of the structure on the (001) BCT plane is shown in Fig. 16a. The main framework of the fourfold structure is formed by the Mn atoms, organized in networks of deformed octagons and squares, in two non-adjacent layers of the structure

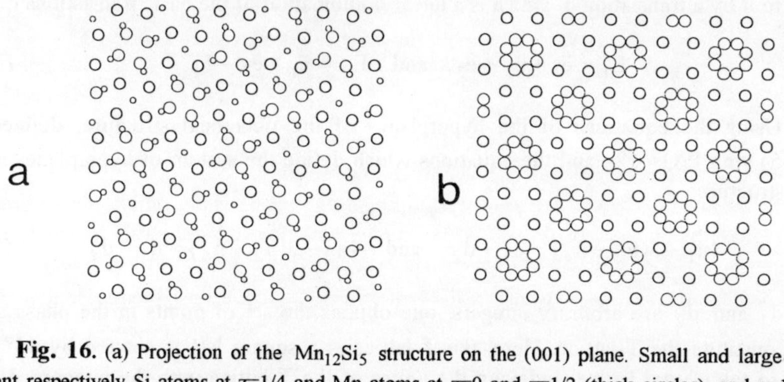

Fig. 16. (a) Projection of the $Mn_{12}Si_5$ structure on the (001) plane. Small and large circles represent respectively Si atoms at z=1/4 and Mn atoms at z=0 and z=1/2 (thick circles), and z=1/4 and z=3/4 (thin circles). (b) Approximant structure of a BCT unit-cell. Large octagons and squares correspond to 4(a) positions, and small octagons to 8(b) positions.

which contain 2/3 of the total number of Mn atoms. In the two intermediate layers, the Mn atoms are arranged in deformed squares, located between the projections of the octagons of the two first layers. The Si atoms are found over the centers of these squares at z=1/4.

Providing a slight displacement field, one can realize the symmetrization of the local structure which gives the ideal approximant. The results of this symmetrization are obvious: the large deformed octagons of Mn atoms become regular octagons, i.e. their longest diagonals become equal and parallel to the diagonals of the squares formed by Mn atoms in 4(a) positions, as represented in Fig.16b. The four Mn atoms belonging to the smallest octagons are randomly distributed in the 8(b) positions. Because the ratio between the sides of the large and small octagons is $(1+\sqrt{2})$ the projection of the Mn positions in the (x, y) plane will form a square lattice of period $a/\sqrt{2}$ with two occupied positions 4(a) and 8(b) respectively located at the two sites : $[0.5, (\sqrt{2}-1)/2]$ and $[(\sqrt{2}-1)/2, (3-2\sqrt{2})/2]$. In addition, only one-fourth of the 8(b) positions are occupied.

From Eq.(5.54) one can deduce that the atomic density distribution in the BCT and octagonal structures are defined in the hyperplanes of the phase space of equations:

$$-2Qn_1 + P(n_2 - n_4) = 0 \quad \text{and} \quad -2Qn_3 + P(n_2 + n_4) = 0 \tag{5.55}$$

with P/Q=3/2 in the BCT structure and P/Q=$\sqrt{2}$ in the octagonal structure.

There exists an infinite set of points of $E_{\parallel}^{q}$ representing the nodes of the T-lattice, in the vicinity of which the octagonal structure can be approximated by the BCT structure. The demonstration of the property follows the same scheme than in the one-dimensional case. We consider the system of hyperplanes which are parallel to the hyperplane corresponding to the crystal structure and passing through the origin of the phase space. Since in this space $\varphi(n_i)$ is periodic, all the hyperplanes are translationally equivalent. They generate structures which are equivalent to the BCT structure, and shifted with respect to it by a translation $\vec{t}$ which is a linear combination of the basic translations:

$$\vec{E}_1 = \vec{e}_1 + \vec{e}_2 + \vec{e}_3, \quad \text{and} \quad \vec{E}_2 = \vec{e}_1 - \vec{e}_3 - \vec{e}_4. \tag{5.56}$$

Using the equations of the hyperplanes of the octagonal structure, defined by Eq.(5.55) with P/Q=$\sqrt{2}$, and the equations which define the system of hyperplanes in the crystal structure:

$$4n_1 - 3(n_2 - n_4) = d_1 + d_2, \quad \text{and} \quad 4n_3 - 3(n_2 + n_4) = d_1 - d_2 \tag{5.57}$$

where d_1 and d_2 are arbitrary integers, one obtains the set of points in the phase space which constitute the T-lattice. Here, the T-lattice is a square lattice rotated by 45° with respect to the crystal lattice. Indexing the nodes of the T-lattice with the integers d_1 and d_2, one can verify that the translation $\vec{t} = d_1\vec{t}_1 + d_2\vec{t}_2$ brings into coincidence the

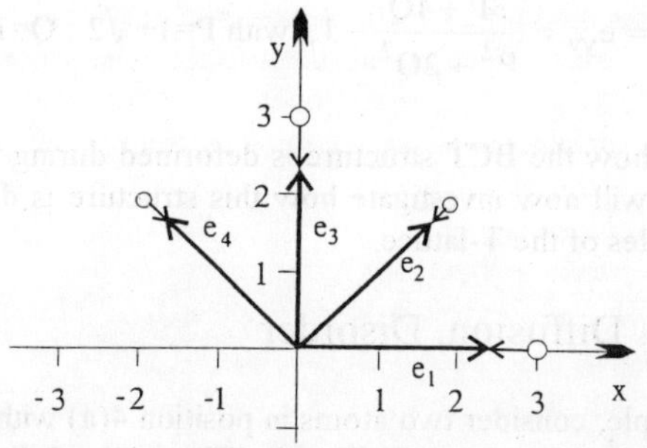

Fig. 17. Change of the reciprocal-lattice vectors at the crystal-octagonal quasicrystal transition. The long thick arrows correspond to the quasicrystal, while the short thin arrows indicate the difference between the initial and final positions of the wave-vectors.

hyperplanes (0, 0) and (d_1, d_2). Thus, in the vicinity of the nodes (d_1, d_2) the crystal and quasicrystal lattice functions $\psi^q(\vec{r}')$ and $\psi^c(\vec{r}')$ are correlated by the condition that *the images of $\vec{r}$ and $\vec{r}'$ differ by $\vec{t}$ in the phase space.*

Fig.17 shows the basis vectors of the reciprocal space of the initial tetragonal and final octagonal structures. It can be noted that the change in the basis vectors $\vec{e}_1$ and $\vec{e}_3$ is expressed entirely in terms of the changes in $\vec{e}_2$ and $\vec{e}_4$:

$$\Delta\vec{e}_1 = \Delta\vec{e}_2 - \Delta\vec{e}_4, \quad \Delta\vec{e}_3 = \Delta\vec{e}_2 + \Delta\vec{e}_4. \tag{5.58}$$

$\Delta\vec{e}_1$ and $\Delta\vec{e}_3$ form an irreducible star belonging to the first-Brillouin-zone of the initial (BCT) crystal, which determines the change in the basic set of density waves at the crystal-quasicrystal transition. It is immediate to show that the vector corresponding to the phase differences $\Delta\vec{e}_1 \cdot \vec{r}$, $\Delta\vec{e}_2 \cdot \vec{r}$, $\Delta\vec{e}_3 \cdot \vec{r}$, $\Delta\vec{e}_4 \cdot \vec{r}$, coincides with the vector $\vec{t}_1$ given by (5.56) at the point $[1/(2-\sqrt{2}), 1/(2-\sqrt{2})]$, and with the vector $\vec{t}_2$ at the point $[-1/(2-\sqrt{2}), 1/(2-\sqrt{2})]$. For other sites of the T-lattice the vector of the phase differences can be expressed as linear integer combinations of $\vec{t}_1$ and $\vec{t}_2$.

Therefore, the quasicrystal lattice function is equal to the crystal lattice function at the sites of the T-lattice: $\psi^q(n_i) = \psi^c(n_i')$. Near each of these sites the *reconstruction* of the lattice at the transition results in a local *homogeneous deformation*, which can be evaluated by considering the distance between two atoms in the BCT and octagonal structure. One finds that the minimal distance between atoms along the X axis, which is a=(1, 0) in the BCT structure and $\vec{t} = 3\vec{e}_1 + 2\vec{e}_2 - 2\vec{e}_4$ in the phase space, undergoes a *dilatation* at the transition to the octagonal structure. This dilatation is characterized by the *local strain tensor*:

$$\frac{\Delta a}{a} = e_{xx} = e_{yy} = \frac{3P+4Q}{P^2+2Q^2} - 1, \text{ with } P=1+\sqrt{2}, Q=1+\frac{1}{\sqrt{2}}. \qquad (5.59)$$

Having clarified how the BCT structure is deformed during the transition, close to the T-lattice nodes, we will now investigate how this structure is deformed in the region located between two nodes of the T-lattice.

5.3. Fusion of sites. Diffusion. Disorder

Let us, for example, consider two atoms in position 4(a) within a layer containing a network of octagons and squares shown in Fig. 18a. The minimal distance between them is the side of the square motif, and each atom has three nearest neighbors. The vectors expressing the difference of coordinates of two nearest neighbor atoms form an orbit of eight vectors which permute under the symmetry operations of the group C_{8v}. This orbit is the projection of $E_{\parallel}^{q}$ of the orbit of the eight vectors of the phase space which are equivalent to the vector $\vec{t}_1 = \vec{e}_1 + \vec{e}_2 + \vec{e}_3$. One can easily deduce that two adjacent regions of the BCT structure, corresponding to nearest neighbor nodes of the T-lattice *overlap* after the transition to the octagonal structure, and the *extent* of the overlapping is determined by the length of the square motif. Midway between two nodes of the T-lattice there is therefore a *fusion of a number of pairs of nearest neighbor positions*.

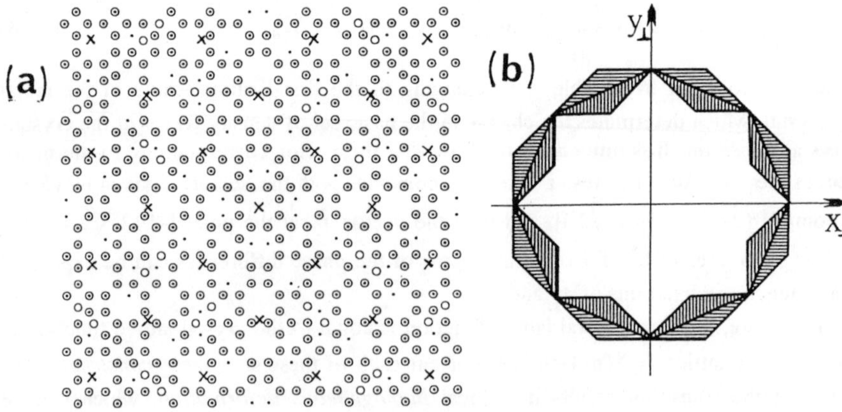

Fig. 18. (a) Structural mechanism proposed for the approximant-octagonal transformation in $Mn_{12}Si_5$. Open circles, crosses and dots represent, respectively, BCT atoms in 4(a) position, T-lattice nodes, and 4(a) positions in the ideal octagonal network. (b) Atomic surface in the $(x_\perp, y_\perp)$ plane showing how the existence of two types of distance between the atomic sites (i.e. the distances between two opposite sides of the external and internal octagons) generates neighboring positions which are too close.

In $Mn_{12}Si_5$ this fusion process involves about 1/6 of the pairs of 4(a) positions, and about 1/10 of the pairs of 8(b) positions. Since the layers occupied by atoms in 8(b) positions have the lowest density, the atoms in excess will occupy the vacancies which preexist in the BCT structure. This *diffusion* of atoms implies *a change in the stoichiometry*, similar to the one observed at the Mn_3Si - $Mn_{12}Si_5$ crystal-quasicrystal transition reported in the same system [56].

Fig. 18a represents our proposed mechanism of the BCT-octagonal transition. One can verify that a number of additional positions appear in the octagonal structure, e.g. a small octagon is surrounded by 8 positions in the BCT lattice, and by 16 positions in the octagonal network. Accordingly, there exists structural disorder in the octagonal structure: each position in this structure may have nearest positions with two different types of coordinations corresponding to different distances, the ratio of which is $2\sin(\pi/8)$ when assuming plane atomic surfaces [60]. Consequently, one finds pairs of positions which are closer to each other, the difference of their coordinates corresponding in phase space to vectors belonging to the orbit of the vector $\vec{t} = \vec{e}_1 + \vec{e}_2$.

Fig. 18b shows how the existence of two types of distances between the atomic sites generates neighboring positions which are too close. The distance between the opposite ends of the outer octagon in Fig. 18b, which determines the boundary of the atomic surface in phase space, equals twice the length of the projection of the vector $\vec{t} = \vec{e}_1 + \vec{e}_2 - \vec{e}_4$ on $E_\perp$. For the inner octagon the distance between the opposite ends equals the length of the projection of the vector $\vec{t} = \vec{e}_1 + \vec{e}_2$ on $E_\perp$. If the plane $E_\parallel$ intersects each atomic surface at a point belonging to the vertically hatched region in Fig. 18b, then it also intersects the atomic surface, displaced from the first surface by a vector belong to the orbit of $\vec{t} = \vec{e}_1 + \vec{e}_2$ in the horizontally hatched region, thereby generating two closely spaced sites.

Since the number of positions in the octagonal structure is larger than the number of atoms and the distance between some of the positions is smaller than the atomic diameter, one can conclude that *there is not more than one atom for each pair of close positions*: Therefore the disordering of the quasicrystalline phase is *symmetry determined*.

The physical mechanism which has been proposed for the crystal-quasicrystal transition in $Mn_{12}Si_5$ can thus be decomposed into the following correlated processes:

1) A displacive mechanism consisting of an inhomogeneous deformation, which results into a fusion of pairs of sites and transforms the crystal structure into an incommensurately modulated structure displaying a periodic system of defects. This structure is formed by an assembly of differently deformed regions possessing a *finite number* of coinciding positions with the initial crystal. For example in $Mn_{12}Si_5$ the ideal approximant and ideal quasicrystal have 16 coinciding positions (the central octagon and 4 neighboring squares). The preceding finite number becomes *infinite* under the effect of a macroscopic homogeneous deformation, which modifies the overall geometry in such a way that the local structure of the sites and their general distribution possess a non-ideal quasicrystal symmetry.

2) A further disordering mechanism, related to the additional positions induced by the increase in the orientational symmetry takes place at the crystal-quasicrystal transition.

The formation of the ideal quasicrystal structure then results from a diffusion of the atoms and a disordering related to the random occupation of the quasicrystal positions.

An essential and general feature of the preceding mechanism is the preservation of a lattice of points - the T-lattice - where the macroscopic deformation is minimal, and in the vicinity of which the quasicrystal structure is dilatated with respect to the crystal structure.

5.4. Return to the density-wave description

Let us formulate in terms of density waves the BCT-octagonal transition mechanism in $Mn_{12}Si_5$ decomposing this mechanism in the two successive steps: 1) the symmetrization of the BCT structure into the crystal approximant structure, 2) the inhomogeneous deformation of the approximant into the octagonal structure.

Taking P=3 and Q=2 in Eqs.(5.52)-(5.54) the lattice function $\psi_c(\vec{r})$ in the BCT structure can be expressed in terms of a basic system of four density waves associated with the basic wave-vectors:

$$\vec{e}_1=(0, 3),\ \vec{e}_2=(2, 2),\ \vec{e}_3=(3, 0),\ \text{and}\ \vec{e}_4=(-2, 2). \tag{5.60}$$

Considering only the Mn atoms (Fig.16a) one can write:

$$\psi_c(\vec{r}) = f_1[\cos 3x + \cos 3y] + f_2[\cos 2(x+y) + \cos 2(x-y)] \tag{5.61}$$

where the structural amplitudes f_1 and f_2 depend on the atomic Mn coordinates:

$$f_1 = \frac{1}{12}\left[\sum_{i=1}^{4} P_1 \cos 3x_i + \sum_{j=1}^{8} P_2 \cos 3x_j\right],$$

$$f_2 = \frac{1}{12}\left[\sum_{i=1}^{4} P_1 \cos 2(x_i + y_i) + \sum_{j=1}^{8} P_2 \cos 2(x_j + y_j)\right].$$

P_1 and P_2 are the respective probabilities of occupancy of 4(a) and 8(b) sites by the atoms. In the approximant structure (Fig.16b) one can assume $f_1 \cong f_2 \cong (f_1+f_2)/2$ (these equalities being strictly verified in the octagonal quasicrystal structure). Therefore the amplitude corresponding to $\psi_c(\vec{r})$ can be approximated as:

$$A_c(\vec{r}_i) \cong \frac{1}{48}\left[\sum_{i=1}^{4} P_1 \psi(\vec{r}_i) + \sum_{j=1}^{8} P_2 \psi(\vec{r}_j)\right] \tag{5.62}$$

with $\psi(\vec{r}_i) = \cos 3x_i + \cos 3y_i + \cos 2(x_i + y_i) + \cos 2(x_i - y_i)$.

In summary, the symmetrization of the BCT structure results in a distribution of Mn atoms, corresponding to the deepest system of extrema of the amplitude (5.62), which is obtained by assuming equal amplitudes for the wave-vectors $\vec{e}_1 - \vec{e}_4$. This distribution is also the closest to the initial BCT distribution of Mn atoms.

In the octagonal structure the basic density waves become equivalent due to the inhomogeneous deformation mechanism, which changes the length of the $\vec{e}_i$ to $\vec{e}_i + \Delta\vec{e}_i$, as shown in Fig. 17. The $\Delta\vec{e}_i$ given by Eq. (5.58) correspond to a critical wave vector having two branches $(\Delta\vec{e}_2, \Delta\vec{e}_4)$, and to a single four dimensional irreducible representation. Writing the lattice function ψ_{oct} in terms of the basic vectors $\vec{q}_i^{ic} = \vec{e}_i + \Delta\vec{e}_i$ and of the non-equilibrium positions $\vec{r}_i$ (see Secs. 2.3 and 3.4) one gets by maximization of the related amplitude $A_{oct}(\vec{r}_i)$ the atomic positions in the octagonal structure.

Fig. 19 decomposes the succession of changes in the structural amplitude from the initial BCT structure (Fig. 19a) to the ideal disordered octagonal structure (Fig. 19d), going through the intermediate approximant structure (Fig. 19b) and incommensurate structure before the deformation process (Fig. 19c). Only the most important contribution of the Mn atoms is taken into account in these simulated diffraction patterns.

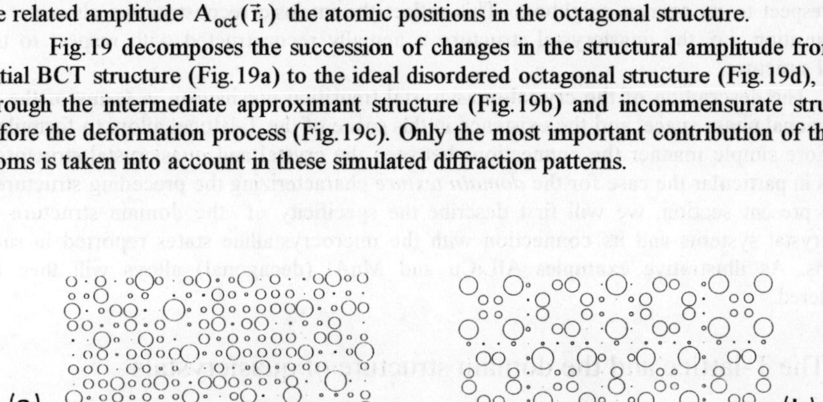

Fig. 19. Simulation of the electronic diffraction patterns of $Mn_{12}Si_5$, showing the variation of the structural amplitudes at the BCT-octagonal transition. (a) BCT structure, (b) Tetragonal approximant, (c) Incommensurate structure before the diffusion process, (d) Octagonal disordered structure.

6. Domain structure and microcrystalline states in quasicrystals

In the preceding sections we have stressed that the absence of bijection between the crystal and quasicrystal structures [61] had to be understood in the following way: One has to distinguish the atoms from the atomic positions and consider, as usually in metallic alloys, that the atomic sites are filled with different probabilities of occupancy. Thus, the number of atoms remains unchanged at the crystal-quasicrystal transition but the number of sites is modified via fusion, splitting etc. mechanisms. In other words, the displacement field relating the two structures is a dissymmetrical field regarding the position of an atom with respect to its nearest neighbors. This reflect the intrinsic reconstructive character of the transition, i.e. the quasicrystal structure is actually reconstructed with respect to the crystal structure.

The description of the crystal-quasicrystal transition mechanism in terms of the n-dimensional phase space, and the existence in this space of the T-lattice, allow to formulate in a more simple manner the connections between the crystal and quasicrystal structures. This is in particular the case for the *domain texture* characterizing the preceding structures. In the present section, we will first describe the specificity of the domain structure of quasicrystal systems and its connection with the microcrystalline states reported in such systems. As illustrative examples AlLiCu and MnAl (decagonal) alloys will then be considered.

6.1. The T-lattice and the domain structure of quasicrystals

Let us first consider the transition *from* a crystal *to* a quasicrystal. Since the translational symmetry of the crystal is lost at the transition one should expect the onset of antiphase domains in the quasicrystal state transforming one into another by the lost crystal translations. Two antiphase domains also transform one into another by the translations of the phase space $E_{\parallel}^{q}$ corresponding to the projections on this subspace of the crystal translations. Shifting from one antiphase domain to another is equivalent to the displacement of a plane of $E_{\parallel}^{q}$ in the perpendicular direction $E_{\perp}^{q}$. If this displacement has a collective character it will be associated to a phason-strain mode [62].

In contrast to antiphase domains which always exist in the quasicrystal state when coming from the crystal state, the existence of *orientational domains* is determined by the relationship between the point-group symmetries of the two structures: Orientational domains will take place in the quasicrystal phase only if the point group of the crystal is not a subgroup of the quasicrystal's point group. However, the rule fixing the number n_d of orientational domains in the quasicrystal phase is not the same as in reconstructive transitions between crystal phases, where one would have:

$$n_d = \frac{\left[\text{order of the crystal point group}\right]}{\left[\begin{array}{c}\text{order of the common subgroup to}\\ \text{the crystal and quasicrystal point groups}\end{array}\right]}. \qquad (5.63)$$

For example in $FeTi_2$ alloys the common subgroup to the crystal (O_h) and quasicrystal (I_h) point groups is C_{3v}. Thus n_d=48/6=8, but only two orientational domains will be found corresponding to the two types of orientations of the 6 basis vectors of the icosahedral structure with respect to the cubic axis. The remaining orientational domains actually become antiphase domains displaced with respect to one another by the translations relating the icosahedral complexes in the cubic unit-cell.

The reversed transition *from* quasicrystal *to* crystal also gives rise to a specific domain texture. One finds in this case: 1) orientational domains transforming one into another by the *non-crystalline* (and eventually crystalline) symmetry elements which are lost at the transition. 2) Antiphase domains *originating* in the loss of the translations belonging to the T-lattice. As each node of the T-lattice may serve as a nucleation center for the crystalline phase, two situations may be met: If the nucleation occurs simultaneously for the different nodes of the T-lattice a single domain will be formed. If, on the contrary, the process is not simultaneous a set of different microdomains will form, which are connected by the translations of the T-lattice. Therefore a microcrystalline state will be produced, corresponding to a coherent twinning of antiphase domains. Let us note that the size of the microdomains may not exceed some basic periods of the T-lattice.

In summary, the formation of defects in the crystal and quasicrystal states is also the result of the mutual formation of orientational and antiphase domains which occur at the transition. A microcrystalline state can be formed when coming from the quasicrystal phase, which consists in a local ordering of the atoms in positions preexisting in the quasicrystal phase, around the nodes of the T-lattice. These conclusions will be now illustrated by two examples: the BCC-icosahedral transition in AlCuLi alloys and the orthorhombic-decagonal transition in Mn-Al alloys.

6.2. The Al_5CuLi_3-Al_6CuLi_3 transition

We have already shown in Sec.2.2 that the BCC (T_h) structure of the R-phase of Al_5CuLi_3 could be described as a packing of distorted triacontahedra and icosahedral complexes. These structural motifs possess a perfect five-fold symmetry in the T_2-phase of the icosahedral Al_6CuLi_3 composition, which is formed by conventional solidification [63]. An intermediate phase has been also reported between the R and T_2 phases which is described as a complex microcrystalline state [64].

The correspondence between the R and T_2 axes is the same as for AlMnSi (relations (5.3)) and nearly identical reflection intensities are found for the respective planes: (035) and (100000), (532)(006) and (110000), (031) and (110001) [65]. From these data one can deduce the following basic vectors of the reciprocal space, for describing the BCC *and* icosahedral structures:

$$\vec{e}_1=(0, Q, P),\ \vec{e}_2=(0, -Q, P),\ \vec{e}_3=(Q, P, 0),\ \vec{e}_4=(-Q, P, 0),$$
$$\vec{e}_5=(p, 0, Q),\ \vec{e}_6=(P, 0, -Q). \tag{5.64}$$

The translational symmetry of the cubic R-phase imposes that P and Q are odd integers and the projection method shows that this phase can be described using the numbers P=5 and Q=3, whereas P/Q=τ in the T_2 phase. Table 5 gives (three first columns) the crystallographic and atomic positions in the R-phase from Ref.[32]. Column (e) of the table lists the coordinates of the positions resulting from the intersection of the atomic surfaces passing by the points of the phase space indicated in column (e) (which are multiplied by a factor 2), and of the $E^c_\parallel$ space. Therefore, position 24g (number 4 in column (a)) corresponds to the atomic surface passing by the origin of the phase space, whereas the positions 16f, 24g (number 3) and 12e (number 7) are associated with atomic surfaces passing by the centers of the unit-cells in phase space. The other positions correspond to atomic surfaces passing by the centers of the edges of the unit-cells.

The components of the position vector $\vec{r}$ can be written in the phase space $E^c_\parallel$ as:

$$r_i = \sum_{j=1}^{6} A_{ij} n_j \quad (i=1\text{-}3) \tag{5.65}$$

where the n_j are the phase coordinates and the matrix A_{ij} is defined by:

$$A_{ij} = \frac{1}{2(3+5\tau)}\begin{bmatrix} 0 & 0 & 1 & -1 & \tau & \tau \\ 1 & -1 & \tau & \tau & 0 & 0 \\ \tau & \tau & 0 & 0 & 1 & -1 \end{bmatrix}. \tag{5.66}$$

The system of equations giving the coordinates of the planes parallel to $E^c_\parallel$ is:

$$\begin{cases} P(n_3 - n_4) - Q(n_5 + n_6) = d_1 \\ P(n_1 - n_2) - Q(n_3 + n_4) = d_2 \\ P(n_5 - n_6) - Q(n_1 + n_2) = d_3 \end{cases} \tag{5.67}$$

where d_1, d_2, and d_3 are integers and $(d_1+d_2+d_3)$ is even.

One can construct the T-lattice in the icosahedral T_2-phase in the following way. At first the equation of $E^q_\parallel$ is obtained by taking in (5.67) P/Q=τ and $d_1=d_2=d_3$=0.

Table 5. (a) and (b): Crystallographic positions corresponding to the BCC (T_h^5) structure of R-Al_5CuLi_3, (c) atomic positions from Ref.[32], (d) concentration (%) of Al, Cu and Li for each type of positions. (e) Coordinates of the positions resulting from the intersection of the atomic surfaces passing by the points of the phase space given in column (e), with the $E_\parallel^c$ space (f).

(a)	(b)	(c)			(d)			(e)			(f)
		x	y	z	Al	Cu	Li	x	y	z	
1	24g	0.0	0.094	0.154	88.6	11.4	-	0.0	0.095	0.155	2 1 1 1 1 -1
2	16f	0.187	0.187	0.187	-	-	100	0.191	0.191	0.191	3 1 3 1 3 1
3	24g	0.0	0.304	0.171	-	-	100	0.0	0.309	0.118	3-1 3 3 1-1
4	24g	0.0	0.180	0.315	53.8	46.2	-	0.0	0.191	0.309	4 2 2 2 2-2
5	48h	0.157	0.188	0.406	88.6	11.4	-	0.155	0.191	0.405	5 3 3 1 4-1
6	12e	0.404	0.0	0.500	100	-	-	0.382	0.0	0.5	5 5 2-2 7 1
7	12e	0.199	0.0	0.500	-	-	100	0.191	0.0	0.5	5 5 1-1 5-1

The equation of the closest plane to $E_\parallel^c$ passing by the origin is also deduced from (5.67) by putting $d_1=0$, $d_2=-1$, $d_3=1$. This plane is deduced from $E_\parallel^c$ by the translation (in the $\vec{e}_i$ basis): $\vec{t}_1=(\tau 1111\text{-}1)$. The projection of $\vec{t}_1$ on $E_\parallel^c$ gives the vector $(0, 1/2\tau^2, 1/2\tau)$ (see line 4 of Table 5). In the $E_\parallel^q$ subspace the coordinates of the node which is the closest to the origin are:

$$\left[0, \frac{-1}{2Q(5-3\tau)}, \frac{1}{2Q(5-3\tau)}\right] \tag{5.68}$$

Taking into account the BCC T_h symmetry of Eqs.(5.67), one thus reaches the result that ***the T-lattice is a face-centered cubic lattice with the primitive translation (5.68).***

The *image* of the T-lattice in the $E_\parallel^c$ space will be determined by the condition that the points corresponding to the coordinates:

$$\left(0, \frac{-\tau^2}{2}, \frac{\tau}{2}\right) + n_1\vec{a}_1 + n_2\vec{a}_2 + n_3\vec{a}_3 \tag{5.69}$$

where n_1, n_2, n_3 are integers and $\vec{a}_1$, $\vec{a}_2$, $\vec{a}_3$ are the BCC lattice vectors:

$$\vec{a}_1 = \left(-\frac{1}{2}, \frac{1}{2}, \frac{1}{2}\right), \quad \vec{a}_2 = \left(\frac{1}{2}, -\frac{1}{2}, \frac{1}{2}\right), \quad \vec{a}_3 = \left(\frac{1}{2}, \frac{1}{2}, -\frac{1}{2}\right) \tag{5.70}$$

should coincide with the point given by (5.68).

From the knowledge of the T-lattice and of its image in $E_{\parallel}^{c}$ one can deduce the geometrical properties of the antiphase domains which appear in the crystal phase when coming from the icosahedral phase. In addition the crystal phase will contain orientational domains transforming into one another by the five-fold symmetry elements lost at the I_h - T_h point-group change. In contrast, since T_h is a subgroup of I_h, the reverse crystal-quasicrystal transition will not generate orientational domains, but only antiphase domains shifted with respect to each other by the cubic translations (5.70).

We will not enter into the detailed description of the transition mechanism [66] but only stress the following points: The inhomogeneous deformation which brings into coincidence the cubic planes with the corresponding icosahedral planes, is equivalent to the transformation bringing into coincidence the T-lattice with its image in $E_{\parallel}^{c}$. One finds that the critical wave-vector associated with this transition is: $\vec{k} = (1/\tau^2, 0, 0)$. It yield a local homogeneous dilatation of the crystal structure corresponding to $e_{xx} = e_{yy} = e_{zz} = = \tau^3/(\tau+2)$, and a fusion of about 6% of the pairs of positions 24g (no.4). Figs.20a and 20b show the simulated diffraction patterns of the Cu-lattices before and after the transition to the icosahedral structure before the diffusion process.

6.3. The $Al_4Mn - Al_{78}Mn_{22}$ (decagonal) transition

As a second example let us consider the orthorhombic-decagonal transition in MnAl alloys, in which the relationship between the structures was established by Schoemaker [67, 68]. The projection of the crystal structure along the five fold directions is shown in Fig.21a, no difference being made between the Al and Mn atoms. The analysis of the transition mechanism follows a scheme very similar to that of the tetragonal-octagonal transition in MnSi alloys, and we restrict ourselves to summarize its main results.

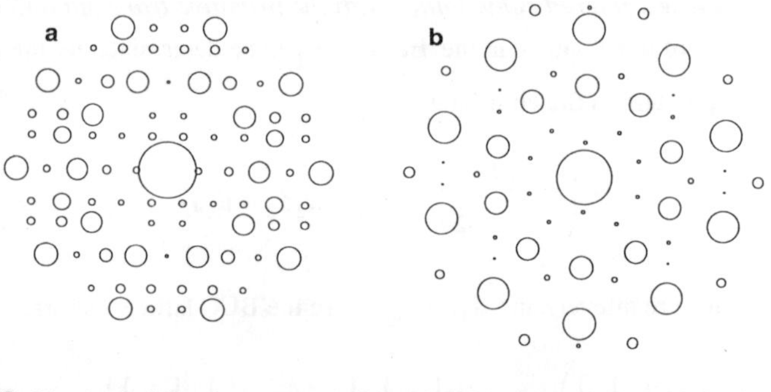

Fig. 20. Simulated diffraction patterns of R- Al_5CuLi_3 (a) and $T_2 - Al_6CuLi_3$ (b) following our proposed transition mechanism.

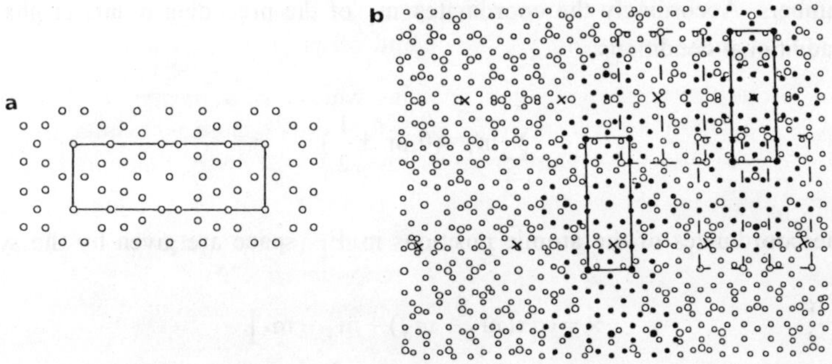

Fig. 21. (a) Projection of the $\pi - MnAl_4$ structure along the 5-fold axis, assuming identical Al and Mn atoms. (b) Ideal decagonal structure in which the T-lattice nodes are represented by crosses. The rectangles correspond to fragments of the crystal structure associated with different crystalline domains.

The basic vectors of the reciprocal space describing the orthorhombic and decagonal structures can be chosen as:

$$\vec{e}_1=(0, 10),\ \ \vec{e}_2=(3, 3),\ \ \vec{e}_3=(-3, 3),\ \ \vec{e}_4=(2, -8),\ \ \vec{e}_5=(-2, -8) \qquad (5.64)$$

from which one can deduce the primitive orthorhombic translations in phase space:

$$(0\ 3\ {-3}\ 2\ {-2}) \quad \text{and} \quad (10\ 3\ 3\ {-8}\ 8). \qquad (5.71)$$

The equations defining $E_{\parallel}^{c}$ are:

$$\sum_{j=1}^{5} A_{ij} n_j = 0 \qquad (5.72)$$

where the n_j are the phases of the density waves and A_{ij} is the 3×5 matrix:

$$\begin{bmatrix} 0 & 2 & -2 & -3 & 3 \\ 22 & 18 & -18 & 7 & 7 \\ 1 & 1 & 1 & 1 & 1 \end{bmatrix}.$$

The vectors projecting the planes (5.72) in the phase space are:

$$(-2\ {-1}\ 0\ 1\ 2\ ) \quad \text{and} \quad (-2\ 0\ {-1}\ 2\ 1). \qquad (5.73)$$

The lattice function in phase space corresponding to Fig.21a can be represented by atomic surfaces passing by the points of the phase space corresponding either to integers or to half integer numbers. Accordingly the coordinates m_i of the preceding points in phase space fulfill the additional conditions:

$$\sum_{i=1}^{5} m_i = (0 \text{ or } \pm\frac{1}{2}).$$

The coordinates of the atomic positions in $E_\parallel$ space are given by the system of equations:

$$x = \alpha\left[\tau(m_2 - m_3) + m_4 - m_5\right]$$
$$y = 2\tau m_1 + m_2 + m_3 - \tau^2(m_4 - m_5) \qquad (5.74)$$

with the normalizing conditions for the translations (a≠b) of the orthorhombic lattice: $a=\alpha(6\tau+4)$ and $b=36\tau+22$.

The two-dimensional T-lattice deduced from the preceding considerations is formed by lozenges with the basic translations (t_x, t_y) and $(-t_x, t_y)$, where:

$$t_x = (3\tau+2)\tau^2\alpha = \tau^2\frac{a}{2}, \text{ and } t_y = (18\tau+11)\tau^2 = \tau^2\frac{b}{2}.$$

Introducing (5.73) in (5.74) one can verify that the microcrystalline domains are displaced one from another by the linear combinations of the 2 vectors:

$$\left(\frac{a}{2\tau^2}, \frac{b}{2\tau^2}\right), \left(-\frac{a}{2\tau^2}, \frac{b}{2\tau^2}\right)$$

as represented in Fig.21b.

7. References

1. T. Janssen, *Acta Cryst.* **A42** (1986) 261.
2. V. Elser, *Acta Cryst.* **A42** (1986) 36.
3. C. Janot and M. de Boissieu, in *Quasicrystals: The State of the Art*, eds. D. P. DiVincenzo and P. J. Steinhardt (World Scientific, Singapore, 1991), p.57.
4. V. P. Dmitriev, S. B. Rochal, Yu. M. Gufan and P. Toledano, *J. de Physique (France)* **51** (1990) 2309.
5. V. P. Dmitriev, S. B. Rochal, V. L. Lorman and P. Toledano, *Sov. Phys. Solid State* **33** (1991) 963.
6. V. P. Dmitriev, S. B. Rochal, V. L. Lorman and P. Toledano, *Crystallogr. Reports* **38** (1993) 224.

7. V. P. Dmitriev, S. B. Rochal, V. L. Lorman and P. Toledano, *Phys. Rev.* **B49** (1994) 838.
8. S. B. Rochal, V. P. Dmitriev, V. L. Lorman and P. Toledano, *Phys. Solid State* **37** (1995) 130.
9. P. Bak, *Phys. Rev. Lett.* **54** (1985) 1517.
10. N. D. Mermin and S. Troian, *Phys. Rev. Lett.* **54** (1985) 1524.
11. S. Sachdev and D. R. Nelson, *Phys. Rev.* **B32** (1985) 4592.
12. T. L. Ho, *Phys. Rev. Lett.* **56** (1986) 468.
13. L. D. Landau, *Phys. Z. Sowiet* **11** (1937) 545, transl. in *The Collected Papers of L. D. Landau*, ed. D. Ter Haar (Pergamon Press, New York, 1965), p. 193.
14. S. Alexander and J. McTague, *Phys. Rev. Lett.* **41** (1978) 702.
15. B. Dubost, J. M. Lang, M. Tanaka, P. Sainfort and M. Audier, *Nature* **324** (1986) 48.
16. W. Ohasi and F. Spaepen, *Nature* **330** (1987) 557.
17. M. K. Sanyal, V. C. Sahni and G. K. Dey, *Nature* **328** (1987) 704.
18. J. L. Robertson, in *Methods of Structural Analysis of Modulated Structures and Quasicrystals*, eds. J. M. Perez-Mato, F. J. Zuniga, G. Madariga and A. Lopez-Echarri (World Scientific, Singapore, 1991), p. 521.
19. R. A. Dunlap, R. C. O'Handley, M. E. Henry and R. Chatterjee, *Phys. Rev.* **B37** (1988) 8484.
20. W. Steurer, *Mater. Science Forum* **150-151** (1994) 15.
21. C. L. Henley, in *Quasicrystals*, eds. T. Fujiwara and T. Ogawa (Academic, Boston, 1988), Chap.3.
22. Y. Ishii, *Phys. Rev.* **B45** (1992) 5228.
23. M. Fettweiss, P. Launois, R. Reich, R. Wittmann and F. Denoyer, *Phys. Rev.* **B51** (1995) 6700.
24. M. Cooper, *Acta Cryst.* **23** (1967) 1106.
25. W. B. Pearson, *A Handbook of Lattice Spacings and Structures of Metals and Alloys* (Pergamon, Oxford, 1987).
26. F. C. Frank and J. S. Kaspers, *Acta Cryst.* **11** (1958) 184.
27. V. Elser and C. L. Henley, *Phys. Rev. Lett.* **55** (1985) 2883.
28. M. Audier and P. Guyot, *Phil. Mag.* **B53** (1986) L43.
29. P. A. Bancel, P. A. Heiney, P. W. Stephens, A. I. Goldman and P. M. Horn, *Phys. Rev. Lett.* **54** (1985) 2422.
30. M. Audier, J. Pannetier, M. Leblanc, C. Janot, J. M. Lang and B. Dubost, *Physica* **B153** (1988) 136.
31. P. A. Heiney, P. Bancel, P. M. Horn, J. L. Jordan, S. LaPlaca, J. Angilello and F. W. Gayle, *Science* **238** (1987) 660.
32. F. Denoyer, G. Heger. M. Lambert, J. M. Lang and P. Sainfort, *J. Physique (France)* **48** (1987) 1357.
33. P. Donnadieu, *J. Physique I (France)* **4** (1994) 791.
34. M. Fettweiss, P. Launois, F. Denoyer and M. Lambert, *Mat. Science Forum* **150-151** (1994) 267.

35. J. Devaud-Rzepski, A. Quivy, Y. Calvayrac, M. Cornier-Quiquandon and D. Gratias, *Phil. Mag.* **B60** (1989) 855.
36. M. Ellner and U. Brukhardt, *Mat. Science Forum* **150-151** (1994) 97.
37. O. V. Kovalev, *Irreducible Representations of the Space Groups* (Gordon and Breach, New York, 1965).
38. *Incommensurate Phases in Dielectrics*, eds. R. Blinc and A. P. Levanyuk (North Holland, Amsterdam, 1986), Vols. 1-2.
39. K. Fujiwara, *J. Phys. Soc. Japan* **12** (1957) 7.
40. M. Cooper and K. Robinson, *Acta Cryst.* **20** (1953) 614.
41. R. N. Corby and P. J. Black, *Acta Cryst.* **B33** (1977) 3468.
42. J. K. Critchley and J. W. Jeffery, *Acta Cryst.* **19** (1965) 674.
43. P. I. Kripjakevich, *Structural Types of Intermetallic Compounds* (Nauka, Moscow, 1977), p. 86.
44. R. Chatterjee and J. W. O'Handley, *Phys. Rev.* **B39** (1989) 8128.
45. C. Dong, Z. K. Hei, L. B. Wang, O. H. Song, Y. K. Wu and K. H. Kuo, *Scrip. Metall.* **20** (1986) 1155.
46. K. F. Kelton, P. C. Gibbons and P. N. Sabes, *Phys. Rev.* **B38** (1988) 7810.
47. D. S. Rokhsar, N. D. Mermin and D. C. Wright, *Phys. Rev.* **B35** (1987) 5847.
48. A. J. Bradley and H. J. Goldschmidt, *J. Inst. Metals* **65** (1939) 403.
49. P. A. Bancel, in *Quasicrystals: The State of the Art*, eds. D. P. DiVincenzo and P. J. Steinhardt (World Scientific, Singapore, 1991), p.17.
50. J. M. Dubois, C. Dong, C. Janot, M. de Boissieu and M. Audier, *Phase Transitions* **32** (1991) 3.
51. Y. Calvayrac, A. Quivy, M. Bessiere, S. Lefebre, M. Cornier-Quiquandon and D. Gratias, *J. Physique (France)* **51** (1990) 417.
52. J. W. Cahn, D. Schechtman and D. Gratias, *J. Mat. Res.* **1** (1986) 13.
53. P. A. Bancel, *Phys. Rev. Lett.* **63** (1989) 25.
54. M. Audier and P. Guyot, *Adriatico Research Conf. on Quasicrystals*, eds. M. Jaric and S. Lundquist (World Scientific, Singapore, 1989), p. 74.
55. M. Audier, Y. Brechet, M. de Boissieu, P. Guyot, C. Janot and J. M. Dubois, *Phil. Mag.* **63** (1991) 1375.
56. W. Cao, H. Q. Ye and K. H. Kuo, *Phys. Stat. Sol. (a)* **107** (1988) 511.
57. P. Bak, *Script. Metall.* **20** (1986) 1199.
58. A. Katz and M. Duneau, *J. Physique (France)* **47** (1986) 181.
59. E. M. Lifshitz, *Z. Eksp. Teor. Fiz.* **11** (1941) 255.
60. L. S. Levitov, in *Quasicrystals: The State of the Art*, eds. D. P. DiVincenzo and P. J. Steinhardt (World Scientific, Singapore, 1991), p.239
61. M. Duneau and S. Oguey, *J. Phys.* **A24** (1991) 461.
62. J. E. S. Socolar, T. C. Lubensky and P. J. Steinhardt, *Phys. Rev.* **B34** (1986) 3345.
63. P. Sainfort, B. Dubost and A. Dubus, *Comptes Rend. Acad. Sci. Paris II* **301** (1985) 689.
64. P. Donnadieu and C. Degand, *Phil. Mag.* **B68** (1993) 310.
65. P. Sainfort and B. Dubost, *J. Phys. Colloq. France* **47** (1986) C3-321.

66. S. B. Rochal, *Preprint*, to appear in *Crystallographic Reports*.
67. C. B. Shoemaker, *Phil. Mag.* **B67** (1993) 869.
68. C. B. Shoemaker, *Mat. Science Forum* **150-151** (1994) 191.

APPENDIX 1

GROUP-THEORETICAL ASPECTS OF THE PHENOMENOLOGICAL THEORY OF STRUCTURAL PHASE TRANSITIONS

1. Crystallographic space-groups

1.1. Geometrical elements composing a space-group

In a crystalline solid the equilibrium configuration of the average positions of the atoms has a three-dimensional periodicity. It means that the preceding configuration is transformed into a configuration indistinguishable by translations of vectors:

$$\vec{t}(m_1, m_2, m_3) = m_1\vec{a}_1 + m_2\vec{a}_2 + m_3\vec{a}_3 \tag{A1}$$

where m_i are arbitrary integers and the $\vec{a}_i$ non-coplanar vectors which are called the *basic translations* of the crystal's structure. The set of the translations $\vec{t}(m_1, m_2, m_3)$, which constitutes an infinite discrete abelian group $\vec{T}$ of geometrical transformations is called the group of *primitive translations* of the crystal's structure. In general, the set of geometrical transformations leaving the structure invariant is not reduced to $\vec{T}$, and there exist *additional displacements* which leave the structure invariant. The set of all such displacements, including $\vec{T}$, form a group G which is called the *space-group* of the crystal.

In addition to $\vec{T}$, the group G generally comprises four types of geometrical displacements (as well as their products):

i) *Rotations* by an angle θ about an axis passing through a point 0 of the crystal.

ii) *Reflections* about a *plane* perpendicular to the direction $\vec{u}$ and passing through a point 0 of the crystal.

iii) The symmetry about a point 0 of the crystal (*inversion*). It is a combination of two of the above defined displacements: e.g. a rotation by $\theta=\pi$ around an axis passing by 0 and parallel to the direction $\vec{u}$, followed by a symmetry in the plane perpendicular to $\vec{u}$ and passing through 0.

Rotations, reflections and the inversion are called *point-symmetry* displacements as they have the common property to leave, at least, one point 0 unmoved in the displacement (unlike translations which displace all the points of the system). We will denote R a point-symmetry operation.

iv) The displacements obtained by applying subsequently a point-symmetry operation R and a translation $\vec{t}$. If R is a rotation of angle θ about an axis $\vec{u}$, the interesting case corresponds to $\vec{t}$ parallel to $\vec{u}$: One obtains a type of displacement called an *helicoidal-*

rotation, or *screw axis*. If R is a reflection, the interesting case corresponds to $\vec{t}$ parallel to the plane: one obtain a displacement called a *glide-plane*.

It is worth pointing out that any space-group containing pure rotations or reflections R will contain elements of type iv) generated by the products of R by the primitive translations $\vec{t}$ of the crystal. However, certain space-groups can contain in addition non-trivial products of the type iv), with $\vec{t}$ a non-primitive translation (i.e. not belonging to $\vec{T}$) denoted $\vec{t}\,*$.

The notation of the set of elements constituting a space-group is customarily denoted by:

$$\left\{R\middle|\vec{t}\right\} \tag{A2}$$

where $\vec{t}$ is either a primitive translation $\vec{t}$ or a non-primitive translation $\vec{t}\,*$. With reference to the same origin, the product of any two elements of the space group is given by the simple mathematical rule:

$$\left\{R_1\middle|\vec{t}_1\right\}\cdot\left\{R_2\middle|\vec{t}_2\right\}=\left\{R_1R_2\middle|\vec{t}_1+R_1\vec{t}_2\right\} \tag{A3}$$

one can see that even if R_1 and R_2 commute, the product of two space-group elements does not commute.

1.2. Bravais lattice and point-group of a crystal

If one choses an arbitrary point 0 in space, and if, with this point as origin, one constructs the infinite discrete set of vectors $\vec{t}(m_1, m_2, m_3)$ specified by Eq.(A1), one obtains as extremities of these vectors a 3-dimensional lattice of points which is called the Bravais lattice associated with the crystal. The Bravais lattice provides a geometrical picture of the translation group of the crystal.

Eq.(A3) shows that the set of all the R_i point-symmetry elements referred to the chosen origin 0, form a group which is called the point-group of the crystal and is denoted $\hat{G}$. When the space-group G contains screw-axes or glide-planes among its elements G is said to be *non-symmorphic*. In this case the R_i associated to these operations do not leave the crystal invariant. Accordingly the point group $\hat{G}$ will not be a group of invariance of the crystal. Conversely if G contains neither screw-axes nor glide-planes G is said to be *symmorphic*, and in this case $\hat{G}$ will be a group of invariance of the crystal. With respect to the same origin 0, the symmorphic group G can be considered as the product of the translation-group $\vec{T}$ by the group of elements $\left\{R_i\middle|\vec{0}\right\}$ isomorphous to $\hat{G}$.

An important property of all space-groups is the invariance of their Bravais lattice by the point-group $\hat{G}$, referred to one of the nodes of the lattice. This property results from the fact that the translation group $\vec{T}$ of the crystal is an *invariant subgroup* of G, i.e.

$\{R|\vec{t}\}\cdot\vec{T}=\vec{T}\cdot\{R|\vec{t}\}$. Accordingly, if we chose a set of coordinate axes along the three basic vectors ($\vec{a}_1, \vec{a}_2, \vec{a}_3$) of the lattice, the 3-dimensional orthogonal matrices representing the R_i, in this frame, will necessarily have all their elements equal to integers (since the R_i exchange the nodes of the Bravais lattice). In particular the *traces* $(1+2\cos\theta_i)$ of the 3-dimensional rotations R_i *have to be integers*. This imposes to the θ_i to be one of the following angle values: θ=0, π, 2π/3, 4π/3, π/2, 3π/2, π/3, 5π/3. Accordingly a crystal will possess exclusively rotations corresponding to the preceding angles. These rotations are respectively denoted C_1, C_2, C_3, C_3^2, C_4, C_4^3, C_6, C_6^5, in the Schoenflies notation, which labels σ a plane of symmetry and I the inversion.

Thus the point-group $\hat{G}$ of a crystal is a subgroup of O(3) composed of the preceding rotations, eventually combined with reflections. Up to a configuration of O(3), the point-groups which can be constructed from such symmetry elements, form a limited set of *32 distinct groups* which are the crystallographic point-groups. Their list can be found in a number of text books [1].

Due to the fact that the Bravais lattice is a configuration of points (the nodes of the lattice), it generally possesses, besides $\hat{G}$, additional point-symmetry elements leaving it invariant. Its invariance point-group $\hat{H}$ is generally larger than $\hat{G}$. Hence, if $\hat{G}$ does not contain the inversion I about the origin (a node of the lattice), $\hat{H}$ necessarily contains this operation. Also, if there exist planes forming lattice nodes and intersecting at right angle, the lattice will be invariant by symmetries about these planes. The groups $\hat{H}$ are part of the 32 crystallographic point-groups and coincide with the groups which contain the inversion, and, if relevant, symmetry planes intersecting at right-angle: There are 7 crystallographic groups, called the *holohedries*, which satisfy these conditions. These groups are the full invariance groups of possible Bravais lattices.

A partial order can be defined among the 7 holohedries by means of the group subgroup relationship. The symmetry of a crystal will be said to pertain to a given *crystal system* (a *syngony*) defined by $\hat{H}$, if $\hat{H}$ is the smallest holohedry containing the point-group $\hat{G}$ of the crystal. The list of the *7 syngonies*, of the symbols for their defining *holohedries*, and of the orders of these groups, are given in Table 1.

It is desirable to define an equivalence between Bravais lattices in order to express the identity of their shapes. Thus two square lattices only differing in the size of the squares should be considered as equivalent. Two Bravais lattices are equivalent if: 1) they have the same invariance group $\hat{H}$, 2) if one can be transformed into the other by continuous deformation of the lattice, without reaching, at any stage of the deformation, a lattice of lower symmetry than $\hat{H}$. Hence, different holohedries will define unequivalent Bravais lattices. However, the specification of $\hat{H}$ does not always define completely, up to an equivalence, a Bravais lattice. An enumeration of unequivalent lattices compatible with the seven holohedries $\hat{H}$ shows that there exist 14 unequivalent types of lattices: 14 distinct types of translation groups $\vec{T}$ which leave invariant a crystalline structure. Their symbols are listed in Table 1.

Table 1. Syngonies, holohedries, and Bravais lattices of crystals.

Syngony	Cubic	Hexagonal	Rhombohedral	Tetragonal	Orthorhombic	Monoclinic	Triclinic
$\hat{H}$	O_h	D_{6h}	D_{3d}	D_{4h}	D_{2h}	C_{2h}	C_i
Order	48	24	12	16	8	4	2
Bravais lattice	P, I, F	P	R	P, I	P, C, I, F	P, C	P

1.3. Space-group of a crystal. Unit-cell

If the invariance group G of a crystal contains screw axes or glide planes $\{R|\vec{t}\}$, the repeated applications of these geometrical transformations must belong to G. This requirement restricts the possible vectors $\vec{t}\,*$ defining the non-primitive translations involved in a screw axis or glade plane. If R is a rotation C_n, $\vec{t}\,*$ can only be of the form $l\vec{t}/n$ with $\vec{t}$ a primitive translation (A1) and l an integer with 0<l<n. If R is a plane-symmetry, one has $\vec{t}\,*=\vec{t}/2$.

In summary the elements $\{R|\vec{t}\}$ composing a space group G comply with the following restrictions: 1) The point-symmetry elements form one of the 32 crystallographic point-groups. 2) The primitive translations $\vec{t}$ constitute, up to an equivalence, one of the 14 translation groups represented by the 14 Bravais lattices. 3) The non-primitive translations associated to a point-symmetry element can be equal to one of a few specified fractions of the primitive translations. As a consequence, the number of distinct space-groups, i.e. differing by either of the preceding characteristics, is limited. 230 space groups, given in [1], can be constructed. Some of the groups have in common the features 1)-3), but the relative orientations of the Bravais lattice basic vectors, and of the elements of the point group are different in the distinct groups. 72 among the 230 space groups are symmorphic.

Two different types of symbols are usually used to identify point groups and space groups. The first one uses the Schoenflies symbol for the point-group $\hat{G}$ and an additional number referring to a standard list of the space-groups having in common this point-group [1]: T_d^3 is the third space-group with point group T_d. The international symbol shows more explicitly the type of Bravais lattice, the point-group, and the possible occurrence of glide planes and screw axes: P4mm has a P Bravais lattice, the point-group denoted 4mm, and no glide plane or screw axes, whereas P4bm has a glide plane with non-primitive translation parallel to the b axes. Both notations have their advantages and disadvantages. In this book we have used the Schoenflies notation.

A primitive unit-cell of a crystal is the smallest volume, which displaced by the translations $\vec{t} \in \vec{T}$, allows an exact filling of space. Clearly, unit-cells can be found by considering the Bravais lattice of the crystal. In general, several primitive unit-cells exist, all having the same volume. One of them, the Wigner-Seitz cell, displays the same symmetry as the entire Bravais lattice. This cell coincides with the one constructed on the basic translations $\vec{a}_1$, $\vec{a}_2$, $\vec{a}_3$, for P-Bravais lattices, while these primitive unit-cells are different for other types of Bravais lattices. For the latter lattices, it is customary to define a multiple-unit-cell whose volume is a multiple of volume of the primitive one, and which, like the Wigner-Seitz cell, displays the full symmetry of the lattice. In this book we call unit-cell the primitive unit-cell and conventional unit-cell the multiple unit-cell.

2. Irreducible representations of the space-groups

2.1. Irreducible representations of a group G

Consider on the x-axis the transformation I(x)=-x. Certain functions are *symmetric* (even) with respect to the action of the inversion I. They satisfy the condition: $f_1(x) = f_1(-x)$. The *anti-symmetric* (odd) functions satisfy the condition $f_2(-x) = -f_2(x)$. Considering the f_i as the basis of a vector space, we note that this vector space is invariant by the action of I, as well as by the action of the powers of I, since the iterated action of I is either the identical transformation E, or is equal to that of I: the vector space f_i is *invariant* by the group G of the two elements (E, I). In the space f_i, G can be *represented* by the set of two matrices expressing the action of the two elements of the group, as shown in Table 2. These matrices are one-dimensional (±1) for the two spaces of symmetric (f_1) and antisymmetric (f_2) functions. Each set of matrices (1, 1) and (1, -1) constitutes an irreducible representation of G, the irreducibility being defined by the fact that the spaces f_i do not contain any smaller subspace itself invariant by the two elements of G_0.

The prominent role of the two types of functions $f_1(x)$ and $f_2(x)$ arises from the fact that there are no irreducible invariant spaces other than the symmetric and antisymmetric ones: any set of functions $\Phi_i(x)$ spanning a vector space *invariant* and *irreducible* with respect to G will give rise to matrices for the set (E, I) identical to one of the two sets listed in Table 2. Besides, any function Φ(x) can be expressed as the sum of two functions, a symmetric one and an antisymmetric one.

The generalization of the preceding example consists in considering a m-dimensional vector space ε, defined on complex numbers, which is left *globally invariant* by the action of every element g_i of a group G: g_i transforms any vector of ε into a vector of ε. Each g_i can be represented by a matrix $M(g_i)$ referred to a certain basis $\{\varphi_k\}$ of the space. The set $M(g_i)$ is an *irreducible representation of G* if ε contains no subspace invariant by *all* the elements of G. Otherwise the set $M(g_i)$ will be said to form a *reducible* representation of G.

In general G will not act in the same manner on different irreducible spaces (e. g. I *preserves* f_1 and *reverses* f_2). In order to classify the different action of G in different spaces, we compare the two sets of matrices $\{M_1(g_i)\}$ and $\{M_2(g_i)\}$ corresponding to

Table 2. Matrices expressing the action of the elements E, I of the group G on the spaces f_1 and f_2.

	E	I
f_1	1	1
f_2	1	-1

the irreducible spaces ε_1 and ε_2. $\{M_1\}$ and $\{M_2\}$ are equivalent if for all the elements g_i of G, it is possible to write:

$$M_2(g_i) = S \cdot M_1(g_i) \cdot S^{-1} \tag{A4}$$

where the same unitary matrix S is used for all the g_i of G. If (A4) is obeyed, the irreducible representations carried by ε_1 and ε_2 are said to be equivalent. Otherwise they are unequivalent. The unequivalent irreducible representations of a group can be enumerated on the sole basis of the structure of the group, i. e. independently from the particular vector spaces ε which carry the representations. In this enumeration each irreducible representation can be specified unambiguously and independently from the choice of a basis, by the set of *traces* $\chi(g_i)$ of the matrices $M(g_i)$ which are termed *characters* of the irreducible representation.

A vector space ε is *invariant* and *irreducible* with respect to G if one can decompose it into subspaces ε_i which are invariant and irreducible. One says that the reducible representation Γ of G carried by ε is the *sum* of irreducible representations carried by the ε_i: $\Gamma = \Sigma \Gamma_i$.

A *finite* group G has a finite number ν of unequivalent irreducible representations. ν is generally smaller than the *order* n of the group (i. e. its number of elements). More precisely, ν is equal to the number of *classes of equivalent elements* g_i of the group G, the equivalence between g_i and g_k being defined by $g_k = g \cdot g_i \cdot g^{-1}$ where g is also an element of G. In this case the enumeration of the irreducible representations of G can be displayed in the form of a *character table* of the group. For abelian groups (for which the product of any two elements commute: $g_i \cdot g_j = g_j \cdot g_i$) ν=n and two unequivalent irreducible representations are carried by one-dimensional spaces: abelian groups of order n possess n one-dimensional irreducible representations. These properties do not hold for non abelian groups for which $\nu < n$ and one has irreducible spaces of dimension $m > 1$. By contrast all groups (abelian or non abelian) possess a one-dimensional irreducible representation for which all the characters satisfy the condition $\chi(g_i)$=1. This irreducible representation (which leaves invariant the unique vector of its one-dimensional carrier space) is termed the *identity* (or *totally symmetric*) representation of G.

As an illustrative example let us consider the group G=C_{3v} of rotations and reflections which preserve the equilateral triangle: rotations by $2\pi/3$ and $4\pi/3$ about the center of gravity of the triangle, denoted C_3 and C_3^2, and reflections σ_i about the three

Table 3. Irreducible representations of the group C_{3v}.
For Γ_3 the matrices are given for C_1, C_3 and σ_1.

	E	(C_3, C_3^2)	$(\sigma_1, \sigma_2, \sigma_3)$
Γ_1	1	1	1
Γ_2	1	1	-1
Γ_3	2	-1	0
	$\begin{bmatrix} 1 & 0 \\ 0 & 1 \end{bmatrix}$	$\begin{bmatrix} -1/2 & -\sqrt{3}/2 \\ \sqrt{3}/2 & -1/2 \end{bmatrix}$	$\begin{bmatrix} 1 & 0 \\ 0 & -1 \end{bmatrix}$

medians of the triangle. The order of G is n=6, while the number of its classes of equivalence is ν=3: $\{C_1\}$, $\{C_3, C_3^2\}$, $\{\sigma_1, \sigma_2, \sigma_3\}$. The character table of the C_{3v} group is given in Table 3, as well as the matrices representing the operations C_1, C_3 and σ_1 for Γ_3. Examples of carrier spaces corresponding to the three unequivalent irreducible representations of G are provided 1) by the Z-axis, perpendicular to the plane of the triangle, which spans the identity representation Γ_1, 2) by the difference of ordered pairs of coordinates (x, y)-(y, x) where x and y are coordinates in the plane (x, y) of the triangle, which span the one-dimensional irreducible representation Γ_2, 3) by the two-dimensional space constituted by the plane (x, y) of the triangle. It spans the two-dimensional irreducible representation Γ_3.

Consider the 3 transformations C_1, C_3, C_3^2 of C_{3v}. They form an abelian subgroup G' of G, which has three irreducible representations Γ_i' given in Table 4. We note that Γ_2' and Γ_3' have characters, and matrices, relative to C_3 and C_3^2, which are complex numbers. These two representations are termed *complex representations*. Their carrier spaces are respectively the complex coordinates x+iy and x-iy. If we consider the *reducible* vector space spanned by the set (x+iy, x-iy) a unitary change to the basis (x, y) will transform the complex diagonal matrices of C_3^2 and C_3 into the real matrices indicated in Table 3 for the same rotations. This *reducible* representation is real, and if we do not allow complex combinations of x and y to be taken as basis of the (x, y) space, this space will be *irreducible*: the representation carried by the space (x, y) will be said to be

Table 4. Irreducible representations of the group C_3.
$\omega=\exp\{2\pi i/3\}$.

	E	C_3	C_3^2
Γ_1'	1	1	1
Γ_2'	1	ω	ω^2
Γ_3'	1	ω^2	ω

irreducible on the real numbers ("physically irreducible") and reducible on the complex numbers.

2.2. Irreducible representation of the Translation group. Brillouin-zone. Star of $\vec{k}$. Small representations of $G(\vec{k})$

Space-groups are discrete infinite groups (since they contain the discrete infinite group $\bar{T}$ of translations) possessing non commuting elements. They possess an infinite set of irreducible representations which cannot be tabulated in a concise manner as for finite groups. Actually, the existing tabulation do not provide explicitly the representations but only elements which are needed to construct these representations according to a well defined procedure [2, 3], which is given in the next section. In the present section we define the main ingredients which are necessary for that purpose.

As the translation group $\bar{T}$ is abelian, its irreducible representations are one-dimensional. Given an arbitrary vector $\vec{k}$, the set of one-dimensional matrices:

$$M(\vec{t}) = e^{-i\vec{k}\vec{t}} \tag{A5}$$

clearly constitutes a one-dimensional irreducible representation, of $\vec{T}$, which can be denoted $\gamma(\vec{k})$. However two distinct vectors $\vec{k}_1$, and $\vec{k}_2$ do not necessarily define unequivalent irreducible representations of $\bar{T}$. It can easily be checked that $\gamma(\vec{k}_1)$ is equivalent to $\gamma(\vec{k}_2)$ if, and only if, one has:

$$\vec{k}_1 - \vec{k}_2 = \vec{K} = m_1\vec{a}_1^* + m_2\vec{a}_2^* + m_3\vec{a}_3^* \tag{A6}$$

where the m_i are positive, negative, or zero integers, and :

$$\vec{a}_1^* = 2\pi\frac{\vec{a}_2 \times \vec{a}_3}{\vec{a}_1 \cdot \vec{a}_2 \times \vec{a}_3}, \quad \vec{a}_2^* = 2\pi\frac{\vec{a}_3 \times \vec{a}_1}{\vec{a}_1 \cdot \vec{a}_2 \times \vec{a}_3}, \quad \vec{a}_3^* = 2\pi\frac{\vec{a}_1 \times \vec{a}_2}{\vec{a}_1 \cdot \vec{a}_2 \times \vec{a}_3}. \tag{A7}$$

The $\vec{a}_i^*$ are the *reciprocal-basic-vectors*. The lattice constructed on the $\vec{k}$-vectors (A6), starting from an arbitrary origin, is the *reciprocal Bravais lattice* (or more usually the reciprocal lattice) associated with $\bar{T}$. It has the same invariance group $\bar{H}$ as the Bravais lattice of $\bar{T}$, but *it is generally unequivalent to it.* Eq.(A6) shows that the $\vec{k}$ vectors contained in one *unit-cell of the reciprocal lattice* allow to index all the unequivalent irreducible representations of $\bar{T}$. Besides, two $\vec{k}$-vectors *inside* a unit-cell correspond necessarily to unequivalent irreducible representations. By contrast, several vectors at the *surface* limiting the unit-cell can correspond to equivalent irreducible representations. Any of the various possible types of unit-cells has this property and, in particular, the Wigner-

Seitz cell of the reciprocal lattice. The lattice cell is called the *Brillouin-zone*. Accordingly, the $\vec{k}$-vectors of the Brillouin-zone constitute a geometrical picture of the set of irreducible representations of the translation group $\vec{T}$ of the crystal. Let us note that the identity representation of $\vec{T}$ corresponds to $\vec{k}=0$.

Let G be a space group and $\vec{k}_1$ a vector of the Brillouin-zone. If we apply to $\vec{k}_1$ the transformations belonging to the point-group $\hat{G}$ (referred to the center of the Brillouin-zone), this vector is transformed into vectors $\vec{k}_2,\ldots,\vec{k}_n$ (where n is the order of $\hat{G}$) of same modulus, and all belonging to the Brillouin-zone. Some of these vectors represent the same irreducible representations of $\vec{T}$ as $\vec{k}_1$, and can be considered as equivalent to $\vec{k}_1$. Their existence implies that certain elements of $\hat{G}$, forming a subgroup of $\hat{G}$, transform $\vec{k}_1$ into an equivalent vector. This subgroup, which we denote $\hat{G}(\vec{k}_1)$, is called the *point group of the vector* $\vec{k}_1$, or the *little point-group of* $\vec{k}_1$. Starting from each element $\hat{G}(\vec{k}_1)$, we can consider all the elements $\{R|\vec{t}\}$ of G which involve R as a point symmetry element. The set of elements $\{R|\vec{t}\}$ associated with all the elements R of $\hat{G}(\vec{k}_1)$ constitute a subgroup G($\vec{k}_1$) of G, which contains the group of translations $\vec{T}$. G($\vec{k}_1$), which is one of the 230 space groups, is called the *space group of the vector* $\vec{k}_1$, or more concretely, *the little group of* $\vec{k}_1$.

The set of *unequivalent* vectors in $(\vec{k}_1,\vec{k}_2,\ldots,\vec{k}_n)$ which are generated by the application of $\hat{G}$, constitute *the star of* $\vec{k}_1$, which we denote $\vec{k}_1^*$. Each $\vec{k}_i$ is an *arm* of $\vec{k}_1^*$. The number of arms in the star of $\vec{k}_1$ is equal to the index of the subgroup $\hat{G}(\vec{k}_1)$ with respect to the group $\hat{G}$. Note that the same star is generated by the application of $\hat{G}$ to any arm of $\vec{k}_1^*$.

Consider a point A of the Brillouin-zone and $\vec{k}_0$ the vector of the reciprocal space forming the Brillouin-zone, joining the center of this zone to A. Similarly, let $\vec{k}$ be the vector associated with a point located in the neighborhood of A. If $\hat{G}(\vec{k})\subset\hat{G}(\vec{k}_0)$ strictly, for any point in the neighborhood of A, then A is a *high-symmetry point* of the Brillouin-zone. One can find analogously *lines* or *planes* of high-symmetry in the Brillouin-zone. A point which does not belong to a symmetry manifold corresponds to a *general direction* in the reciprocal space and has $\hat{G}(\vec{k})$ reduced to the identity E. In this case the corresponding vector $\vec{k}$ defines a star $\vec{k}^*$ whose number of arms equals the order of the point-group $\hat{G}$ of the crystal. Customarily, high-symmetry points, and points on high-symmetry manifolds are labeled by letters (e. g. Γ, A, Z, Λ, Δ etc.) the meaning of which are precised in Ref.[2]. Kovalev's table [3] possesses its own labeling. We have used in the book, the notations given by Zak [2] and Kovalev [3].

The little group $G(\vec{k}_1)$ has irreducible representations which play an essential role in the construction of the irreducible representations of the entire space group G. Though $G(\vec{k}_1)$ is like G, a space-group, its irreducible representations are simpler than those of G, because it preserves, up to an equivalence, the vector $\vec{k}_1$ of the Brillouin-zone. Thus, its irreducible representations associated with $\vec{k}_1$ are in finite number and called the small representations. The i-th irreducible representation of $G(\vec{k}_1)$ is denoted $\tau_i(\vec{k}_1)$ or more simply τ_i. The matrices of τ_i for the elements of $G(\vec{k}_1)$ are denoted: $\tau_i\left(\left\{R|\vec{t}\right\}\right)$. Starting from these matrices one can define the following set of matrices:

$$\tau_i'(R) = \tau_i\left(\left\{R|\vec{t}\right\}\right)\cdot e^{i\vec{k}_1\cdot\vec{t}}. \tag{A8}$$

With respect to the nature of the τ_i', 3 different situations can be distinguished.

i) $\vec{k}_1$ is a vector whose extremity lies *inside* the Brillouin-zone. $\tau_i'(R)$ coincides with the set of matrices of an irreducible representation of the finite point group $\hat{G}(\vec{k}_1)$. Equation (A8) provides us with the matrices of the small representation $\tau_i\left(\left\{R|\vec{t}\right\}\right)$ deduced from the irreducible representations of $\hat{G}(\vec{k}_1)$.

ii) $\vec{k}_1$ is any vector of the Brillouin-zone, but the group G is symmorphic. The same rule as in case i) appears, with the particularity that only primitive translations $\vec{t}$ are involved in the elements of G.

iii) $\vec{k}_1$ is a vector whose extremity lies on the limiting surfaces of the Brillouin-zone and G is non-symmorphic. In this case τ_i' is not an irreducible representation of $\hat{G}(\vec{k}_1)$ in the usual sense. It is the so-called "weighted" irreducible representation of the finite point-group $\hat{G}(\vec{k}_1)$. A weighted irreducible representation does not only depend on the structure of the point-group $\hat{G}(\vec{k}_1)$ but also on the $\vec{k}_1$ vector considered and on the nature of the translations $\vec{t}$ existing in the space-group G. The weighted irreducible representations are in finite number and have been tabulated for all the space groups and relevant $\vec{k}_1$ vectors [3]. Using these tables and Eq.(A8), one can construct the small representations τ_i associated to each weighted irreducible representations τ_i'.

2.3. Construction of the irreducible representations of a space group

A space-group G has an infinite number of irreducible representations. A given representation can be specified by two elements: 1) a star k*, generated by applying $\hat{G}$ to a given vector $\vec{k}$ of the Brillouin-zone. Unequivalent stars will define unequivalent irreducible representations of G. 2) One of the small representations $\tau_i(k)$ of the group

$G(\vec{k})$. Unequivalent τ_i define unequivalent irreducible representations of G. Hence an irreducible representations of G can be denoted $\Gamma_i(\vec{k})$ (or $\Gamma_i(\vec{k}^*)$). There is an infinite set of $\vec{k}$-vectors in the Brillouin-zone, and for each $\vec{k}$-vector, a finite set of possible values for the index i_k. The dimension of an irreducible representation $\Gamma_i(\vec{k})$ is n=r×s where r is the dimension of the small representation τ_i, and s is the number of arms of the star $\vec{k}^*$.

In order to construct the matrices of $\Gamma_i(\vec{k}_1)$ the following sequence of operations must be performed:

1) Consider the vector $\vec{k}_1$ and find the group $\hat{G}(\vec{k}_1)\subset\hat{G}$ which leaves $\vec{k}_1$ invariant, up to an equivalence.

2) Determine the star $\vec{k}_1^*$. In this view, apply to $\vec{k}_1$ the elements of $\hat{G}$ not belonging to $\hat{G}(\vec{k}_1)$ and determine the s unequivalent vectors $\vec{k}_2$, $\vec{k}_3,\ldots,\vec{k}_s$ generated in this manner. Check that s is equal to the index of $\hat{G}(\vec{k}_1)$ in $\hat{G}$.

3) Chose a set (E, $R_1,\ldots,R_s$) of elements of $\hat{G}$ satisfying $\vec{k}_P = R_P\vec{k}_1$. For each R_P, chose an element $g_P = \{R_P|\vec{t}_P\}$ of the space-group G.

4) Find the form of the matrices of the small representation τ_i of $G(\vec{k}_1)$. The realization of this stage of the procedure depends if the situation considered pertains to the cases i), ii) or iii) indicated in Sec.2.2. In case i) or ii) the matrices τ_i are obtained through Eq.(A8) from the matrices τ_i' of an irreducible representation of the point-group $\hat{G}(\vec{k}_1)$. The latter-group being one of the 32 crystallographic groups, its irreducible representations are tabulated in a number of text books [2, 3]. In case iii), τ_i is still obtained through Eq.(A8), but, this time, τ_i' is a weighted-representation of $\hat{G}$, the matrices of which have been tabulated by Kovalev [3].

5) Construct the matrices of the irreducible representations of G for any element $\{R|\vec{t}\}$ of G. This last stage can be realized as follows: Assume that the carrier space of $\tau_i(\vec{k})$ is $\varepsilon(\vec{k})$, spanned by vectors $\Phi_1(\vec{k}),\ldots,\Phi_r(\vec{k})$. Let, on the other hand $g_P = \{R_P|\vec{t}_P\}$ be s elements of the space-group G, chosen in such a way that the P-th vector in the star $\vec{k}_1^*$ is:

$$\vec{k}_P = \hat{g}_P\cdot\vec{k}_1 = R_P\cdot\vec{k}_1. \tag{A9}$$

A standard basis of r×s vectors for the representation $\Gamma_i(\vec{k}_1)$ is provided by the set of vectors:

$$\Phi_l(\vec{k}_P) = g_P\cdot\Phi_l(\vec{k}). \tag{A10}$$

The vectors of this basis can be ordered as follows:

$$\Phi_1(\vec{k}_1),..., \Phi_r(\vec{k}_1);\ \Phi_1(\vec{k}_2),..., \Phi_r(\vec{k}_2);\ \Phi_1(\vec{k}_s),..., \Phi_r(\vec{k}_s).$$

In this case the matrices $M(\{R|\vec{t}\})$ of $\Gamma_i(\vec{k})$ can be considered as formed of s×s blocks, each block being related to a pair $(\vec{k}_{m_1}, \vec{k}_{m_2})$ of vectors of $\vec{k}^*$. Let $\mu(\vec{k}_{m_1}, \vec{k}_{m_2}|\{R|\vec{t}\})$ be the r×r matrix constituting such a block. It can be shown [1-3] that:

$$\mu(\vec{k}_{m_1}, \vec{k}_{m_2}|\{R|\vec{t}\})=0 \quad \text{if } R\cdot\vec{k}_{m_1} \neq k_{m_2},$$

$$\mu(\vec{k}_{m_1}, \vec{k}_{m_2}|\{R|\vec{t}\})=\tau_i\left[g_{m_1}^{-1}\cdot\{R|\vec{t}\}\cdot g_{m_2}\right] \quad \text{if } R\cdot\vec{k}_{m_1} = k_{m_2} \tag{A11}$$

where g_{m_1} and g_{m_2} are defined by (A9) and τ_i is the matrix of the *small representation*, corresponding to the element between brackets (which is an element of $G(\vec{k})$). Eq.(A11) allows to determine the blocks $\mu(\vec{k}_{m_1}, \vec{k}_{m_2}|\{R|\vec{t}\})$ constituting the matrices $M(\{R|\vec{t}\})$.

3. Group theoretical properties used in the formulation of the Landau theory

3.1. Decomposition of $f(\vec{r})$ into a sum of irreducible functions

Let G be a group of transformations acting in the 3-dimensional space. G will also act on any function $f(\vec{r})$ defined in the preceding space, and one can write: $f(g_i(\vec{r}))=f_i(\vec{r})$. Any function $f(\vec{r})$ can be written:

$$f(\vec{r}) = \sum_k\left[\lambda_{kr}\varphi_{kr}(\vec{r})\right] \tag{A12}$$

where the set of functions $\varphi_{kr}(\vec{r})$ carries the irreducible representations of the group G. The origin of this property can be understood by the following considerations: If G is finite, one can construct all the transformations $f_i(\vec{r})=g_i f(\vec{r})$ ((i=1, 2,...,n) where n is the order of G) of $f(\vec{r})$ by the elements g_i of G. *The set* $f_i(\vec{r})$ is globally invariant by G and can be considered as generating an invariant vector space. If this space is not irreducible, one can decompose it into irreducible spaces. Each irreducible space will be spanned by functions $\varphi_{kr}(\vec{r})$ which are linear combinations of the $f_i(\vec{r})$. Conversely each $f_i(\vec{r})$ can be expressed as a linear combination of the $\varphi_{kr}(\vec{r})$. In particular $f(\vec{r})$ can be expressed as in Eq.(A12).

3.2. Invariant polynomials

Consider a vector space ε left globally invariant by G. The elements of G act linearly on the coordinates x_j in this space. We can consider functions $f(x_1,...,x_n)$ and their transforms $f_i = g_i f = f(g_i x_j)$. An important case is constituted by functions which are homogeneous polynomials of a given degree P. These polynomials are linear combinations of C^P_{P+n-1} monomials of degree P. Thus, they can be considered to generate a vector space ε^P of dimension C^P_{P+n-1}. Since G acts linearly on the x_j, a polynomial of degree P is transformed into a polynomial of the same degree. The group G therefore acts linearly on the space ε^P and leaves this space invariant. ε^P carries a representation of G which, in general, will be reducible. If we decompose ε^P into irreducible vector spaces with respect to G, there can be in this decomposition, one or several spaces which carry the totally symmetric representation Γ_1 of G. Each of the latter one-dimensional spaces is spanned by a polynomial of degree P which has the property to be left invariant by the elements of G. Such polynomials are termed *invariant polynomials* of the x_j coordinates.

Consider, for instance, the space $\varepsilon=(x, y)$ which, with respect to the group $G=C_{3v}$ (see Table 3) carries the irreducible representation Γ_3. Homogeneous polynomials of degree 2 form a 3-dimensional space ε^2 spanned by (x^2, y^2, xy). The decomposition of this space into irreducible spaces show that ε^2 contains one totally-symmetric subspace spanned by the polynomials $((x^2, y^2)$. Likewise, ε^3 is generated by (x^3, y^3, x^2y, xy^2) which contains a single totally symmetric subspace spanned by the polynomial $x(x^2 - 3y^2)$. Finally ε^4 is generated by the 5 monomials $(x^4, y^4, x^3y, xy^3, x^2y^2)$. It also contains a single Γ_1-space spanned by $(x^2 + y^2)$.

3.3. Reduction of an invariant space into irreducible spaces

Consider a space ε, spanned by the vectors $\vec{\phi}_i$, and carrying a reducible representation Γ of the group G. ε can be decomposed into subspaces ε_k carrying irreducible representations Γ_k of G. Knowledge of the irreducible representations involved into the decomposition of Γ requires knowledge of the characters $\chi(g_i)$ of Γ. The practical rule is provided by the formula [1, 2]:

$$n_k = \frac{1}{n_G} \sum_{g_i} \bar{\chi}(g_i) \cdot \chi^{(k)}(g_i) \tag{A13}$$

where n_G is the order of the group G and n_k the number of irreducible spaces of type Γ_k contained in ε. $\chi^{(k)}(g_i)$ are the characters of Γ_k, to be found in the character tables of G.

3.4. n-th power of a representation

Let ε_1 and ε_2 be two spaces invariant under the action of G and respectively spanned by the vectors $\vec{\varphi}_i^1$ and $\vec{\varphi}_j^2$. ε_1 and ε_2 respectively carry representations Γ_1 and Γ_2 of G which are reducible or irreducible. The tensorial product $\varepsilon_1 \times \varepsilon_2$ of the two spaces (taken in this order) is the vector space generated by the set of elements $(\vec{\phi}_i^1, \phi_j^2) = \vec{\varphi}_{ij}$ where $\vec{\varphi}_{ij}$ is assumed to have a linear dependence with respect to each member of the pair. We define, on the other hand, the action of $g_i \in G$ on the space $\varepsilon^1 \times \varepsilon^2$ by:

$$g_i \vec{\varphi}_{ij} = \left(g_i \vec{\varphi}_i^1, g_i \vec{\varphi}_j^2 \right). \tag{A14}$$

It results from this definition that $\varepsilon^1 \times \varepsilon^2$ is invariant by G and that each $\vec{\varphi}_{ij}$ is transformed linearly into the $\vec{\varphi}_{i'j'}$. This space is a carrier space for a representation Γ_{12} of G, which is called the product of the two representations Γ_1 and Γ_2. This representation is reducible if either Γ_1 or Γ_2 are reducible. It can also be reducible if Γ_1 and Γ_2 are irreducible. The characters $\chi^{12}(g_i)$ of Γ_{12} are

$$\chi^{12}(g_i) = \chi^1(g_i) \cdot \chi^2(g_i). \tag{A15}$$

The definition of a tensorial product applies to the case where ε^1 and ε^2 are identical. Γ_{11} is then called the square of the representation Γ_1. It is straightforward to extend this definition to the n-th power of a representation Γ_1 which is denoted Γ^n. Eq.(A15) shows that the characters of Γ^n will be $[\chi(g_i)]^n$ where $\chi(g_i)$ are the characters of Γ.

4. References

1. *International Tables for X-ray Crystallography.* Vol.1 (Kynoch Press, Birmingham, 1952).
2. C. J. Bradley and A. P. Cracknell, *The Mathematical Theory of Symmetry in Solids* (Clarendon Press, Oxford, 1972);
   J. Zak, A. Casher, H. Glück and Y. Gur, *The Irreducible Representations of Space Groups* (Benjamin, New York, 1969);

G. Ya. Lyubarskii, *The Application of Group Theory in Physics* (Pergamon, London, 1960).
3. O.V. Kovalev, *Irreducible Representations of the Space Groups* (Gordon and Breach, New York, 1965).

Subject Index